TOPOLOGICAL SOLITONS

Topological solitons occur in many nonlinear classical field theories. They are stable, particle-like objects, with finite mass and a smooth structure. Examples are monopoles and Skyrmions, Ginzburg–Landau vortices and sigma-model lumps, and Yang–Mills instantons. This book is a comprehensive survey of static topological solitons and their dynamical interactions. Particular emphasis is placed on the solitons that satisfy first-order Bogomolny equations. For these, the soliton dynamics can be investigated by finding the geodesics on the moduli space of static multi-soliton solutions. Remarkable scattering processes can be understood this way.

NICHOLAS MANTON received his Ph.D. from the University of Cambridge in 1978. Following postdoctoral positions at the Ecole Normale Supérieure in Paris; Massachusetts Institute of Technology; and University of California, Santa Barbara, he returned to Cambridge and is now Professor of Mathematical Physics in the Department of Applied Mathematics and Theoretical Physics. He is also head of the department's High Energy Physics group, and a fellow of St John's College. He introduced and helped develop the method of modelling topological soliton dynamics by geodesic motion on soliton moduli spaces. He also discovered, with Frans Klinkhamer, the unstable sphaleron solution in the electroweak theory of elementary particles. Professor Manton was awarded the London Mathematical Society's Whitehead Prize in 1991, and he was elected a fellow of the Royal Society in 1996.

PAUL SUTCLIFFE received his Ph.D. from the University of Durham in 1992. Following postdoctoral appointments at Heriot-Watt, Orsay and Cambridge, he moved to the University of Kent, where he held an EPSRC Advanced Fellowship and later a Chair. In 2006 he returned to Durham and is now Professor of Mathematical Physics in the Department of Mathematical Sciences. He has researched widely on topological solitons, especially multi-soliton solutions and soliton dynamics, and has found surprising relations between different kinds of soliton. One of his principal research contributions was to reveal the symmetric structures formed by Skyrmions and monopoles, their links with fullerenes in carbon chemistry, and finding associated novel scattering processes. He also discovered, with Richard Battye, the first stable knotted soliton solution in classical field theory. Professor Sutcliffe was awarded the London Mathematical Society's Whitehead Prize in 2006.

T0340054

CAMBRIDGE MONOGRAPHS ON MATHEMATICAL PHYSICS

General editors: P. V. Landshoff, D. R. Nelson, S. Weinberg

[†] Issued as a paperback

Topological Solitons

NICHOLAS MANTON

University of Cambridge

PAUL SUTCLIFFE

University of Durham

CAMBRIDGE UNIVERSITY PRESS
Cambridge, New York, Melbourne, Madrid, Cape Town, Singapore, São Paulo

Cambridge University Press
The Edinburgh Building, Cambridge CB2 8RU, UK

Published in the United States of America by Cambridge University Press, New York

www.cambridge.org
Information on this title: www.cambridge.org/9780521838368

First published 2004
This digitally printed version (with corrections) 2007

A catalogue record for this publication is available from the British Library

Library of Congress Cataloguing in Publication data

Manton, Nicholas, 1952–
Topological solitons / Nicholas Manton, Paul Sutcliffe.
p.   cm. – (Cambridge monographs on mathematical physics)
Includes bibliographical references and index.
ISBN 0 521 83836 3
1. Solitons.   I. Sutcliffe, Paul (Paul M.)   II. Title.   III. Series.
QC174.26.W28M36 2004
530.14 – dc22   2003069072

ISBN 978-0-521-83836-8 hardback
ISBN 978-0-521-04096-9 paperback

Contents

Preface

Topological solitons have been investigated by theoretical physicists and mathematicians for more than a quarter of a century, and it is now a good time to survey the progress that has been made. Many types of soliton have been understood in detail, both analytically and geometrically, and also numerically, and various links between them have been discovered.

This book introduces the main examples of topological solitons in classical field theories, discusses the forces between solitons, and surveys in detail both static and dynamic multi-soliton solutions. Kinks in one dimension, lumps and vortices in two dimensions, monopoles and Skyrmions in three dimensions, and instantons in four dimensions, are all discussed. In some field theories, there are no static forces between solitons, and there is a large class of static multi-soliton solutions satisfying an equation of the Bogomolny type. Deep mathematical methods can be used to investigate these. The manifold of solutions is known as moduli space, and its dimension increases with the soliton number. We survey the results in this area. We also discuss the idea of geodesic dynamics on moduli space, which is an adiabatic theory of multi-soliton motion at modest speeds when the static forces vanish, or almost vanish.

Some variants of the solitons mentioned above are considered, but we do not consider the coupling of fermions to solitons, nor solitons in supersymmetric theories, where there are sometimes remarkable dualities between the solitons and elementary particles, nor solitons coupled to gravity, although all these topics are interesting. Also not discussed are the solitons of string theory, known as branes, and the related noncommutative solitons. Much recent work has been on these, but we are not knowledgable enough to write about them. There is some discussion of Skyrmion quantization, because this is essential for the physical interpretation of Skyrmions, but nothing else on soliton quantization. At the end we discuss the unstable analogues of solitons, known as sphalerons.

To make this book reasonably self-contained, we start with an introduction and three general chapters on classical field theory and the mathematical tools useful for understanding various solitons. For those new to the subject we recommend a quick read through these chapters, and then a careful study of the subsequent chapters on kinks and lumps. Here, gauge fields do not occur. The later chapters are longer and some of the material technically harder. The reader can return to the earlier chapters for some of the necessary background material while reading these.

We have tried to make the discussion mathematically sound, at the level customary in theoretical physics, and many calculations are given in detail. But the analysis and topology should be regarded as heuristic. Fortunately, many aspects of the theory we present have been given a rigorous analytical basis through the work of Taubes, Uhlenbeck and Stuart, among others. The geometrical and topological aspects have been put on a firm basis by Atiyah, Hitchin and their collaborators.

Many numerical results concerning solitons have been obtained over the past decades, and some are presented here. To ensure the accuracy of what we present we have recalculated and plotted afresh almost everything. This was partly to achieve consistency with our conventions and notation, partly to take advantage of up-to-date computational power, and partly to avoid the need to copy graphs from other publications.

We would like to record here our thanks to many friends and colleagues who have shared our interest in solitons.

N. S. M. especially thanks Peter Goddard for introducing him to the subject, for supervising his Ph.D. thesis and for later support, and Michael Atiyah (now Sir Michael) for inspirational guidance on many topics at the interface of mathematics and theoretical physics. He thanks those who have collaborated on joint papers in this area: Peter Forgács, Roman Jackiw, Ian Affleck, Orlando Alvarez, Frans Klinkhamer, Gary Gibbons, Peter Ruback, Fred Goldhaber, Andy Jackson, Andreas Wirzba, Michael Atiyah, Robert Leese, Nigel Hitchin, Michael Murray, Houari Merabet, Bernard Piette, and Martin Speight. He would also like to warmly thank his Ph.D. students who have worked in the area of topological solitons, and in some cases collaborated on papers: Mark Temple-Raston, Trevor Samols, Bernd Schroers, Margaret James, Paul Shah, Kim Baskerville, Conor Houghton, Sazzad Nasir, Patrick Irwin, Steffen Krusch, Nuno Romão, João Baptista and Anne Schunck. Many results and ideas presented in this book are due to them.

N. S. M. is grateful to CERN for a Scientific Associateship during part of 2001, which allowed significant progress on this book. He is also grateful to Julia Blackwell for typing drafts of many of the chapters. He particularly thanks Anneli and Ben for their love and understanding while enduring many evenings and weekends of book writing and rewriting.

P. M. S. was fortunate to be an undergraduate and Ph.D. student at Durham during a time when there were many soliton experts around. He thanks Ed Corrigan, David Fairlie, Robert Leese, Bernard Piette and Ian Strachan for many valuable discussions, and in particular Wojtek Zakrzewski and Richard Ward, his Ph.D. supervisor. As a research fellow in Cambridge he benefited greatly from conversations with Sir Michael Atiyah, Gary Gibbons, Nigel Hitchin, Trevor Samols and Paul Shellard. He would like to thank all his collaborators, and particularly Richard Battye, Conor Houghton and Theodora Ioannidou. P. M. S. acknowledges the EPSRC for an advanced fellowship held at the time of writing this book. He thanks Zoë for her love and motivation (through continually asking "isn't it finished yet?"), and Steven and Jonathan for warning that a book with no wizards in it will never sell.

Together, we especially thank Wojtek Zakrzewski. The writing of this book emerged from an earlier project he initiated, which unfortunately did not come to fruition.

1

Introduction

1.1 Solitons as particles

In the 1960s and early 1970s a novel approach to quantum field theory developed and became popular. Physicists and mathematicians began to seriously study the classical field equations in their fully nonlinear form, and to interpret some of the solutions as candidates for particles of the theory. These particles had not been recognized before – they are different from the elementary particles that arise from the quantization of the wave-like excitations of the fields. Their properties are largely determined by the classical equations, although a systematic treatment of quantum corrections is possible.

A characteristic feature of the new, particle-like solutions is their topological structure, which differs from the vacuum. If one supposes that quantum excitations about the vacuum are associated with smooth deformations of the field, then such excitations do not change the topology. So the usual elementary particles of quantum field theory, e.g. the photon, have no topological structure. The new particles owe their stability to their topological distinctiveness. Although they are often of large energy, they can not simply decay into a number of elementary particles.

In many cases, the topological character of the field is captured by a single integer N, called the topological charge. This is usually a topological degree, or generalized winding number of the field. The topological charge N can be identified as the net number of the new type of particle, with the energy increasing as $|N|$ increases. The basic particle has $N = 1$; the minimal energy field configuration with $N = 1$ is a classically stable solution, as it can not decay into a topologically trivial field. The energy density is smooth, and concentrated in some finite region of space. Such a field configuration is called a topological soliton – or just soliton. The ending "-on" indicates the particle-like nature of the solution. There is usually a reflection symmetry reversing the sign of N, and hence there is an antisoliton with $N = -1$. Soliton-antisoliton pairs can annihilate or be pair-produced. Field configurations with $N > 1$ are interpreted as multi-soliton states. Sometimes it is energetically favourable for these to decay into N well separated charge 1 solitons; alternatively they can relax to a classical bound state of N solitons.

The length scale and energy of a soliton depend on the coupling constants appearing in the Lagrangian and field equations. In a Lorentz invariant theory, and in units where the speed of light is unity, the energy of a soliton is identified as its rest mass. In contrast, the elementary particles have a mass proportional to Planck's constant $\hbar$ (this is sometimes not recognized, because of the choice of units). Quantum effects become small in the limit $\hbar \to 0$. In this limit, the topological soliton has finite mass, but the elementary particle mass goes to zero. Furthermore, the quantum corrections to the soliton mass go to zero.

There are important relationships between the solitons of a theory and the wave-like fields which satisfy the linearized field equations. It is the quantization of the latter that gives the elementary particle states, with the nonlinear terms being responsible for interactions between these particles. First, in the region of space far from the soliton, the field approaches the vacuum, and the rate of approach is determined by the linearized field equation. Thus, if the linearized equation has no mass term, making the elementary particles massless, then the soliton's tail will be long-range, falling off with an inverse power of distance. If the linearized equation has a positive mass coefficient m, implying that the elementary particle mass is $m\hbar$ to first approximation, then the soliton field approaches the vacuum exponentially fast, the difference being e^{-mr} corrected by powers of r, at a distance r from the soliton core. This is called a Yukawa tail. Typically, the constant coefficient multiplying this is not determined by the linearized equation, but must be calculated using the full field equations solved throughout space.

Secondly, when one has two well separated solitons, their interaction energy depends on their separation in rather a simple way, completely determined by the linearized, asymptotic field of the solitons. The derivative of the interaction energy with respect to the separation is the force between the solitons.

Thirdly, the linearized theory can be used to describe the scattering of waves off the soliton. Here the equations are linearized around the soliton solution, rather than the vacuum. In two or three dimensions this involves partial wave analysis which shows that incoming plane waves emerge after collision with the soliton as radially scattered waves. This has the quantum interpretation [343]:*

Soliton + elementary particle $\longrightarrow$ Soliton + elementary particle.

There are simplifications in one space dimension.

Finally, the linearized waves are important in soliton-soliton scattering. Although topological charge conservation implies that solitons do not disappear, nevertheless some of the soliton kinetic energy can be converted into radiation during the scattering process, especially in a high energy, relativistic collision. The radiation disperses into space, and at low amplitude is described by the linearized field equations. Sometimes, the total amount of radiation can be estimated using the linearized theory, treating the moving solitons as sources, but this only works at modest collision speeds. The interpretation in the quantized field theory is that soliton-soliton collisions can produce elementary particles:

Soliton + Soliton $\longrightarrow$ Soliton + Soliton + elementary particles.

1.2 A brief history of topological solitons

There were a number of antecedents to the discovery of particle-like topological solitons in field theory. One of the first was Kelvin's vortex model of atoms [400]. Kelvin suggested that these could be represented by knotted structures in an ideal fluid. The topology of the knot would be unchanging, corresponding to the chemical immutability of atoms; the many distinct knot types would classify the many naturally occurring elements. The dynamics of the fluid, leading to vibrations of the knot shape, would explain atomic spectra. However, constructing partial differential equations (PDEs) with knot-like solutions is not easy. The Skyrme-Faddeev model, discussed at the end of Chapter 9, is a particularly successful example.

With the discovery of the electron, and later the constituents of the nucleus, our modern view of atoms emerged. Now the problem was to understand the subatomic particles. A point-like electron has an infinite Coulomb energy, classically. Abraham and Lorentz [269], and later Born and Infeld [59] made various attempts to give finite structure to the electron. The Abraham-Lorentz model gives the electron a distributed charge density, and requires short-range scalar interactions to stabilize the electric repulsion. The structure is mathematically rather arbitrary, and there are difficulties maintaining relativistic invariance. However, as

---

* Most references are postponed until the later, more specialized chapters, starting with Chapter 5. We only give references before then to key ideas and those that are not discussed later.

we shall see, the balancing of two types of force is characteristic of several types of topological soliton. The Born-Infeld model is a nonlinear variant of Maxwell's electrodynamics. Here, an electrically charged point source gives rise to finite field strengths and finite total energy. However, the fields are not smooth, having a discontinuous gradient at the source. The solution is not really a soliton. In both types of model, the length scale is the "classical electron radius", which is chosen such that the electrostatic field energy outside this radius, for a Coulomb field, is of order the electron mass.

One further antecedent was the Dirac monopole. This is a singular solution of the usual electromagnetic equations, with a net magnetic charge. It has a point singularity, and has infinite energy. However, the fields at a fixed distance r from the source are topologically interesting, and their topology is related to the magnetic charge, which can only occur in integer multiples of the quantum $2\pi\hbar/q$, where q is the basic unit of electric charge. We shall describe the Dirac monopole in detail in Chapter 8 prior to a discussion of soliton-like magnetic monopoles.

Historically, the first example of a topological soliton model of a particle was the Skyrmion. For a survey of Skyrme's pioneering work, see the book compiled by Brown [68]. The Skyrmion emerged from the Yukawa model, a field theory for spin $\frac{1}{2}$ nucleons (protons and neutrons) and the three types of spinless pion (π^+, π^-, π^0), with the relatively heavy nucleons interacting through pion exchange. Skyrme believed that the nucleons in a nucleus were moving in a nonlinear, classical pion medium. This made him reconsider the pion interaction terms. Symmetry arguments led to a particular form of Lagrangian for the three-component pion field, with a topological structure which allowed a topologically stable soliton solution of the classical field equation, distinct from the vacuum. This Skyrmion has rotational degrees of freedom, and Skyrme had the insight to see that when these were quantized it was quite permissible for the state to have spin $\frac{1}{2}$. Thus a purely bosonic field theory could lead to spin $\frac{1}{2}$ fermionic states, which could be identified as nucleons. Within Skyrme's model it therefore became unnecessary to include independent nucleons coupled to the pion fields. They emerge naturally as the soliton states of the theory. Subsequent work has shown that multi-Skyrmion solutions have some relation to nuclei, and recently there has been considerable progress finding classical multi-Skyrmion solutions with Skyrmion numbers up to 20 and beyond.

Skyrme at first found it challenging to analyse his pion field theory in three space dimensions. As a toy model he proposed a Lorentz invariant field theory in one space dimension, where the field has values on a circle. This is the sine-Gordon theory. Here there is also a topological soliton. Developing Skyrme's work, Coleman [85] and Mandelstam [274] later showed that an exact quantization of the sine-Gordon theory is possible, and it has both elementary meson states, analogous to pions, and solitons. The solitons behave to a certain extent as fermions, although in one dimension there is no possibility of spin.

Another, slightly earlier strand in the historical development came from condensed matter theory. Condensed matter systems are fundamentally quantal in nature, and non-relativistic, and involve complicated, many electron states. Against this background, a number of phenomenological approaches were developed based on classical field theory. The basic field usually represents a density of fermions, and is assumed to be slowly varying in space and time. This field carries sufficient information about the quantum state that one can write down an energy function for the field. This is the Ginzburg-Landau (GL) approach. Its most famous use is in describing superconductors, where a complex scalar field ϕ represents the density and phase of the superconducting paired electrons. GL theory is superceded by the more fundamental BCS theory [33] in certain circumstances, but is still valuable for studying spatially varying states, and types of superconductor where a more fundamental theory is lacking.

In the GL theory the field ϕ is coupled to the electromagnetic gauge potential a_μ. One basic feature is that in the lowest energy state, ϕ is a non-zero constant ϕ_0. The magnitude of ϕ_0 is determined by the GL energy function, but the phase is arbitrary. It may be assumed that ϕ_0 is real and positive, but this is a gauge choice. One says that the phase symmetry $\phi \mapsto e^{i\alpha}\phi$ is spontaneously broken. A consequence is that the electromagnetic field acquires a length scale, which accounts for the finite penetration depth of the magnetic field in a superconductor. In the relativistic generalization of GL theory, the photon acquires a mass.

It was discovered by Abrikosov in 1957 that the GL energy function has topological solitons. The topology arises from the fact that the vacuum manifold is a circle, since the phase of ϕ_0 is arbitrary. The solitons really only exist in the version of the theory in two space dimensions – they are called magnetic flux vortices, or simply vortices. Along a large circle surrounding a basic vortex, the phase of ϕ changes by 2π. In three dimensions these vortices extend into tubes, and they carry magnetic flux through the superconductor. These magnetic flux vortices persist in the relativistic theory, as shown by Nielsen and Olesen, and can be interpreted either as particles in two dimensions or as massive, relativistic strings in three dimensions. If they were present at very large scales in the universe, they would be called cosmic strings.

The GL theory is an example of a gauge theory with spontaneously broken gauge symmetry. The gauge group is the abelian group $U(1)$. In the late 1960s such models with non-abelian gauge symmetry were proposed to unify the electromagnetic and weak interactions. The famous electroweak theory of Glashow-Weinberg-Salam, with $SU(2) \times U(1)$ gauge symmetry spontaneously broken to $U(1)$ became established as part of the standard model of elementary particle physics, being confirmed by the subsequent discovery of the massive $W^{\pm}$ and Z gauge bosons in the 1980s. The breaking is produced by a Higgs scalar field, whose vacuum manifold is the orbit space $SU(2) \times U(1)/U(1)$, which is a 3-sphere.

The non-trivial topology of the 3-sphere encouraged a search for stable topological solitons in the electroweak theory, but it now appears fairly certain that none exist. However, the theory does have some non-trivial unstable solutions related to the topology. These solutions are called sphalerons (sphaleros ≡ unstable).

Until the Glashow-Weinberg-Salam theory was experimentally established, other gauge theories with spontaneously broken symmetry were considered. Particularly interesting among these is the Georgi-Glashow model, with $SO(3)$ symmetry broken to $U(1)$. The Higgs vacuum manifold here is $SO(3)/U(1)$, a 2-sphere. This is just the right structure to permit soliton solutions in three dimensions. The Higgs field defines a map from the 2-sphere at spatial infinity to the 2-sphere of the Higgs vacuum manifold, whose degree is the topological charge. As for the GL vortex, the topological structure is associated with a non-trivial magnetic field, which in this case points radially inwards or outwards. The Georgi-Glashow model therefore has magnetic monopoles as topological solitons. This was discovered independently by 't Hooft and Polyakov in 1974.

An important ingredient of the standard model of particle physics is the strongly interacting sector, described by quantum chromodynamics (QCD). QCD is a pure Yang-Mills gauge theory, without Higgs fields, coupled to fermionic quark fields. This is the theory to which the Skyrme model is possibly a low energy approximation [428]. If the quarks are ignored, there is just a pure gauge theory, with gauge group $SU(3)$, whose classical field equation is the Yang-Mills equation. This does not have soliton solutions in three space dimensions. However, the Yang-Mills equation in four space dimensions does have topological soliton solutions, known as instantons. The name arises because four-dimensional space can be regarded as Euclideanized space-time, where the Minkowski metric $ds^2 = dt^2 - d\mathbf{x} \cdot d\mathbf{x}$ is replaced by the Euclidean metric $ds^2 = dt^2 + d\mathbf{x} \cdot d\mathbf{x}$. A solution localized in four-dimensional space can therefore be interpreted as simultaneously localized in three-dimensional space and in (Euclidean) time. It therefore corresponds to a spatially localized event occurring in an instant.

In many circumstances, the amplitudes for quantum processes can be treated as Euclidean functional integrals. In the semi-classical approximation, the integrals are dominated by the classical solutions. Serious attempts have been made to understand quantum Yang-Mills theory, especially for gauge group $SU(2)$, by assuming that the functional integral is dominated by instantons [71]. This approach is hard to implement in pure Yang-Mills theory, because the integration must involve multi-instanton/multi-anti-instanton contributions. However, in supersymmetric Yang-Mills theory with certain geometries, just the multi-instantons contribute, and a precise calculation of quantum correlation functions can be made taking this into account [112].

1.3 Bogomolny equations and moduli spaces

A key discovery, which has aided the study of topological solitons in many field theories, is that the field equations can be reduced from second to first order PDEs, provided the coupling constants take special values. Several examples were exposed in a seminal paper of Bogomolny in 1976, and many others are now known. Generally, the first order equations are called Bogomolny equations. Bogomolny equations never involve time derivatives, and their solutions are static soliton or multi-soliton configurations.

Bogomolny showed that in these special field theories, the energy is bounded below by a numerical multiple of the modulus of the topological charge N, with equality if the field satisfies the Bogomolny equation. Thus, solutions of the Bogomolny equation of a given charge all have the same energy; and since the fields *minimize* the energy, they are automatically stable. In general the Hessian, or second variation of the energy, in the background of a static solution, has a spectrum consisting of a finite number of negative eigenvalues and a finite number of zero eigenvalues. The corresponding (normalizable) eigenfunctions are called negative and zero modes, respectively. There are also infinitely many positive eigenvalues. Bogomolny solitons are stable in the sense that they have no negative modes; although zero modes may still lead to rolling instabilities. They also automatically satisfy the Euler-Lagrange equations, which normally only imply that a static solution is a stationary point of the energy.

Kinks in one dimension are solutions of a Bogomolny equation, although this is rather a trivial case. The reduction of the GL equations for gauged vortices to a coupled pair of Bogomolny equations occurs at the critical value of the coupling separating the Type I and Type II superconducting regimes. Monopoles satisfy a Bogomolny equation if the Higgs field is massless. Instantons satisfy the self-dual Yang-Mills equation, which is like a Bogomolny equation in four-dimensional space.

The set of solutions to the Bogomolny equation of a particular theory is often large. The N-soliton solution space, with any gauge freedom quotiented out, is called the N-soliton moduli space, and denoted $\mathcal{M}_N$. It is a smooth manifold. Originally, index theorems were used to establish the dimension of $\mathcal{M}_N$. The dimension is a small integer multiple of the number of solitons, the multiple just counting the number of degrees of freedom of each soliton. Subsequently the global structure of $\mathcal{M}_N$ has been clarified.

The existence of non-trivial solutions of Bogomolny equations mathematically explains why the force between solitons is sometimes zero. For example, for two well separated monopoles, the total force between them is the sum of a magnetic repulsion and also a scalar attraction (because of the Higgs field). When the Higgs field is massless, both forces are long range, and in fact cancel. So far this only implies that the leading order $1/r^2$ force vanishes as $r \to \infty$. Without the Bogomolny equation, it would be much more difficult to understand the exact cancellation of forces. But Taubes proved that static solutions exist with monopoles at (essentially) any separation. The nature of such solutions is now understood in great detail. The fact that the energy of these solutions is independent of separation implies that the forces exactly cancel.

1.4 Soliton dynamics

So far, we have discussed static solitons in various field theories, and their interpretation as particles. An important issue is to understand the dynamics and interaction of these solitons. In a relativistic field theory, the dynamical equations are essentially uniquely determined as the relativistic generalization of the Euler-Lagrange equations for static fields. A soliton can be boosted to move at an arbitrary speed less than the speed of light. In a non-relativistic theory, like the GL theory of superconductors, it is not so easy to determine the correct equations for time dependent fields, and experimental input is needed. The dynamics may or may not be dissipative. However, we shall show that in almost all situations, solitons behave like ordinary particles.

When solitons are well separated they can be approximated as point-like objects carrying charges, or perhaps more complicated internal structure. The charges are defined in terms of the asymptotic form of the fields surrounding the soliton. One can calculate the forces between well separated solitons, and their relative motion, and interpret the result in terms of the charges. For example, Bogomolny monopoles carry a magnetic charge and a scalar charge. As we mentioned previously, for monopoles at rest the corresponding forces exactly cancel; however, for monopoles in relative motion, they do not cancel, and this results in velocity dependent forces, and hence accelerations. Also there is a net force between a monopole and antimonopole. The forces can be calculated directly from the time dependent field equations, or by considering integrals of the energy-momentum tensor. There is no need to postulate a force law for solitons, as one needs to do for electrically charged point particles, where the force

$$\mathbf{F} = \frac{q^2 \mathbf{r}}{4\pi r^3} \tag{1.1}$$

between two charges q at separation $\mathbf{r}$ is a basic postulate of electromagnetic theory, supplementing Maxwell's equations.

Thus topological solitons realize a dream of theoretical physics, which is to give a unified understanding of the existence and internal structure of particles, and of the dynamics and interactions of particles. All these things follow from the nonlinear dynamical field equations.

The treatment of solitons as point-like objects breaks down when the solitons come close together. If solitons collide at high speed, then the scattering behaviour can be very complicated, the only certainty being the conservation of topological charge. Thus if a soliton and antisoliton collide, they may annihilate and the energy emerge as wave-like radiation – the field pattern is generally complicated and can only be found numerically. Alternatively, the soliton and antisoliton may survive and separate, with a smaller amount of radiation being generated.

In a high energy soliton-soliton collision, there must be at least two solitons surviving the collision, but there can again be a complicated radiation pattern carrying away part of the energy, and part of this may convert into soliton-antisoliton pairs.

However, there are circumstances when soliton-soliton collisions occur rather gently, and adiabatically. The number of outgoing solitons equals the number of incoming solitons, and there is little accompanying radiation. This occurs generally in a theory with a Bogomolny equation, where the initial data are a field configuration close to a static multi-soliton solution, but perturbed a little to give the solitons some relative motion. An example is the collision (perhaps head-on) of two monopoles, where their initial velocities are small compared to the speed of light. The net force, being velocity dependent and vanishing at zero velocity, is small. One might imagine that in these circumstances the monopoles hardly interact at all, and just pass through each other, preserving their momenta. This is very far from the case. We shall show later that monopoles, and also vortices, in such a gentle head-on collision, usually scatter through a right angle.

The adiabatic dynamics of solitons can be approximated by a finite-dimensional dynamical system – the dynamics on moduli space. If the original field theory is second order in time derivatives of the fields, then the motion on moduli space reduces to geodesic motion. The metric on moduli space is not flat, and this is responsible for the non-trivial scattering. If the original field theory has couplings close to, but not exactly, the critical ones for the Bogomolny equation to be valid, then there is a residual potential energy defined on moduli space, and the adiabatic motion of solitons is approximately given by geodesic dynamics on moduli space modified by this potential. It is remarkable that the moduli space dynamics smoothly extends the asymptotic dynamics of solitons, where the solitons can be approximated as point-like, into the region where the solitons are close together relative to their size, and strongly deformed.

We shall discuss in detail the forces between most types of soliton, including kinks, vortices, monopoles and Skyrmions. We shall also give a detailed discussion of how the moduli space for vortices and monopoles is constructed, how the metric on moduli space can be found, and how one can study the geodesic dynamics on moduli space modelling second order adiabatic soliton motion. A remarkable feature of the metrics on moduli space is that they are often Kähler or hyperkähler. We shall also discuss some examples of field theories which are first order in time derivatives. These lead to first order dynamics on moduli space, where the initial soliton configuration, but not the soliton velocities, must be specified as initial data.

1.5 Solitons and integrable systems

The notion of soliton has often been used in recent decades in a somewhat different sense [2]. Certain partial differential equations have localized, smooth soliton solutions which do not disperse. Moreover if a number of these solitons are superposed at large separations, and set in motion, then there is a collision, but they emerge from the collision almost unchanged. The number of solitons is unchanged, and the momenta (and energies) before and after are all the same. If one could label the solitons, then one would say that the momenta had been permuted. The solitons just experience a time delay or time advance due to the collisions. Such behaviour is very interesting and it occurs in integrable PDEs. The conservation of number and momenta of solitons is a consequence of an infinite number of conservation laws, which means that after a suitable transformation of the field one can find an infinite set of decoupled dynamical variables each obeying rather simple dynamics. This kind of integrable soliton dynamics is most often studied in one space dimension. Classic examples are the KdV and sine-Gordon solitons. The latter is particularly interesting as a model of a particle, because the sine-Gordon theory is Lorentz invariant.

Integrable evolution equations do exist in more than one space dimension, such as the KP equation, which is a planar example. However, such multi-dimensional integrable systems usually break the spatial symmetry of the domain, so that the properties of a soliton depend upon the direction in which it travels.

The topological solitons that we shall consider in this book have the feature of being smooth and localized, and having a conserved particle number. They exist in theories in space dimensions up to four and beyond. But, with the exception of the sine-Gordon model, they do not exhibit integrability. There is inelastic and non-trivial scattering of topological solitons in collisions. If there were a Lorentz invariant, integrable Lagrangian field theory in two or more space dimensions, with solitons, that would be very interesting. But so far, despite considerable searching, no such model has been found. The models that we shall consider, which have topological soliton solutions, are sometimes Lorentz invariant. Even those that are not Lorentz invariant still possess the symmetry of the underlying spatial domain. Thus for a model defined in $\mathbb{R}^d$, flat d-dimensional space, there is at least Euclidean symmetry $E(d)$. Broadly speaking, theoretical physicists, though perhaps not all mathematicians, think that there are sufficient features of the particles we consider to call them solitons, even if there is no integrability.

In several cases, the first order, time independent Bogomolny equation for topological solitons is an integrable system. Although this does not imply that the explicit construction of multi-soliton solutions is always possible, it nevertheless implies that strong results can be derived concerning the solutions and the nature of the moduli spaces. This is a rather curious situation. It means that *static* multi-solitons are related to integrable systems, but the dynamics is not integrable.

The geodesic dynamics on moduli space is an approximation to the full field theory dynamics. One may ask if this reduced dynamics is integrable. It appears that in many cases, for example for vortices and for monopoles, it is not. Indeed, the metric on the two-monopole moduli space is known explicitly, because it is hyperkähler and because of symmetry. However, analytic and numerical investigations show that the geodesic motion on this moduli space is a non-integrable Hamiltonian dynamical system.

It is worthwhile to compare the radiation aspects of integrable soliton dynamics in one dimension with topological soliton dynamics. Systems like KdV are infinite-dimensional dynamical systems, where the N-soliton dynamics can be separated off as a finite-dimensional subsystem. The subsystem is integrable, and hence one can determine the soliton dynamics (e.g. time delays). Because the whole system is integrable, small amplitude radiation degrees of freedom are also integrable. These are the wave-like modes.

In the topological soliton situation, the N-soliton dynamics can also be separated off as a finite-dimensional dynamical system – the motion on moduli space. The latter system is exactly defined, because of the integrability of the Bogomolny equation. Thus, so far, the analogy with the integrable solitons is really close. The difference is that the remaining radiation modes do not completely decouple. The field theory dynamics of topological solitons does excite some radiation. Moreover, the radiation is not integrable, so exact results can not be obtained.

Atiyah and Hitchin have made the following comparison between integrable soliton dynamics, and the dynamics of topological solitons where there is a Bogomolny equation for static fields. They point out [17] that the classical equations of field theory (e.g. for monopoles) may be regarded as an exact description of at least a simplified model of nature. The moduli space of static solitons is a precise finite-dimensional truncation of great mathematical beauty, but one must accept that the geodesic dynamics on moduli space only approximately describes the soliton dynamics. On the other hand, equations like KdV are not exact. KdV emerges from an approximation to the equations for real fluid waves after making a number of assumptions about the amount of nonlinearity, the depth of fluid, and the directionality of the waves. However, KdV is an essentially solvable PDE, and soliton scattering properties can be calculated exactly. So moduli space dynamics of topological solitons is an approximate treatment of an exact equation, whereas integrable soliton dynamics is an exact treatment of an approximate equation. Both are only an approximation to the truth.

1.6 Solitons – experimental status

Let us conclude this introduction with some remarks on the physical status of the ideas presented in this book. So far, there are rather limited experimental tests of many of the ideas and mathematical results.

There are a number of physical systems which carry one-dimensional solitons, e.g. optical fibres, and narrow water channels [115]. The solitons are sometimes described by an integrable model, or a near-integrable variant. Sigma model lumps occur as solitons in certain idealized, planar ferromagnetic and antiferromagnetic systems in the continuum approximation [138]. They are mathematically interesting because the static soliton solutions are rational functions of a single complex variable, and thus can easily be written down explicitly. Moreover, as we shall see, rational maps play an important role in the theory of monopoles and Skyrmions. The two-dimensional Ginzburg-Landau vortices are observed as solitons in thin superconductors, and as extended vortices in three-dimensional superconductors [326, 231]. A dissipative dynamical equation is relevant for most superconductors, so there has not yet been the possibility of firing vortices at one another and observing the right-angle scattering. Very few superconductors are close to the critical coupling separating Type I/Type II; the nearest are niobium and certain lead alloys. So the

Bogomolny equations for static multi-vortex solutions have not been very important. Some versions of first order Chern-Simons vortex dynamics may be relevant to quantum Hall systems, but this needs clarification [439, 348]. Global vortices have recently been created experimentally as extended strings in three-dimensional Bose-Einstein condensates composed of trapped dilute alkali gases [298]. Relativistic, cosmic strings have not been observed [202].

Models which give the electron a finite structure are not supported experimentally. Quantum field theory, and in particular quantum electrodynamics, can deal with a fundamental point-like electron through the renormalization programme, although ultimately this may be unsatisfactory.

Magnetic monopoles have not been observed, despite a long history of searching [161]. Fortunately, the standard model of elementary particles has no monopole solutions. Certain Grand Unified Theories have them, and the non-observance of monopoles severely constrains these models [438, 341]. The standard electroweak model does have unstable, sphaleron solutions, and they are in principle formed at the 10 TeV energy scale, which may be accessible in future particle accelerators. The crucial challenge before this is to find the Higgs particle – without the Higgs the whole mathematical structure leading to monopoles and sphalerons is in doubt.

Skyrmions remain an interesting possible model for nucleons and nuclei. For a single nucleon, the model works well, and the model also gives a reasonable description of the deuteron – the bound state of a proton and a neutron. However, the Skyrme model predictions for larger nuclei need further analysis. There is some evidence that the spin states of multi-Skyrmions match those of a number of nuclei, but there is no evidence yet that the surprisingly symmetric, classical multi-Skyrmion shapes give new insight into nuclear structure.

The study of pure quantized Yang-Mills theory using instantons is not yet finalized. Maybe the theory of multi-instantons, and their moduli spaces, will play an important role. Instantons are relevant to the dynamics, as is clear from lattice QCD studies. However, the whole problem of understanding quantum Yang-Mills theory, and quark confinement, remains open.

Should supersymmetry be discovered there would be much new interest in solitons, and especially in solitons described by Bogomolny equations, as the critical couplings that make Bogomolny equations possible are the same as those that arise in supersymmetric theories [429, 188]. Solitons in superstring theory [335] are currently being enthusiastically studied, where they are known as "branes".

1.7 Outline of this book

We shall explore all the examples of topological solitons that have been mentioned so far, and a few more. The organization is as follows. We shall present in Chapter 2 some essential background ideas, concerning Lagrangians and field equations, the role of symmetries, and the structure of gauge theories and their physical content. In Chapter 3 we present ideas and calculational methods from topology that are needed to understand topological solitons. In Chapter 4 we discuss some further general ideas and methods that can be applied to study several types of soliton.

Chapter 5 is about solitons in one-dimensional field theories, especially the ϕ^4 kink and sine-Gordon soliton. Chapters 6 and 7 are about solitons in two space dimensions. The former deals with sigma model lumps and two-dimensional (Baby) Skyrmions whereas the latter is concerned with GL vortices both in gauged and ungauged theories. Chapter 8 is on monopoles, and Chapter 9 on Skyrmions, both three-dimensional solitons. Chapter 10 is about Yang-Mills instantons, and Chapter 11 is on sphaleron solutions.

This book focusses on the classical solutions describing solitons and their dynamics. Readers may well be interested in soliton quantization – the treatment of solitons and their dynamics in quantum field theory. The coupling of solitons to additional bosonic and fermionic quantum fields is also interesting, and leads to ideas of fractional charge, and to the study of solitons in supersymmetric theories. However, these topics would require a further book (with other authors) to survey them in depth, and with the exception of Skyrmion quantization, they are not discussed here.

2

Lagrangians and fields

In this book we shall be dealing with classical field dynamics, and also classical particle dynamics. We shall be showing how topological solitons in field theory behave like particles, and this will involve reducing the field equations to an effective particle dynamics. Field theory has an infinite number of dynamical degrees of freedom, whereas particle dynamics has a finite number proportional to the number of particles. We need to understand the structure of field and particle dynamics separately before looking at the relationship between them. Particle dynamics, being a finite-dimensional system, is conceptually more basic, so we look at that first. If the dynamics is non-dissipative, we shall use the Lagrangian formalism. The simplest dissipative equation of motion, the gradient flow, can easily be deduced from a Lagrangian structure.

2.1 Finite-dimensional systems

In Lagrangian dynamics, the configuration space is a smooth manifold M, of dimension D, say. The system is represented by one point in M, varying with time along a smooth trajectory. Let $\mathbf{q} = (q^1, \ldots, q^D)$ denote local coordinates in M. The trajectory is expressed either as $\mathbf{q}(t)$, or in component form as $q^i(t)$, $1 \leq i \leq D$.

To define the Lagrangian function, we need further structure on M. We consider three types, though they need not all be present at once. The first is a potential energy function – a scalar function on M given locally as $V(\mathbf{q})$. The second is a Riemannian metric on M. This is given locally by a symmetric non-degenerate rank 2 tensor $g_{ij}(\mathbf{q})$. The last type of structure is an abelian gauge potential, locally a covariant rank 1 tensor or 1-form on M, with components denoted by $a_i(\mathbf{q})$. We shall denote, in the standard way, the inverse of the metric by g^{ij}, so $g^{ij}g_{jk} = \delta^i_k$ with δ^i_k the Kronecker delta symbol ($\delta^i_k = 1$ if $i = k$, 0 otherwise). Here and below, we use the summation convention; if an index is repeated, it is summed over.

Given these data on M, and a trajectory $\mathbf{q}(t)$, we define the Lagrangian

$$L = \frac{1}{2}g_{ij}(\mathbf{q})\dot{q}^i\dot{q}^j - a_i(\mathbf{q})\dot{q}^i - V(\mathbf{q}) \, . \tag{2.1}$$

At a given time t, L depends on the position $\mathbf{q}$ and velocity $\dot{\mathbf{q}} = \frac{d\mathbf{q}}{dt}$, and is assumed to be no more than quadratic in the velocity. We refer to the first two terms in L as kinetic terms. Although the Lagrangian is defined locally in terms of some coordinate system, it is actually coordinate invariant if we suppose that g_{ij} and a_i transform as covariant tensor fields on M.

The dynamical principle determining the trajectory $\mathbf{q}(t)$ is the principle of stationary action. We consider all trajectories which begin at $\mathbf{q}^{(1)}$ at $t = t_1$ and end at $\mathbf{q}^{(2)}$ at $t = t_2$ $(t_2 > t_1)$. (The velocities at t_1 and t_2 do not need to be specified.) We define the action

$$S = \int_{t_1}^{t_2} L(t)\,dt \, , \tag{2.2}$$

for any trajectory satisfying the initial and final condition. The actual motion is one for which S is stationary.

The standard result in the calculus of variations implies that the true motion satisfies the Euler-Lagrange equation

$$\frac{d}{dt}\left(\frac{\partial L}{\partial \dot{q}^i}\right) - \frac{\partial L}{\partial q^i} = 0 \, , \tag{2.3}$$

which here takes the form

$$\frac{d}{dt}\left(g_{ij}\dot{q}^j - a_i\right) - \frac{\partial}{\partial q^i}\left(\frac{1}{2}g_{jk}\dot{q}^j\dot{q}^k - a_j\dot{q}^j - V\right) = 0 \, . \tag{2.4}$$

This equation of motion can be reexpressed as

$$g_{ij}(\ddot{q}^j + \Gamma^j_{kl}\dot{q}^k\dot{q}^l) + f_{ij}\dot{q}^j + \partial_i V = 0 \, , \tag{2.5}$$

where ∂_i denotes $\frac{\partial}{\partial q^i}$ and

$$f_{ij} = \partial_i a_j - \partial_j a_i \tag{2.6}$$

is the gauge field strength, an antisymmetric tensor or 2-form, and

$$\Gamma^j_{kl} = \frac{1}{2}g^{ij}(\partial_k g_{li} + \partial_l g_{ki} - \partial_i g_{kl}) \tag{2.7}$$

is the Levi-Civita connection on M.

The Euler-Lagrange equation appears to be more fundamental than the action principle that gives rise to it, since other variational principles give the same equation. We shall regard any solution of the Euler-Lagrange equation as a physically acceptable motion, without worrying about initial and final data. The solution is unique if we specify the initial position and initial velocity.

Suppose, locally, we replace $a_i(\mathbf{q})$ by $a_i(\mathbf{q}) + \partial_i \alpha(\mathbf{q})$ where α is a scalar function on M. Then L changes by the subtraction of

$$\partial_i \alpha \, \dot{q}^i = \dot{\alpha} \,, \tag{2.8}$$

a total time derivative. The action S changes by

$$\alpha(\mathbf{q}^{(1)}) - \alpha(\mathbf{q}^{(2)}) \,, \tag{2.9}$$

which is independent of the trajectory joining $\mathbf{q}^{(1)}$ to $\mathbf{q}^{(2)}$. Thus, this change of a_i has no effect on the equation of motion, and can be regarded as having no effect at all (at least classically). Such a change of a_i is a gauge transformation. Notice that the equation of motion only involves f_{ij}, which is gauge invariant. In fact, because of this gauge invariance, a_i need not be a globally defined 1-form; it need only be a local 1-form representing a connection on a $U(1)$ bundle over M. We expand on this in Section 3.4 below.

The law of conservation of energy is obtained by multiplying (2.5) by $\dot{q}^i$ and summing over i. One finds that

$$\frac{d}{dt}\left(\frac{1}{2}g_{ij}\dot{q}^i\dot{q}^j + V\right) = 0 \,, \tag{2.10}$$

and hence

$$\frac{1}{2}g_{ij}\dot{q}^i\dot{q}^j + V = E \,, \tag{2.11}$$

where the constant E is the conserved total energy. To derive this we have used the fact that f_{ij} is antisymmetric. Although $\dot{\mathbf{q}}$ is the coordinate velocity, on a Riemannian manifold the geometrical speed v – the distance moved per unit time – depends on the metric, and is given by the formula $v = \sqrt{g_{ij}\dot{q}^i\dot{q}^j}$. Therefore, the first term in E is the kinetic energy $\frac{1}{2}v^2$.

In the simplest case of a constant potential V, and no gauge field, the Euler-Lagrange equation is

$$\ddot{q}^j + \Gamma^j_{kl}\dot{q}^k\dot{q}^l = 0 \,, \tag{2.12}$$

which is the equation for geodesic motion on M. Conservation of energy here implies that the motion along the geodesic is at constant speed. Note that solutions of the Euler-Lagrange equation with given initial and final data need not be unique. For example, there are infinitely many geodesic trajectories connecting two points on a sphere; provided the points are non-coincident and non-antipodal, these geodesics all lie on the same great circle, but involve different numbers of rotations around it, at different speeds.

A special case of the above Lagrangian formalism is the Newtonian dynamics of one or more particles. For one particle of mass m moving in d dimensions, the manifold M is the Cartesian space $\mathbb{R}^d$ with the Euclidean metric $g_{ij} = m\delta_{ij}$. We denote the (Cartesian) trajectory of the particle by $\mathbf{x}(t)$. The Lagrangian is

$$L = \frac{1}{2}m\dot{x}^i\dot{x}^i - a_i(\mathbf{x})\dot{x}^i - V(\mathbf{x})\,, \tag{2.13}$$

and the equation of motion

$$m\ddot{x}^i = -f_{ij}\dot{x}^j - \partial_i V\,. \tag{2.14}$$

There is a static force due to V, and a velocity dependent force due to the gauge field. In the absence of forces, the equation of motion is simply

$$m\ddot{x}^i = 0\,, \tag{2.15}$$

whose solutions are straight line trajectories at constant speed.

In three dimensions one may define the axial vector magnetic field $b_i = -\frac{1}{2}\varepsilon_{ijk}f_{jk}$, where ε_{ijk} is the alternating tensor: $\varepsilon_{ijk} = 1\,(-1)$ if (i,j,k) is an even (odd) permutation of $(1,2,3)$, and 0 otherwise. ($\mathbf{a}$ is a covariant vector or 1-form, and the usual vector potential of electromagnetic theory, being the spatial part of a contravariant 4-vector, is $-\mathbf{a}$. Hence $\mathbf{b} = \nabla \times (-\mathbf{a})$.) One may also regard V as the electrostatic potential, and $-\partial_i V$ as the electric field e_i. Then (2.14) becomes, in vector notation

$$m\ddot{\mathbf{x}} = \dot{\mathbf{x}} \times \mathbf{b} + \mathbf{e}\,, \tag{2.16}$$

which is the Lorentz force law for a non-relativistic particle of mass m, and unit electric charge. But note that interpretations of Eq. (2.14), other than this electromagnetic one, may arise.

For N similar particles, $M = \mathbb{R}^d \otimes \cdots \otimes \mathbb{R}^d = \mathbb{R}^{Nd}$ with its Euclidean metric. The trajectory of the N particles is $(\mathbf{x}_1(t),\ldots,\mathbf{x}_N(t))$, and the Newtonian form of the Lagrangian is

$$L = \sum_{r=1}^{N}\frac{1}{2}m\dot{x}^i_r\dot{x}^i_r - \sum_{r=1}^{N}a_i^{(r)}(\mathbf{x}_1,\ldots,\mathbf{x}_N)\dot{x}^i_r - V(\mathbf{x}_1,\ldots,\mathbf{x}_N)\,. \tag{2.17}$$

The potential V is a single function on $\mathbb{R}^{Nd}$, and the functions $\{a_i^{(r)}\}$ may be thought of as Nd components of a single 1-form on $\mathbb{R}^{Nd}$. If there are no background fields on $\mathbb{R}^d$, the particles are influenced only by each other. In this case there is translational and rotational symmetry. In particular $\{\mathbf{a}^{(r)}\}$ and V depend only on the differences in the particle positions $\mathbf{x}_r - \mathbf{x}_s$. We shall discuss the consequences of symmetries further in Section 2.2.

A small generalization of the Newtonian dynamics is where the N particles are moving on a given d-dimensional Riemannian manifold X, with local coordinates $\mathbf{x}$ and metric $h_{ij}(\mathbf{x})$. The manifold M is X^N, the Nth Cartesian power of X, and the Lagrangian is

$$L = \sum_{r=1}^{N} \frac{1}{2} h_{ij}(\mathbf{x}_r) \dot{x}_r^i \dot{x}_r^j - \sum_{r=1}^{N} a_i^{(r)}(\mathbf{x}_1, \dots, \mathbf{x}_N) \dot{x}_r^i - V(\mathbf{x}_1, \dots, \mathbf{x}_N). \quad (2.18)$$

Again, if the particles are only interacting with each other, then the symmetries of X will constrain the form of $\{\mathbf{a}^{(r)}\}$ and V.

An interesting issue is whether the permutation symmetry between the N particles has any significance. Acting on X^N, the permutation group S_N permutes $\{\mathbf{x}_1, \dots, \mathbf{x}_N\}$ in all possible ways. Since the particles are similar, the functions $\{\mathbf{a}^{(r)}\}$ and V are invariant under permutations. Thus

$$\mathbf{a}^{(r)}(\mathbf{x}_{1'}, \dots, \mathbf{x}_{N'}) = \mathbf{a}^{(r')}(\mathbf{x}_1, \dots, \mathbf{x}_N) \quad (2.19)$$

$$V(\mathbf{x}_{1'}, \dots, \mathbf{x}_{N'}) = V(\mathbf{x}_1, \dots, \mathbf{x}_N), \quad (2.20)$$

where $(1', \dots, N')$ is a permutation of $(1, \dots, N)$. This just ensures that if the labels of the particles are permuted, the trajectories are unchanged.

Provided the particles are at N distinct points, one may quotient X^N by S_N and obtain a Lagrangian system on the quotient. Locally, the Lagrangian and equations of motion look the same. However, this quotient generally leads to singularities where two or more particles become coincident. The dynamics on the quotient does not unambiguously tell us how particles emerge from a collision. However, the original dynamics on X^N is unproblematic. If the potential V has no singularity when particles collide, then particles simply pass through each other if they have sufficient energy. One may label the particles 1 to N at some instant, and then follow their trajectories, which are always smooth. One does not lose track of the labels, even in a collision.

Later, we shall be investigating field theory defined on a Riemannian space X (often $X = \mathbb{R}^d$) which admits topological solitons. We shall show that N-soliton dynamics can be approximated by a reduced Lagrangian system of type (2.1), where M is a manifold of dimension proportional to N (in the simplest case $\dim M = Nd$). We shall also show that the solitons, when they are well separated, behave as similar, independent particles. In this well separated regime, the reduced dynamics is of the form (2.18), with the functions $\{\mathbf{a}^{(r)}\}$ and V being small and only affecting the dynamics slightly. To first approximation *each* soliton follows a geodesic on X. So M has the local asymptotic form X^N, with the corresponding product metric. Now, because the solitons are well separated, we can safely take the quotient, and regard M as asymptotic to X^N/S_N. It turns out that this quotienting is actually the required thing to do in field theory. The solitons can not be labelled in a natural way. When soliton positions are permuted, the *field* is *identical* before and after the permutation. However, this seems to produce a difficulty, because X^N/S_N is potentially singular when points in X coincide. What happens is that the finite size of the solitons prevents a singularity. M is not globally X^N/S_N, but only approaches it in the asymptotic regime. M is a manifold which smooths out the singularities of X^N/S_N. Even the topology of M is sometimes different from that of X^N/S_N or X^N. This is very remarkable. It means that a Newtonian description of N solitons is possible while they are far apart, but it breaks down completely when they are close together. Solitons can not consistently be labelled along trajectories. Indeed the trajectory on M does not generally have an interpretation as a set of N unlabelled trajectories in X.

Two first order dynamical systems are related to the Lagrangian (2.1). One of these is the Lagrangian with no quadratic terms in velocity

$$L = -a_i(\mathbf{q})\dot{q}^i - V(\mathbf{q}) \,. \tag{2.21}$$

The equation of motion is

$$f_{ij}\dot{q}^j + \partial_i V = 0 \,. \tag{2.22}$$

This determines $\dot{q}^i$ and hence the trajectory $\mathbf{q}(t)$, given initial data $\mathbf{q}(t_1) = \mathbf{q}^{(1)}$, provided f_{ij} is invertible. In this situation, M can be regarded as a phase space, with $-f_{ij}$ the symplectic 2-form and V the Hamiltonian, and (2.22) Hamilton's equations. V is conserved, since

$$\frac{dV}{dt} = \partial_i V \dot{q}^i = -f_{ij}\dot{q}^j \dot{q}^i = 0 \,, \tag{2.23}$$

the last equality following from the antisymmetry of f_{ij}.

The other first order system is the gradient flow associated to the Riemannian manifold M with potential V. This is

$$\kappa g_{ij}\dot{q}^j + \partial_i V = 0 \,, \tag{2.24}$$

where κ is a fixed positive constant. If $M = \mathbb{R}^d$ with the Euclidean metric, then (2.24) simplifies to

$$\kappa \dot{x}^i + \partial_i V = 0 \,. \tag{2.25}$$

The only difference from the Newtonian equation

$$m\ddot{x}^i + \partial_i V = 0 \tag{2.26}$$

is that second time derivatives are replaced by first time derivatives (and the constant m replaced by κ for dimensional reasons).

The interpretation of the gradient flow equation on M is as follows. At any point $\mathbf{q}$ of M, $g^{ij}\partial_j V$ is a vector orthogonal to the contour of V (hypersurface of constant V) through $\mathbf{q}$, pointing in the direction of increasing V, and of magnitude $\frac{dV}{ds}$ where s is the distance in the direction orthogonal to the contour. Equation (2.24) states that the velocity is the negative of this vector, divided by κ. The gradient flow equation implies that

$$\frac{dV}{dt} = \partial_i V \dot{q}^i = \begin{cases} -\kappa g_{ij}\dot{q}^i\dot{q}^j \\ -\frac{1}{\kappa}g^{ij}\partial_i V\partial_j V \, . \end{cases} \tag{2.27}$$

Both expressions show that V decreases along a gradient flow trajectory. In fact, a gradient flow trajectory is a path of steepest descent.

Suppose $\mathbf{q}(t)$ ($-\infty < t < \infty$) is a complete trajectory, satisfying Eq. (2.24). If $\mathbf{q}_{\pm\infty} = \lim_{t\to\pm\infty}\mathbf{q}(t)$ both exist, then $\lim_{t\to\pm\infty}\dot{\mathbf{q}}(t) = 0$, so $\mathbf{q}_{\pm\infty}$ are stationary points of V. The gradient flow joins these two stationary points. If M is compact, and V a non-singular function, then all non-trivial gradient flow trajectories have good limits as $t \to \pm\infty$, and connect a higher to a lower stationary point of V.

2.2 Symmetries and conservation laws

Suppose a Lie group G of symmetries acts on M, leaving invariant the metric g_{ij}, the 1-form a_i, and the potential V. Then there is a set of conservation laws for solutions of the equation of motion (2.5), one for each generator of the group.

Let $\mathbf{q} \mapsto \mathbf{q} + \varepsilon\boldsymbol{\xi}(\mathbf{q})$ be the infinitesimal action of one of the generators of G, where $\boldsymbol{\xi}$ is a vector field on M, and ε is an infinitesimal parameter. The metric is invariant if

$$\mathcal{L}_{\boldsymbol{\xi}} g_{ij} = 0 \, , \tag{2.28}$$

where $\mathcal{L}_{\boldsymbol{\xi}}$ denotes the Lie derivative in the direction of $\boldsymbol{\xi}$. Explicitly, the Lie derivative of the metric is

$$\mathcal{L}_{\boldsymbol{\xi}} g_{ij} = \xi^k \partial_k g_{ij} + \partial_i \xi^k g_{kj} + \partial_j \xi^k g_{ik} \, . \tag{2.29}$$

This can be reexpressed using the Levi-Civita connection, so that (2.28) reduces to the Killing equation

$$\xi_{i;j} + \xi_{j;i} = 0 \, , \tag{2.30}$$

where $\xi_i = g_{ij}\xi^j$ and $\xi_{i;j} = \partial_j\xi_i - \Gamma^k_{ij}\xi_k$ is its covariant derivative. A group G leaving the metric on M invariant is called a group of isometries of M and an individual generating vector field $\boldsymbol{\xi}$ satisfying (2.30) is called a Killing vector. The gauge potential a_i is invariant if

$$\mathcal{L}_{\boldsymbol{\xi}}a_i = \partial_i\alpha_{\boldsymbol{\xi}}\,,\qquad(2.31)$$

where $\alpha_{\boldsymbol{\xi}}$ is a scalar function on M, also linearly dependent on $\boldsymbol{\xi}$. The interpretation of (2.31) is that the Lie derivative of a_i does not need to be strictly zero although it could be; it is sufficient that the Lie derivative equals a gauge transformation, with generator $\alpha_{\boldsymbol{\xi}}$. Explicitly,

$$\xi^j\partial_j a_i + \partial_i\xi^j a_j = \partial_i\alpha_{\boldsymbol{\xi}}\,.\qquad(2.32)$$

Under a general gauge transformation, $a_i \mapsto a_i + \partial_i\alpha$, the left-hand side of this equation changes by the addition of $\partial_i(\xi^j\partial_j\alpha)$, so there needs to be an associated gauge transformation

$$\alpha_{\boldsymbol{\xi}} \mapsto \alpha_{\boldsymbol{\xi}} + \xi^j\partial_j\alpha\,.\qquad(2.33)$$

Finally, the potential V is invariant if

$$\mathcal{L}_{\boldsymbol{\xi}}V = 0\,.\qquad(2.34)$$

Explicitly,

$$\xi^i\partial_i V = 0\,.\qquad(2.35)$$

We now investigate the effect on the Lagrangian (2.1) of a shift in the trajectory $\mathbf{q}(t)$ to $\mathbf{q}(t)+\varepsilon\boldsymbol{\xi}(\mathbf{q}(t))$. The velocity $\dot{q}^i$ changes to $\dot{q}^i + \varepsilon\partial_k\xi^i\dot{q}^k$, so

$$\begin{aligned}
\delta L &= \varepsilon\Big(\frac{1}{2}(\partial_k g_{ij}\xi^k\dot{q}^i\dot{q}^j + g_{ij}\partial_k\xi^i\dot{q}^k\dot{q}^j + g_{ij}\dot{q}^i\partial_k\xi^j\dot{q}^k)\\
&\qquad -(\partial_k a_i\xi^k\dot{q}^i + a_i\partial_k\xi^i\dot{q}^k) - \partial_i V\xi^i\Big)\\
&= -\varepsilon(\partial_i\alpha_{\boldsymbol{\xi}})\dot{q}^i\\
&= -\varepsilon\dot{\alpha}_{\boldsymbol{\xi}}\,,
\end{aligned}\qquad(2.36)$$

using the invariance conditions given above. Since δL is a total time derivative, Noether's theorem asserts that there is a conserved quantity [214]

$$\begin{aligned}
Q_{\boldsymbol{\xi}} &= \xi^i\frac{\partial L}{\partial\dot{q}^i} + \alpha_{\boldsymbol{\xi}}\\
&= g_{ij}\xi^i\dot{q}^j - \xi^i a_i + \alpha_{\boldsymbol{\xi}}\\
&= g_{ij}\xi^i\dot{q}^j + \psi_{\boldsymbol{\xi}}\,.
\end{aligned}\qquad(2.37)$$

The first term is the inner product of $\boldsymbol{\xi}$ with the velocity vector $\dot{\mathbf{q}}$. The quantity $\psi_{\boldsymbol{\xi}} \equiv -\xi^i a_i + \alpha_{\boldsymbol{\xi}}$ is gauge invariant, and it is directly related to the field tensor f_{ij}. Using (2.31), one finds that

$$\xi^i f_{ij} = \partial_j \psi_{\boldsymbol{\xi}} \,. \tag{2.38}$$

One may prove that $Q_{\boldsymbol{\xi}}$ is independent of time using a variant of the argument given for field theory in Section 2.4. Alternatively, one may verify this directly, using the equation of motion (2.5) and the invariance conditions.

For the Lagrangian with no quadratic term in velocities (2.21), the conserved quantity $Q_{\boldsymbol{\xi}}$ associated with $\boldsymbol{\xi}$ is simply $\psi_{\boldsymbol{\xi}}$. This is easily verified by contracting the equation of motion (2.22) with ξ_i.

For the gradient flow, given by equation (2.24), $Q_{\boldsymbol{\xi}} = g_{ij}\xi^i \dot{q}^j$. One can verify not only that $Q_{\boldsymbol{\xi}}$ is conserved, but that $Q_{\boldsymbol{\xi}} = 0$, since in this case the equation of motion (2.24) implies that

$$g_{ij}\xi^i \dot{q}^j = -\frac{1}{\kappa}\xi^i \partial_i V = 0 \,. \tag{2.39}$$

Thus the gradient flow is orthogonal to the vector field $\boldsymbol{\xi}$. More generally, the flow is orthogonal to orbits of G. Since G preserves V, this is not surprising, since the orbits of G are submanifolds of the hypersurfaces $V = \text{const}$, and the flow is orthogonal to these hypersurfaces.

2.3 Field theory

Classical Lagrangian field theory is concerned with the dynamics of one or more fields defined throughout space and evolving in time. Let us first suppose space is $\mathbb{R}^d$. Space-time is $\mathbb{R} \times \mathbb{R}^d$. Local coordinates are $x = (t, \mathbf{x})$, and we shall often identify $x^0 = t$. The simplest field is a scalar field ϕ, a function on $\mathbb{R} \times \mathbb{R}^d$, denoted locally by $\phi(t, \mathbf{x})$.

One may regard the field as representing an infinite number of dynamical degrees of freedom. Formally, the value of ϕ at each spatial point is one degree of freedom, which evolves in time. Generally, the field values at distinct points are coupled together, because the Lagrangian depends not just on ϕ and its time derivative $\partial_0 \phi$, but also on its space derivatives $\boldsymbol{\nabla}\phi$. The components of $\boldsymbol{\nabla}\phi$ are $\partial_i \phi$, where ∂_i now denotes $\frac{\partial}{\partial x^i}$. As in the finite-dimensional systems, we suppose that the Lagrangian depends on $\partial_0 \phi$ polynomially, with no higher than quadratic terms. Similarly, we suppose that the dependence on $\boldsymbol{\nabla}\phi$ is polynomial.

The simplest type of Lagrangian for the field ϕ is

$$L = \int \left(\frac{1}{2}(\partial_0 \phi)^2 - \frac{1}{2}\boldsymbol{\nabla}\phi \cdot \boldsymbol{\nabla}\phi - U(\phi)\right) d^d x \,. \tag{2.40}$$

The Lagrangian density $\mathcal{L}$, the integrand here, is a local quantity depending isotropically on $\nabla\phi$. The potential U is some function of ϕ (and not explicitly dependent on $\mathbf{x}$), often taken to be a polynomial.

There is a natural splitting of this Lagrangian into kinetic energy and potential energy terms, $L = T - V$. We call

$$T = \int \frac{1}{2}(\partial_0\phi)^2 \, d^d x \tag{2.41}$$

the kinetic energy of the field, and

$$V = \int \left(\frac{1}{2}\nabla\phi \cdot \nabla\phi + U(\phi) \right) d^d x \tag{2.42}$$

the potential energy. Note that the potential energy is defined for a field at a given time, and it has a gradient energy contribution in addition to that of the potential U. It is important that the potential energy is bounded below, otherwise the dynamics is liable to produce singular fields. This justifies the choice of sign in front of the gradient term. Also U should be bounded below, which means that if it is a polynomial its leading term should be an even power with positive coefficient. We shall almost always arrange that the minimal value of U is zero.

For a field theory defined in $\mathbb{R} \times \mathbb{R}^d$ we shall always insist on Euclidean invariance in $\mathbb{R}^d$, and time translation invariance. The Euclidean group $E(d)$ combines spatial translations and rotations. Translational symmetry is ensured by having no explicit dependence on $\mathbf{x}$ in the Lagrangian, and integrating over the whole of $\mathbb{R}^d$ using the standard measure. Rotational invariance requires combining the gradient terms into a scalar, as in (2.40). Acceptable generalizations of (2.40) could involve a kinetic term

$$\int (\nabla\phi \cdot \nabla\phi)(\partial_0\phi)^2 \, d^d x , \tag{2.43}$$

and a term similar to this occurs in the Skyrme model. Terms linear in $\partial_0\phi$ are also possible, but must not just be total time derivatives. One of the simplest possibilities occurs in a theory with two fields ϕ_1 and ϕ_2, where the kinetic term could include

$$\int (\phi_1\partial_0\phi_2 - \phi_2\partial_0\phi_1) \, d^d x . \tag{2.44}$$

The potential energy could include further terms like

$$\int (\nabla\phi \cdot \nabla\phi)^2 \, d^d x , \tag{2.45}$$

or

$$\int W(\phi)\nabla\phi \cdot \nabla\phi \, d^d x , \tag{2.46}$$

for some positive function W.

The field theory with Lagrangian (2.40) possesses more than the Euclidean symmetry $E(d)$. Because the time and space derivatives of ϕ both occur quadratically and with a relative minus sign, the theory is Lorentz invariant (the speed of light is unity). Lorentz invariance allows additions to (2.40) of the form

$$\int W(\phi) \left((\partial_0 \phi)^2 - \boldsymbol{\nabla}\phi \cdot \boldsymbol{\nabla}\phi \right) d^d x. \tag{2.47}$$

Lorentz invariance is vital for theories purporting to describe elementary particles in Minkowski space-time $\mathbb{R} \times \mathbb{R}^d$. Many of the field theories we shall consider have Lorentz invariance. In such theories, the soliton dynamics will be Lorentz invariant. However, some theories are intended to describe condensed matter systems, and Lorentz invariance is then not required. In Lorentz invariant theories we shall often use the more condensed notation $\partial_\mu \phi \partial^\mu \phi$ to denote $(\partial_0 \phi)^2 - \boldsymbol{\nabla}\phi \cdot \boldsymbol{\nabla}\phi$. Generally, Greek indices will run from 0 to d in $(d+1)$-dimensional Minkowski space-time, and will be raised or lowered using the Minkowski metric $\eta^{\mu\nu}$, with signature $(1, -1, \ldots, -1)$.

The action associated with a Lagrangian density $\mathcal{L}(\partial_0 \phi, \boldsymbol{\nabla}\phi, \phi)$ is

$$S = \int_{t_1}^{t_2} L \, dt = \int_{t_1}^{t_2} \int \mathcal{L}(\partial_0 \phi, \boldsymbol{\nabla}\phi, \phi) \, d^d x \, dt. \tag{2.48}$$

The action principle is that S is stationary for given initial and final data defined throughout $\mathbb{R}^d$; $\phi(t_1, \mathbf{x}) = \phi^{(1)}(\mathbf{x})$, $\phi(t_2, \mathbf{x}) = \phi^{(2)}(\mathbf{x})$. So consider a variation of the trajectory $\phi(t, \mathbf{x})$ to $\phi(t, \mathbf{x}) + \delta\phi(t, \mathbf{x})$ where $\delta\phi \to 0$ as $|\mathbf{x}| \to \infty$, and $\delta\phi = 0$ at times t_1 and t_2. The variation of S is

$$\delta S = \int_{t_1}^{t_2} \int \left(\frac{\partial \mathcal{L}}{\partial(\partial_0 \phi)} \partial_0 \delta\phi + \frac{\partial \mathcal{L}}{\partial(\boldsymbol{\nabla}\phi)} \cdot \boldsymbol{\nabla}\delta\phi + \frac{\partial \mathcal{L}}{\partial \phi} \delta\phi \right) d^d x \, dt. \tag{2.49}$$

Integrating by parts,

$$\delta S = \int_{t_1}^{t_2} \int \left\{ \left(-\partial_0 \frac{\partial \mathcal{L}}{\partial(\partial_0 \phi)} - \boldsymbol{\nabla} \cdot \left(\frac{\partial \mathcal{L}}{\partial(\boldsymbol{\nabla}\phi)} \right) + \frac{\partial \mathcal{L}}{\partial \phi} \right) \delta\phi \right\} d^d x \, dt. \tag{2.50}$$

For δS to vanish for all $\delta\phi$, we require

$$\partial_0 \frac{\partial \mathcal{L}}{\partial(\partial_0 \phi)} + \boldsymbol{\nabla} \cdot \left(\frac{\partial \mathcal{L}}{\partial(\boldsymbol{\nabla}\phi)} \right) - \frac{\partial \mathcal{L}}{\partial \phi} = 0, \tag{2.51}$$

and this is the Euler-Lagrange equation satisfied by the dynamical field. Note that the last two terms together (at $\mathbf{x}$) can be regarded formally as (minus) the derivative of $\mathcal{L}$ with respect to $\phi(\mathbf{x})$.

For the basic Lagrangian (2.40), Eq. (2.51) is

$$\partial_0 \partial_0 \phi - \nabla^2 \phi + \frac{dU}{d\phi} = 0 \,, \tag{2.52}$$

which is a Lorentz invariant nonlinear wave equation.

For a static field, the field equation simplifies to the nonlinear Laplace equation

$$\nabla^2 \phi = \frac{dU}{d\phi} \,, \tag{2.53}$$

and this is the condition for ϕ to be a stationary point of the potential energy function V. A solution ϕ is stable if it is a minimum of V.

The Lagrangian formalism can easily be extended to a theory of n scalar fields $\phi = (\phi_1, \ldots, \phi_n)$. A dynamical field configuration is a map $\phi : \mathbb{R} \times \mathbb{R}^d \mapsto \mathbb{R}^n$. The Lagrangian now depends on all n component fields and their derivatives, and the action is stationary with respect to independent variations of each of them. The Euler-Lagrange equations are thus

$$\partial_0 \frac{\partial \mathcal{L}}{\partial(\partial_0 \phi_l)} + \nabla \cdot \left(\frac{\partial \mathcal{L}}{\partial(\nabla \phi_l)} \right) - \frac{\partial \mathcal{L}}{\partial \phi_l} = 0 \,, \qquad 1 \le l \le n \,. \tag{2.54}$$

Because the field takes values in $\mathbb{R}^n$, this type of theory is called a linear scalar field theory, even though the field equations (2.54) are nonlinear.

The fields can also have a tensorial character, which requires additional use of the Minkowski metric to produce a Lorentz invariant Lagrangian density. We shall need to deal with $U(1)$ gauge fields below, which entails a 1-form potential a_μ with time and space components $\{a_0, a_i\}$. From the point of view of deriving the Euler-Lagrange equations, each component of a tensor field can be treated as an independent field. (Caution: Here a_μ is a dynamical field, and not a fixed background 1-form coupled to a dynamical particle, as in Section 2.1.)

An important phenomenon in field theory is the possibility of internal symmetries, unrelated to the isometries of space. For example, suppose the Lagrangian density is

$$\mathcal{L} = \frac{1}{2} \partial_0 \phi_l \partial_0 \phi_l - \frac{1}{2} \nabla \phi_l \cdot \nabla \phi_l - U(\phi_l \phi_l) \,, \tag{2.55}$$

where the repeated index l labelling the fields is to be summed over from 1 to n, and U is a function of just the single quantity $\phi_l \phi_l$. This is invariant under internal rotations

$$\phi_l \mapsto R_{lm} \phi_m \,, \tag{2.56}$$

with R_{lm} an $SO(n)$ matrix. The symmetry leads to conservation laws for the dynamics.

Let us next consider the Lagrangian for a scalar field $\phi(t, \mathbf{x})$ defined on the space X with Riemannian metric $h_{ij}(\mathbf{x})$. Space-time is $\mathbb{R} \times X$, and the metric on X is extended trivially to a metric of Lorentzian signature on space-time

$$ds^2 = dt^2 - h_{ij}(\mathbf{x})dx^i dx^j . \tag{2.57}$$

It is possible to define field theories on more general curved space-time backgrounds, and this is important if gravitational effects are significant. We do not consider this possibility here. If X is curved, we think of this as due not to gravity, but to material constraints. For example vortices can be considered on a 2-sphere, where the sphere is a thin, curved layer of superconductor in flat three-dimensional space. Einstein's equations play no role.

We wish the dynamics of the field to depend just on the intrinsic geometry of X, and to respect the symmetries of X. This means that the integration measure becomes $\sqrt{\det h}\, d^d x$, the natural measure on X, and the expression $\boldsymbol{\nabla}\phi \cdot \boldsymbol{\nabla}\phi$ is replaced by $h^{ij}\partial_i \phi \partial_j \phi$. $\det h$ denotes the determinant of h_{ij}. No other changes are allowed. Thus the basic Lagrangian (2.40) becomes

$$L = \int_X \left(\frac{1}{2}(\partial_0 \phi)^2 - \frac{1}{2}h^{ij}\partial_i \phi \partial_j \phi - U(\phi) \right) \sqrt{\det h}\, d^d x , \tag{2.58}$$

and the action is

$$S = \int_{t_1}^{t_2} L\, dt . \tag{2.59}$$

The field equation is now

$$\partial_0 \partial_0 \phi - \frac{1}{\sqrt{\det h}}\partial_i(\sqrt{\det h}\, h^{ij}\partial_j \phi) + \frac{dU}{d\phi} = 0 , \tag{2.60}$$

where the only significant change from (2.52) is that the ordinary Laplacian operator is replaced by the covariant Laplacian on X.

Yet another variant of field theory is where the field ϕ takes values in a non-trivial manifold Y, so

$$\phi : \mathbb{R} \times X \mapsto Y . \tag{2.61}$$

Locally, the field is represented by $\phi(t, \mathbf{x}) = (\phi_1(t, \mathbf{x}), \ldots, \phi_n(t, \mathbf{x}))$, where $n = \dim Y$. Abstractly, $(\phi_1, \ldots, \phi_n)$ are coordinates on Y. One needs a metric $H_{lm}(\phi_1, \ldots, \phi_n)$ on Y to define the Lagrangian. A particular example is the sigma model Lagrangian, which depends quadratically on the time and space derivatives of the field, and has no potential term. The Lagrangian is

$$L = \frac{1}{2}\int_X \left(\partial_0 \phi_l \partial_0 \phi_m H^{lm} - h^{ij}\partial_i \phi_l \partial_j \phi_m H^{lm} \right) \sqrt{\det h}\, d^d x . \tag{2.62}$$

The symmetry group of this Lagrangian is the product of the isometry groups of X and Y, together with time translation invariance.

Such a theory is called a nonlinear scalar theory. Sometimes it is formulated as a linear theory, with the scalar field subject to a nonlinear constraint. This may be convenient if Y, with its metric, is a simple submanifold of a Euclidean space. For example, in the basic sigma model, and also the Skyrme model, Y is a round sphere, and this sits conveniently in a Euclidean space of one higher dimension.

We conclude with a remark about the relationship between field theory and the finite-dimensional dynamical systems we considered earlier. Let us consider the sigma model example with field $\phi = (\phi_1, \ldots, \phi_n)$. A field configuration is a multiplet of specific functions $(\phi_1(\mathbf{x}), \ldots, \phi_n(\mathbf{x}))$ defined throughout X, at a given time. A configuration is not necessarily a static solution of the Euler-Lagrange field equations, but it could be the instantaneous form of a dynamical field. Then we can think of the function space $\mathcal{C}$, whose points are field configurations $(\phi_1(\mathbf{x}), \ldots, \phi_n(\mathbf{x}))$, as the infinite-dimensional analogue of the finite-dimensional manifold M with points $\mathbf{q}$, the configuration space of a finite-dimensional dynamical system. The field potential energy

$$V = \frac{1}{2} \int_X h^{ij} \partial_i \phi_l \partial_j \phi_m H^{lm} \sqrt{\det h} \, d^d x \qquad (2.63)$$

depends only on the configuration $(\phi_1(\mathbf{x}), \ldots, \phi_n(\mathbf{x}))$, so is a scalar function on $\mathcal{C}$.

There is a natural Riemannian distance in $\mathcal{C}$ between two infinitesimally close configurations $(\phi_1(\mathbf{x}), \ldots, \phi_n(\mathbf{x}))$ and $(\phi_1(\mathbf{x}) + \delta\phi_1(\mathbf{x}), \ldots, \phi_n(\mathbf{x}) + \delta\phi_n(\mathbf{x}))$, namely

$$ds^2 = \int_X \delta\phi_l(\mathbf{x}) \delta\phi_m(\mathbf{x}) H^{lm} \sqrt{\det h} \, d^d x \,, \qquad (2.64)$$

the natural squared norm of $(\delta\phi_1, \ldots, \delta\phi_n)$. If the fields are time dependent, then the kinetic energy T is obtained from ds^2 by replacing $\delta\phi_l$ by $\partial_0\phi_l$, and dividing by 2, giving the first term in the expression (2.62).

Similar arguments apply to all the other examples of field theory we have presented so far. So a field theory is simply a dynamical system of a certain type, defined on an infinite-dimensional Riemannian configuration space $\mathcal{C}$.

2.4 Noether's theorem in field theory

If a Lagrangian field theory has an infinitesimal symmetry then there is an associated current $J^\mu(x)$ which is conserved: $\partial_\mu J^\mu = \partial_0 J^0 + \nabla \cdot \mathbf{J} = 0$.

Both space-time symmetries and internal symmetries lead to conservation laws.

Consider a theory in $(d+1)$-dimensional Minkowski space, for the field ϕ. Let the infinitesimal variation of ϕ be

$$\phi(x) \mapsto \phi(x) + \varepsilon \Delta \phi(x) \,, \tag{2.65}$$

with ε infinitesimal. This variation is a symmetry if one can show, without using the field equation, that the corresponding variation of the Lagrangian density $\mathcal{L}$ is a total divergence,

$$\mathcal{L}(x) \mapsto \mathcal{L}(x) + \varepsilon \partial_\mu K^\mu(x) \,. \tag{2.66}$$

The action then varies only by a surface term. Sometimes, K^μ will be zero, and the action strictly invariant.

Now let us calculate the change in $\mathcal{L}$ more explicitly:

$$
\begin{aligned}
\mathcal{L}(x) \quad &\mapsto \quad \mathcal{L}(x) + \varepsilon \frac{\partial \mathcal{L}}{\partial(\partial_\mu \phi)} \partial_\mu(\Delta \phi) + \varepsilon \frac{\partial \mathcal{L}}{\partial \phi} \Delta \phi \tag{2.67} \\
&= \quad \mathcal{L}(x) + \varepsilon \partial_\mu \left(\frac{\partial \mathcal{L}}{\partial(\partial_\mu \phi)} \Delta \phi \right) + \varepsilon \left(\frac{\partial \mathcal{L}}{\partial \phi} - \partial_\mu \frac{\partial \mathcal{L}}{\partial(\partial_\mu \phi)} \right) \Delta \phi \,.
\end{aligned}
$$

Using the field equation, the last parentheses vanish. Identifying (2.66) and (2.67), we see that the current

$$J^\mu \equiv \frac{\partial \mathcal{L}}{\partial(\partial_\mu \phi)} \Delta \phi - K^\mu \tag{2.68}$$

is conserved. This is Noether's theorem. If there are several fields, the first term in J^μ becomes a sum over terms, one for each field.

Current conservation $\partial_\mu J^\mu = 0$ implies the conservation of the Noether charge

$$Q = \int J^0 \, d^d x \,. \tag{2.69}$$

Q is time independent because

$$\frac{dQ}{dt} = \int \partial_0 J^0 \, d^d x = - \int \boldsymbol{\nabla} \cdot \mathbf{J} \, d^d x = 0 \,. \tag{2.70}$$

Here we have used the divergence theorem, and assumed that $\mathbf{J} \to \mathbf{0}$ as $|\mathbf{x}| \to \infty$.

As a first example, consider the complex Klein-Gordon field $\phi(x)$, with Lagrangian density

$$\mathcal{L} = \frac{1}{2} \partial_\mu \bar{\phi} \partial^\mu \phi - \frac{1}{2} m^2 \bar{\phi} \phi \,. \tag{2.71}$$

The phase rotation $\phi \mapsto e^{i\alpha}\phi$, $\bar{\phi} \mapsto e^{-i\alpha}\bar{\phi}$ is a $U(1)$ symmetry, leaving $\mathcal{L}$ invariant. Infinitesimally, $\Delta\phi = i\phi$, $\Delta\bar{\phi} = -i\bar{\phi}$. Here $K^\mu = 0$, so the conserved current is

$$J^\mu = -\frac{i}{2}(\bar{\phi}\partial^\mu\phi - \phi\partial^\mu\bar{\phi}).\tag{2.72}$$

One can verify $\partial_\mu J^\mu = 0$ directly, using the Klein-Gordon equation

$$\partial_\mu\partial^\mu\phi + m^2\phi = 0\tag{2.73}$$

and its complex conjugate. J^μ is interpreted as the electric current, and $Q = \int J^0 \, d^d x$ as the total electric charge.

As a second, rather general example, consider infinitesimal translations in Minkowski space-time

$$x^\mu \mapsto x^\mu + \varepsilon^\mu,\tag{2.74}$$

with ε infinitesmal, for an arbitrary Lagrangian not depending explicitly on the space-time coordinates. The effect on a field ϕ is

$$\phi(x) \mapsto \phi(x + \varepsilon) = \phi(x) + \varepsilon^\nu\partial_\nu\phi(x),\tag{2.75}$$

and similarly for derivatives of ϕ. The effect on the Lagrangian density, no matter what the details of its structure, is

$$\mathcal{L} \mapsto \mathcal{L} + \varepsilon^\nu\partial_\nu\mathcal{L} = \mathcal{L} + \varepsilon^\nu\partial_\mu(\delta^\mu_\nu\mathcal{L}).\tag{2.76}$$

Since the infinitesimal parameter ε^ν is a space-time vector, the conserved current is a tensor

$$T^\mu_\nu = \frac{\partial\mathcal{L}}{\partial(\partial_\mu\phi)}\partial_\nu\phi - \delta^\mu_\nu\mathcal{L}.\tag{2.77}$$

Naively, there is one current for each component of ε^ν. T^μ_ν is the energy-momentum tensor, and it satisfies $\partial_\mu T^\mu_\nu = 0$.

The conserved charge associated with time translation symmetry is the energy

$$E = \int T^0_0 \, d^d x = \int \left(\frac{\partial\mathcal{L}}{\partial(\partial_0\phi)}\partial_0\phi - \mathcal{L}\right) d^d x.\tag{2.78}$$

The conserved charge associated with spatial translations is the momentum vector

$$P_i = -\int T^0_i \, d^d x = -\int \frac{\partial\mathcal{L}}{\partial(\partial_0\phi)}\partial_i\phi \, d^d x.\tag{2.79}$$

2.5 Vacua and spontaneous symmetry breaking

Consider a Lagrangian describing a multiplet of n real scalar fields $\phi = (\phi_1, \ldots, \phi_n)$ in Minkowski space-time $\mathbb{R} \times \mathbb{R}^d$, with no explicit time or space dependence. Let the potential $U(\phi)$ have minimal value $U_{\min} = 0$. We denote by $\mathcal{V}$ the submanifold of $\mathbb{R}^n$ where U attains its minimum, and call this the vacuum manifold of the theory. If the field takes its value in $\mathcal{V}$, and is constant throughout space (and time), we call this a vacuum configuration, or vacuum for short; it is a stable solution of the field equations and its total energy is zero. The effect of internal symmetries on the possible vacua is the topic of this section. The internal symmetry group can be discrete or a Lie group.

For example, for the Lagrangian density (2.55), with $SO(n)$ symmetry, the energy is

$$\int \left(\frac{1}{2} \partial_0 \phi_l \partial_0 \phi_l + \frac{1}{2} \nabla \phi_l \cdot \nabla \phi_l + U(\phi_l \phi_l) \right) d^d x . \tag{2.80}$$

This is minimized by a field configuration which is independent of t and $\mathbf{x}$, and minimizes U. The $SO(n)$ symmetry means that constant configurations lying on any given orbit of $SO(n)$ have the same energy, so the vacuum is not necessarily unique. We assume here that the minimum of U is attained on just a single orbit of $SO(n)$. There are two possibilities. If this orbit consists of just one point, the vacuum is unique, and is invariant under the symmetry group $SO(n)$. One says that the symmetry is unbroken in the vacuum. If the orbit is non-trivial, then the vacuum is not uniquely determined. The vacuum is a (random) choice of a point on the orbit. One says that the symmetry is spontaneously broken.

Quartic potentials illustrate the two possibilities. If

$$U = (c + \phi_l \phi_l)^2 - c^2 , \tag{2.81}$$

with $c \geq 0$, then the minimum of U occurs at $\phi_l = 0$, and the symmetry is unbroken. If

$$U = (c - \phi_l \phi_l)^2 , \tag{2.82}$$

with $c > 0$, the minimum occurs on the orbit $\phi_l \phi_l = c$. The vacuum is a choice of a point on the orbit, e.g. the n-vector

$$(\phi_1, \phi_2, \ldots, \phi_n) = (0, 0, \ldots, \sqrt{c}), \tag{2.83}$$

and this spontaneously breaks the $SO(n)$ symmetry. The subgroup of $SO(n)$ whose action leaves the n-vector fixed is $SO(n-1)$. Geometrically, this is called the isotropy group of the particular vector. Physically, it is the unbroken subgroup of the original symmetry group, leaving the vacuum fixed. More generally, if the symmetry group of the potential U

is the group G, and the choice of vacuum leaves the group H unbroken, then the vacuum orbit of the symmetry group is G/H, with H the isotropy group of the chosen vacuum.

The occurrence of spontaneous breaking of an internal symmetry group G has important consequences for the dynamics of the field, particularly if G is a Lie group. Small amplitude oscillations around a chosen vacuum can be decomposed into the directions orthogonal to the orbit G/H and tangent to the orbit. The tangent directions are "flat" directions, since the potential function U is unchanging in these directions. Generally, the potential is not flat in the orthogonal directions, but increases quadratically. In the flat directions, the oscillating field components ψ satisfy the wave equation

$$\partial_0\partial_0\psi - \nabla^2\psi = 0\,, \qquad (2.84)$$

to linear order in ψ. The plane wave solutions are $\psi = \psi_0 e^{-i(\mathbf{k}\cdot\mathbf{x}-\omega t)}$, with the relation between frequency ω and wave-vector $\mathbf{k}$

$$\omega = |\mathbf{k}|\,, \qquad (2.85)$$

so ω is arbitrarily close to zero. In the quantized theory there are massless, elementary scalar particles associated with such waves. This is Goldstone's theorem [162]. Its proof is not completely straightforward, as it does not assume the small amplitude approximation that we have just made. The number of distinct Goldstone particles is $\dim G - \dim H$, which is the dimension of the orbit G/H. In the example (2.82), the orbit is $SO(n)/SO(n-1)$, which is the $(n-1)$-sphere, and the number of Goldstone particles is $n-1$. In classical field theory, the consequence of spontaneous symmetry breaking is the presence of long-range interactions corresponding to the exchange of the massless Goldstone particles. By contrast, in a theory of scalar fields with a symmetry group G which is unbroken, all particles are generally massive and interactions are short-range.

2.6 Gauge theory

Since the 1970s, the standard model of elementary particle physics has been a gauge field theory. A gauge theory is one where an internal Lie symmetry group G acts locally, that is, independently at each space-time point. Field configurations which differ only by a gauge transformation are to be regarded as physically the same. This means that the true configuration space of a gauge theory is smaller than the naive space of all field configurations, $\mathcal{A}$. Typically, $\mathcal{A}$ is an infinite-dimensional linear space; the group of gauge transformations $\mathcal{G}$ is the space of maps from space into the Lie group G, which is an infinite-dimensional curved space;

and the true configuration space $\mathcal{C}$ is the quotient $\mathcal{A}/\mathcal{G}$, which is also curved. The existence of solitons in gauge theory is related to the nonlinear nature of $\mathcal{C}$. We shall explore this later, but here we just describe the basic dynamical structure of a gauge theory, and how to interpret such a theory as a dynamical system on $\mathcal{C}$.

The simplest type of gauge theory is based on the group $G = U(1)$. Since $U(1)$ is abelian (commutative), this is called an abelian gauge theory. Physically, the theory describes the electromagnetic field interacting with other fields. We shall suppose these other fields are scalars. The basic example with one complex scalar field is scalar electrodynamics. Let us start with the ungauged theory in space-time $\mathbb{R} \times \mathbb{R}^d$ with Lagrangian

$$L = \int \left(\frac{1}{2} \partial_\mu \bar{\phi} \partial^\mu \phi - U(\bar{\phi}\phi) \right) d^d x \,. \tag{2.86}$$

Here $\phi(x)$ is the complex-valued scalar field, which can be expressed in terms of two real fields as $\phi = \phi_1 + i\phi_2$. U only depends on $|\phi|^2 = \bar{\phi}\phi$. This Lagrangian, expressed in terms of ϕ_1 and ϕ_2, is of the type (2.55) with $n = 2$. There is an internal symmetry $U(1)$, or equivalently $SO(2)$, as the global phase rotation

$$\phi \mapsto e^{i\alpha} \phi \tag{2.87}$$

leaves the Lagrangian L invariant. The conserved current is given, as before, by (2.72).

To obtain a $U(1)$ gauge theory, one requires the Lagrangian to be invariant under

$$\phi(x) \mapsto e^{i\alpha(x)} \phi(x) \,, \tag{2.88}$$

where $\alpha(x)$ is an arbitrary function of space and time. The term $U(\bar{\phi}\phi)$ is already invariant, but the terms involving derivatives of ϕ are not. To remedy this one needs to introduce the electromagnetic gauge potential $a_\mu(x)$, with time and space components $\{a_0, \mathbf{a}\}$. These are new, independent fields. One defines the gauge covariant derivative of ϕ, with components

$$D_\mu \phi = \partial_\mu \phi - i a_\mu \phi \,, \tag{2.89}$$

and postulates that a_μ transforms under the gauge transformation (2.88) to

$$a_\mu \mapsto a_\mu + \partial_\mu \alpha \,. \tag{2.90}$$

$D_\mu \phi$ gauge transforms in the same way (covariantly) as ϕ itself:

$$
\begin{aligned}
\partial_\mu \phi - i a_\mu \phi \quad \mapsto \quad & \partial_\mu (e^{i\alpha} \phi) - i(a_\mu + \partial_\mu \alpha) e^{i\alpha} \phi \\
= \quad & i \partial_\mu \alpha \, e^{i\alpha} \phi + e^{i\alpha} \partial_\mu \phi - i a_\mu e^{i\alpha} \phi - i \partial_\mu \alpha \, e^{i\alpha} \phi \\
= \quad & e^{i\alpha} (\partial_\mu \phi - i a_\mu \phi) \,.
\end{aligned}
\tag{2.91}
$$

The covariant derivative of $\bar\phi$ is $\overline{D_\mu\phi} = \partial_\mu\bar\phi + ia_\mu\bar\phi$, the complex conjugate of the covariant derivative of ϕ, since a_μ are real fields. Under a gauge transformation,

$$\overline{D_\mu\phi} \mapsto e^{-i\alpha}\overline{D_\mu\phi}. \qquad (2.92)$$

Thus the expression $\overline{D_\mu\phi}D^\mu\phi$ is gauge invariant, and may appear in the Lagrangian.

If α is infinitesimal, the gauge transformations of ϕ and a_μ reduce to

$$\phi \mapsto \phi + i\alpha\phi \qquad (2.93)$$
$$a_\mu \mapsto a_\mu + \partial_\mu\alpha. \qquad (2.94)$$

For the fields a_μ to be dynamical, we need to include terms involving their derivatives in the Lagrangian. This is done using the field tensor

$$f_{\mu\nu} = \partial_\mu a_\nu - \partial_\nu a_\mu. \qquad (2.95)$$

This is gauge invariant, since under the gauge transformation (2.90),

$$\begin{aligned} f_{\mu\nu} &\mapsto \partial_\mu(a_\nu + \partial_\nu\alpha) - \partial_\nu(a_\mu + \partial_\mu\alpha) \\ &= \partial_\mu a_\nu - \partial_\nu a_\mu \\ &= f_{\mu\nu}, \end{aligned} \qquad (2.96)$$

using the symmetry property of double partial derivatives. The components of the field tensor are the electric components

$$e_i = f_{0i} = \partial_0 a_i - \partial_i a_0, \qquad (2.97)$$

or in vector notation $\mathbf{e} = \partial_0\mathbf{a} - \boldsymbol{\nabla}a_0$, and the magnetic components

$$f_{ij} = \partial_i a_j - \partial_j a_i. \qquad (2.98)$$

The electric components of $f_{\mu\nu}$ comprise a 1-form in space, the magnetic components a 2-form.

Using all these ingredients one can construct a Lorentz invariant Lagrangian density for scalar electrodynamics,

$$\mathcal{L} = -\frac{1}{4}f_{\mu\nu}f^{\mu\nu} + \frac{1}{2}\overline{D_\mu\phi}D^\mu\phi - U(\bar\phi\phi). \qquad (2.99)$$

Explicitly separating space and time parts, we obtain

$$\mathcal{L} = \frac{1}{2}e_i e_i + \frac{1}{2}\overline{D_0\phi}D_0\phi - \frac{1}{4}f_{ij}f_{ij} - \frac{1}{2}\overline{D_i\phi}D_i\phi - U(\bar\phi\phi), \qquad (2.100)$$

where both ∂_0 and a_0 terms are included as "time" parts. We define the kinetic and potential energies

$$T = \int\left(\frac{1}{2}e_i e_i + \frac{1}{2}\overline{D_0\phi}D_0\phi\right)d^d x \qquad (2.101)$$

$$V = \int\left(\frac{1}{4}f_{ij}f_{ij} + \frac{1}{2}\overline{D_i\phi}D_i\phi + U(\bar\phi\phi)\right)d^d x, \qquad (2.102)$$

although we will need to say more precisely what a_0 is before we can actually call T the kinetic energy. Notice that the choice of signs in (2.99) ensures that T is positive definite.

There is no difficulty extending this theory to a space-time $\mathbb{R} \times X$. We just need to use the metric h_{ij} on X, to contract tensor indices. Thus the Lagrangian density becomes

$$\mathcal{L} = \frac{1}{2} h^{ij} e_i e_j + \frac{1}{2} \overline{D_0 \phi} D_0 \phi$$
$$- \frac{1}{4} h^{ik} h^{jl} f_{ij} f_{kl} - \frac{1}{2} h^{ij} \overline{D_i \phi} D_j \phi - U(\bar{\phi}\phi), \quad (2.103)$$

and the integration measure is $\sqrt{\det h}\, d^d x$. The further generalization, where ϕ is a section of a $U(1)$ bundle over X and a_i are components of a connection 1-form on this bundle, is described in Chapter 3.

We do not necessarily require the theory to be Lorentz invariant. On $\mathbb{R} \times \mathbb{R}^d$, it is usual for the potential energy expression of scalar electrodynamics to be taken to be (2.102). This is called the gauged Ginzburg-Landau energy. However, there are alternatives for the kinetic term. In particular, in Chapter 7, we shall consider a kinetic energy in two spatial dimensions where the Maxwell term $\frac{1}{2} e_i e_i$ is dropped in favour of a Chern-Simons term, and where the covariant time derivative of ϕ appears in the Lagrangian density as $i(\bar{\phi} D_0 \phi - \phi \overline{D_0 \phi})$, which is real and gauge invariant. The field equations of Lorentz invariant scalar electrodynamics with Lagrangian density (2.99) are obtained, as usual, by requiring the action

$$S = \int_{t_1}^{t_2} \mathcal{L}\, d^d x\, dt \quad (2.104)$$

to be stationary under variations of $\{\phi, a_\mu\}$. The variations $\{\delta\phi, \delta a_\mu\}$ are assumed to vanish at $t = t_1, t = t_2$, and as $|\mathbf{x}| \to \infty$. We use $\overline{\delta\phi} = \delta\bar{\phi}$. The Euler-Lagrange equations are

$$D_\mu D^\mu \phi = -2U'(\bar{\phi}\phi)\phi \quad (2.105)$$

$$\partial_\mu f^{\mu\nu} = -\frac{i}{2}(\bar{\phi} D^\nu \phi - \phi \overline{D^\nu \phi}), \quad (2.106)$$

where U' denotes the derivative of U with respect to its single argument $\bar{\phi}\phi$. The right-hand side of (2.106) is the Noether current J^ν associated with the $U(1)$ global symmetry $\phi \mapsto e^{i\alpha}\phi$, and its conservation, $\partial_\nu J^\nu = 0$, is consistent with (2.106). Most of the component equations of (2.105) and (2.106) are evolution equations for ϕ and $\mathbf{a}$. However, the equation with $\nu = 0$, which comes from varying a_0 in the action, is rather different. This is Gauss' law, which reads

$$\partial_i e_i = -\frac{i}{2}(\bar{\phi} D_0 \phi - \phi \overline{D_0 \phi}). \quad (2.107)$$

Expanding out, this becomes

$$\partial_i(\partial_0 a_i - \partial_i a_0) = -\frac{i}{2}(\bar{\phi}\partial_0\phi - \phi\partial_0\bar{\phi}) - a_0\bar{\phi}\phi, \qquad (2.108)$$

which can be rearranged as

$$(\nabla^2 - \bar{\phi}\phi)a_0 = \partial_i\partial_0 a_i + \frac{i}{2}(\bar{\phi}\partial_0\phi - \phi\partial_0\bar{\phi}). \qquad (2.109)$$

This equation for a_0 can, in principle, be solved at a given time if the time derivatives of a_i and ϕ are known. a_0 is thus not an independent dynamical field, but may be eliminated, although the question of existence and uniqueness of solutions for a_0 is rather subtle, and depends on fixing boundary conditions. Notice that the equation (2.107) is gauge invariant. Sometimes it is possible to fix the gauge so that the right-hand side of (2.109) vanishes. Then one may choose $a_0 = 0$.

This discussion of the role of a_0 in a $U(1)$ gauge theory can be given a more geometrical flavour. As a dynamical theory, one should regard the configuration space of scalar electrodynamics as $\mathcal{C} = \mathcal{A}/\mathcal{G}$, where $\mathcal{A}$ is the set of spatial field configurations $\{\phi(\mathbf{x}), \mathbf{a}(\mathbf{x})\}$ and $\mathcal{G} = \{e^{i\alpha(\mathbf{x})}\}$ is the group of position dependent gauge transformations (see Fig. 2.1).

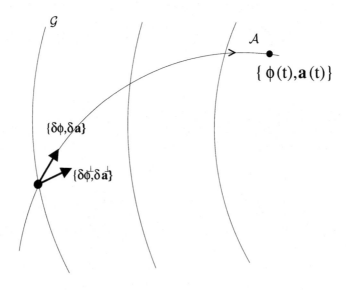

Fig. 2.1. Sketch of the configuration space illustrating variations of the fields and their projection orthogonal to gauge orbits.

The orbits of $\mathcal{G}$ are generically all similar, and $\mathcal{C}$ is the space of orbits.

A dynamical trajectory, whether or not it satisfies the field equations, is a trajectory in $\mathcal{C}$.

How do we calculate the kinetic and potential energies of a trajectory in $\mathcal{C}$? We need to lift the trajectory in $\mathcal{C}$ to a trajectory $\{\phi(t, \mathbf{x}), \mathbf{a}(t, \mathbf{x})\}$ in $\mathcal{A}$, but we want to extract information that does not depend on the choice of lift but only depends on the projection down to $\mathcal{C}$. The potential energy V is gauge invariant, and does not involve time derivatives, so can be thought of as defined, at each instant, on $\mathcal{C}$.

For the kinetic energy T one needs to make a genuine projection. Suppose at time t the fields are $\{\phi, \mathbf{a}\}$ and at $t + \delta t$ they are $\{\phi + \delta\phi, \mathbf{a} + \delta\mathbf{a}\}$. The naive contribution to the kinetic energy would be

$$\frac{1}{2} \int \frac{1}{(\delta t)^2} \left(\delta\mathbf{a} \cdot \delta\mathbf{a} + \overline{\delta\phi}\, \delta\phi \right) d^d x \,. \tag{2.110}$$

However, this would be non-zero if $\{\delta\phi, \delta\mathbf{a}\}$ were simply an infinitesimal gauge transformation

$$\delta\phi = i\alpha\phi \tag{2.111}$$
$$\delta\mathbf{a} = \boldsymbol{\nabla}\alpha \,, \tag{2.112}$$

which represents motion along a gauge orbit. Such motion does not correspond to a physical change and should have no associated kinetic energy.

One deals with this by projecting $\{\delta\phi, \delta\mathbf{a}\}$ orthogonally to the gauge orbit through $\{\phi, \mathbf{a}\}$. Thus we define

$$\delta\phi^{\perp} = \delta\phi - i\beta\phi \tag{2.113}$$
$$\delta\mathbf{a}^{\perp} = \delta\mathbf{a} - \boldsymbol{\nabla}\beta \,, \tag{2.114}$$

where β is chosen so that

$$\int \left(\delta\mathbf{a}^{\perp} \cdot \boldsymbol{\nabla}\alpha + \frac{1}{2}(\delta\phi^{\perp}\overline{(i\alpha\phi)} + \overline{(\delta\phi^{\perp})}i\alpha\phi) \right) d^d x = 0 \tag{2.115}$$

for all (infinitesimal) functions $\alpha(\mathbf{x})$. Integrating by parts, and discarding boundary terms, we see that

$$\int \left(\boldsymbol{\nabla} \cdot \delta\mathbf{a}^{\perp} + \frac{i}{2}(\bar{\phi}\delta\phi^{\perp} - \phi\overline{\delta\phi^{\perp}}) \right) \alpha \, d^d x = 0 \,, \tag{2.116}$$

so

$$\boldsymbol{\nabla} \cdot \delta\mathbf{a}^{\perp} + \frac{i}{2}(\bar{\phi}\delta\phi^{\perp} - \phi\overline{\delta\phi^{\perp}}) = 0 \,. \tag{2.117}$$

Equation (2.117) is the defining equation of $(\delta\phi^{\perp}, \delta\mathbf{a}^{\perp})$ and it is an equation for β. Expanding out, we find

$$(\nabla^2 - \bar{\phi}\phi)\beta = \boldsymbol{\nabla} \cdot \delta\mathbf{a} + \frac{i}{2}(\bar{\phi}\delta\phi - \phi\overline{\delta\phi}) \,. \tag{2.118}$$

Now notice that this is precisely the equation (2.109) for a_0, if we replace $(\delta\phi, \delta\mathbf{a})$ by $(\partial_0\phi, \partial_0\mathbf{a})$. Therefore, we can interpret

$$D_0\phi = \partial_0\phi - ia_0\phi \qquad (2.119)$$

$$e_i = \partial_0 a_i - \partial_i a_0 \qquad (2.120)$$

as the projection of $\partial_0\phi$ and $\partial_0 a_i$ orthogonal to gauge orbits, provided we impose Gauss' law (2.109) on a_0.

We deduce that if Gauss' law is satisfied, the expression

$$\frac{1}{2}\int \left(e_i e_i + \overline{D_0\phi}D_0\phi\right) d^d x \qquad (2.121)$$

can be interpreted as

$$\frac{1}{2}\int \left(\partial_0 a_i^{\perp}\partial_0 a_i^{\perp} + \overline{\partial_0\phi^{\perp}}\partial_0\phi^{\perp}\right) d^d x\,, \qquad (2.122)$$

which is one half the (speed)^2 of the projected motion to $\mathcal{C}$, and thus the natural gauge invariant kinetic energy for a trajectory in $\mathcal{C}$. The corresponding expression for the metric on $\mathcal{C}$ is

$$ds^2 = \int \left(\delta a_i^{\perp}\delta a_i^{\perp} + \overline{\delta\phi^{\perp}}\delta\phi^{\perp}\right) d^d x\,. \qquad (2.123)$$

Our conclusion is that T and V are well defined in a $U(1)$ gauge theory for any trajectory, and gauge invariant, provided one imposes Gauss' law to determine a_0. If one can arrange the gauge choice

$$\nabla \cdot \partial_0\mathbf{a} + \frac{i}{2}(\bar{\phi}\partial_0\phi - \phi\partial_0\bar{\phi}) = 0\,, \qquad (2.124)$$

then it is satisfactory to set $a_0 = 0$. These considerations will be important when we discuss the dynamics of solitons in $U(1)$ gauge theories.

We now briefly describe the extension of the gauge theory formalism to a non-abelian Lie group G [434]. Let the identity element of G be denoted by I. We assume that G is finite-dimensional to have a finite number of independent fields. We also need to assume that G is compact, to ensure that the gauge invariant kinetic energy expression is positive definite. Such a group G can always be identified with a group of unitary matrices (i.e. $U(n)$ for some n, or a Lie subgroup of this). The Lie algebra of G is then a vector space of antihermitian $n \times n$ matrices. Generally, we denote the Lie algebra of a group G by Lie(G); however, for specific matrix groups, like $U(n)$ or $SO(n)$, their Lie algebras are denoted by $u(n)$ and $so(n)$. Let $\{t^a : 1 \leq a \leq \dim G\}$ be an orthonormalized basis of Lie(G), satisfying $\text{Tr}(t^a t^b) = C\delta^{ab}$ for some fixed negative constant C. The Lie algebra structure is

$$[t^a, t^b] = f^{abc}t^c\,, \qquad (2.125)$$

and f^{abc} are the structure constants. The normalization condition implies that f^{abc} are totally antisymmetric in their indices.

Let us consider a field theory with a multiplet of complex scalar fields $\Phi = (\Phi_1, \ldots, \Phi_n)$ acted on by G. Suppressing indices, the global action of G is

$$\Phi \mapsto g\Phi \, , \; g \in G \, . \tag{2.126}$$

We desire a theory invariant under space-time dependent gauge transformations

$$\Phi(x) \mapsto g(x)\Phi(x) \, . \tag{2.127}$$

To construct a gauge invariant Lagrangian we need to have a covariant derivative of Φ. This requires the introduction of a gauge potential $A_\mu(x)$, taking values in Lie(G). In terms of the basis $\{t^a\}$, A_μ has a component expansion $A_\mu = A_\mu^a t^a$. The covariant derivative is

$$D_\mu \Phi = \partial_\mu \Phi + A_\mu \Phi \, . \tag{2.128}$$

If we now postulate that A_μ gauge transforms as

$$A_\mu \mapsto g A_\mu g^{-1} - \partial_\mu g \, g^{-1} \, , \tag{2.129}$$

then

$$
\begin{aligned}
D_\mu \Phi \; &\mapsto \; \partial_\mu(g\Phi) + (g A_\mu g^{-1} - \partial_\mu g g^{-1})g\Phi \\
&= \; \partial_\mu g \Phi + g \partial_\mu \Phi + g A_\mu \Phi - \partial_\mu g \Phi \\
&= \; g D_\mu \Phi \, ,
\end{aligned}
\tag{2.130}
$$

so $D_\mu \Phi$ transforms like Φ, i.e. covariantly. Occasionally, we will denote the covariant derivative operator by D_μ^A if we wish to draw attention to a particular gauge field configuration $A_\mu(x)$.

The final ingredient of the Lagrangian is the Yang-Mills field tensor $F_{\mu\nu}$, defined as the commutator of covariant derivatives D_μ and D_ν. When one expands out the defining equation

$$[D_\mu, D_\nu]\Phi = F_{\mu\nu}\Phi \, , \tag{2.131}$$

one obtains the explicit formula

$$F_{\mu\nu} = \partial_\mu A_\nu - \partial_\nu A_\mu + [A_\mu, A_\nu] \, , \tag{2.132}$$

which is valued in Lie(G). (In the abelian case, $[D_\mu, D_\nu] = -i f_{\mu\nu}$.) Under a gauge transformation,

$$
\begin{aligned}
F_{\mu\nu} \; &\mapsto \; \partial_\mu(g A_\nu g^{-1} - \partial_\nu g g^{-1}) - \partial_\nu(g A_\mu g^{-1} - \partial_\mu g g^{-1}) \\
&\qquad + [g A_\mu g^{-1} - \partial_\mu g g^{-1}, g A_\nu g^{-1} - \partial_\nu g g^{-1}] \\
&= \; g F_{\mu\nu} g^{-1}
\end{aligned}
\tag{2.133}
$$

(using $\partial_\mu g^{-1} = -g^{-1}\partial_\mu g g^{-1}$, which ensures $\partial_\mu(g^{-1}g) = \partial_\mu I = 0$). This is also clear from (2.131).

The Yang-Mills Lagrangian density is a Lorentz invariant combination of these ingredients,

$$\mathcal{L} = \frac{1}{8}\text{Tr}(F_{\mu\nu}F^{\mu\nu}) + \frac{1}{2}(D_\mu\Phi)^\dagger D^\mu\Phi - U(\Phi^\dagger\Phi). \qquad (2.134)$$

Under a gauge transformation,

$$\text{Tr}(F_{\mu\nu}F^{\mu\nu}) \mapsto \text{Tr}(gF_{\mu\nu}g^{-1}gF^{\mu\nu}g^{-1}) = \text{Tr}(F_{\mu\nu}F^{\mu\nu}), \qquad (2.135)$$

using the cyclicity of the trace. Also

$$\Phi^\dagger\Phi \mapsto (g\Phi)^\dagger g\Phi = \Phi^\dagger g^\dagger g\Phi = \Phi^\dagger\Phi, \qquad (2.136)$$

because g is unitary, and similarly

$$(D_\mu\Phi)^\dagger D^\mu\Phi \mapsto (gD_\mu\Phi)^\dagger gD^\mu\Phi = (D_\mu\Phi)^\dagger g^\dagger gD^\mu\Phi = (D_\mu\Phi)^\dagger D^\mu\Phi. \qquad (2.137)$$

So $\mathcal{L}$ is gauge invariant.

The Euler-Lagrange field equations for the Yang-Mills theory are

$$D_\mu D^\mu\Phi = -2U'(\Phi^\dagger\Phi)\Phi \qquad (2.138)$$
$$D_\mu F^{\mu\nu} = J^\nu, \qquad (2.139)$$

where the Lie(G)-valued current is

$$J^\nu = -\frac{1}{C}\left(\Phi^\dagger t^a D^\nu\Phi - (D^\nu\Phi)^\dagger t^a\Phi\right)t^a, \qquad (2.140)$$

and the field tensor (and similarly any other Lie(G)-valued quantity) has covariant derivative $D_\lambda F^{\mu\nu} = \partial_\lambda F^{\mu\nu} + [A_\lambda, F^{\mu\nu}]$. Like the $U(1)$ Lagrangian, the Yang-Mills Lagrangian can be split into kinetic and potential parts

$$T = \int\left(-\frac{1}{4}\text{Tr}(E_i E_i) + \frac{1}{2}(D_0\Phi)^\dagger D_0\Phi\right)d^d x \qquad (2.141)$$

$$V = \int\left(-\frac{1}{8}\text{Tr}(F_{ij}F_{ij}) + \frac{1}{2}(D_i\Phi)^\dagger D_i\Phi + U(\Phi^\dagger\Phi)\right)d^d x, \qquad (2.142)$$

where $E_i = F_{0i}$, but the expression T only describes the physical kinetic energy if one imposes Gauss' law

$$D_i E_i = -\frac{1}{C}(\Phi^\dagger t^a D_0\Phi - (D_0\Phi)^\dagger t^a\Phi)t^a, \qquad (2.143)$$

which reduces to an equation for A_0.

Note that an infinitesimal non-abelian gauge transformation has the form $g = e^{\alpha} = I + \alpha$, with α an infinitesimal antihermitian matrix. Under such a transformation,

$$\delta \Phi = \alpha \Phi \tag{2.144}$$

$$\delta A_{\mu} = -D_{\mu}\alpha. \tag{2.145}$$

Therefore $D_0 \Phi = \partial_0 \Phi + A_0 \Phi$ and $E_i = \partial_0 A_i - D_i A_0$ are related to $\partial_0 \Phi$ and $\partial_0 A_i$ by the addition of a gauge transformation with parameter A_0. Gauss' law determines A_0 in such a way that $\{D_0\Phi, E_i\}$ is the projection of $\{\partial_0\Phi, \partial_0 A_i\}$ orthogonal to gauge orbits, i.e. in such a way that the orthogonality condition

$$\int \left(\mathrm{Tr}(E_i D_i \alpha) + (D_0\Phi)^{\dagger}\alpha\Phi + (\alpha\Phi)^{\dagger} D_0\Phi \right) d^d x = 0 \tag{2.146}$$

is satisfied for all functions α which vanish at infinity. One verifies this by multiplying (2.143) by α, taking the trace, and using the divergence theorem. One also needs the covariant divergence identity

$$\partial_i(\mathrm{Tr}(E_i\alpha)) = \mathrm{Tr}(D_i E_i\, \alpha) + \mathrm{Tr}(E_i\, D_i\alpha), \tag{2.147}$$

which follows from the identity $\partial_i(\mathrm{Tr}(E_i\alpha)) = \mathrm{Tr}(\partial_i E_i\,\alpha) + \mathrm{Tr}(E_i\,\partial_i\alpha)$ because the cyclic property of a trace implies $\mathrm{Tr}([A_i, E_i]\alpha) + \mathrm{Tr}(E_i[A_i, \alpha]) = 0$. Thus, when Gauss' law is satisfied, T is the natural kinetic energy for a trajectory in the true configuration space $\mathcal{C}$.

There are two important variants of Yang-Mills theory that we will be considering later. In the first, we take the scalar field Φ to be valued in the Lie algebra, $\mathrm{Lie}(G)$. That is, Φ is an $n \times n$ matrix field, which has a component expansion $\Phi = \Phi^a t^a$. The gauge transformation of Φ is now by conjugation (the adjoint action of G),

$$\Phi \mapsto g\Phi g^{-1}. \tag{2.148}$$

This transformation keeps Φ in $\mathrm{Lie}(G)$. The covariant derivative of Φ is

$$D_{\mu}\Phi = \partial_{\mu}\Phi + [A_{\mu}, \Phi], \tag{2.149}$$

which gauge transforms as

$$\begin{aligned} D_{\mu}\Phi &\mapsto \partial_{\mu}(g\Phi g^{-1}) + [gA_{\mu}g^{-1} - \partial_{\mu}g g^{-1},\ g\Phi g^{-1}] \\ &= gD_{\mu}\Phi g^{-1}. \end{aligned} \tag{2.150}$$

The Lagrangian density is

$$\mathcal{L} = \frac{1}{8}\mathrm{Tr}\left(F_{\mu\nu}F^{\mu\nu}\right) - \frac{1}{4}\mathrm{Tr}\left(D_{\mu}\Phi D^{\mu}\Phi\right) - U(\mathrm{Tr}\,\Phi^2), \tag{2.151}$$

and is gauge invariant because of the cyclic property of the trace. The second variant is pure Yang-Mills theory. Here there is no scalar field, but only the gauge field A_μ and its field tensor $F_{\mu\nu}$, with Lagrangian density

$$\mathcal{L} = \frac{1}{8}\mathrm{Tr}\left(F_{\mu\nu}F^{\mu\nu}\right), \qquad (2.152)$$

and field equation

$$D_\mu F^{\mu\nu} = 0. \qquad (2.153)$$

Each variant of gauge theory has its own version of Gauss' law, but the interpretation that it projects the time derivatives of the fields orthogonal to gauge orbits persists.

Naively, the vacuum of pure Yang-Mills theory is the field configuration $A_\mu = 0$. However, a gauge transformation of this, $A_\mu = -\partial_\mu g g^{-1}$, is equally well a vacuum configuration. Such a configuration is called a pure gauge. Its field tensor vanishes, so the field equation (2.153) is trivially satisfied.

We described earlier how space-time and internal symmetries lead to conserved quantities in scalar field theory. The space-time and global symmetries lead in the same way to conserved quantities in gauge theories. One might imagine that the very much larger group of local gauge symmetries leads to yet further local conservation laws. However, this is not the case, because the conserved charges vanish identically if Gauss' law is satisfied at each space-time point.

Although gauge transformations by themselves do not give new conservation laws, it is interesting that suitable gauge transformations can be used to improve familiar conservation laws [214]. Thus, consider pure Yang-Mills theory. An infinitesimal translation in the direction ε^ν is a symmetry. The change in the field, naively, is

$$A_\lambda \mapsto A_\lambda + \varepsilon^\nu \partial_\nu A_\lambda. \qquad (2.154)$$

As usual, the change in the Lagrangian density is $\varepsilon^\nu \partial_\nu \mathcal{L} = \varepsilon^\nu \partial_\mu(\delta^\mu_\nu \mathcal{L})$, and hence the conserved energy-momentum tensor is

$$T^\mu_\nu = \frac{1}{8}\mathrm{Tr}\left(4F^{\mu\lambda}\partial_\nu A_\lambda - \delta^\mu_\nu F^{\kappa\lambda}F_{\kappa\lambda}\right). \qquad (2.155)$$

This satisfies $\partial_\mu T^\mu_\nu = 0$, but it is not gauge invariant. It can be improved by adding to the original variation of the gauge potential an infinitesimal gauge transformation with parameter $\varepsilon^\nu A_\nu$. Thus

$$\begin{aligned} A_\lambda &\mapsto A_\lambda + \varepsilon^\nu \partial_\nu A_\lambda - D_\lambda(\varepsilon^\nu A_\nu) \\ &= A_\lambda + \varepsilon^\nu F_{\nu\lambda}, \end{aligned} \qquad (2.156)$$

which already looks more covariant. The improved energy-momentum tensor is

$$\widetilde{T}^\mu_\nu = \frac{1}{8}\mathrm{Tr}\left(4F^{\mu\lambda}F_{\nu\lambda} - \delta^\mu_\nu F^{\kappa\lambda}F_{\kappa\lambda}\right).$$ (2.157)

This is conserved and gauge invariant, and $\widetilde{T}_{\mu\nu}$ is symmetric under interchange of μ and ν. We shall apply this improvement technique in later sections.

2.7 The Higgs mechanism

It is important to understand the field content of a gauge theory linearized around the vacuum, and whether the fields are effectively massive or massless. This will determine the asymptotic nature of the fields of any soliton, and the type of interactions to expect between well separated solitons.

The pure Maxwell theory is the prototype. Here the field equation is $\partial_\mu f^{\mu\nu} = 0$, and is already linear. We can impose the transverse gauge condition $\nabla \cdot \mathbf{a} = 0$. Gauss' law now allows $a_0 = 0$. The remaining field equation is the massless wave equation

$$(\partial_0\partial_0 - \nabla^2)\mathbf{a} = 0.$$ (2.158)

This has plane wave solutions $\mathbf{a} = \boldsymbol{\varepsilon}e^{i(\mathbf{k}\cdot\mathbf{x}-\omega t)}$ with dispersion relation $\omega = |\mathbf{k}|$, and the polarization $\boldsymbol{\varepsilon}$ must be transverse to the wave-vector, that is $\boldsymbol{\varepsilon}\cdot\mathbf{k} = 0$, to satisfy $\nabla \cdot \mathbf{a} = 0$. The particle arising from quantization of the waves is the massless photon, which has two independent polarization states.

Consider now scalar electrodynamics, with Lagrangian (2.99). If U has its minimum at $\phi = 0$, then the $U(1)$ gauge symmetry is unbroken, and the linearized equations for $\mathbf{a}$ and ϕ decouple. Generally, ϕ will be a massive field with two real components. The mass depends on the second derivative $\partial_{\bar\phi}\partial_\phi U$, evaluated at $\phi = 0$. More explicitly, if the Taylor expansion of U is $U = U_0 + \frac{1}{2}m^2\bar\phi\phi + \cdots$, then the scalar fields both have mass m. However, the photon is massless as before, and is long-range.

When there is spontaneous symmetry breaking, things are different. Suppose $U(\bar\phi\phi)$ has its minimum at $|\phi| \neq 0$. Because of the $U(1)$ symmetry, there is a whole circular $U(1)$ orbit that minimizes U. In the absence of the gauge fields, there was a scalar Goldstone particle. However, now it makes sense to fix the gauge so that ϕ is real, and the vacuum is $\phi = \phi_0 > 0$. The linearized equation for ϕ involves only oscillations of the magnitude of ϕ, but its phase remains zero. Generally, U has a quadratic minimum in this direction, so there is a single massive scalar field. On quantization this becomes the Higgs particle of scalar electrodynamics.

Provided that ϕ is at or close to its vacuum value, the part of the Lagrangian quadratic in the gauge field a_μ is

$$-\frac{1}{4}f_{\mu\nu}f^{\mu\nu} + \frac{1}{2}\phi_0^2 a_\mu a^\mu, \qquad (2.159)$$

so the positive coefficient ϕ_0 acts as a non-zero mass parameter. There is no remaining gauge freedom. The field equation associated with the Lagrangian (2.159) is

$$\partial_\mu f^{\mu\nu} + \phi_0^2 a^\nu = 0. \qquad (2.160)$$

Expanding out, this becomes

$$(\partial_\mu\partial^\mu + \phi_0^2)a^\nu - \partial^\nu\partial_\mu a^\mu = 0. \qquad (2.161)$$

Acting with ∂_ν, the triple derivative terms cancel, and it follows that

$$\partial_\nu a^\nu = 0. \qquad (2.162)$$

So (2.161) simplifies to the massive Klein-Gordon equation

$$(\partial_\mu\partial^\mu + \phi_0^2)a^\nu = 0. \qquad (2.163)$$

Because of the auxillary condition, $\partial_\nu a^\nu = 0$, one may regard the spatial components of the gauge field, $\mathbf{a}$, as independently satisfying (2.163), but a_0 is dependent on these. The plane wave solutions of (2.163) and (2.162) are of the form

$$a_0 = \frac{\mathbf{k}\cdot\boldsymbol{\varepsilon}}{\omega}e^{i(\mathbf{k}\cdot\mathbf{x}-\omega t)}, \quad \mathbf{a} = \boldsymbol{\varepsilon}e^{i(\mathbf{k}\cdot\mathbf{x}-\omega t)}, \qquad (2.164)$$

with $\omega = \sqrt{|\mathbf{k}|^2 + \phi_0^2}$, and $\boldsymbol{\varepsilon}$ an unconstrained polarization vector. On quantization, one gets a massive photon, with three independent polarization states, and no scalar Goldstone particle. This is the Higgs mechanism [180, 120, 233].

The Yang-Mills analogue is as follows. In pure Yang-Mills theory, with gauge group G, there are dim G massless gauge fields (gluons). When scalar fields Φ are coupled to the Yang-Mills field, and if the potential function $U(\Phi^\dagger\Phi)$ has its minimum at $\Phi = 0$ (the unbroken case), then the gluons are still massless, and there are n massive scalar fields, all with the same mass. The gauge symmetry is spontaneously broken if U has its minimum at $\Phi_0 \neq 0$. Let the isotropy subgroup of Φ_0 be H. Then U takes the same minimal value on the whole orbit of Φ_0, which is G/H. If G is a sufficiently large subgroup of $U(n)$, one can fix the gauge so that $\Phi = \Phi_0(1 + \eta)$ with η real. There is then just one massive scalar Higgs field, associated with η. All components of the gauge field need also to be considered. There is a mass term

$$\frac{1}{2}(A_\mu \Phi_0)^\dagger A^\mu \Phi_0 , \tag{2.165}$$

coming from the part of the Lagrangian density $\frac{1}{2}(D_\mu \Phi)^\dagger D^\mu \Phi$. This can be rewritten as $\frac{1}{2}A_\mu^a M^{ab} A^{\mu b}$, where $M^{ab} = (t^a \Phi_0)^\dagger t^b \Phi_0 = -\Phi_0^\dagger t^a t^b \Phi_0$ is the mass matrix, a quadratic form on $\mathrm{Lie}(G)$. It has zero eigenvalues in the directions of $\mathrm{Lie}(H)$, because the generators of H annihilate Φ_0, and non-zero eigenvalues in the directions of the orthogonal subspace $\mathrm{Lie}(G)$ $- \mathrm{Lie}(H)$. Thus there are dim H massless gauge fields associated with the unbroken group H, but the remaining gauge fields become massive. There are no Goldstone particles if the gauge group G acts transitively on the vacuum manifold of U, that is, if the orbit of Φ_0 is the entire set minimizing U.

We shall see the Higgs mechanism in action when we discuss vortices and monopoles in Chapters 7 and 8.

2.8 Gradient flow in field theory

Recall from Section 2.1 that the gradient flow equation associated with a second order dynamical system on a Euclidean manifold is obtained by replacing second by first time derivatives. If the background metric is non-trivial, it needs to be included, as in (2.24).

Similarly, in scalar field theory, which has an infinite-dimensional Euclidean configuration space, the gradient flow equation is obtained by replacing the second order nonlinear wave equation (2.52) by

$$\kappa \partial_0 \phi = \nabla^2 \phi - \frac{dU}{d\phi} , \tag{2.166}$$

with κ positive, as before. This is a nonlinear diffusion equation. The flow is in the direction where the potential energy V decreases most steeply. Short wavelength fields decay the most rapidly for equation (2.166), and the field is smoothed out. For arbitrary initial data, at time t_0, the field ϕ is infinitely differentiable at any time $t > t_0$. An important contrast between (2.166) and the analogous equation (2.25) for a finite-dimensional system, related to this, is that in the field theory time can not be reversed for generic initial data, because the shortest wavelengths would blow up arbitrarily fast.

However, there are solutions $\phi(t, \mathbf{x})$ of (2.166) which are defined for all time. If $\lim_{t \to -\infty} \phi(t, \mathbf{x}) = \phi_{-\infty}(\mathbf{x})$ and $\lim_{t \to \infty} \phi(t, \mathbf{x}) = \phi_\infty(\mathbf{x})$ both exist, then $\phi_{-\infty}$ and ϕ_∞ are static solutions of the equation, and hence stationary points of V.

The gradient flow equation has a number of uses, one of which is simply to search for static solutions (generally minima of V). Physically, it describes a dissipative relaxation of the field towards equilibrium.

In a gauge theory there is also a gradient flow. One version is gauge invariant. Thus in scalar electrodynamics, the gradient flow equations are

$$\kappa D_0 \phi \; = \; D_i D_i \phi - 2U'(\bar\phi \phi)\phi \qquad (2.167)$$

$$\kappa e_i \; = \; \partial_j f_{ji} - \frac{i}{2}(\bar\phi D_i \phi - \phi \overline{D_i \phi}), \qquad (2.168)$$

which are related to equations (2.105) and (2.106), but one time derivative ∂_0 or covariant time derivative D_0 has been dropped. Perhaps surprisingly, Gauss' law (2.107) is automatically satisfied. One verifies this by taking the divergence of (2.168), and using (2.167) to replace the terms involving $D_i D_i \phi$. This calculation explains the necessity for the coefficient κ to be the same in (2.167) and (2.168). The interpretation is that in gradient flow, the projected motion $\{D_0\phi, e_i\}$ is automatically orthogonal to gauge orbits. In fact, this is not so surprising, because the flow is orthogonal to hypersurfaces of the potential energy V, and these hypersurfaces include all gauge orbits since V is gauge invariant. It follows that one can set $a_0 = 0$. The gauge-fixed gradient flow equations are then

$$\kappa \partial_0 \phi \; = \; D_i D_i \phi - 2U'(\bar\phi \phi)\phi \qquad (2.169)$$

$$\kappa \partial_0 a_i \; = \; \partial_j f_{ji} - \frac{i}{2}(\bar\phi D_i \phi - \phi \overline{D_i \phi}), \qquad (2.170)$$

and these are consistent with the version of Gauss' law with $a_0 = 0$.

Similar considerations apply to non-abelian gauge theory. The gradient flow equation of pure Yang-Mills theory is

$$\kappa F_{0i} = D_j F_{ji}. \qquad (2.171)$$

Gauss' law is $D_i F_{0i} = 0$, and this is satisfied since

$$\kappa D_i F_{0i} = D_i D_j F_{ji} = \frac{1}{2}[D_i, D_j]F_{ji} = -\frac{1}{2}[F_{ij}, F_{ij}] = 0. \qquad (2.172)$$

The gauge-fixed gradient flow equation, with $A_0 = 0$, is

$$\kappa \partial_0 A_i = D_j F_{ji}. \qquad (2.173)$$

Thus the gradient flow in gauge theory is a flow in the full configuration space $\mathcal{A}$, which may or may not be gauge-fixed; and because Gauss' law is satisfied, there is a consistent interpretation as a flow in the true configuration space $\mathcal{C}$ with its natural metric and potential.

3

Topology in field theory

We shall need some ideas from topology in order to understand the classification of solitons in field theory, and especially to understand their stability. However, this is not a textbook on topology, and our discussion will be somewhat heuristic. For a rigorous discussion of these topics we recommend the books [181] and [60]. In this chapter, and in a few places later in the book, it will be very helpful to use differential forms. We assume the reader has some familiarity with these (an elementary introduction can be found in the book [133]).

There are two basic techniques for classifying solitons in theories with scalar fields. The first is homotopy theory, and the second is topological degree theory, which can sometimes be used to calculate a homotopy class. Topological degree is a special case of homology ideas.

In gauge theories, the Chern numbers classify solitons. These are found by integrating Chern forms over space. The Chern forms are gauge invariant differential forms of even degree constructed algebraically from the field tensor. The simplest are 2-forms and 4-forms. The former can be integrated over a plane or surface, the latter over $\mathbb{R}^4$ or another 4-manifold. Since we are interested in the application to solitons in dimensions up to four, we do not discuss higher degree Chern forms. After discussing Chern forms and Chern numbers, we shall consider the related Chern-Simons forms and their integrals.

3.1 Homotopy theory

Let X and Y be two manifolds without boundary, and consider the continuous maps between them, $\Psi : X \mapsto Y$. Often it is helpful to identify base points $\mathbf{x}_0 \in X$ and $\mathbf{y}_0 \in Y$ and require $\Psi(\mathbf{x}_0) = \mathbf{y}_0$; then Ψ is a based map. A based map $\Psi_0 : X \mapsto Y$ is said to be homotopic to another such map Ψ_1 if Ψ_0 can be continuously deformed into Ψ_1. Precisely, Ψ_0

is homotopic to Ψ_1 if there is a continuous map

$$\widetilde{\Psi} : X \times [0, 1] \mapsto Y , \qquad (3.1)$$

with "time" τ parametrizing the interval $[0, 1]$, such that $\widetilde{\Psi}\,|_{\tau=0} = \Psi_0$ and $\widetilde{\Psi}\,|_{\tau=1} = \Psi_1$, and $\widetilde{\Psi}(\mathbf{x}_0; \tau) = \mathbf{y}_0$ for all τ.

"Homotopic" is an equivalence relation; it is symmetric (Ψ_0 homotopic to Ψ_1 implies Ψ_1 homotopic to Ψ_0) because the time flow can be reversed; it is transitive (Ψ_0 homotopic to Ψ_1 and Ψ_1 homotopic to Ψ_2 implies Ψ_0 homotopic to Ψ_2) because time intervals can be adjoined, and then rescaled; and is obviously reflexive (Ψ_0 homotopic to itself). Thus the maps Ψ can be classified into homotopy classes. One class is the constant class, consisting of maps homotopic to the constant map Ψ for which $\Psi(\mathbf{x}) = \mathbf{y}_0$ for all $\mathbf{x}$.

One can say more about homotopy classes if X is a sphere. The n-sphere S^n is the set of points in $\mathbb{R}^{n+1}$ at unit distance from the origin. We shall be especially interested in the cases S^1 the circle, which is also the manifold of the group $U(1)$, S^2 the usual sphere, and S^3 the unit sphere in four dimensions, which is also the manifold of the group $SU(2)$.

The set of homotopy classes of based maps $\Psi : S^n \mapsto Y$ is denoted by $\pi_n(Y)$. We take as base points the North pole $\mathbf{p}$ in S^n, that is, the point $(0, 0, \ldots, 1) \in \mathbb{R}^{n+1}$, and some chosen point $\mathbf{y}_0 \in Y$. (In $\mathbb{R}^2$, the usual choice of base point is $(1, 0)$.) For $n \geq 1$, the set $\pi_n(Y)$ forms a group; the nth homotopy group of Y. To understand this one needs to define the composition of two classes of maps, and show that the usual group axioms are satisfied. In practice, one composes two maps and shows that the composition is independent of the choice within the homotopy class.

The construction, for $\pi_1(Y)$, is schematically as in Fig. 3.1.

A map $S^1 \mapsto Y$, and also its image in Y, is called a loop. Ψ_0 and Ψ_1 are

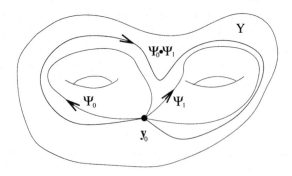

Fig. 3.1. Sketch illustrating the composition of two maps involved in the construction of $\pi_1(Y)$.

two based loops in Y. Their composition $\Psi_0 \cdot \Psi_1$ is the loop Ψ obtained by following Ψ_0 by Ψ_1. The homotopy class of Ψ depends only on the classes of Ψ_0 and Ψ_1. The composition is associative,

$$\Psi_0 \cdot (\Psi_1 \cdot \Psi_2) = (\Psi_0 \cdot \Psi_1) \cdot \Psi_2, \tag{3.2}$$

because each of these is the loop Ψ_0 followed by Ψ_1 followed by Ψ_2. The class of the constant map $S^1 \mapsto \mathbf{y}_0$ is the identity element of the group $\pi_1(Y)$. When composed with another map Ψ, the class of Ψ is unchanged. The inverse of Ψ is Ψ traversed in the opposite direction, which composes with Ψ to give a loop in the constant class. Note that $\pi_1(Y)$ is generally non-abelian, since the composition of loops $\Psi_0 \cdot \Psi_1$ is not necessarily homotopic to $\Psi_1 \cdot \Psi_0$. $\pi_1(Y)$ is known as the fundamental group of Y. If Y is connected and $\pi_1(Y) = I$, where I denotes the trivial group with just the identity element, then the space Y is said to be simply connected. In this case, every loop is contractible, i.e. homotopic to the trivial loop. Consider $\mathbb{R}^d$ with the origin as base point. Any loop $\Psi_0 : S^1 \mapsto \mathbb{R}^d$ is contractible (parametrize S^1 by $\theta \in [0, 2\pi]$, and define $\widetilde{\Psi}(\theta; \tau) = (1 - \tau)\Psi_0(\theta)$), therefore $\pi_1(\mathbb{R}^d) = I$. The same construction works on many other spaces, after defining a suitable origin and local Cartesian coordinates. Thus, for all $d \geq 2$, loops in S^d are contractible, so $\pi_1(S^d) = I$.

One can show that $\pi_1(S^1) = \mathbb{Z}$. A map $S^1 \mapsto S^1$ is defined by a continuous function $f(\theta)$ on $[0, 2\pi]$, where $f(\theta)$ is the angle on the target. The map is based if $f(0) = 0$, and continuity of the map requires that $f(2\pi) = 2\pi k$ for some $k \in \mathbb{Z}$. k is called the winding number of the map. It is the net number of times that the image $f(\theta)$ winds around the target as θ goes once around the domain. k, being an integer, can not change under a homotopy. Conversely, maps f_0 and f_1 with the same k *are* homotopic, as one sees from the formula $f = (1 - \tau)f_0 + \tau f_1$, which continuously deforms f_0 into f_1. Thus the homotopy classes of $\pi_1(S^1)$ are labelled by the integers. Moreover, as a group, $\pi_1(S^1)$ is $\mathbb{Z}$ with the usual addition. This is checked by composing maps f_0 and f_1, in the sense above, and noting that the winding numbers k_0 and k_1 add. In this example, the fundamental group is abelian. More complicated is the d-dimensional torus $\mathbb{T}^d$. $\mathbb{T}^d$ can be regarded as $\mathbb{R}^d$, with base point the origin, modulo a lattice Λ with d independent basis vectors. Then $\pi_1(\mathbb{T}^d) = \Lambda$, which is isomorphic as a group to $\mathbb{Z}^d$, again abelian.

The fundamental group $\pi_1(Y)$ of a closed Riemann surface Y of genus $g > 1$ is non-abelian, being generated by $2g$ loops $a_1, \ldots, a_g, b_1, \ldots, b_g$ and their inverses, starting and ending at the base point $\mathbf{y}_0$. They are subject to the single relation

$$a_1 b_1 a_1^{-1} b_1^{-1} a_2 b_2 a_2^{-1} b_2^{-1} \cdots a_g b_g a_g^{-1} b_g^{-1} = I. \tag{3.3}$$

By cutting Y along all the generating loops one can represent Y as a polygon, every vertex of which corresponds to $\mathbf{y}_0$, and whose edges are identified in pairs; see Fig. 3.2. The surface is reconstructed by gluing the edge pairs a_s and a_s^{-1} together, and similarly b_s and b_s^{-1}, in such a way that the arrows match up.

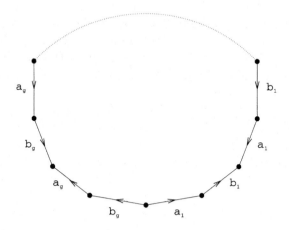

Fig. 3.2. Polygon representation of a genus g Riemann surface.

Let us next consider maps from S^n to a general manifold Y, for $n \geq 2$. We can represent the n-sphere S^n as $\mathbb{R}^n$ with all points at infinity identified, using stereographic projection from the North pole $\mathbf{p}$; see Fig. 3.3.

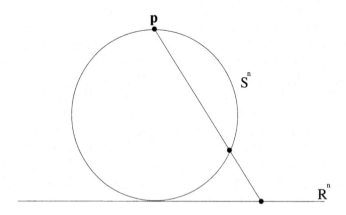

Fig. 3.3. Pictorial representation of stereographic projection from S^n to $\mathbb{R}^n$.

Stereographic projection is a one-to-one correspondence between $\{S^n - \mathbf{p}\}$

and $\mathbb{R}^n$. If we adjoin a single point at infinity to $\mathbb{R}^n$, and regard this as the image of $\mathbf{p}$, then we have our desired representation of S^n. A based map $\Psi : S^n \mapsto Y$ is therefore homotopically equivalent to a map $\Psi : \mathbb{R}^n \mapsto Y$, provided $\lim_{\mathbf{x} \to \infty} \Psi(\mathbf{x}) = \mathbf{y}_0$. One says that the point at infinity is mapped to $\mathbf{y}_0$.

Two alternative representations are helpful. We can contract $\mathbb{R}^n$ onto the interior of an n-ball. This is explicitly achieved by the map $\rho \mapsto \tan^{-1} \rho$, with ρ the distance from the origin. It follows that a continuous map from $\mathbb{R}^n$ to Y, where the boundary of the ball and all exterior points map to $\mathbf{y}_0$, is equivalent to a based map from S^n to Y. Similarly, we can contract $\mathbb{R}^n$ onto an n-dimensional hypercube in $\mathbb{R}^n$. A continuous map from $\mathbb{R}^n$ to Y, where the boundary of the hypercube and all exterior points map to $\mathbf{y}_0$, is again equivalent to a based map from S^n to Y.

Using the hypercube representation one can see how $\pi_n(Y)$ forms a group for $n \geq 2$. Maps Ψ_0 and Ψ_1 from S^n to Y, representing two homotopy classes, are composed schematically as in Fig. 3.4.

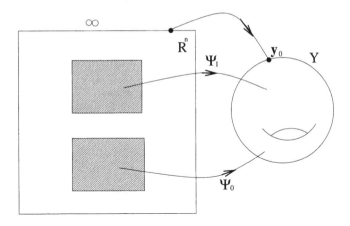

Fig. 3.4. Schematic representation of how $\pi_n(Y)$ forms a group using the hypercube representation.

The result $\Psi_0 \cdot \Psi_1$ has the correct behaviour as $\mathbf{x} \to \infty$ to be regarded as a map from S^n to Y. The identity element of $\pi_n(Y)$ is the class of the constant map $S^n \mapsto \mathbf{y}_0$. The homotopy inverse of a map Ψ (or rather, its class) is obtained, in the hypercube representation, by making a single reflection in $\mathbb{R}^n$, i.e. defining $\Psi^{-1}(x^1, \ldots, x^n) = \Psi(-x^1, \ldots, x^n)$. It can be checked that the composition of Ψ and Ψ^{-1} is homotopic to the constant map. Three maps combine associatively, as before. Thus $\pi_n(Y)$

is a group.

There is a crucial difference between this case with $n \geq 2$, and the case $n = 1$. Figure 3.5 indicates that by a homotopy, the order in which two maps (classes) are composed can be reversed if $n \geq 2$.

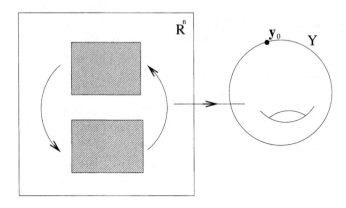

Fig. 3.5.   Sketch to illustrate that for $n \geq 2$ the order in which two maps are composed can be reversed.

The group $\pi_n(Y)$, for $n \geq 2$, is therefore abelian.

Calculation of homotopy groups $\pi_n(Y)$ is a major task of algebraic topology, and is not easy. We quote some of the most important results for us. A basic result is

$$\pi_n(S^n) = \mathbb{Z} \qquad \forall n \geq 1. \tag{3.4}$$

The generator of the group is the class of the identity map from S^n to itself. A representative of the kth homotopy class is where there is a k-fold winding in a 2-plane, e.g. the map

$$(r\cos\theta, r\sin\theta, x^3, x^4, \ldots, x^{n+1}) \mapsto (r\cos k\theta, r\sin k\theta, x^3, x^4, \ldots, x^{n+1}) \tag{3.5}$$

with $r^2 + (x^3)^2 + (x^4)^2 + \cdots + (x^{n+1})^2 = 1$. For $1 \leq n < d$, $\pi_n(S^d) = I$ because all maps $S^n \mapsto S^d$ are contractible to a constant map. This is because the image excludes at least one point of S^d; removing this, one obtains an image of S^n in $\mathbb{R}^d$, which can be linearly contracted to a point. The groups $\pi_n(S^d)$ for $n > d$ are increasingly difficult to compute as n and d increase. Some examples are

$$\pi_n(S^1) = I \quad \forall n \geq 2, \qquad \pi_3(S^2) = \mathbb{Z},$$
$$\pi_{n+1}(S^n) = \mathbb{Z}_2 \quad \forall n \geq 3, \qquad \pi_{n+2}(S^n) = \mathbb{Z}_2 \quad \forall n \geq 2. \tag{3.6}$$

The higher homotopy groups of a torus are trivial; $\pi_n(\mathbb{T}^d) = I$ if $n \geq 2$.

One further homotopy notion is the set of unbased homotopy classes $\pi_0(Y)$. These classes are maps from a single point to Y, up to homotopy equivalence. (Note, we regard the 0-sphere here as one point, not the two points $\pm 1 \in \mathbb{R}^1$.) Maps with image points in the same connected component of Y are homotopic. Thus $\pi_0(Y)$ is the set of distinct, connected components of Y. If Y is connected, then $\pi_0(Y)$ has just one element. $\pi_0(Y)$ is generally not a group.

We shall need some results on the homotopy groups of coset spaces in Chapter 8. Let G be a Lie group with subgroup H. The crucial isomorphism results are

$$\pi_2(G/H) = \pi_1(H), \quad \pi_1(G/H) = \pi_0(H), \tag{3.7}$$

which hold provided that G is both connected and simply connected, i.e. $\pi_0(G) = \pi_1(G) = I$. Note that $\pi_0(G)$ and $\pi_0(H)$ are groups because one can use the group multiplication of G to compose elements. The isomorphisms (3.7) follow from the existence of a series of homomorphisms between homotopy groups, called the exact homotopy sequence. Its proof involves a consideration of the obstructions to lifting a continuous map $\Psi : S^2 \mapsto G/H$ to a continuous map $\tilde{\Psi} : S^2 \mapsto G$. We refer the interested reader to ref. [181] for a detailed proof.

A simple but relevant example is

$$\pi_2(SU(2)/U(1)) = \pi_1(U(1)) = \pi_1(S^1) = \mathbb{Z}, \tag{3.8}$$

which we already knew because the coset space $SU(2)/U(1)$ may be identified with S^2. Two useful generalizations of this example are

$$\pi_2\left(\frac{SU(m)}{U(1)^{m-1}}\right) = \pi_1(U(1)^{m-1}) = \mathbb{Z}^{m-1} \tag{3.9}$$

$$\pi_2\left(\frac{SU(m)}{U(m-1)}\right) = \pi_1(U(m-1)) = \mathbb{Z}, \tag{3.10}$$

where the final equality in the last line follows because $U(m-1) = U(1) \times SU(m-1)/\mathbb{Z}_{m-1}$ and $SU(m-1)$ is simply connected, so the integer is associated with a winding around the $U(1)$ factor.

Homotopy theory can be applied directly to a scalar field theory of the type governed by the Lagrangian (2.62), where the field at a given time is a map $\phi : X \mapsto Y$. If the field (strongly) satisfies the dynamical field equation then it is continuous in space and time, so its homotopy class is well defined and unchanging with time. The homotopy class is a topological, conserved quantity. Homotopy theory can also be applied to field theories defined in $\mathbb{R}^d$, but here the boundary conditions play a crucial role. These applications are discussed further in Section 4.1, and in later chapters.

So far, we have considered the homotopy groups of a finite-dimensional space Y. One may also consider homotopy groups of a space of based continuous functions, or a space of based continuous maps, from X to Y, say. We shall call this space of maps $\text{Maps}(X \mapsto Y)$. This space is generally disconnected, with each connected component being one homotopy class of maps. We denote by $\text{Maps}_0(X \mapsto Y)$ the class of based maps which contains the constant map $X \mapsto \mathbf{y}_0$. Let us limit the discussion to maps from S^m to Y, and let W be the space of maps

$$W = \text{Maps}_0(S^m \mapsto Y). \tag{3.11}$$

W is connected, and the base point of W is the constant map itself, $S^m \mapsto \mathbf{y}_0$. Then it is easy to see that for $n \geq 1$,

$$\pi_n(W) = \pi_n(\text{Maps}_0(S^m \mapsto Y)) = \pi_{n+m}(Y). \tag{3.12}$$

This general result becomes clear using the hypercube representations of S^m and S^n introduced above. An n-sphere's worth of based maps in the constant class, from $S^m \mapsto Y$, may be represented as in Fig. 3.6, where the hypercube is now $(n + m)$-dimensional, each point on its boundary is mapped to $\mathbf{y}_0$, and each slice (fixing n of the coordinates) gives a map $S^m \mapsto Y$. An example is

$$W = \text{Maps}_0(S^1 \mapsto S^2). \tag{3.13}$$

Since $\pi_1(S^2) = I$, every map is in the class of the constant map. Here $\pi_1(W) = \mathbb{Z}$, because $\pi_2(S^2) = \mathbb{Z}$. Heuristically, each map is contractible, but a whole loop of maps is not simultaneously contractible (in general). The picture is as in Fig. 3.7, where we show the images of a 1-parameter family of maps from $S^1 \mapsto S^2$.

Similarly, if $W = \text{Maps}_0(S^2 \mapsto S^3)$, then $\pi_1(W) = \mathbb{Z}$.

In the context of field theory, the space $\text{Maps}_0(X \mapsto Y)$ is the vacuum component of the configuration space $\mathcal{C}$ of the scalar field theory with Lagrangian (2.62), so $\pi_n(\text{Maps}_0(X \mapsto Y))$ is the nth homotopy group of this component of $\mathcal{C}$. The base point condition is natural if one point of X is special (for example, represents the point at infinity). There are other applications too, to scalar field theories and to gauge theories. For example, the fact that $\pi_1(\text{Maps}_0(S^2 \mapsto S^3)) = \mathbb{Z}$ underlies the construction of sphalerons in the electroweak theory, which we describe in Chapter 11.

3.2 Topological degree

Topological degree is a more limited, but also more refined tool than homotopy theory, and it allows the calculation of the homotopy class of

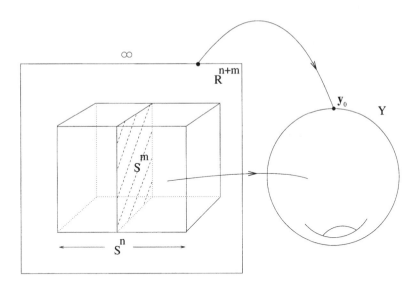

Fig. 3.6. An n-sphere's worth of based maps, from $S^m \mapsto Y$, illustrated using the hypercube representation.

a map in certain circumstances. It is useful, because it occurs in various ways in field theories with solitons. Often, the topological aspect of a soliton is entirely captured by the degree of a map directly related to the soliton field. However, the more general homotopy theory ideas are in the background, and can be brought into action where necessary.

The topological degree is defined for a map Ψ between two closed manifolds of the same dimension, $\Psi : X \mapsto Y$. Let $\dim X = \dim Y = d$. Both X and Y must be oriented, and the map should be differentiable everywhere, with continuous derivatives. To avoid trivial difficulties, we suppose X is connected. We may as well suppose Y is connected too, since the image of X will always lie in one of the connected components of Y.

We need next to suppose that a normalized volume form Ω is defined on Y. Locally, this maps an oriented frame of tangent vectors at each point of Y to the reals, and preferably the positive reals. If Y is a Riemannian manifold, the Hodge dual of a positive function on Y is such a volume form. The normalization condition is

$$\int_Y \Omega = 1 \,. \tag{3.14}$$

Now consider $\Psi^*(\Omega)$, the pull-back of Ω to X using the map Ψ. In terms of local coordinates, if $\Omega = \beta(\mathbf{y}) dy^1 \wedge dy^2 \wedge \cdots \wedge dy^d$, and Ψ is

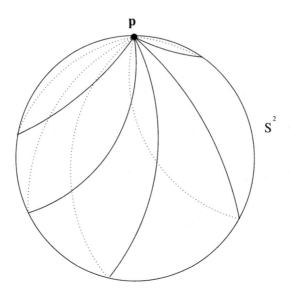

Fig. 3.7.   The images of a 1-parameter family of maps from $S^1 \mapsto S^2$.

represented by functions $\mathbf{y}(\mathbf{x})$, then

$$
\begin{aligned}
\Psi^*(\Omega) &= \beta(\mathbf{y}(\mathbf{x}))\frac{\partial y^1}{\partial x^j}dx^j \wedge \frac{\partial y^2}{\partial x^k}dx^k \wedge \cdots \wedge \frac{\partial y^d}{\partial x^l}dx^l \\
&= \beta(\mathbf{y}(\mathbf{x}))\det\left(\frac{\partial y^i}{\partial x^j}\right)dx^1 \wedge dx^2 \wedge \cdots \wedge dx^d, \qquad (3.15)
\end{aligned}
$$

where $J(\mathbf{x}) = \det\left(\frac{\partial y^i}{\partial x^j}\right)$ is the Jacobian of the map at $\mathbf{x}$.
     Now define

$$
\deg \Psi = \int_X \Psi^*(\Omega). \qquad (3.16)
$$

This integral occurs naturally in various field theories. $\deg \Psi$ is called the topological degree of the map Ψ, and is an integer, as we shall show below. The topological degree is a homotopy invariant of Ψ, simply because an integer can not change under a continuous deformation. It is also independent of the choice of Ω, because the difference of two normalized volume forms on Y is a d-form whose integral is zero, and hence an exact form. The pull-back of the difference is therefore exact on X, and integrates to zero.
     The most important example is for a map $\Psi : S^n \mapsto S^n$. Suppose Ψ is in the kth homotopy class of $\pi_n(S^n)$. Then $\deg \Psi = k$. This is clear for the map (3.5), which is a representative of the class. If one pulls back the standard, rotationally invariant, normalized volume form on the target

S^n, one gets k times the standard normalized volume form on the domain S^n. In particular, for a map $\Psi : S^1 \mapsto S^1$, the degree is equal to the winding number. This is verified by choosing the volume form $\frac{1}{2\pi} d\theta$ on S^1, and noting that for a map given by a function $f(\theta)$, the formula (3.16) reduces to

$$\deg \Psi = \frac{1}{2\pi} \int_0^{2\pi} \frac{df}{d\theta} \, d\theta = \frac{1}{2\pi} (f(2\pi) - f(0)) = k \,. \tag{3.17}$$

Sometimes the degree of a more general map, between higher-dimensional manifolds, is also called a winding number.

Another important example for us is the degree of a map from a three-dimensional manifold X to $SU(2)$. Elements of $SU(2)$ can be written as

$$g = c_0 1_2 + i\mathbf{c} \cdot \boldsymbol{\tau} \,, \tag{3.18}$$

where 1_2 is the unit 2×2 matrix and τ_1, τ_2, τ_3 are the Pauli matrices, and where $c_0^2 + \mathbf{c} \cdot \mathbf{c} = 1$. Geometrically, $SU(2)$ is a 3-sphere. The standard normalized volume form on $SU(2)$ can be expressed as

$$\Omega = \frac{1}{24\pi^2} \operatorname{Tr} \left(dgg^{-1} \wedge dgg^{-1} \wedge dgg^{-1} \right) . \tag{3.19}$$

To understand this, note that Ω is invariant under left and right multiplication by fixed elements of $SU(2)$, $g \mapsto g_1 g g_2$, and since $SU(2) \times SU(2)/\mathbb{Z}_2 = SO(4)$, Ω is rotationally invariant. To understand the normalization factor, consider g close to 1_2, where c_0 is essentially constant. Then (3.19) simplifies to

$$\Omega = \frac{-6i}{24\pi^2} \operatorname{Tr} \left(\tau_1 \tau_2 \tau_3 \right) dc_1 \wedge dc_2 \wedge dc_3 = \frac{1}{2\pi^2} dc_1 \wedge dc_2 \wedge dc_3 \,. \tag{3.20}$$

This is the desired normalization, because the unit 3-sphere has total volume $2\pi^2$ and has volume element $dc_1 \wedge dc_2 \wedge dc_3$ close to $c_0 = 1$.

If Ψ is a map from X to $SU(2)$, represented by a function $g(\mathbf{x})$, then

$$\deg \Psi = \frac{1}{24\pi^2} \int_X \operatorname{Tr} \left(dgg^{-1} \wedge dgg^{-1} \wedge dgg^{-1} \right) , \tag{3.21}$$

where dg now denotes $\partial_i g \, dx^i$. This is the integral of the pull-back of Ω to X.

A very useful feature of the topological degree of a map $\Psi : X \mapsto Y$ is that there is a second, apparently independent way to compute it. Choose a point $\mathbf{y}$ on Y, such that the set of preimages of $\mathbf{y}$, the points on X mapped to $\mathbf{y}$, is a set (possibly empty) of isolated points $\{\mathbf{x}^{(1)}, \ldots, \mathbf{x}^{(M)}\}$ at each of which the Jacobian of the map is non-zero. Such points $\mathbf{y}$ occur almost everywhere on Y. Let

$$\widetilde{\deg \Psi} = \sum_{m=1}^{M} \text{sign}\left(J(\mathbf{x}^{(m)})\right), \tag{3.22}$$

where $\text{sign}\left(J(\mathbf{x}^{(m)})\right)$ is the sign of the Jacobian at $\mathbf{x}^{(m)}$. One says that $\widetilde{\deg \Psi}$ counts the preimages of $\mathbf{y}$ with their multiplicity, which is 1 or -1, depending on whether Ψ is locally orientation preserving or orientation reversing. Clearly $\widetilde{\deg \Psi}$ is an integer. It is a theorem that $\widetilde{\deg \Psi} = \deg \Psi$, and hence is independent of the choice of $\mathbf{y}$.

To prove that $\widetilde{\deg \Psi} = \deg \Psi$, we proceed as follows. Deform the volume form Ω on Y so that it is concentrated on a small neighbourhood of the point $\mathbf{y}$, and still normalized. $\deg \Psi$ is unaffected, because, as we argued earlier, it doesn't depend on the choice of Ω. $\Psi^*(\Omega)$ is now concentrated on small neighbourhoods of each of the preimages $\{\mathbf{x}^{(1)}, \ldots, \mathbf{x}^{(M)}\}$. Moreover, the integral of $\Psi^*(\Omega)$ over one of these neighbourhoods is simply ± 1, which can be understood by a naive local change of coordinates from $\{x^i\}$ back to $\{y^i\}$, which introduces a factor $|J|^{-1}$, and reproduces the volume form Ω up to an orientation preserving/reversing sign. The integral is then unity, by the normalization condition. Summing over the preimages, we see that the formula (3.16) for $\deg \Psi$ reduces to the expression (3.22) for $\widetilde{\deg \Psi}$. $\deg \Psi$ is consequently an integer.

The standard map (3.5) from S^n to S^n again provides an example. A suitable choice for $\mathbf{y}$ is any point with $r \neq 0$. The preimages of

$$(r \cos \psi, \ r \sin \psi, \ x^3, \ldots, x^{n+1}) \tag{3.23}$$

are the points

$$(r \cos \theta, \ r \sin \theta, x^3, \ldots, x^{n+1}), \tag{3.24}$$

where $\cos k\theta = \cos \psi$ and $\sin k\theta = \sin \psi$. This requires $k\theta = \psi \bmod 2\pi$. If we choose both ψ and θ to be in the range $[0, 2\pi)$ then $\theta = \frac{\psi}{k}$, $\frac{\psi + 2\pi}{k}, \ldots, \frac{\psi + 2\pi(k-1)}{k}$. There are k preimages, and it is easy to see that the Jacobian is positive at each. So the degree of the map, by counting preimages, is k.

Preimage counting is often the easiest way of determining the degree of a map. A good example is the degree of a rational map from the Riemann sphere to itself. The Riemann sphere is the complex plane with one point at infinity adjoined, $\mathbb{C} \cup \{\infty\}$. Topologically it is S^2. We denote a point on the Riemann sphere by a complex number z, which can take any value, including infinity. A rational map is a function

$$R(z) = \frac{p(z)}{q(z)}, \tag{3.25}$$

where p and q are polynomials in z. p and q must have no common roots, otherwise factors can be cancelled between them. q can be a non-zero constant, in which case R is just a polynomial. For finite z, $R(z)$ may have any complex value, including infinity. The value is infinity where q vanishes. $R(\infty)$ is the limit as $z \to \infty$ of $p(z)/q(z)$, and can either be finite or infinity. Functions of the type (3.25) are smooth maps from S^2 to S^2.

The algebraic degree k_{alg} of R is the larger of the degrees of the polynomials p and q. For example, the maps $(z - a)/(z - b)$, $1/z^2$ and $z^3 + a$ have algebraic degrees 1, 2 and 3, respectively.

The topological degree of R is the number of preimages of a given point c, counted with multiplicity. These are found by solving $R(z) = c$, or equivalently

$$p(z) - cq(z) = 0, \tag{3.26}$$

and this is a polynomial equation, generally of degree k_{alg} and with k_{alg} simple roots. At each of these isolated roots, the complex derivative dR/dz is non-zero. By expanding in real and imaginary parts, we find that as a real map between 2-spheres, R has Jacobian $|dR/dz|^2$, which is positive. More geometrically, this is because the map, being holomorphic, locally preserves orientation. Thus each preimage of c occurs with positive multiplicity. Therefore, the topological degree of R equals the algebraic degree.

Some values of c are exceptional. As c varies, the roots of $p - cq$ will sometimes coalesce, but the net number of preimages doesn't change if one defines their multiplicities with care. Also $p - cq$ may sometimes have one or more leading powers of z missing. But then the missing finite roots of (3.26) are regarded as being at infinity. This becomes clear if one changes c a little. For example, the equation

$$\frac{1}{z^2} = c, \tag{3.27}$$

with c small, has roots at $z = \pm\sqrt{\frac{1}{c}}$ near infinity, so the equation

$$\frac{1}{z^2} = 0, \tag{3.28}$$

which degenerates if expressed in the form (3.26), is regarded as having a double root at $z = \infty$. From either viewpoint, the map $R(z) = 1/z^2$ has algebraic and topological degree 2.

Rational maps have several important applications in soliton theory, and we shall encounter them more than once in the following chapters.

3.3 Gauge fields as differential forms

For studying the topological properties of gauge fields, it is convenient to express the gauge potential and field tensor as differential forms.

Let us start with an abelian gauge theory defined on X, a manifold with local coordinates $(x^1, \ldots, x^d)$, and let us ignore the time dependence. We postulated in Chapter 2 that a gauge potential is a covariant rank 1 tensor. That means that under a coordinate transformation $x^i \mapsto x'^i$, the components a_i of the gauge potential transform to

$$a'_i = \frac{\partial x^j}{\partial x'^i} a_j \,. \tag{3.29}$$

This is natural, since it implies that the gauge covariant derivative of a scalar field, $\partial_i \phi - i a_i \phi$, transforms in the same way as $\partial_i \phi$ under a coordinate transformation. It follows that the gauge potential components can be combined into a differential 1-form

$$a = a_i \, dx^i = a_1 \, dx^1 + a_2 \, dx^2 + \cdots + a_d \, dx^d \,, \tag{3.30}$$

and the covariant derivative of ϕ becomes the 1-form $d\phi - i a \phi$. The 1-form a is coordinate invariant, since the transformation rule (3.29) is equivalent to the equation

$$a_i \, dx^i = a'_i \, dx'^i \,. \tag{3.31}$$

The field tensor components combine into the 2-form field strength

$$f = da = \sum_{i<j} (\partial_i a_j - \partial_j a_i) \, dx^i \wedge dx^j \,, \tag{3.32}$$

the exterior derivative of a. This is also coordinate invariant. Two basic properties of the exterior derivative operator d are the Leibniz rule $d(u \wedge v) = du \wedge v + (-1)^r u \wedge dv$, if u is an r-form, and that the operator gives zero when applied twice, that is, $dd = 0$. Because of the latter property,

$$df = d(da) = 0 \,, \tag{3.33}$$

so f is closed. Under a gauge transformation,

$$a \mapsto a + d\alpha \,. \tag{3.34}$$

f is gauge invariant, since

$$d(a + d\alpha) = da + d(d\alpha) = da \,. \tag{3.35}$$

The gauge potential in a non-abelian gauge theory defined on X has spatial components A_i. These are each valued in $\mathrm{Lie}(G)$, the Lie algebra of the gauge group G, which as before we take to be a vector space of antihermitian matrices. The components can be combined into $A = A_i \, dx^i$. This is a matrix of the type occurring in $\text{Lie}(G)$, whose entries are 1-forms, and generally complex. A is referred to as a $\text{Lie}(G)$-valued 1-form. Thus in an $SU(2)$ theory, the gauge potential is a 2×2 traceless antihermitian matrix of 1-forms. This can be written as

$$A = \begin{pmatrix} iA^3 & iA^1 + A^2 \\ iA^1 - A^2 & -iA^3 \end{pmatrix} \tag{3.36}$$

or equivalently $A = A^a(i\tau^a)$, where A^1, A^2 and A^3 are ordinary, real 1-forms, and $\{i\tau^a\}$ is a Pauli matrix basis of $su(2)$. The covariant derivative of a scalar field Φ is now $D\Phi = d\Phi + A\Phi$.

The field strength is

$$\begin{aligned} F = dA + A \wedge A &= \sum_{i,j}(\partial_i A_j + A_i A_j) \, dx^i \wedge dx^j \\ &= \sum_{i<j}(\partial_i A_j - \partial_j A_i + [A_i, A_j]) \, dx^i \wedge dx^j \\ &= \sum_{i<j} F_{ij} \, dx^i \wedge dx^j \,, \end{aligned} \tag{3.37}$$

a $\text{Lie}(G)$-valued 2-form. The exterior derivative operator d acts on each matrix entry of A in the usual way. $A \wedge A$ means that the matrix A is multiplied by itself in the usual way, with individual entries being multiplied using the wedge product of 1-forms. Generally, $A \wedge A$ is not zero. In the $SU(2)$ case,

$$F = \begin{pmatrix} i\Delta A^3 & i\Delta A^1 + \Delta A^2 \\ i\Delta A^1 - \Delta A^2 & -i\Delta A^3 \end{pmatrix}, \tag{3.38}$$

where $\Delta A^1 = dA^1 - 2A^2 \wedge A^3$, and cyclically.

The field strength F is not gauge invariant. Under a gauge transformation,

$$\begin{aligned} A &\mapsto gAg^{-1} - dgg^{-1} \\ F &\mapsto gFg^{-1} . \end{aligned} \tag{3.39} \tag{3.40}$$

Also, F is not a closed 2-form. However, it satisfies the Bianchi identity

$$dF + A \wedge F - F \wedge A = 0 \tag{3.41}$$

because, from (3.37),

$$dF = dA \wedge A - A \wedge dA = (dA + A \wedge A) \wedge A - A \wedge (dA + A \wedge A). \tag{3.42}$$

3.4 Chern numbers of abelian gauge fields

The first Chern form of an abelian gauge field is defined to be the 2-form

$$C_1 = \frac{1}{2\pi} f. \tag{3.43}$$

The factor $\frac{1}{2\pi}$ will be seen to be useful later.

Let us first consider a gauge field in the plane, $\mathbb{R}^2$. The first Chern number c_1 is the integral of the first Chern form C_1,

$$c_1 = \frac{1}{2\pi} \int_{\mathbb{R}^2} f. \tag{3.44}$$

If f is smooth and decays to zero as $|\mathbf{x}| \to \infty$ more rapidly than $|\mathbf{x}|^{-2}$, then c_1 is finite. In Cartesian coordinates,

$$c_1 = \frac{1}{2\pi} \int_{-\infty}^{\infty} \int_{-\infty}^{\infty} f_{12} \, dx^1 \, dx^2, \tag{3.45}$$

where $f_{12} = \partial_1 a_2 - \partial_2 a_1$ is the magnetic field in the plane. Therefore the first Chern number is the total magnetic flux through the plane, divided by 2π. (If the plane is thought of as embedded in $\mathbb{R}^3$, then $b = -f_{12}$ is the component of the magnetic field pointing in the x^3 direction, recalling again that $\mathbf{a}$ is minus the usual vector potential.)

By Stokes' theorem for differential forms, c_1 can be expressed as a line integral along the circle at infinity

$$c_1 = \frac{1}{2\pi} \int_{S_\infty^1} a = \frac{1}{2\pi} \int_0^{2\pi} a_\theta \, d\theta \bigg|_{\rho=\infty}, \tag{3.46}$$

where (ρ, θ) are polar coordinates.

Finally, because $f \to 0$ as $|\mathbf{x}| \to \infty$, the gauge potential for large $|\mathbf{x}|$ can be expressed as a pure gauge, that is $a = d\alpha$. In particular, on the circle at infinity, $a_\theta = \partial_\theta \alpha$. Therefore

$$c_1 = \frac{1}{2\pi} \int_0^{2\pi} \frac{\partial \alpha}{\partial \theta} \, d\theta = \frac{1}{2\pi} (\alpha(2\pi) - \alpha(0)). \tag{3.47}$$

We have three expressions for c_1, namely (3.44), (3.46) and (3.47). However, there is no reason for c_1 to take any particular value. c_1 is not necessarily an integer for a pure abelian gauge field in $\mathbb{R}^2$. This is because α is not necessarily single-valued, nor does it need to increase by an integer multiple of 2π around the circle at infinity. However, we shall see later, in our discussion of vortices, that the coupling of the gauge potential to a scalar field ϕ does bring in further restrictions, and then c_1 must be an integer.

Now let X be a compact two-dimensional surface, without a boundary. Let us suppose a scalar field and a $U(1)$ gauge field are defined on X, represented locally by a complex-valued function ϕ and a real 1-form a. The field strength is $f = da$. If we required $\{\phi, a\}$ to be a globally defined function and 1-form on X, then f would be a globally defined *exact* 2-form on X (the exterior derivative of a 1-form). In this case, Stokes' theorem for forms would imply that

$$\int_X f = \int_X da = \int_{\partial X} a = 0, \tag{3.48}$$

since ∂X, which denotes the boundary of X, is here absent. Thus, the net magnetic flux through X would be zero, and this is rather uninteresting.

One obtains a more interesting situation by relaxing one's view of the global nature of the fields $\{\phi, a\}$. They need only be regarded as a section and connection on a $U(1)$ bundle over the surface X. We will not present the mathematically careful definition of these concepts. For that, see [60]. Instead, we shall give the theoretical physicist's picture and justification, following ref. [432]. The idea is that, physically, ϕ and a are not observable, but only their gauge equivalence class is. So we require ϕ and a to be globally well defined only up to a gauge transformation. Gauge invariant quantities are globally well defined.

To proceed, we imagine X to be divided up into a (finite) number of overlapping, contractible regions (patches) $\{U_p\}$. A pair of such regions is illustrated in Fig. 3.8.

We assume that the section and connection are represented by a scalar field and 1-form $\{\phi^{(1)}, a^{(1)}\}$ on region U_1, and similarly by $\{\phi^{(2)}, a^{(2)}\}$ on region U_2. On the overlap of these regions, $U_{21} = U_2 \cap U_1$, we require that $\{\phi^{(1)}, a^{(1)}\}$ and $\{\phi^{(2)}, a^{(2)}\}$ are related by a gauge transformation

$$\phi^{(2)} = e^{-i\alpha^{(21)}}\phi^{(1)} \tag{3.49}$$
$$a^{(2)} = a^{(1)} - d\alpha^{(21)}, \tag{3.50}$$

where $e^{-i\alpha^{(21)}} \in U(1)$ need only be defined on U_{21}. Note that we may have used different coordinate systems on U_1 and U_2, but the formulae (3.49) and (3.50) are coordinate independent, so this doesn't matter. Note also that we still have the freedom to make independent gauge transformations on U_1 and U_2, so that the fields on these regions could be replaced by

$$\phi'^{(1)} = e^{i\alpha^{(1)}}\phi^{(1)} \tag{3.51}$$
$$a'^{(1)} = a^{(1)} + d\alpha^{(1)}, \tag{3.52}$$

and similarly for $\{\phi^{(2)}, a^{(2)}\}$. The primed fields still satisfy equations (3.49) and (3.50), but the transition function $e^{-i\alpha^{(21)}}$ needs to be replaced

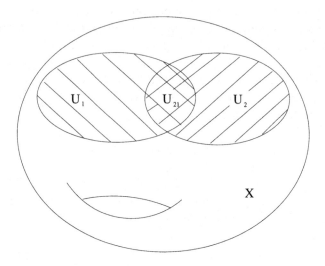

Fig. 3.8.   Two overlapping patches on X.

by $e^{i\alpha^{(2)}}e^{-i\alpha^{(21)}}e^{-i\alpha^{(1)}}$. It turns out that for a well defined bundle on X, there is a further constraint on the transition functions. On triply overlapping patches $U_3 \cap U_2 \cap U_1$, the transition functions must obey

$$e^{-i\alpha^{(32)}}e^{-i\alpha^{(21)}}e^{-i\alpha^{(13)}} = 1. \qquad (3.53)$$

(Here, $e^{-i\alpha^{(pq)}}$ is always defined to equal $e^{i\alpha^{(qp)}}$.)

The formula (3.49) implies that on U_{21}, the gauge invariant quantities $|\phi^{(1)}|^2$ and $|\phi^{(2)}|^2$ are equal, so $|\phi|^2$ is globally well defined. More importantly, the field strength f is globally well defined, since

$$f^{(1)} = da^{(1)} = da^{(2)} = f^{(2)}, \qquad (3.54)$$

using (3.50).

Since $f = da$ locally, f is a closed 2-form, satisfying $df = 0$. Because a is not globally well defined, f is not necessarily an exact 2-form. This by itself tells us nothing about the integral of f over X, except that it can be non-zero. However, the fact that the transition between regions is as given by (3.49) and (3.50) leads to an important constraint on the integral. The easiest example that shows this is where X is the 2-sphere, S^2.

Two regions, U_1 and U_2, are sufficient to cover S^2. Suppose U_1 covers most of the sphere, except a disc surrounding the South pole, and U_2 most of the sphere, except a disc surrounding the North pole. We use spherical polar coordinates (θ, φ). (The reader might be worried here, because polar coordinates are ill defined at the North and South poles, and φ is multivalued, but it turns out this is no problem.) Let $\{\phi^{(1)}, a^{(1)}\}$ and $\{\phi^{(2)}, a^{(2)}\}$ be the fields on U_1 and U_2. U_1 and U_2 overlap on a region U_{21} including the equator. Let $e^{-i\alpha^{(21)}(\theta,\varphi)}$ be the transition function, defined over the whole 2π range of φ, and over some range of θ including $\frac{\pi}{2}$ but not extending to 0 or π. Then

$$a^{(2)} = a^{(1)} - d\alpha^{(21)} . \tag{3.55}$$

Since $\phi^{(2)} = e^{-i\alpha^{(21)}} \phi^{(1)}$, and $\phi^{(1)}, \phi^{(2)}$ are themselves well defined on their respective regions, $e^{-i\alpha^{(21)}}$ must be single-valued, i.e.

$$e^{-i\alpha^{(21)}(\theta,2\pi)} = e^{-i\alpha^{(21)}(\theta,0)} . \tag{3.56}$$

Even if $\phi^{(1)}, \phi^{(2)}$ were absent (or had value zero) we would require this.

Now let us calculate the integral of f over S^2. We split the 2-sphere into hemispheres, with boundary the equator, and use $a^{(1)}, a^{(2)}$ in the Northern and Southern hemispheres, respectively. Then

$$\int_{S^2} f = \int_{\text{Northern hemisphere}} da^{(1)} + \int_{\text{Southern hemisphere}} da^{(2)}$$

$$= \int_{\text{Equator}} a^{(1)} - \int_{\text{Equator}} a^{(2)} . \tag{3.57}$$

The step from the first to the second line uses Stokes' theorem for each hemisphere. The integral around the equator is taken in the usual sense (φ increasing). Next, using (3.55), we deduce that

$$\int_{S^2} f = \int_{\text{Equator}} d\alpha^{(21)} = \alpha^{(21)} \left(\frac{\pi}{2}, 2\pi \right) - \alpha^{(21)} \left(\frac{\pi}{2}, 0 \right) , \tag{3.58}$$

i.e. the magnetic flux equals the increase in $\alpha^{(21)}$ going round the equator once. Because of Eq. (3.56), this is some integer multiple of 2π, say $2\pi N$. For example, $\alpha^{(21)}$ could be of the simple form $\alpha^{(21)}(\theta, \varphi) = N\varphi$, which gives this result. Thus we have the quantization of flux

$$\int_{S^2} f = 2\pi N \tag{3.59}$$

for some integer N, or equivalently

$$c_1 = \frac{1}{2\pi} \int_{S^2} f = N , \tag{3.60}$$

so the first Chern number is an integer.

Rather remarkably, for a compact Riemann surface X without boundary, of any genus g, one obtains the same constraint on the first Chern number

$$c_1 = \frac{1}{2\pi} \int_X f = N \,, \tag{3.61}$$

with N an arbitrary integer. One can understand this by using the representation of the surface by a polygon with edges identified, as in Fig. 3.2. The gauge potential a can be extended smoothly to the whole polygon, but there are constraints. The gauge potential in a neighbourhood of the edge a_s must agree up to a gauge transformation $e^{i\alpha^{(s)}}$ with the gauge potential in a neighbourhood of the edge a_s^{-1}. This gauge transformation is single-valued on the loop in X corresponding to a_s. Now, by Stokes' theorem, c_1 can be expressed as $\frac{1}{2\pi} \int a$, where the integral is along the boundary of the polygon. The contributions from edges a_s and a_s^{-1} would cancel if the gauge transformation $e^{i\alpha^{(s)}}$ were unity, because of the arrows being oppositely oriented. The total contribution from a_s and a_s^{-1} is in fact $\frac{1}{2\pi} \int d\alpha^{(s)}$, where the integral is along a_s, and this equals an integer N_{a_s}. Similarly the pair of edges b_s and b_s^{-1} contribute an integer N_{b_s}. Summing these contributions, we get the integer result (3.61).

There is one further characterization of the first Chern number of a $U(1)$ bundle over a surface X. It can be demonstrated that the number of zeros of a section ϕ of the bundle, counted with multiplicity, equals c_1.

The second Chern form for an abelian gauge field is

$$C_2 = \frac{1}{8\pi^2} f \wedge f \,. \tag{3.62}$$

This 4-form is closed, because $df = 0$, and is locally exact, since

$$C_2 = \frac{1}{8\pi^2} d(f \wedge a) \,. \tag{3.63}$$

On $\mathbb{R}^4$, the second Chern number of an abelian field is therefore

$$c_2 = \int_{\mathbb{R}^4} C_2 = \frac{1}{8\pi^2} \int_{S^3_\infty} f \wedge a \,. \tag{3.64}$$

If $f \to 0$ as $|x| \to \infty$, then this integral vanishes. So the second Chern number of an abelian field on $\mathbb{R}^4$, with this boundary condition, is not interesting.

For an abelian field on a closed 4-manifold X, c_2 can be non-zero if X has topologically non-trivial closed two-dimensional submanifolds, and it is always an integer. For example, if X is $X^{(1)} \times X^{(2)}$, where each factor is a compact surface, then

$$c_2 = \frac{1}{8\pi^2} \int_{X^{(1)} \times X^{(2)}} f \wedge f = \left(\frac{1}{2\pi} \int_{X^{(1)}} f \right) \left(\frac{1}{2\pi} \int_{X^{(2)}} f \right) = c_1^{(1)} c_1^{(2)} \,.$$
$$\tag{3.65}$$

$c_1^{(1)}$ and $c_1^{(2)}$ are the first Chern numbers of the field on $X^{(1)}$ and $X^{(2)}$. One factor of 2 disappears, because if (for simplicity) $f = f^{(1)} + f^{(2)}$, where $f^{(1)}$ is non-zero on $X^{(1)}$, and $f^{(2)}$ on $X^{(2)}$, then $f \wedge f = 2f^{(1)} \wedge f^{(2)}$. This illustrates why the second Chern form has a prefactor $\frac{1}{8\pi^2}$.

3.5 Chern numbers for non-abelian gauge fields

Because the field strength F is not gauge invariant, it can not be used to construct a direct analogue of the first Chern form $C_1 = \frac{1}{2\pi} f$ of the abelian theory. However, for a $U(n)$ gauge theory, or if the gauge group G is a subgroup of the $n \times n$ unitary matrices, one defines

$$C_1 = \frac{i}{2\pi} \operatorname{Tr} F, \tag{3.66}$$

where Tr denotes the trace. C_1 is gauge invariant. At the Lie algebra level, $u(n) = su(n) \oplus u(1)$, and the trace picks out the $U(1)$ part of the field strength, which commutes (as a matrix) with the $SU(n)$ part. For an $SU(2)$ gauge field, with field strength (3.38), C_1 vanishes.

The second Chern form, C_2, is defined as

$$C_2 = \frac{1}{8\pi^2} \left(\operatorname{Tr} (F \wedge F) - \operatorname{Tr} F \wedge \operatorname{Tr} F \right). \tag{3.67}$$

Let us assume that F has no $U(1)$ part; then only the term $\operatorname{Tr}(F \wedge F)$ contributes. $F \wedge F$ is the matrix product of F with itself, with the 2-form entries being combined by wedge product. The result, after taking the trace, is a 4-form. C_2 is gauge invariant, since

$$\operatorname{Tr} (gFg^{-1} \wedge gFg^{-1}) = \operatorname{Tr} (gF \wedge Fg^{-1}) = \operatorname{Tr} (F \wedge F). \tag{3.68}$$

(The cyclic property of a trace $\operatorname{Tr}(\alpha\beta\ldots\gamma) = \operatorname{Tr}(\beta\ldots\gamma\alpha)$ is valid if the entries of the matrices $\alpha, \beta, \ldots, \gamma$ are commuting objects, and here the entries of both g and F, which are 0-forms and 2-forms, commute.)

A key property of C_2 is that it is a closed 4-form, because

$$
\begin{aligned}
dC_2 &= \frac{1}{4\pi^2} \operatorname{Tr} (dF \wedge F) \\
&= \frac{1}{4\pi^2} \left(\operatorname{Tr} (F \wedge A \wedge F) - \operatorname{Tr} (A \wedge F \wedge F) \right) \\
&= 0,
\end{aligned}
\tag{3.69}
$$

using the Bianchi identity (3.41) and the cyclicity of the trace (and the fact that 2-forms commute with r-forms for any r). Locally, C_2 can be explicitly written as an exact form

$$C_2 = d \left(\frac{1}{8\pi^2} \operatorname{Tr} \left(F \wedge A - \frac{1}{3} A \wedge A \wedge A \right) \right). \tag{3.70}$$

Let us now specialize to an $SU(2)$ gauge field defined in $\mathbb{R}^4$, regarded as space-time, and define the second Chern number

$$c_2 = \int_{\mathbb{R}^4} C_2 \,. \tag{3.71}$$

The importance of c_2 is that it is integer-valued if the field strength F approaches zero sufficiently fast as $|x| \to \infty$. (This is different from the situation for c_1 in a $U(1)$ gauge theory in $\mathbb{R}^2$.) To see this, we assume this fast decay of F, writing this concisely as $F^\infty = 0$, and fix the radial gauge $x^\mu A_\mu = 0$. Then there is a limiting gauge potential which is pure gauge,

$$A^\infty = -dg^\infty (g^\infty)^{-1} \,. \tag{3.72}$$

g^∞ is defined on the 3-sphere at infinity S^3_∞, and takes values in $SU(2)$, also a 3-sphere. Therefore, we have a map

$$g^\infty : S^3_\infty \mapsto SU(2) \,. \tag{3.73}$$

If one doubts that g^∞ is globally defined on S^3_∞, one may start with two functions g^∞_N and g^∞_S defined on the Northern and Southern hemispheres of S^3_∞, with

$$A^\infty_N = -dg^\infty_N (g^\infty_N)^{-1}, \quad A^\infty_S = -dg^\infty_S (g^\infty_S)^{-1} \,. \tag{3.74}$$

On the equator where they overlap, there must be a gauge transformation g^∞_E connecting them,

$$g^\infty_N = g^\infty_E \, g^\infty_S \,. \tag{3.75}$$

g^∞_E is a map from $S^2 \mapsto SU(2)$. Since $\pi_2(S^3) = I$, such a map is homotopic to the constant map, $S^2 \mapsto 1_2$, so g^∞_E can be continuously extended to a map over the Northern hemisphere. Then $(g^\infty_E)^{-1}$ can be used to change g^∞_N over its hemisphere keeping g^∞_S fixed, such that after the change, $g^\infty_N = g^\infty_S$ on the equator, and hence $A^\infty_N = A^\infty_S$. Then g^∞ is continuously defined over all of S^3_∞, and A^∞ is given by (3.72).

Now recall the formula (3.70), which is globally valid on $\mathbb{R}^4$. Stokes' theorem implies that

$$c_2 = \int_{\mathbb{R}^4} C_2 = \frac{1}{8\pi^2} \int_{S^3_\infty} \text{Tr}\left(F \wedge A - \frac{1}{3} A \wedge A \wedge A \right) \,. \tag{3.76}$$

Since $F^\infty = 0$, the first term vanishes, so

$$c_2 = -\frac{1}{24\pi^2} \int_{S^3_\infty} \text{Tr}\left(A^\infty \wedge A^\infty \wedge A^\infty \right), \tag{3.77}$$

which can be reexpressed in terms of g^∞ as

$$c_2 = \frac{1}{24\pi^2} \int_{S^3_\infty} \text{Tr}\left(dg^\infty (g^\infty)^{-1} \wedge dg^\infty (g^\infty)^{-1} \wedge dg^\infty (g^\infty)^{-1} \right) \,. \tag{3.78}$$

This is precisely the formula (3.21) we obtained earlier for the degree of a map from S^3 to $SU(2)$. Therefore the second Chern number c_2 is the degree of g^∞, associated with the pure gauge at infinity, and hence an integer. For reasons that will be explained in Chapter 10, this integer is also called the instanton number of the field.

Non-abelian gauge fields can be defined on a general closed 4-manifold X (without boundary), for example a 4-sphere or a 4-torus. As in the abelian theory on surfaces, A is a connection, this time on a complex vector bundle. X must be covered in patches $\{U_p\}$, and the connection A is a separately defined Lie(G)-valued 1-form $A^{(p)}$ on each patch. On the overlap U_{qp} of two patches U_p and U_q there is a gauge transformation $g^{(qp)} \in G$ relating $A^{(q)}$ and $A^{(p)}$. The Chern form C_2 is defined locally as before, but because it is gauge invariant, it is a global 4-form on X. Because of the (local) Bianchi identity, C_2 is closed. C_2 is not exact unless A is globally defined over X (which is possible only if the bundle is trivial). The second Chern number is

$$c_2 = \int_X C_2 . \tag{3.79}$$

If C_2 is exact, then $c_2 = 0$, since X has no boundary. c_2 is generally non-zero, and remarkably, it is always an integer. Moreover it depends only on the transition functions $\{g^{(qp)}\}$, so it is a topological invariant of the bundle over X. If X has dimension higher than four, we may integrate C_2 over any four-dimensional closed submanifold of X. c_2 is again an integer.

For applications to solitons, one usually thinks of the Chern number as a topological invariant of the field, but it is actually an invariant of the underlying bundle.

3.6 Chern-Simons forms

Locally, Chern forms can be expressed as exact forms. The expression constructed from the gauge field whose exterior derivative gives the Chern form is called a Chern-Simons form. The first Chern form of an abelian field is locally the 2-form $C_1 = \frac{1}{2\pi} da$, so

$$Y_1 = \frac{1}{2\pi} a \tag{3.80}$$

is the Chern-Simons 1-form. For a more general gauge group, the Chern-Simons 1-form is

$$Y_1 = \frac{i}{2\pi} \text{Tr}\, A . \tag{3.81}$$

From (3.70) we see that the Chern-Simons 3-form of a non-abelian field is

$$Y_3 = \frac{1}{8\pi^2} \text{Tr} \left(F \wedge A - \frac{1}{3} A \wedge A \wedge A \right) . \tag{3.82}$$

The abelian version is

$$Y_3 = \frac{1}{8\pi^2} f \wedge a . \tag{3.83}$$

None of these Chern-Simons forms are gauge invariant; however, their integrals often are.

Consider the example of an $SU(2)$ gauge field configuration in $\mathbb{R}^4$. Let us explicitly distinguish $\mathbb{R}^3$ spatial variables $\mathbf{x}$ and the time t. (Since we are integrating differential forms over manifolds, the metric plays no role, so the time could be Euclidean or Minkowskian.) Let $F \to 0$ as $|\mathbf{x}| \to \infty$. We do not at the moment restrict how F behaves as $t \to \pm\infty$. The 1-form gauge potential approaches a pure gauge $A^\infty = -dg^\infty(g^\infty)^{-1}$ on S^2_∞, the 2-sphere at spatial infinity, if we impose the radial gauge condition $A_r = 0$ ($r = |\mathbf{x}|$). There is now no problem finding a gauge transformation that makes $A^\infty = 0$. We need to smoothly extend the map $g^\infty : S^2_\infty \mapsto SU(2)$ to a map $g(\mathbf{x}) : \mathbb{R}^3 \mapsto SU(2)$, but there is no obstruction to this since $\pi_2(S^3) = I$, so g^∞ is contractible. The required gauge transformation is $g(\mathbf{x})^{-1}$. This construction can be extended for any interval of time.

Now consider the equation $C_2 = dY_3$, and its integral over $[t_0, t_1] \times \mathbb{R}^3$. Using Stokes' theorem,

$$\int_{[t_0,t_1] \times \mathbb{R}^3} C_2 = \int_{\mathbb{R}^3} Y_3 \bigg|_{t=t_1} - \int_{\mathbb{R}^3} Y_3 \bigg|_{t=t_0} . \tag{3.84}$$

There is no contribution from the boundary at spatial infinity because of our gauge choice, which makes not only F, but all spatial components of A vanish there.

Let us define the Chern-Simons number y_3 as the integral of Y_3,

$$y_3 = \int_{\mathbb{R}^3} Y_3 = \frac{1}{8\pi^2} \int_{\mathbb{R}^3} \text{Tr} \left(F \wedge A - \frac{1}{3} A \wedge A \wedge A \right) . \tag{3.85}$$

Only the spatial components of A and F contribute. Equation (3.84) states that the integral of the Chern form C_2 over space, and from time t_0 to t_1, is the change in Chern-Simons number,

$$\int_{[t_0,t_1] \times \mathbb{R}^3} C_2 = y_3(t_1) - y_3(t_0) . \tag{3.86}$$

Next, let us study the gauge invariance of y_3. We may perform a further gauge transformation $g(\mathbf{x})$, but it must satisfy $\lim_{|\mathbf{x}| \to \infty} g(\mathbf{x}) = 1_2$, so as to preserve the condition that A vanishes at infinity. A map $g : \mathbb{R}^3 \mapsto SU(2)$, subject to this limiting behaviour, is equivalent to a map $S^3 \mapsto SU(2)$, and it has integer degree, deg g. Under the gauge transformation $g(\mathbf{x})$,

$$
\begin{aligned}
Y_3 \quad &\mapsto \quad \frac{1}{8\pi^2}\mathrm{Tr}\left(gFg^{-1} \wedge (gAg^{-1} - dgg^{-1}) - \frac{1}{3}(gAg^{-1} - dgg^{-1})^3\right) \\
&= \quad Y_3 + \frac{1}{8\pi^2}dT_2 + \frac{1}{24\pi^2}\mathrm{Tr}\,(dgg^{-1})^3,
\end{aligned} \tag{3.87}
$$

where $T_2 = \mathrm{Tr}\,(dgg^{-1} \wedge gAg^{-1})$, and the superscript 3 is shorthand for the triple wedge product. T_2 vanishes at spatial infinity, so dT_2 integrates to zero. Therefore,

$$
\begin{aligned}
y_3 \quad &\mapsto \quad y_3 + \frac{1}{24\pi^2}\int_{\mathbb{R}^3} \mathrm{Tr}\,(dgg^{-1})^3 \\
&= \quad y_3 + \deg g\,.
\end{aligned} \tag{3.88}
$$

Thus, the Chern-Simons number is not gauge invariant, but a gauge transformation can only change it by an integer. The fractional part of y_3, which is usually taken to lie in the interval $[0, 1)$, is gauge invariant. y_3 is strictly gauge invariant under "small" gauge transformations, for which deg $g = 0$. However, one doesn't want to restrict to these.

The further gauge transformation g could depend on $\mathbf{x}$ and t. However, deg g being an integer must be time independent, by continuity. Thus the difference in y_3 between times t_0 and t_1 is completely gauge invariant, even though y_3 itself can be shifted by an integer. This is consistent with (3.86), since C_2 is gauge invariant.

Let us now also suppose that $F \to 0$ as $t \to \pm\infty$. We maintain our choice of gauge, for which $A = 0$ at spatial infinity. (Note that this is a different gauge choice from that of the previous section.) As $t \to -\infty$, the gauge potential approaches a pure gauge throughout $\mathbb{R}^3$

$$
A = -dg^-(g^-)^{-1}, \tag{3.89}
$$

where $g^- = 1_2$ at spatial infinity. Performing the gauge transformation $(g^-(\mathbf{x}))^{-1}$ (possibly of non-zero degree) makes $A = 0$ at $t = -\infty$. This is the naive vacuum, and has Chern-Simons number zero. The same gauge transformation must be made for all t, to avoid discontinuities.

Now consider the field as $t \to \infty$. Here the gauge potential is another pure gauge

$$
A = -dg^+(g^+)^{-1}. \tag{3.90}
$$

Because of the boundary condition $g^+ = 1_2$ at spatial infinity, the map $g^+ : \mathbb{R}^3 \mapsto SU(2)$ is again equivalent to a map $g^+ : S^3 \mapsto SU(2)$, with integer degree, deg g^+. deg g^+ is the interesting quantity. First of all it is the Chern-Simons number y_3 as $t \to \infty$. But it is also a gauge invariant integer associated with the whole field configuration in $\mathbb{R}^4$. It can not be changed without spoiling the naive vacuum at $t = -\infty$, or the boundary condition at spatial infinity. Since $y_3 = 0$ at $t = -\infty$, Eq. (3.88) implies that

$$c_2 = \lim_{t \to \infty} y_3(t) = \deg g^+ . \tag{3.91}$$

Thus, in the gauge we have chosen, the second Chern number of a field configuration for which $F \to 0$ at infinity equals the Chern-Simons number of its limiting vacuum configuration at $t = \infty$. In the language of instantons, a gauge field with instanton number N interpolates between the naive vacuum and a vacuum with degree N.

In Chapter 11 we shall describe the electroweak sphaleron, and show that its Chern-Simons number is $\frac{1}{2}$.

Let us now turn to abelian gauge fields in lower dimensions. Consider a cylindrical space-time where space is a circle S^1, parametrized by a coordinate x in the range $[0, 2\pi]$, with the ends identified, and time is the usual linear variable. The circle could have length $2\pi L$, but the value of L is here irrelevant. Let the 1-form gauge potential be $a = a_0\, dt + a_x\, dx$. The field strength is $f = f_{0x}\, dt \wedge dx$, where $f_{0x} = \partial_0 a_x - \partial_x a_0$ is the electric field on the circle. The Chern-Simons 1-form is $Y_1 = \frac{1}{2\pi}a$. Integrating this around the circle gives us the Chern-Simons number

$$y_1 = \frac{1}{2\pi}\int_{S^1} a = \frac{1}{2\pi}\int_0^{2\pi} a_x\, dx . \tag{3.92}$$

The allowed gauge transformations $g(t,x) \in U(1)$ are those which are periodic around the circle. Such gauge transformations are time dependent maps $S^1 \mapsto S^1$. Let us write

$$g(t,x) = e^{i\alpha(t,x)} \tag{3.93}$$

where α is periodic in x, mod 2π. Under this gauge transformation,

$$a_x \mapsto a_x + \partial_x \alpha \tag{3.94}$$

so

$$\begin{aligned}
y_1 \quad &\mapsto \quad y_1 + \frac{1}{2\pi}\int_0^{2\pi} \partial_x \alpha\, dx \\
&= \quad y_1 + \frac{1}{2\pi}(\alpha(2\pi) - \alpha(0)) \\
&= \quad y_1 + k
\end{aligned} \tag{3.95}$$

where k is the winding number of $g(t,x)$, which by continuity is independent of time.

Again, the fractional part of y_1 is gauge invariant. The change of y_1 over time is

$$y_1(t_1) - y_1(t_0) = \frac{1}{2\pi} \int_{t_0}^{t_1} \int_0^{2\pi} \partial_0 a_x \, dx \, dt = \frac{1}{2\pi} \int_{t_0}^{t_1} \int_0^{2\pi} f_{0x} \, dx \, dt, \quad (3.96)$$

provided a_0 is single-valued, so that the spatial integral of $\partial_x a_0$ vanishes. Therefore

$$y_1(t_1) - y_1(t_0) = \int_{[t_0, t_1] \times S^1} C_1, \quad (3.97)$$

which is gauge invariant, and the analogue of (3.86).

If space is the line $\mathbb{R}$, we still define the Chern-Simons number as

$$y_1 = \frac{1}{2\pi} \int_{-\infty}^{\infty} a_x \, dx, \quad (3.98)$$

and the fractional part of this is gauge invariant if we insist that only gauge transformations $g(t, x)$ which approach 1 as $x \to \pm\infty$ are allowed. This Chern-Simons number plays a role in the discussion of gauged kinks in Chapter 11.

In Chapter 7, we shall be considering field theories in $2+1$ dimensions, with vortices, where part of the action is the integral of the abelian Chern-Simons 3-form. The variational principle requires us to fix the fields at an initial and final time t_0 and t_1. The Chern-Simons part of the action is a constant multiple of

$$y_3 = \frac{1}{8\pi^2} \int_{t_0}^{t_1} \int_{\mathbb{R}^2} f \wedge a, \quad (3.99)$$

where we assume also that $f \to 0$ at spatial infinity and therefore a approaches a pure gauge. For smooth f with this boundary condition, y_3 is convergent.

A gauge transformation $g(t, \mathbf{x})$ must have the property $g = 1$ throughout $\mathbb{R}^2$ at $t = t_0$ and $t = t_1$. By continuity, we can suppose $g = 1$ at spatial infinity for all t. (g can have no winding on the circle at spatial infinity, because for all t, it has to extend to $\mathbb{R}^2$.) Write $g(t, \mathbf{x}) = e^{i\alpha(t, \mathbf{x})}$, with $\alpha = 0$ on the entire boundary of $[t_0, t_1] \times \mathbb{R}^2$. Under this gauge transformation,

$$
\begin{aligned}
y_3 \;\longmapsto\; & y_3 + \frac{1}{8\pi^2} \int_{t_0}^{t_1} \int_{\mathbb{R}^2} f \wedge d\alpha \\
= \; & y_3 + \frac{1}{8\pi^2} \int_{t_0}^{t_1} \int_{\mathbb{R}^2} d(f \wedge \alpha) \\
= \; & y_3 + \frac{1}{8\pi^2} \int_{\text{boundary}} f \wedge \alpha \\
= \; & y_3.
\end{aligned}
\quad (3.100)
$$

Thus the Chern-Simons action is gauge invariant.

It is interesting to determine the variation of y_3 under a general variation of the gauge field that vanishes on the boundary. Under $a \mapsto a + \delta a$, $f \mapsto f + d(\delta a)$, so

$$
\begin{aligned}
\delta y_3 &= \frac{1}{8\pi^2} \int \left(d(\delta a) \wedge a + f \wedge \delta a \right) \\
&= \frac{1}{8\pi^2} \int \left(d(\delta a \wedge a) + 2 f \wedge \delta a \right) \\
&= \frac{1}{4\pi^2} \int f \wedge \delta a ,
\end{aligned}
\tag{3.101}
$$

where the integrals are over $[t_0, t_1] \times \mathbb{R}^2$, as before. This is gauge invariant, since a but not δa changes under a gauge transformation. For δy_3 to vanish for all δa,

$$
f = 0 .
\tag{3.102}
$$

This is the field equation for an abelian theory in $2+1$ dimensions whose action is just a Chern-Simons term. The magnetic and electric fields must both vanish.

4
Solitons – general theory

This chapter is concerned with methods for deciding if a particular Lagrangian field theory can have topological soliton solutions, and with one general method for finding them. It is also concerned with soliton dynamics. Ultimately, to find solitons, one must solve the field equations, either analytically or numerically. However, it is very helpful to know beforehand if a theory is likely to have solitons or not, how the solitons are topologically classified, and what symmetries the solitons may have. The topological data are intimately tied up with boundary conditions.

The topics we shall discuss are i) the topological structure and classification of solitons, ii) the Derrick scaling argument for the existence or non-existence of solitons, iii) symmetries, and the reduction of Euler-Lagrange field equations to ordinary differential equations (ODEs), iv) the modelling of soliton dynamics at slow speeds by a finite-dimensional dynamical system – the dynamics on moduli space.

4.1 Topology and solitons

Here we shall apply results from Chapter 3 to investigate whether particular field theories have the required topological structure for topological solitons to exist. A key point is that topological and energetic considerations need to be combined. We shall first consider theories with a multiplet of scalar fields, possibly coupled to gauge fields, and then discuss pure gauge theories.

In Chapter 3 we often stressed the base point condition when discussing the homotopy class of a map. This is important if one wants to have a homotopy group structure, but usually, if one just wants to know whether maps can or can not be deformed into each other, then the base point condition can be dropped. The classification of homotopy classes is unaffected provided the domain X of the map is connected. In the context of field theory, we shall often drop the base point requirement if it plays no role.

Our main aim is to elucidate the topological aspects of fields defined on a flat space $\mathbb{R}^d$. Just assuming the fields are continuous is not sufficient. In the absence of further structure, linear fields are topologically trivial. For example, any field configuration $\phi(\mathbf{x})$ can be replaced by $(1-\tau)\phi(\mathbf{x})$. If τ runs from 0 to 1, then that is a homotopy, taking the initial configuration to the trivial one, $\phi = 0$. The nonlinear case is a bit different. Suppose the field is a map from $\mathbb{R}^d$ to a target manifold Y. Since $\mathbb{R}^d$ is contractible to a point, the only topological invariant is the component of Y where the field takes its value. So field configurations would be classified by $\pi_0(Y)$.

The topological classification becomes more interesting if we assume the energy density decays rapidly as $\rho \to \infty$, where ρ is the distance from the origin. In fact, most solitons have finite total energy, which is a stronger property, but there are exceptions, like global vortices. The requirement that the energy density is zero at infinity imposes boundary conditions on the fields, crucial for the topological classification. We have already seen an example of this. We showed that a pure $SU(2)$ gauge field in $\mathbb{R}^4$ is classified by its second Chern number, provided the field strength F decays rapidly towards infinity.

Consider a multiplet of n scalar fields, $\phi = (\phi_1, \ldots, \phi_n)$, with an energy functional of the form

$$E = \int \left(\frac{1}{2} \boldsymbol{\nabla}\phi_l \cdot \boldsymbol{\nabla}\phi_l + U(\phi_1, \ldots, \phi_n) \right) d^d x \,. \tag{4.1}$$

For time independent fields, E is the total energy. Assume the potential function $U(\phi_1, \ldots, \phi_n)$ takes its minimal value $U_{\min} = 0$ on a submanifold $\mathcal{V} \subset \mathbb{R}^n$, the vacuum manifold of the theory. There is no constraint on the value of ϕ at any finite point $\mathbf{x}$. However, at spatial infinity ϕ must take its values in $\mathcal{V}$, to ensure zero energy density there. Its value can be different in different directions. A field configuration therefore defines a map from S^{d-1}_∞, the sphere at infinity in $\mathbb{R}^d$, to $\mathcal{V}$.

Topologically, we lose no information if we just retain these asymptotic data

$$\phi^\infty : S^{d-1}_\infty \mapsto \mathcal{V} \,. \tag{4.2}$$

In a linear theory, two field configurations with the same asymptotic data are homotopic. Moreover, field configurations $\phi, \tilde{\phi}$ with distinct asymptotic data $\phi^\infty, \tilde{\phi}^\infty$ are still homotopic if ϕ^∞ is homotopic to $\tilde{\phi}^\infty$. The topological character of the configuration $\phi(\mathbf{x})$ is therefore determined by the homotopy class of the map ϕ^∞, which is an element of $\pi_{d-1}(\mathcal{V})$.

In many examples, there is an $SO(n)$ internal symmetry, and the potential is of the form $U(\phi_l \phi_l)$. If $\mathcal{V}$ is a single orbit of $SO(n)$, then $\mathcal{V}$ is either a single point, or a sphere S^{n-1}, depending on the parameters in U. If $\mathcal{V}$ is one point, which means that the symmetry is unbroken in the vacuum, then there are no homotopy classes beyond the trivial one, and no topological solitons. If $\mathcal{V} = S^{n-1}$, there is spontaneous breaking of the internal symmetry. The asymptotic field is a map

$$\phi^\infty : S_\infty^{d-1} \mapsto S^{n-1}, \tag{4.3}$$

so fields are classified by elements of the homotopy group $\pi_{d-1}(S^{n-1})$.

Let us restrict attention to theories in dimensions $1, 2$ and 3, and consider these in turn:

$d = 1$: Here S_∞^{d-1} consists of the points $\pm\infty$ in $\mathbb{R}$, so ϕ^∞ is a map from two points to $\mathcal{V}$. The components of $\mathcal{V}$ are classified by $\pi_0(\mathcal{V})$, and this is the set of topologically distinct vacua. The topological class of a field configuration $\phi(x)$ is therefore an element of $\pi_0(\mathcal{V}) \times \pi_0(\mathcal{V})$. In particular, if $\mathcal{V}$ consists of p points in $\mathbb{R}^n$, then there are p^2 topologically distinct types of field. Suppose the field is characterized by $(v_1, v_2) \in \pi_0(\mathcal{V}) \times \pi_0(\mathcal{V})$. If $v_1 = v_2$ then the field is in the class of the vacuum v_1. If $v_1 \neq v_2$ then the field is kink-like, and connects the vacuum v_1 at $-\infty$ to the vacuum v_2 at ∞.

$d = 2$: Here S_∞^{d-1} is a circle. The field is topologically characterized by an element of $\pi_1(\mathcal{V})$, the fundamental group of $\mathcal{V}$. A field configuration for which this element is the identity is in the vacuum sector, since the field can be deformed to take a constant value in $\mathcal{V}$ on the circle at infinity, and then to a constant field throughout $\mathbb{R}^2$. A field configuration characterized by a non-trivial element of $\pi_1(\mathcal{V})$ has the character of a vortex, and the element of $\pi_1(\mathcal{V})$ gives its "winding at infinity".

If $\mathcal{V}$ is S^{n-1}, with $n > 2$, then $\pi_1(\mathcal{V})$ is trivial so there are no vortices. If $n = 2$, then field configurations are classified by $\pi_1(S^1) = \mathbb{Z}$, and the winding number N is the vortex number. Of course there are plenty of manifolds $\mathcal{V}$ with non-trivial fundamental group, but we want $\mathcal{V}$ to be a submanifold of $\mathbb{R}^n$, minimizing a potential U. Then it is not so easy to construct natural examples. One example, somewhat artificial, is to consider $\phi = (\phi_1, \ldots, \phi_6)$, with a potential U of the form

$$
\begin{aligned}
U(\phi) \;=\; & (\phi_1^2 + \phi_2^2 + \phi_3^2 + 2\phi_4^2 + 2\phi_5^2 + 2\phi_6^2 - 1)^2 \\
& + (\phi_1\phi_2 - \phi_4^2)^2 + (\phi_1\phi_3 - \phi_5^2)^2 + (\phi_2\phi_3 - \phi_6^2)^2 \\
& + (\phi_1\phi_6 - \phi_4\phi_5)^2 + (\phi_2\phi_5 - \phi_4\phi_6)^2 + (\phi_3\phi_4 - \phi_5\phi_6)^2 .
\end{aligned}
\tag{4.4}
$$

This is minimized when all brackets vanish. If we organize ϕ as a symmetric matrix

$$
M = \begin{pmatrix} \phi_1 & \phi_4 & \phi_5 \\ \phi_4 & \phi_2 & \phi_6 \\ \phi_5 & \phi_6 & \phi_3 \end{pmatrix},
\tag{4.5}
$$

then the vanishing of U implies that each 2×2 determinant of M is zero, and the sum of the squares of the entries of M is unity. Therefore M is of rank 1. The general matrix satisfying these conditions can be expressed in terms of a unit 3-vector $\psi = (\psi_1, \psi_2, \psi_3)$ as

$$M = \begin{pmatrix} \psi_1^2 & \psi_1\psi_2 & \psi_1\psi_3 \\ \psi_1\psi_2 & \psi_2^2 & \psi_2\psi_3 \\ \psi_1\psi_3 & \psi_2\psi_3 & \psi_3^2 \end{pmatrix}, \tag{4.6}$$

with each row and each column proportional to ψ. Note that if ψ is replaced by $-\psi$ then M is unaltered, so the vacuum manifold is a 2-sphere with opposite points identified, i.e. the real projective plane $\mathbb{RP}^2$. The equations $\phi_1 = \psi_1^2$, $\phi_2 = \psi_2^2$, $\phi_3 = \psi_3^2$, $\phi_4 = \psi_1\psi_2$, $\phi_5 = \psi_1\psi_3$ and $\phi_6 = \psi_2\psi_3$ define the Veronese embedding of $\mathbb{RP}^2$ in $\mathbb{R}^6$. The image in fact lies in the 4-sphere which is at the intersection of the 5-sphere (after rescaling ϕ_4, ϕ_5 and ϕ_6 by $\sqrt{2}$)

$$\phi_1^2 + \phi_2^2 + \phi_3^2 + 2\phi_4^2 + 2\phi_5^2 + 2\phi_6^2 = 1 \tag{4.7}$$

and the hyperplane

$$\phi_1 + \phi_2 + \phi_3 = 1. \tag{4.8}$$

Note also that the potential U is $SO(3)$-symmetric, with $SO(3)$ acting by conjugation on the matrix M. At the minima of U, $SO(3)$ acts by rotating ψ. A choice of ψ spontaneously breaks the symmetry to $O(2)$, the subgroup of rotations that either preserve ψ or rotate ψ to $-\psi$.

Since $\pi_1(\mathcal{V}) = \pi_1(\mathbb{RP}^2) = \mathbb{Z}_2$, a theory based on this potential can have $\mathbb{Z}_2$ vortices. Disclinations in uniaxial nematic liquid crystals are examples of such vortices [105]. There is a vacuum sector, and a vortex sector, and a field configuration which is made from two vortices is topologically equivalent to the vacuum.

$d = 3$: This is the dimension of physical space. Spatial infinity is S_∞^2, a 2-sphere, so the homotopy classes of linear fields are elements of the group $\pi_2(\mathcal{V})$. If $\mathcal{V}$ is simply a discrete set of points or curves, then $\pi_2(\mathcal{V})$ is trivial, and there are no topological solitons. The simplest non-trivial example is where $\mathcal{V} = S^2$. Since $\pi_2(S^2) = \mathbb{Z}$, a field configuration is labelled by an integer N, called the monopole number. If $\mathcal{V} = S^{n-1}$, with $n > 3$, then $\pi_2(\mathcal{V})$ is again trivial, and there are no solitons.

It is possible for $\mathcal{V}$ to be a manifold other than S^2, with $\pi_2(\mathcal{V})$ non-trivial. An example would be $\mathbb{CP}^2$. It is not so easy, however, to realize such examples as minima of a potential in a linear space.

If a linear scalar field is the only field in the theory, then the corresponding topological objects in dimensions $1, 2$ and 3 (assuming they exist) are known as kinks, global vortices, and global monopoles. Global vortices and global monopoles, which do exist as solutions of the field equation in certain theories, have a divergent total energy, due to gradients of the field at infinity. Suppose the field at infinity is a topologically non-trivial map from S_∞^{d-1} to $\mathcal{V}$, with $d \geq 2$. Then (in polar coordinates) the angular derivatives of the field remain finite as the radius ρ tends to infinity. The Cartesian components of the gradient of the field therefore decay as ρ^{-1}, and the gradient energy density decays as ρ^{-2}. This can be integrated over the angular coordinates, leaving a radial integral of order

$$\int^\infty \rho^{d-3}\,d\rho. \tag{4.9}$$

This is logarithmically divergent if $d = 2$, and linearly divergent if $d = 3$.

The presence of a gauge field does not change the topological classification, though gauge fields have several important effects, which we shall discuss more carefully later. Perhaps most importantly, the gradient terms are replaced by covariant gradient terms, which can vanish rapidly as $\rho \to \infty$, even though the scalar field at infinity is topologically non-trivial. The corresponding solitons in two or three dimensions then have finite energy, and for this reason we regard them as more interesting, and more truly topological solitons. We call them simply vortices and monopoles.

We turn now to nonlinear scalar fields $\phi : \mathbb{R}^d \mapsto Y$, where Y is a closed manifold, and $d \geq 2$. Let us assume there are no gauge fields, as these make the classification complicated. There may be a potential $U(\phi)$ with its minimum occurring on a non-trivial submanifold $\mathcal{V} \subset Y$. The simplest type of theory is where there is no potential, and the energy depends just on the gradient of ϕ. This is called a sigma model if the energy depends quadratically on the gradient, and a Skyrme model if the energy has a more complicated structure. Let us now assume that the field ϕ has *finite* energy. To avoid a divergent gradient energy, the field should tend to a constant value at infinity, independent of direction. Thus $\phi^\infty : S_\infty^{d-1} \mapsto Y$ must be a constant map, with value $\mathbf{y}_0$, say. (In the absence of a potential, the choice of $\mathbf{y}_0$ is arbitrary, but once it is made we take it to be the base point of Y. If a potential is present, then $\mathbf{y}_0$ must be in $\mathcal{V}$.) This boundary condition allows a topological compactification of space $\mathbb{R}^d$ to S^d. (Stereographic projection achieves this.) A single point at spatial infinity is added, and this is taken to be the base point of S^d. $\phi : \mathbb{R}^d \mapsto Y$ then extends to a continuous, based map, $\phi : S^d \mapsto Y$. The topological class of ϕ is therefore given by an element of the homotopy group $\pi_d(Y)$.

Consider the cases $d = 2$ and 3, and finally $d = 1$.

$d = 2$: Here, field configurations are labelled by elements of $\pi_2(Y)$. The simplest possibility is with $Y = S^2$. Since $\pi_2(S^2) = \mathbb{Z}$, there is an integer topological label. Solitons classified by this integer are $O(3)$ sigma model lumps, $O(3)$ being the symmetry group of S^2, and Baby Skyrmions. The most interesting generalization is for $Y = \mathbb{CP}^n$ or Y another compact Kähler manifold. (Note that, topologically, $\mathbb{CP}^1 = S^2$.) Since $\pi_2(\mathbb{CP}^n) = \mathbb{Z}$, lumps with an integer topological charge are possible in the $\mathbb{CP}^n$ sigma model. Generally, lump solutions are holomorphic functions in $\mathbb{C}$, the complexified spatial plane.

$d = 3$: Here the relevant homotopy group is $\pi_3(Y)$. The simplest non-trivial case is $Y = S^3$, with $\pi_3(S^3) = \mathbb{Z}$. Topological solitons labelled by elements of this group occur in Skyrme models and are called Skyrmions, the integer label being the Skyrmion number or baryon number. Note that if G is any compact, non-abelian simple Lie group, then $\pi_3(G) = \mathbb{Z}$ (essentially because there is a canonical $SU(2)$ subgroup in G, and $SU(2) = S^3$ topologically). Thus Skyrmions are possible in theories with a scalar field $\phi : \mathbb{R}^3 \mapsto G$.

Since $\pi_3(S^2) = \mathbb{Z}$, a theory with a scalar field $\phi : \mathbb{R}^3 \mapsto S^2$ can have topological solitons. The Skyrme-Faddeev model has a suitable Lagrangian for these to exist. A unit charge soliton is related to the Hopf fibration

$$S^3$$
$$\Downarrow$$
$$S^2 .$$

Generally, the integer label is called the Hopf charge.

$d = 1$: This case is rather different, because "infinity" consists of two points, $\pm\infty$. We call a topological soliton on $\mathbb{R}$, with a field ϕ taking values in a closed manifold Y, a nonlinear kink. If we insist that $\phi(\infty) = \mathbf{y}_0 = \phi(-\infty)$, or if the potential U has its minimum just at the one point $\mathbf{y}_0$, then $\mathbb{R}$ can be compactified to a circle, and solitons classified by elements of $\pi_1(Y)$. But if $\phi(\infty)$ and $\phi(-\infty)$ are distinct, then the topological classification is a bit more complicated. An example is the sine-Gordon kink. Normally, the field is regarded as linear, and the potential U periodic. However, it is possible to regard the field as taking values on a circle, and U then has a unique minimum on the circle. This interpretation makes the sine-Gordon kink into a nonlinear kink.

Coupling a nonlinear scalar field ϕ to a gauge field can have a significant topological effect, because it can allow the field to be non-constant on S^{d-1}_∞, while the energy remains finite. The field is a map $\phi : \mathbb{R}^d \mapsto Y$, and its asymptotic form is a map $\phi^\infty : S^{d-1}_\infty \mapsto Y$ where the image is restricted to a single orbit of the gauge group. The topological classification of solitons is now in terms of relative homotopy groups. A model with solitons of this type has been investigated by Yang [435].

Table 4.1. Classification table of possible solitons in linear and nonlinear scalar field theories.

Linear	d	Nonlinear
Kink	1	Nonlinear kink
$\pi_0(\mathcal{V}) \times \pi_0(\mathcal{V})$		$\pi_1(Y)$
Vortex	2	Sigma model lump
		Baby Skyrmion
$\pi_1(\mathcal{V})$		$\pi_2(Y)$
Monopole	3	Skyrmion
$\pi_2(\mathcal{V})$		$\pi_3(Y)$

Table 4.1 is a classification table of solitons in linear and nonlinear scalar field theories in one, two or three space dimensions, and the homotopy group (or set) that classifies them.

The nonlinear types of soliton that we refer to as Skyrmions are sometimes called textures.

Consider next a scalar field defined on a closed manifold X, rather than flat space. We remarked already that for a scalar field which is a map from X to a target manifold Y, the homotopy classes of maps $X \mapsto Y$ directly classify the possible field configurations. If the energy of a static field is a positive expression involving the gradient of the field, as in (4.1), then the lowest energy configuration, the vacuum, is a constant map. If there is a homotopy class of maps distinct from the class of the constant map, then one may seek solutions of the field equation there, and in particular, a solution of minimal energy.

Such solutions may be thought of as solitons if the energy density is localized on a small part of X. This may or may not be the case. In Chapter 6 we shall consider lumps defined on a 2-sphere of finite radius. These are maps from S^2 to S^2, whose topological charge N is the degree of the map. We shall see that minimal energy solutions exist for all non-zero values of N, and they have a large number of parameters. As these parameters vary, the solutions can change from being concentrated on small regions of the sphere to being rather evenly spread over it. Lumps are therefore not very good examples of solitons. On the other hand, Skyrmions are. They are energy minimizing maps from a Riemannian 3-manifold X to $SU(2)$. They have a built-in length scale, because the energy involves both quadratic and quartic terms in derivatives, and at the characteristic scale these are comparable in magnitude. The Skyrmion number N is the topological degree of the map $X \mapsto SU(2)$. There is evidence that Skyrmion solutions exist for any N, and if X is a sufficiently large manifold, then the Skyrmion is localized, so it is a topological soliton.

Let us now briefly discuss pure gauge fields. In $\mathbb{R}^2$, the only relevant quantity is the first Chern number c_1, which doesn't have to be an integer, so there are no topological solitons. On a closed surface X, c_1 is an arbitrary integer for an abelian field (or abelian part of a non-abelian field). It is the total magnetic flux through X, divided by 2π, and is a conserved topological quantity. But, for a solution of the field equation, the magnetic flux and energy are not localized, so the solution should not be interpreted as one or more two-dimensional solitons. We shall see below that if X is a smoothly embedded surface in $\mathbb{R}^3$, and $c_1 \neq 0$ on X, then the gauge field has magnetic monopole singularities inside X, which are the sources of the magnetic flux. These Dirac monopoles in three dimensions are effectively the topological solitons of pure $U(1)$ gauge theory, although they have infinite energy.

For smooth gauge fields in $\mathbb{R}^3$, the quantity that naturally occurs is the Chern-Simons number y_3, which can be non-zero for both abelian and non-abelian fields. However, it can take any real value, and does not classify solitons.

Finally, in $\mathbb{R}^4$, non-abelian gauge fields of finite energy are classified by the second Chern number c_2, which is an integer. Corresponding solitons – instantons – do exist, and c_2 is the topological charge, the instanton number.

4.2 Scaling arguments

In this section, we shall only consider time independent field configurations with finite energy. Our discussion so far has shown that these can be classified by their homotopy class. The vacuum, which is spatially constant and has the minimal energy of all fields, lies in the trivial class. It is natural to ask whether there exist minima of the energy in other homotopy classes. Such minima are usually stable solitons. More generally, one may ask if there are any non-minimal stationary points of the energy. Such stationary points are often rather like solitons, but unstable.

A simple and important *non-existence* theorem is due to Derrick [107]. It applies to field theories defined in flat space. Derrick noted that in many theories, the energy functional for static fields has the property that its variation with respect to a spatial rescaling is never zero for *any* non-vacuum field configuration. But a field configuration which is a stationary point of the energy should be stationary against all variations including spatial rescaling. So in these theories, there can be no static finite energy solutions of the field equation in any homotopy class, except the vacuum. In particular, there are no topological solitons.

More precisely, in $\mathbb{R}^d$ a spatial rescaling is a map $\mathbf{x} \mapsto \mu\mathbf{x}$, with $\mu > 0$. Let $\Psi(\mathbf{x})$ be a finite energy field configuration, with Ψ any kind of field or multiplet of fields, and let $\Psi^{(\mu)}(\mathbf{x})$, $0 < \mu < \infty$, be the 1-parameter family of field configurations obtained from $\Psi(\mathbf{x})$ by applying the map $\mathbf{x} \mapsto \mu\mathbf{x}$. We shall clarify how $\Psi^{(\mu)}(\mathbf{x})$ is related to $\Psi(\mathbf{x})$ below. Let

$$e(\mu) = E(\Psi^{(\mu)}) \tag{4.10}$$

denote the energy of the field configuration $\Psi^{(\mu)}(\mathbf{x})$, as a function of μ. Then we have Derrick's theorem:

Suppose that for an arbitrary, finite energy field configuration $\Psi(\mathbf{x})$, which is not the vacuum, the function $e(\mu)$ has no stationary point. Then the theory has no static solutions of the field equation with finite energy, other than the vacuum.

The usefulness of this non-existence theorem depends on defining $\Psi^{(\mu)}$ in an appropriate way so that it is easy to determine $e(\mu)$. For a scalar field configuration $\phi(\mathbf{x})$ one defines simply

$$\phi^{(\mu)}(\mathbf{x}) = \phi(\mu\mathbf{x}) \,. \tag{4.11}$$

The gradient of $\phi^{(\mu)}$ is then

$$\nabla\phi^{(\mu)}(\mathbf{x}) = \nabla(\phi(\mu\mathbf{x})) = \mu\nabla\phi(\mu\mathbf{x}) \,. \tag{4.12}$$

For a 1-form gauge potential A, possibly coupled to a scalar field Φ, one defines

$$A^{(\mu)}(\mathbf{x}) = \mu A(\mu\mathbf{x}) \,. \tag{4.13}$$

The additional factor of μ is natural for a 1-form. Its effect is to give the same scaling behaviour to the covariant derivative of $\Phi^{(\mu)}$ as to the ordinary derivative,

$$D^{A^{(\mu)}}\Phi^{(\mu)}(\mathbf{x}) = (d\Phi^{(\mu)} + A^{(\mu)}\Phi^{(\mu)})(\mathbf{x}) = \mu D^A\Phi(\mu\mathbf{x}) \,. \tag{4.14}$$

The field strength involves one further derivative, so

$$F^{(\mu)}(\mathbf{x}) = \mu^2 F(\mu\mathbf{x}) \,. \tag{4.15}$$

(The notation here is for a non-abelian theory, but the abelian case is similar.) Other kinds of field would be rescaled in the appropriate way depending on their geometrical character. For example a vector field $\mathbf{V}$ would scale as

$$\mathbf{V}^{(\mu)}(\mathbf{x}) = \frac{1}{\mu}\mathbf{V}(\mu\mathbf{x}) \,. \tag{4.16}$$

Note that the boundary conditions, $\phi \in \mathcal{V}$, $D\Phi = 0$, $F = 0$ on S_∞^{d-1} are preserved by rescaling according to these rules. Therefore, as μ varies, the energy remains finite and the topological class of the field does not change. Also, if ϕ is a nonlinear scalar field then the rescaling is consistent; if $\phi(\mathbf{x}) \in Y$ then $\phi^{(\mu)}(\mathbf{x}) \in Y$.

In a theory with just a scalar field ϕ, the energy is often of the form

$$\begin{aligned}
E(\phi) &= \int \Big(W(\phi) \boldsymbol{\nabla}\phi \cdot \boldsymbol{\nabla}\phi + U(\phi) \Big)\, d^d x \\
&\equiv E_2 + E_0 \,,
\end{aligned} \tag{4.17}$$

where we have decomposed the energy into its component parts, and the subscripts indicate the explicit powers of μ that occur when the integrand is rescaled. Then

$$\begin{aligned}
e(\mu) = E(\phi^{(\mu)}) &= \int \Big(W(\phi^{(\mu)}) \boldsymbol{\nabla}\phi^{(\mu)} \cdot \boldsymbol{\nabla}\phi^{(\mu)} + U(\phi^{(\mu)}) \Big)\, d^d x \\
&= \int \Big(\mu^2 W(\phi(\mu \mathbf{x})) \boldsymbol{\nabla}\phi(\mu \mathbf{x}) \cdot \boldsymbol{\nabla}\phi(\mu \mathbf{x}) + U(\phi(\mu \mathbf{x})) \Big)\, d^d x \\
&= \mu^{2-d} E_2 + \mu^{-d} E_0 \,,
\end{aligned} \tag{4.18}$$

where the last step follows by a change of variables from $\mathbf{x}$ to $\mu \mathbf{x}$. Thus $e(\mu)$ is a simple function of μ, with the coefficients E_2 and E_0 depending on the initial choice of field configuration $\phi(\mathbf{x})$.

Generally E_2 and E_0 are both positive. Then the nature of $e(\mu)$ depends crucially on the spatial dimension d. If $d = 3$ or $d = 2$,

$$e(\mu) = \begin{cases} \frac{1}{\mu} E_2 + \frac{1}{\mu^3} E_0 & d = 3 \\[2mm] E_2 + \frac{1}{\mu^2} E_0 & d = 2 \end{cases} \tag{4.19}$$

so $e(\mu)$ decreases monotonically as μ increases. There is no stationary point, so no non-trivial solutions of the field equation are possible. If $d = 1$,

$$e(\mu) = \mu E_2 + \frac{1}{\mu} E_0 \,, \tag{4.20}$$

which is stationary at $\mu = \sqrt{E_0 / E_2}$, so in this case solutions are not ruled out. Thus, finite energy topological solitons in purely scalar theories with an energy of the type (4.17) are possible in one dimension, but not in higher dimensions. We shall discuss these one-dimensional kink solutions in Chapter 5.

Note that the vacuum solution evades Derrick's theorem in all dimensions, because, by definition, the vacuum is a field that is constant in space and where the potential takes its minimal value, so $E_2 = E_0 = 0$. There is a possibility to evade the theorem in two dimensions if $E_0 = 0$, for example, if the potential term is absent. In this case $e(\mu) = E_2$ is independent of μ. We shall discuss sigma model lump solutions, which arise in this way, in Chapter 6. The other way to evade Derrick's theorem in a scalar field theory is to include terms in the Lagrangian and energy involving higher powers of the derivatives of ϕ, or higher order derivatives. This leads to Baby Skyrmions and Skyrmions, discussed in Chapters 6 and 9, respectively. Global vortices evade the theorem because they have infinite energy.

In a gauge theory with a scalar field, the general form of the energy functional (simplifying the algebraic structure, and ignoring numerical coefficients) is

$$E = \int \Big(|F|^2 + |D\Phi|^2 + U(\Phi) \Big) d^d x$$
$$\equiv E_4 + E_2 + E_0 \,. \tag{4.21}$$

Generally, each term contributes positively to the energy. Replacing $\{\Phi, A\}$ by rescaled fields $\{\Phi^{(\mu)}, A^{(\mu)}\}$ gives an energy

$$e(\mu) = \mu^{4-d} E_4 + \mu^{2-d} E_2 + \mu^{-d} E_0 \,. \tag{4.22}$$

Derrick's argument has now lost most of its teeth. If $d = 2$ or $d = 3$, $e(\mu)$ has a minimum for some μ in the range $0 < \mu < \infty$. This is because $e(\mu)$ is a continuous function bounded below by zero, which tends to infinity both as $\mu \to 0$ (the E_0 term) and as $\mu \to \infty$ (the E_4 term). Thus solutions with E_0, E_2 and E_4 all positive are *not ruled out* in two or three spatial dimensions, and there are indeed gauged vortices in two dimensions, and gauged monopoles in three dimensions. (In one dimension, the field strength of a static gauge field vanishes, and the gauge potential can be locally gauge transformed away.) If $d = 4$,

$$e(\mu) = E_4 + \frac{1}{\mu^2} E_2 + \frac{1}{\mu^4} E_0 \,, \tag{4.23}$$

which has no stationary point. So a gauge theory with scalars has *no* non-trivial solutions in four-dimensional Euclidean space-time.

In a pure Yang-Mills gauge theory, the terms involving Φ are absent, and

$$e(\mu) = \mu^{4-d} E_4 \,. \tag{4.24}$$

This has no stationary point if $d < 4$ and $E_4 > 0$, so there is only the vacuum solution $F = 0$ (which is the only type of field with $E_4 = 0$). However, if $d = 4$, $e(\mu)$ is scale independent, and non-vacuum solutions are possible. Indeed instantons of pure Yang-Mills theory exist in four dimensions, and are discussed in Chapter 10.

We have so far used the scaling argument of Derrick to rule out the existence of solutions other than the vacuum in a range of field theories. However, it can be used in two positive ways. The first is to use the condition that the energy of a solution is stationary under rescaling to find relations between the various contributions to the energy. These relations are called virial theorems.

For example, suppose that $d = 1$, and that $\phi(x)$ is a *solution* of the field equation of the theory with energy (4.17). (Previously $\phi(x)$ was just a finite energy field configuration.) Then

$$e(\mu) = \mu E_2 + \frac{1}{\mu} E_0 \,, \tag{4.25}$$

so

$$\frac{de}{d\mu} = E_2 - \frac{1}{\mu^2} E_0 \,. \tag{4.26}$$

This derivative must be zero at $\mu = 1$. Therefore $E_2 = E_0$, so the gradient term and the potential term (integrated over $\mathbb{R}$) each contribute half of the total energy.

Similarly, if $d = 2$ and $\{\Phi, A\}$ is a solution of the field equations for the theory with energy (4.21), then

$$e(\mu) = \mu^2 E_4 + E_2 + \frac{1}{\mu^2} E_0 \,. \tag{4.27}$$

$\frac{de}{d\mu}$ vanishes at $\mu = 1$ only if $E_4 = E_0$. Thus the Yang-Mills (or Maxwell) energy and the energy from the potential U contribute equally to the total energy (and less than half, because of E_2).

A second use (or perhaps, misuse) of Derrick's theorem is to suggest that if the theorem does not rule out a topological soliton solution, i.e. if $e(\mu)$ has a minimum for finite μ, and $e(\mu) \to \infty$ if either $\mu \to 0$ or $\mu \to \infty$, then such a soliton probably exists. Suppose that the homotopy group classifying fields is $\mathbb{Z}$, so that the class $1 \in \mathbb{Z}$ is the sector which potentially has the basic, stable soliton solution. The energy E certainly has an infimum for field configurations in this class, and this is non-negative. Consider a sequence of field configurations whose energy approaches the infimum. The sequence may fail to converge for various reasons. One is that the centre of mass (moment of the energy density) drifts out to infinity – but this can be prevented by centring each element of the sequence at the origin. Another is that the energy density concentrates into a spike over one point – but if this is essentially a local rescaling of the field, with $\mu \to \infty$, then this is ruled out if $e(\mu)$ has its minimum at finite μ. Similarly, failure to converge because the field spreads out and the energy spreads thinly throughout space – again if this is essentially a rescaling with $\mu \to 0$, then it is ruled out.

We can not rigorously conclude that a soliton exists by this argument, because other types of singularity in the field could develop as one approaches the infimum of E. However, in several examples, where solitons of unit topological charge have been rigorously proved to exist (vortices, monopoles), the proof does depend in an important way on understanding that fields which are rescaled by μ have divergent energy, both as $\mu \to 0$ and as $\mu \to \infty$.

The same argument does not work so simply for higher charge solitons. A sequence of configurations in a higher homotopy class can fail to converge because the configuration splits into soliton clusters of lower charge, which separate to infinity as the energy approaches its infimum. To show that this does not occur, one has to show that soliton clusters attract each other at large separation, which is true for some kinds of soliton and not for others.

4.3 Symmetry and reduction of dimension

We have already seen that the space-time and internal symmetries of a Lagrangian field theory have important consequences. They lead to conservation laws for the dynamics. Symmetries have another important role, especially in the study of solitons. It turns out that solitons are frequently of a symmetric form, and recognizing this helps to find and understand them.

The maximal spatial symmetry that a time independent field can have is the full Euclidean symmetry, that is, invariance under translations and rotations. Usually, the vacuum is the only finite energy field of this type. Any field configuration with a positive energy density would have infinite energy if it were translation invariant in even one direction. Solitons are localized solutions whose energy density vanishes at spatial infinity, so they can not have any translational symmetry. They can have, at most, full rotational symmetry.

In certain variational problems, there is a rigorous proof that the configuration that optimizes the "energy" also has maximal symmetry. For example, in the plane, the closed curve of given length with maximal enclosed area is the circle. There are no such results for solitons, except in some very special cases. For example, the basic Skyrmion in $\mathbb{R}^3$ is believed to be rotationally symmetric, but there is no proof that a field configuration with less symmetry can not have lower energy.

To explore the possible symmetries of solitons, one generally makes an "ansatz" for the field. One *assumes* that the field is invariant under a group of symmetries which is some subgroup of the complete symmetry group of the energy functional, and then seeks solutions with this symmetry. It is important to write down the most general field satisfying the invariance conditions. For example, if one imposes time independence and $SO(d)$ rotational symmetry in $\mathbb{R}^d$ ($d \geq 2$), then the symmetry determines the angular behaviour of the fields, and the most general field with the symmetry can be expressed in terms of a number of functions depending only on the radial variable. It is important not to accidentally or deliberately suppress any of these functions, unless there is a further symmetry condition, for example a reflection symmetry, to justify it.

Imposing symmetries on a field often restricts the topological class of the field. For example, for a multiplet of scalar fields, rotational symmetry restricts the homotopy class of the map $S_\infty^{d-1} \mapsto \mathcal{V}$ associated with the field at infinity.

There are now two routes that one can follow. One can take the ansatz for the field, which involves the unknown radial functions, and substitute it into the field equation. One will find that the equation reduces to a number of ordinary differential equations, involving just the radial derivatives of the remaining functions. It is much easier to solve this set of ODEs than the original PDE in $\mathbb{R}^d$. Solutions of finite energy, satisfying the appropriate boundary conditions, are candidate soliton solutions.

The second route, often slightly easier to implement, is to take the ansatz for the field and substitute into the energy functional. The integral over the angular variables can then be done trivially, because the rotational symmetry implies that the energy density is independent of the angles. There remains a radial integral of a simplified energy density, which depends only on the radial functions in the ansatz, and their derivatives. This simplified energy functional can be regarded as that of a dimensionally reduced theory. It is the energy for a field theory defined in one spatial dimension (actually on a half-line, because the radius is non-negative). One may calculate the Euler-Lagrange equations for this theory in the usual way. These turn out to be identical to the equations obtained by the first route, where the ansatz was substituted in the d-dimensional field equation. As before, these dimensionally reduced equations can be solved, and the solutions are candidate solitons.

The fact that these two routes lead to the same equations and solutions is a consequence of the principle of symmetric criticality. We shall give a proof of this in the next section.

Let us analyse in more detail the example of rotational symmetry in $\mathbb{R}^d$, as it exemplifies many features of more general symmetries. The action of an element $R \in SO(d)$ on a point $\mathbf{x} \in \mathbb{R}^d$,

$$\mathbf{x} \mapsto R\mathbf{x}, \tag{4.28}$$

preserves the length of $\mathbf{x}$, that is, $\mathbf{x} \cdot \mathbf{x} = (R\mathbf{x}) \cdot (R\mathbf{x})$ for all R and $\mathbf{x}$. Moreover the length is the only invariant, so the orbit of $\mathbf{x}$ under the action of $SO(d)$ is the complete sphere S^{d-1}, centred at the origin, of radius $|\mathbf{x}|$. $\mathbb{R}^d$ is thus foliated into a 1-parameter family of spheres, labelled by the radius. Each sphere can be identified with the coset space $SO(d)/SO(d-1)$ where $SO(d-1)$ is the isotropy group, the subgroup of $SO(d)$ which leaves a point on the sphere fixed. For example, the point

$$\mathbf{x}_0 = (0, \ldots, 0, \rho), \quad \rho > 0 \tag{4.29}$$

remains fixed under the action of the $SO(d-1)$ subgroup of $SO(d)$ consisting of matrices of the form

$$R = \left(\begin{array}{c|c} r & 0 \\ \hline 0 & 1 \end{array} \right) \tag{4.30}$$

where r is a $(d-1) \times (d-1)$ matrix.* Any other point in the orbit of $\mathbf{x}_0$ is fixed by an $SO(d-1)$ subgroup conjugate to this.

In $\mathbb{R}^d$ there is one exceptional orbit of $SO(d)$, the origin. Here the isotropy group is the whole of $SO(d)$.

For a single scalar field ϕ, $SO(d)$ invariance requires that

$$\phi(R\mathbf{x}) = \phi(\mathbf{x}) \tag{4.31}$$

for all $\mathbf{x}$ and all R. ϕ is constant on the orbits of $SO(d)$, and is determined by its values on a curve that intersects each orbit once. This we can choose to be the half-line whose points are of the form (4.29), extended to $\rho = 0$. ϕ reduces to a function f of the radial variable ρ alone,

$$\phi(\mathbf{x}) = f(\rho), \quad \rho = |\mathbf{x}|. \tag{4.32}$$

ϕ is continuous if and only if f is continuous, and $f(0)$ can take any value. Differentiability of ϕ imposes a stronger condition. By symmetry,

$$\phi(0, \ldots, 0, \rho) = \phi(0, \ldots, 0, -\rho) \tag{4.33}$$

(a rotation connects these points), so $\frac{\partial \phi}{\partial x^d} = 0$ at $\mathbf{x} = \mathbf{0}$. Therefore f must be differentiable and satisfy

$$\left. \frac{df}{d\rho} \right|_{\rho=0} = 0. \tag{4.34}$$

Suppose now that we have a field theory in $\mathbb{R}^d$ with a multiplet of n scalar fields $\phi = (\phi_1, \ldots, \phi_n)$, and that there is an $SO(n)$ internal symmetry. There may now be more than one way to impose rotational symmetry,

---

* Here, $\mathbf{x}_0$ should be a column vector, acted on from the left by the matrix R. However, it is notationally more convenient to present components as a row. We shall treat some other vectors similarly, not distinguishing the row and column forms.

involving combined rotations and internal rotations. The full symmetry group of the energy functional for static fields, ignoring translations and reflections, is $SO(d) \times SO(n)$. We can require the field to be invariant under an $SO(d)$ subgroup whose elements are of the form $(R, D(R))$ where $D : SO(d) \mapsto SO(n)$ associates an $SO(n)$ matrix $D(R)$ with each rotation matrix $R \in SO(d)$. The group multiplication law

$$(R_1, D(R_1)) \cdot (R_2, D(R_2)) = (R_1 R_2, D(R_1)D(R_2)) \qquad (4.35)$$

is consistent only if

$$D(R_1 R_2) = D(R_1)D(R_2). \qquad (4.36)$$

Thus D is a homomorphism. The invariance condition is now

$$\phi(R\mathbf{x}) = D(R)\phi(\mathbf{x}). \qquad (4.37)$$

Any choice of homomorphism will give a consistent ansatz for the field.

One possibility is that the homomorphism D is trivial, and that $D(R) = 1_n$ for all R. In this case, the invariance condition is essentially the same as for a single scalar field; each component of ϕ is rotationally invariant, and just depends on the radial variable ρ. A more interesting possibility is where D maps $SO(d)$ isomorphically onto a subgroup of $SO(n)$, which is only possible if $n \geq d$. Let us consider the simplest case, where

$$D(R) = \left(\begin{array}{c|c} R & 0 \\ \hline 0 & 1_{n-d} \end{array} \right). \qquad (4.38)$$

ϕ splits into $(\phi_1, \ldots, \phi_d, \phi_{d+1}, \ldots, \phi_n)$. The last $n - d$ components are again rotationally invariant in the sense of just being functions of ρ. Let us ignore these, and assume that $n = d$. Then $D(R) = R$.

Our invariance condition is now

$$\phi(R\mathbf{x}) = R\phi(\mathbf{x}). \qquad (4.39)$$

Thinking about the general solution of this equation gives insight into the construction of symmetric fields in almost any situation. Note that the equation determines ϕ at $R\mathbf{x}$ in terms of its value at $\mathbf{x}$. Thus the field on a whole orbit of $SO(d)$ is determined by its value at one point of the orbit. On the other hand, the values on distinct orbits are not algebraically related.

Let us assume for the moment that $d > 2$, so that both $SO(d)$ and $SO(d - 1)$ are non-trivial. The key point is to consider the action of the isotropy group of $\mathbf{x}$. Let R be any element of this isotropy group. Then $R\mathbf{x} = \mathbf{x}$, so

$$\phi(\mathbf{x}) = R\phi(\mathbf{x}), \qquad (4.40)$$

which is an algebraic constraint on the value of ϕ at $\mathbf{x}$. We see that $\phi(\mathbf{x})$ must be invariant under the "internal" action of the isotropy group. At the origin, where the isotropy group is $SO(d)$, ϕ must vanish. At $\mathbf{x}_0 = (0, 0, \ldots, \rho)$, with $\rho > 0$, the isotropy group $SO(d-1)$ consists of matrices of the form (4.30), and (4.40) reduces to

$$
\begin{pmatrix} \phi_1 \\ \cdot \\ \cdot \\ \cdot \\ \phi_{d-1} \\ \phi_d \end{pmatrix} = \left(\begin{array}{c|c} r & 0 \\ \hline 0 & 1 \end{array} \right) \begin{pmatrix} \phi_1 \\ \cdot \\ \cdot \\ \cdot \\ \phi_{d-1} \\ \phi_d \end{pmatrix}. \tag{4.41}
$$

This is satisfied for all r only if $\phi_1 = \phi_2 = \cdots = \phi_{d-1} = 0$. The value of the remaining component, ϕ_d, is arbitrary. Thus we can write

$$
\phi(\mathbf{x}_0) = \begin{pmatrix} 0 \\ \cdot \\ \cdot \\ \cdot \\ 0 \\ f(\rho) \end{pmatrix}. \tag{4.42}
$$

A general point can be expressed as $\mathbf{x} = R(0, \ldots, 0, \rho)$ for some $R \in SO(d)$, and here $\phi = R(0, \ldots, 0, f(\rho))$, by (4.39). It follows that $\phi(\mathbf{x})$ has the form

$$
\phi(\mathbf{x}) = f(\rho) \frac{\mathbf{x}}{\rho}. \tag{4.43}
$$

This is called the hedgehog ansatz for a multiplet of d scalar fields [336]. The set of d functions of d variables is reduced to a single function of one variable, $f(\rho)$, because of rotational symmetry. Substituting the ansatz into the field equation gives an ODE for the function $f(\rho)$.

Let us rewrite the hedgehog ansatz as $\phi(\mathbf{x}) = g(\rho)\mathbf{x}$, where $g(\rho) = f(\rho)/\rho$. Continuity of ϕ at $\mathbf{x} = \mathbf{0}$ requires that g has a finite limit as $\rho \to 0$, and therefore $f(\rho) = O(\rho)$. Provided g is differentiable, ϕ is differentiable. These conditions are rather different from the conditions we found earlier for a rotationally invariant, one-component scalar field.

Suppose the d-component field ϕ has a potential term $U(\phi) = (c - \phi_l \phi_l)^2$ with $c > 0$, leading to spontaneous breaking of the $SO(d)$ internal symmetry. For the energy density to go to zero at infinity, a field of hedgehog form must satisfy the boundary condition $f(\rho) \to \pm\sqrt{c}$ as $\rho \to \infty$. The rotational symmetry and the boundary condition determine the homotopy class of the field. If $f(\infty) = \sqrt{c}$, then the field at infinity is the identity map $S_\infty^{d-1} \mapsto S^{d-1}$, which is in the homotopy class 1 of the group

$\pi_{d-1}(S^{d-1}) = \mathbb{Z}$. If $f(\infty) = -\sqrt{c}$ then the field at infinity is the antipodal map, which is in the class -1 if d is odd, but the class 1 if d is even (since for d even, the antipodal map can be obtained by a continuous rotation of the identity map). Thus, for linear fields in three spatial dimensions, the basic soliton or antisoliton can be of hedgehog type, but multi-solitons can not be.

The hedgehog ansatz is a rather special consequence of the groups acting here. The basic principle that applies to any symmetric field is that i) the field value at any point on an orbit of the symmetry group is determined by its value at one base point on the orbit, ii) the field at the base point is constrained because it must be invariant under the isotropy group there.

In two dimensions, the rotation group is $SO(2)$, which is abelian. Its orbits in the plane are circles, with the origin as an exceptional orbit. The invariance condition for a multiplet of real scalar fields is

$$\phi(R\mathbf{x}) = D(R)\phi(\mathbf{x}), \qquad (4.44)$$

where D is a choice of homomorphism from $SO(2)$ to the internal symmetry group. The basic example is where ϕ is a two-component field, with internal symmetry group $SO(2)$. The homomorphisms $D : SO(2) \mapsto SO(2)$ are labelled by an integer j, and are given by the formulae

$$D\begin{pmatrix} \cos\theta & -\sin\theta \\ \sin\theta & \cos\theta \end{pmatrix} = \begin{pmatrix} \cos j\theta & -\sin j\theta \\ \sin j\theta & \cos j\theta \end{pmatrix}. \qquad (4.45)$$

A scalar doublet field satisfying (4.44) has the form (in polar coordinates)

$$\phi(\rho,\theta) = f(\rho)\begin{pmatrix} \cos j\theta \\ \sin j\theta \end{pmatrix}, \qquad (4.46)$$

or what can be obtained from this by a further constant internal rotation. Since the isotropy group is trivial for $\rho > 0$, there are no further constraints on f here. However, continuity and rotational invariance at $\rho = 0$ requires that $f \to 0$ as $\rho \to 0$, if $j \neq 0$. If there is spontaneous symmetry breaking, and f tends to a non-zero value as $\rho \to \infty$, then this ansatz for the field has winding number j. If the scalar field has more than two components, then it splits into a number of doublets with this behaviour (not all j necessarily the same), plus singlets with no angular dependence.

We have dealt at some length with symmetries of linear scalar fields. However, nonlinear scalar fields can be treated rather similarly, as we shall see when studying Skyrmions. The internal symmetry group is now the symmetry group of the manifold Y where the nonlinear field takes its values, or a subgroup of this if there is a potential.

It is not difficult to extend the analysis to symmetric vector fields and other tensor fields in $\mathbb{R}^d$. Consider, as before, rotations in $\mathbb{R}^d$, but ignore any internal symmetry group. Under a rotation R, the value of a vector field $\mathbf{V}$ at $\mathbf{x}$ is carried to $R\mathbf{x}$. However, also important is how the neighbourhood of $\mathbf{x}$ is mapped to the neighbourhood of $R\mathbf{x}$. This is calculated by noting that, under R,

$$\mathbf{x} + \delta\mathbf{x} \mapsto R(\mathbf{x} + \delta\mathbf{x}) = R\mathbf{x} + R\delta\mathbf{x} \tag{4.47}$$

so $\delta\mathbf{x}$ goes to $R\delta\mathbf{x}$. Thus the condition of rotational invariance for a vector field is

$$\mathbf{V}(R\mathbf{x}) = R\mathbf{V}(\mathbf{x}). \tag{4.48}$$

This equation is similar to that for a multiplet of d scalar fields, but here there is no choice for the homomorphism D. One must have $D(R) = R$. For $d > 2$, the only fields satisfying (4.48) are of the hedgehog form

$$\mathbf{V}(\mathbf{x}) = g(\rho)\mathbf{x}, \tag{4.49}$$

with $g(0)$ finite. One can see this by splitting $\mathbf{V}$, at any point $\mathbf{x}$ other than the origin, into radial and tangential components. The radial component has a magnitude which depends only on ρ. The tangential component is transformed non-trivially by the isotropy group, so invariance requires it to vanish.

A bit more abstractly, the tangent space at $\mathbf{x}$ is d-dimensional, and under $SO(d-1)$ splits into the direct sum of irreducible modules

$$\underline{1} \oplus \underline{d-1}. \tag{4.50}$$

A rotationally invariant vector field is associated with the singlet, which transforms trivially under $SO(d-1)$, and therefore there is just one function of ρ in the ansatz (4.49).

In two dimensions, the isotropy group is trivial except at the origin, so a rotationally invariant vector field has both radial and tangential components. The ansatz is

$$\begin{pmatrix} V^1 \\ V^2 \end{pmatrix} = \begin{pmatrix} g(\rho)x^1 - h(\rho)x^2 \\ g(\rho)x^2 + h(\rho)x^1 \end{pmatrix}. \tag{4.51}$$

Similar considerations apply to tensors. If $d > 2$, a rotationally invariant, rank 2 symmetric tensor field has the form

$$g^{ij}(\mathbf{x}) = f(\rho)\delta^{ij} + g(\rho)x^i x^j. \tag{4.52}$$

The representation theory of the isotropy group again explains the presence of two functions of ρ. The action of $SO(d-1)$ on the symmetrized tensor product of the tangent space of a point is given by the tensor product representation

$$(\underline{1} \oplus \underline{d-1}) \otimes_S (\underline{1} \oplus \underline{d-1}),   (4.53)$$

whose decomposition into irreducibles has two singlet pieces.

We come finally to symmetric gauge fields. This is potentially quite complicated. For a more substantial analysis see refs. [243, 346, 136]. Again, let us consider rotations in $\mathbb{R}^d$, with $d > 2$, and let the gauge group be the non-abelian group G. (The abelian case is not very different.) A 1-form gauge potential $A(\mathbf{x})$ is rotationally symmetric if each rotation R combined with a suitable gauge transformation leaves the field unchanged. One says that the rotation leaves the field invariant "up to a gauge transformation". Since a gauge transformation has no physical effect, we have in a geometrical sense invariance under the rotation. In terms of spatial components, A is invariant if

$$R_{ji}A_j(R\mathbf{x}) = g_R(\mathbf{x})A_i(\mathbf{x})g_R^{-1}(\mathbf{x}) - \partial_i g_R(\mathbf{x})g_R^{-1}(\mathbf{x}).   (4.54)$$

(This says that R has the same effect as the gauge transformation g_R; or equivalently that the combined effect of R and g_R^{-1} leaves A invariant.) If the gauge field A is coupled to a scalar field Φ, gauge transforming under the fundamental representation of G as in Eq. (2.127), then Φ is also invariant under the rotation if

$$\Phi(R\mathbf{x}) = g_R(\mathbf{x})\Phi(\mathbf{x}).   (4.55)$$

There is a condition on the gauge transformations g_R, coming from the composition rule for rotations. This can be derived from (4.54), but more easily from (4.55). We have $\Phi(R_1 R_2 \mathbf{x}) = g_{R_1 R_2}(\mathbf{x})\Phi(\mathbf{x})$ and also

$$\Phi(R_1 R_2 \mathbf{x}) = g_{R_1}(R_2\mathbf{x})\Phi(R_2\mathbf{x}) = g_{R_1}(R_2\mathbf{x})g_{R_2}(\mathbf{x})\Phi(\mathbf{x}).   (4.56)$$

For consistency, and to avoid an unnecessary constraint on Φ, $g_R(\mathbf{x})$ must satisfy

$$g_{R_1 R_2}(\mathbf{x}) = g_{R_1}(R_2\mathbf{x})g_{R_2}(\mathbf{x}).   (4.57)$$

This is a "cocycle" condition, which arises in several contexts. It is more subtle than the previous "homomorphism" condition (4.36). The interpretation of (4.57) is that the pairs $\{R, g_R\}$ lift the action of the rotation group on $\mathbb{R}^d$ to the bundle over $\mathbb{R}^d$ of which Φ is a section, and on which A is a connection 1-form. Logically, (4.57) should come first; subsequently one can impose the symmetry conditions (4.54) and (4.55) on the fields.

These symmetry conditions for a gauge field coupled to a scalar field are themselves gauge invariant. If $\Phi \mapsto g\Phi$ and $A \mapsto gAg^{-1} - dgg^{-1}$, then the transformed fields still satisfy (4.54) and (4.55) but $g_R(\mathbf{x})$ must be replaced by $g(R\mathbf{x})g_R(\mathbf{x})g^{-1}(\mathbf{x})$. The cocycle condition remains satisfied.

The solution of (4.57) is not difficult. Note first that (4.57) implies that $g_I(\mathbf{x}) = I$ and $g_{R^{-1}}(R\mathbf{x}) = g_R(\mathbf{x})^{-1}$, for all $\mathbf{x}$ and R. Next, consider the point $\mathbf{x}_0 = (0, \ldots, 0, \rho)$, with $\rho > 0$. Let R_1 and R_2, and hence $R_1 R_2$, lie in the isotropy group $SO(d-1)$ of $\mathbf{x}_0$. Since $R_2 \mathbf{x}_0 = \mathbf{x}_0$, Eq. (4.57) simplifies to

$$g_{R_1 R_2}(\mathbf{x}_0) = g_{R_1}(\mathbf{x}_0) g_{R_2}(\mathbf{x}_0). \tag{4.58}$$

This is a homomorphism condition. It is solved by *choosing* a homomorphism

$$\lambda : SO(d-1) \mapsto G. \tag{4.59}$$

λ can be chosen to be independent of ρ. We denote the image of this homomorphism, which is a subgroup of G, by G_λ, and we denote the centralizer of G_λ in G by H. H is the subgroup of G whose elements commute with all elements of G_λ.

Given λ, one can solve (4.57) as follows. Fix a neighbourhood of $\mathbf{x}_0$ on the sphere of radius ρ (actually, the whole sphere except the point $(0, \ldots, 0, -\rho)$). For each point $\mathbf{x}$ in this neighbourhood, there is a special rotation $R_\mathbf{x}$ that takes $\mathbf{x}_0$ to $\mathbf{x}$. It is defined by decomposing the Lie algebra of $SO(d)$ as

$$so(d) = so(d-1) \oplus m. \tag{4.60}$$

m is the orthogonal complement of $so(d-1)$ with respect to the Killing form (or trace), and is also invariant under conjugation by any element of $SO(d-1)$. The dimension of m is the dimension of the orbit space $S^{d-1} = SO(d)/SO(d-1)$. The exponential map, acting on a neighbourhood of the origin in m, in fact an open ball of radius π, gives the desired rotations. For each point $\mathbf{x}$, there is a unique element of m in this ball, whose exponential $R_\mathbf{x}$ rotates $\mathbf{x}_0$ to $\mathbf{x}$.

We may use the gauge freedom to choose, for all $\mathbf{x}$,

$$g_{R_\mathbf{x}}(\mathbf{x}_0) = I, \tag{4.61}$$

where $R_\mathbf{x}$ is the special rotation. Now let R be a general rotation, that sends $\mathbf{x}$ to $R\mathbf{x}$. The rotations $R_\mathbf{x}$, R, and $R_{R\mathbf{x}}^{-1}$ send $\mathbf{x}_0$ successively to $\mathbf{x}$, $R\mathbf{x}$ and back to $\mathbf{x}_0$, so $R_{R\mathbf{x}}^{-1} R R_\mathbf{x}$ belongs to the isotropy group of $\mathbf{x}_0$. It is Wigner's "little group element" associated with R and $\mathbf{x}$. The cocycle condition implies that

$$g_{R_{R\mathbf{x}}^{-1} R R_\mathbf{x}}(\mathbf{x}_0) = g_{R_{R\mathbf{x}}^{-1}}(R\mathbf{x}) g_R(\mathbf{x}) g_{R_\mathbf{x}}(\mathbf{x}_0). \tag{4.62}$$

The third factor on the right-hand side is the identity, because of our gauge choice (4.61), and so is the first factor, because it is equal to $g_{RR\mathbf{x}}(\mathbf{x}_0)^{-1}$. Therefore we obtain the solution of the cocycle condition

$$g_R(\mathbf{x}) = g_{R_{R\mathbf{x}}^{-1}RR\mathbf{x}}(\mathbf{x}_0) = \lambda(R_{R\mathbf{x}}^{-1}RR_\mathbf{x}). \tag{4.63}$$

For all R and $\mathbf{x}$, $g_R(\mathbf{x})$ lies in G_λ.

This solution of the cocycle condition has the following nice property. Let R be an element of the isotropy group of $\mathbf{x}_0$, and $\mathbf{x}$ a general point in the orbit of $\mathbf{x}_0$. $R_\mathbf{x}$ as before denotes the special rotation sending $\mathbf{x}_0$ to $\mathbf{x}$. Observe that

$$RR_\mathbf{x}R^{-1}\mathbf{x}_0 = R\mathbf{x}, \tag{4.64}$$

so $RR_\mathbf{x}R^{-1}$ sends $\mathbf{x}_0$ to $R\mathbf{x}$. Recall also that conjugation by R maps m to itself. Therefore $RR_\mathbf{x}R^{-1}$, like $R_\mathbf{x}$, is the exponential of an element of m. Indeed, if $R_\mathbf{x} = \exp(w)$ then $RR_\mathbf{x}R^{-1} = \exp(RwR^{-1})$, and if w is in the ball of radius π, so is RwR^{-1}. So $RR_\mathbf{x}R^{-1}$ is the special rotation $R_{R\mathbf{x}}$. The formula (4.63) therefore simplifies to

$$g_R(\mathbf{x}) = \lambda(RR_\mathbf{x}^{-1}R^{-1}RR_\mathbf{x}) = \lambda(R) \tag{4.65}$$

for R an element of the isotropy group of $\mathbf{x}_0$. g_R is independent of $\mathbf{x}$ for such R.

Now let us consider the invariance conditions for the scalar field and gauge field at $\mathbf{x}_0$. For R in the isotropy group $SO(d-1)$, Eq. (4.55) reduces to

$$\Phi(\mathbf{x}_0) = \lambda(R)\Phi(\mathbf{x}_0). \tag{4.66}$$

Thus $\Phi(\mathbf{x}_0)$ must be invariant under the subgroup G_λ. This condition means that if one decomposes the fundamental module (representation) of G into irreducible modules of G_λ, then the invariant singlets in the decomposition are the surviving components of Φ, and each contributes one function of ρ. The remaining components are zero. The non-zero components combine into one or more multiplets of H, the centralizer of G_λ.

Similarly, the invariance condition (4.54) at $\mathbf{x}_0$ reduces to

$$R_{ji}A_j(\mathbf{x}_0) = \lambda(R)A_i(\mathbf{x}_0)\lambda(R)^{-1} \tag{4.67}$$

for $R \in SO(d-1)$. (The final, derivative term vanishes because g_R is independent of $\mathbf{x}$.) The equation (4.67) is a linear algebraic constraint and can be solved using Schur's lemma. Generally, $A(\mathbf{x}_0)$ lies in the module of $SO(d) \times G$

$$\underline{d} \otimes \mathrm{Lie}(G), \tag{4.68}$$

where $\underline{d}$ is the cotangent space at $\mathbf{x}_0$, and G acts on $\mathrm{Lie}(G)$ by conjugation. The action of the elements $R \in SO(d-1)$ decomposes $\underline{d}$ into irreducible

$SO(d-1)$ modules $\underline{1} \oplus \underline{d-1}$, where the first factor corresponds to the radial direction, and the second to the tangent space to the sphere at $\mathbf{x}_0$. The corresponding decomposition of the gauge potential is into its radial part A^{rad} and its tangential part A^{tan}. Similarly, the action of $\lambda(R)$ by conjugation turns $\mathrm{Lie}(G)$ into a module of $SO(d-1)$. Equation (4.67) implies that $A(\mathbf{x}_0)$ lies in the submodule invariant under the action of the $SO(d-1)$ subgroup defined by the pairs $(R, \lambda(R))$, where the first factor acts on $\underline{d}$, and the second factor acts by conjugation.

R acts trivially on A^{rad}, so (4.67) implies that

$$A^{\mathrm{rad}}(\mathbf{x}_0) = \lambda(R) A^{\mathrm{rad}}(\mathbf{x}_0) \lambda(R)^{-1}, \tag{4.69}$$

which constrains A^{rad} to lie in the subspace of $\mathrm{Lie}(G)$ consisting of the $SO(d-1)$ singlets in the decomposition of $\mathrm{Lie}(G)$. This subspace is in fact the subalgebra $\mathrm{Lie}(H)$. Within this subspace, A^{rad} is an arbitrary function of ρ. Therefore, the part A^{rad} of a rotationally symmetric gauge potential in $\mathbb{R}^d$ reduces to a gauge potential for the gauge group H on the radial half-line.

The tangential components $A^{\mathrm{tan}}(\mathbf{x}_0)$ are constrained by (4.67) in the following way. Under the $SO(d-1)$ spatial rotations these form the module $\underline{d-1}$. Each module $\underline{d-1}$ in the decomposition of $\mathrm{Lie}(G)$ into $SO(d-1)$ irreducibles can be paired with this, and by Schur's lemma, gives a non-zero component of A^{tan}. These are arbitrary functions of ρ, that behave as scalar fields from the point of view of the reduced gauge theory on the radial half-line. Like the scalars coming from Φ, they can be combined into one or more multiplets of H.

Having determined the components of A and Φ which can be non-vanishing at $\mathbf{x}_0$, one can calculate completely the form of A and Φ over the orbit of $\mathbf{x}_0$, using equations (4.54) and (4.55). Using the special rotations $R_{\mathbf{x}}$, and the fact that $g_{R_{\mathbf{x}}}(\mathbf{x}_0) = I$, one finds

$$\Phi(\mathbf{x}) \;=\; \Phi(\mathbf{x}_0) \tag{4.70}$$

$$(R_{\mathbf{x}})_{ji} A_j(\mathbf{x}) \;=\; A_i(\mathbf{x}_0) - \partial_i g_{R_{\mathbf{x}}}(\tilde{\mathbf{x}}) \Big|_{\tilde{\mathbf{x}}=\mathbf{x}_0}$$

$$=\; A_i(\mathbf{x}_0) - \partial_i \lambda(R_{R_{\mathbf{x}}\tilde{\mathbf{x}}}^{-1} R_{\mathbf{x}} R_{\tilde{\mathbf{x}}}) \Big|_{\tilde{\mathbf{x}}=\mathbf{x}_0}, \tag{4.71}$$

where the derivatives are with respect to $\tilde{\mathbf{x}}$. The first of these equations, which says that Φ is constant on spheres, looks very different from the hedgehog ansatz for a scalar field, but this is mainly a consequence of our gauge choice.

We should not forget here the excluded points of the form $(0, \ldots, 0, -\rho)$. The fields can be extended smoothly and symmetrically to include these points, but sometimes only by introducing another patch of $\mathbb{R}^d$, together with a non-trivial transition function to relate the fields on the different patches. There can also be constraints on the choice of homomorphism λ to make this possible.

There is another important point that is relevant for rotationally symmetric gauge fields in $\mathbb{R}^d$. Remember that there is a special orbit, the origin, whose isotropy group is $SO(d)$. A similar analysis as above applies there. One needs to choose a homomorphism

$$\Lambda : SO(d) \mapsto G \,, \qquad (4.72)$$

and the homomorphism $\lambda : SO(d-1) \mapsto G$ must be the restriction of Λ to the appropriate $SO(d-1)$ subgroup, otherwise there will be discontinuities at the origin. Because of this, one may choose a different gauge than (4.61), namely

$$g_{R_\mathbf{x}}(\mathbf{x}_0) = \Lambda(R_\mathbf{x}) \,. \qquad (4.73)$$

Much of the theory above now simplifies. One finds that

$$g_R(\mathbf{x}) = \Lambda(R) \qquad (4.74)$$

for all $\mathbf{x}$ and R, i.e. g_R is independent of $\mathbf{x}$. This clearly satisfies (4.57), and one can prove that any other choice for $g_R(\mathbf{x})$ (with the desired limit $\Lambda(R)$ at the origin) is gauge equivalent to this.

This observation justifies the assumption made in much of the literature on spherically symmetric non-abelian monopoles, that the fields should be invariant under combined rotations and global (**x**-independent) gauge transformations. However, the earlier approach, using just λ, clarifies the structure of the reduced gauge theory on the radial half-line, and explains why it has gauge group H.

A Dirac monopole in a $U(1)$ gauge theory is singular at the origin, and one can not impose a rotational invariance condition there. The Dirac monopole is spherically symmetric in the earlier sense, involving the choice of a homomorphism λ from the isotropy group $SO(2)$ (in $\mathbb{R}^3$) to the gauge group $U(1)$. A consequence of this is that Dirac monopoles have an infinite range of possible magnetic charges. In contrast, spherically symmetric non-abelian monopoles, which smoothly extend to the origin, have far more restricted charges.

This completes our description of the ansatz for a rotationally symmetric gauge field. The ideas can be applied to more general continuous symmetries than just rotations. The key points are i) the symmetric gauge fields are completely described in terms of dimensionally reduced gauge and scalar fields defined on the parameter space of orbits, ii) one must choose a homomorphism λ from the isotropy group of a point on a generic orbit to the gauge group G, iii) H, the centralizer of its image G_λ, is the reduced gauge group, iv) the invariant gauge potential splits into parts tangential and normal to the orbits. The normal part acts as a gauge potential of the dimensionally reduced gauge theory, whereas the tangential part contributes scalar fields to the reduced theory, transforming under some representation of H, v) the global topology, and the existence of special orbits, can put constraints on the choice of homomorphism λ.

We have assumed, for simplicity, that the fields are time independent, but it is a trivial matter to relax this assumption.

For dynamical scalar and gauge fields, rotational symmetry leads to the same ansatz, but all functions depend on the radius and on the time, and the reduced gauge potential has both radial and time components. The field equations reduce from PDEs in $(d + 1)$-dimensional space-time to PDEs in $(1 + 1)$ dimensions. The latter are the field equations of a gauge theory with Higgs scalar fields, and gauge group H.

We can have a smaller rotational symmetry, e.g. $SO(2)$ symmetry in $\mathbb{R}^3$, or $SO(3)$ symmetry in $\mathbb{R}^4$. In both cases the field equations reduce to those for a theory in two space dimensions. We shall also be interested in fields which are invariant under a discrete subgroup K of $SO(d)$. For example, in three dimensions, K could be the symmetry group of a Platonic solid. Higher charge solitons, both monopoles and Skyrmions, sometimes have these symmetries. However, discrete symmetry groups do not lead to a reduction in dimension of the field equations.

Rather interesting is the possibility of higher symmetry. Pure Yang-Mills theory in $\mathbb{R}^4$ is invariant under the conformal group $SO(5, 1)$. The basic instanton solution is not just rotationally invariant under $SO(4)$, but is actually invariant under an $SO(5)$ subgroup of the conformal group which acts transitively on $\mathbb{R}^4$. The fields are completely determined, algebraically, by this symmetry.

4.4 Principle of symmetric criticality

Let Ψ be the (generic) fields of some Lagrangian field theory with action $S(\Psi)$, and let $\mathcal{C}$ temporarily denote the space of all field configurations $\Psi(x)$ depending on both space and time. If a field configuration is transformed by any element of the symmetry group of the theory, then the action is unchanged.

Let K be a subgroup of the symmetry group, and let $\mathcal{C}_K \subset \mathcal{C}$ denote the configuration space of all K-invariant fields, that is, field configurations Ψ satisfying $k(\Psi) = \Psi$ for all $k \in K$. (The dependence on x is suppressed in this somewhat compact notation.) Let $S_K(\Psi)$ denote the action of the theory restricted to $\mathcal{C}_K$. As we have seen, this is often a theory defined in a lower-dimensional space-time.

Let $\Psi_0(x)$ be a K-invariant field configuration which is a stationary point of the restricted action S_K. Then the principle of symmetric criticality states that $\Psi_0(x)$ is automatically a stationary point of the full action S. The principle undoubtedly has a long history. In the context of solitons in field theory it was enunciated and given a brief proof by Coleman [88]. Palais gave a more careful discussion, with various proofs adapted to varying assumptions [322]. He also showed that the principle is not universally valid, and gave some counterexamples. We shall outline the proof that is valid if K is a finite, discrete group (e.g. a group of reflections, or a Platonic symmetry group) or if K is a compact Lie group (e.g. a rotation group).

We need to consider infinitesimal variations $\Psi_0 + \delta\Psi$ of Ψ_0. $\delta\Psi$ lies in $T\mathcal{C}$, the tangent space to $\mathcal{C}$ at Ψ_0. The group K acts on $\mathcal{C}$, and an element $k \in K$ transforms $\Psi_0 + \delta\Psi$ to

$$k(\Psi_0 + \delta\Psi) = k(\Psi_0) + k'(\delta\Psi) = \Psi_0 + k'(\delta\Psi), \qquad (4.75)$$

where k' denotes the derivative. Because Ψ_0 is invariant under K, K acts linearly on $T\mathcal{C}$ through the derivative; and the map $k \mapsto k'$ is a representation of K. This infinite-dimensional representation can be completely decomposed into finite-dimensional irreducible representations if K is finite or compact. Some subspace $T\mathcal{C}^{\parallel}$ of $T\mathcal{C}$ transforms trivially under K, and a complementary space $T\mathcal{C}^{\perp}$ transforms non-trivially (i.e. when decomposed into irreducibles, all the non-trivial modules lie in $T\mathcal{C}^{\perp}$, and all the invariant singlets lie in $T\mathcal{C}^{\parallel}$).

Now $\mathcal{C}_K$ consists of *all* K-invariant fields. If $\delta\Psi$ lies in $T\mathcal{C}^{\parallel}$ then it is invariant under K. Thus $\Psi_0 + \delta\Psi$ is in $\mathcal{C}_K$, to linearized approximation. We conclude that $T\mathcal{C}^{\parallel} = T\mathcal{C}_K$, the tangent space to $\mathcal{C}_K$ at Ψ_0.

Consider next the action S evaluated on $\Psi_0 + \delta\Psi$. By the functional Taylor series, we have

$$S(\Psi_0 + \delta\Psi) = S(\Psi_0) + S'(\delta\Psi) + O(\delta\Psi)^2. \qquad (4.76)$$

S' is the functional derivative, or first variation, of S at Ψ_0, and is a linear map from $T\mathcal{C}$ to $\mathbb{R}$. To show that Ψ_0 is a stationary point of S we need to show that $S'(\delta\Psi)$ is zero for any $\delta\Psi$.

On the subspace $T\mathcal{C}^{\parallel}$, S' vanishes. This is because, by assumption, Ψ_0 is a stationary point of the restricted action S_K, so S' vanishes for all $\delta\Psi$ in $T\mathcal{C}_K$, which is the same as $T\mathcal{C}^{\parallel}$.

The non-trivial part of the proof is to show that S' vanishes on $T\mathcal{C}^{\perp}$. Let V be an irreducible K-module in $T\mathcal{C}^{\perp}$, and suppose $v \in V$. Consider the orbit of v under K,

$$\{v^{(1)}, v^{(2)}, \ldots, v^{(n)}\}, \qquad (4.77)$$

with $v^{(1)} = v$. (This is written as a finite set, assuming that K is a finite group.) By K-invariance, $S'(v^{(i)})$ has the same value for all i. Therefore $S'(v)$ can be expressed as the average

$$S'(v) = \frac{1}{n} \sum_{i=1}^{n} S'(v^{(i)}). \tag{4.78}$$

But S' is a linear function, so

$$S'(v) = S' \left(\frac{1}{n} \sum_{i=1}^{n} v^{(i)} \right). \tag{4.79}$$

Now $\frac{1}{n} \sum_{i=1}^{n} v^{(i)}$ is invariant under K, because K acts by permuting points in the orbit. But the only invariant element of a vector space V on which K acts irreducibly and non-trivially is the zero vector (otherwise V would contain a proper subspace invariant under K – the subspace of individually invariant vectors). Thus the right-hand side of (4.79) is zero, so

$$S'(v) = 0. \tag{4.80}$$

Since $v \in V$ was arbitrary, S' must be zero on all of V, and by running through all irreducible pieces of $TC^{\perp}$ we conclude that S' is zero on this whole space. Thus S' vanishes on both $TC^{\|}$ and $TC^{\perp}$ and hence vanishes identically. This completes the proof.

For a compact Lie group K the set (4.77) becomes a continuous orbit of vectors $\{k'(v) : k \in K\}$. $S'(v)$ can again be expressed as an average by using a normalized K-invariant measure $d\Omega_K$, which exists on K. Then

$$S'(v) = \int_K S'(k'(v))\, d\Omega_K = S' \left(\int_K k'(v)\, d\Omega_K \right), \tag{4.81}$$

using the invariance and linearity of S'. The argument of S' in the last expression is again K-invariant, and must vanish to avoid a contradiction with the irreducibility of V. The proof is completed as before.

The interpretation of the principle of symmetric criticality is that the solutions of the reduced field equations for symmetric fields are in fact solutions of the full field equation of the theory.

We are particularly interested in static fields of finite energy. Let $\mathcal{C}$ revert to denoting the space of field configurations at a given time. The principle of symmetric criticality applies here too. A stationary point of the reduced energy functional for K-symmetric fields is a solution of the full field equation. In particular, a *minimal energy* field configuration for the restricted problem in $\mathcal{C}_K$ is a *stationary point* of the energy in $\mathcal{C}$. It is not necessarily a minimum in the full configuration space. One needs to investigate the second variation of the energy in the directions $TC^\perp$ to see if the field is a minimum or saddle point.

To establish if a K-symmetric solution minimizes the energy among all K-symmetric fields is often straightforward. One needs to use an analytic argument, or perhaps study an eigenvalue problem for an ODE and show that all eigenvalues are positive. To show that this solution minimizes the energy among *all* fields, one may need to investigate an infinite set of eigenvalue problems, one associated with each irreducible module of K in $TC^\perp$. Such an investigation has been successfully carried out for the spherically symmetric $SU(2)$ monopole (and not just in the Bogomolny case). The monopole is a minimum of the energy in its topological class.

Even these arguments only establish that a solution is a *local* minimum of the energy. The proof that it is a global minimum is still usually lacking.

The fact that symmetric minima are sometimes saddle points in the complete theory is actually a virtue, for this is a way to find saddle point solutions. One should choose a symmetry group which is distinct from, and not a subgroup of, the complete symmetry group of the minimal energy solution. If one finds a solution with this symmetry, then it can not be the minimal energy solution, and is likely to be a saddle point, although it could be a local minimum. Many saddle points of the Skyrme energy function have been found this way.

4.5 Moduli spaces and soliton dynamics

It is a challenging problem, when studying any complicated physical system or a mathematical model of it, to reduce the number of degrees of freedom to those that are essential. For example, consider an elastic body pivoted at its centre of mass and free to rotate. It is experimentally verified that a body like this behaves as a "rigid" body whose essential degrees of freedom are the Euler angles specifying its orientation. However, rigid body motion is only an approximation. It is valid provided the frequencies of the motion are small compared with the elastic vibration frequencies of the body.

Let us look at this in more detail. There is an elastic potential energy function for the body, and a kinetic energy obtained as the integral of the kinetic energy density of the constituent matter. The minimum of the potential is attained when the body is in its "rigid" equilibrium shape. This minimum is not unique. It occurs on a copy of the manifold $SO(3)$ of possible orientations, embedded in the infinite-dimensional space of shapes.

For the reduced dynamics of the body one restricts the full Lagrangian to the $SO(3)$ of minima of the potential. Equivalently, one supposes that the body has its static equilibrium shape, with orientational angles that vary with time. The restriction of the kinetic energy function gives the kinetic energy expression on $SO(3)$ for the rigid body. The coefficient matrix is a left-invariant metric on $SO(3)$, whose exact form depends on the moments of inertia of the body. The potential function on $SO(3)$ is simply constant, by rotational symmetry, so does not contribute to the reduced dynamics. The reduced Lagrangian is therefore purely kinetic, and the rigid body motion is geodesic motion on $SO(3)$, given by an equation of the form (2.12).

The potential energy is a positive quadratic form for small elastic deformations orthogonal to the manifold of minima. The vibrational frequencies depend on this quadratic form and the kinetic energy expression. For very slow rotational motion, there is a rescaling of time which brings the rotational motion back to an angular speed of order 1, and makes the vibrational frequencies large. The limit of negligibly slow rotation is equivalent to rotation at finite speed, but with the vibrational frequencies becoming infinite, so the potential is infinitely steep away from the minimum. In the limit, the kinetic energy is insufficient to deform the body, and the potential becomes effectively a constraint forcing the body to be in its equilibrium shape. This role for the potential explains why the body behaves as rigid, with only three dynamical degrees of freedom, when it is rotating slowly.

This formal argument has been rigorously justified mathematically [356, 49, 139]. At any non-zero speed of rotation (in the original unscaled time) there is some excitation of the transverse, elastic modes of the body. However, this is an adiabatic effect, and small at slow speeds. The body's shape smoothly adjusts to accommodate the rotation, for example, by a centrifugal stretching. Over modest time intervals, there is negligible transfer of the energy of rotational motion to genuine vibrations of the elastic body, provided there are no elastic vibrations initially. There can be significant transfer of energy only over a time scale of order $\exp(\omega_{\mathrm{vib}}/\omega_{\mathrm{rot}})^{\nu}$ where $\omega_{\mathrm{vib}}/\omega_{\mathrm{rot}}$ is the ratio of the lowest vibrational frequency (determined by the properties of the body) to the frequency associated with rotational motion (determined by the initial state of motion of the body), and ν is some positive power.

Another example to keep in mind is a slow bobsleigh on a frictionless bobsleigh track. Assume this is on the level. To first approximation, the dynamics reduces to a one-dimensional motion along the bottom of the track, at constant speed. One knows that, as the bobsleigh enters a curve it has to rise up from the bottom of the track, but as it exits the curve it returns to the bottom, and its initial speed is almost exactly regained. This adiabatic effect modifies, but does not invalidate the reduction of the motion to one dimension. In addition, there is a small residual transverse vibration after the bobsleigh has exited the curve, but provided the curve is completely smooth, and connects segments that are asymptotically straight, the amplitude of vibration is exponentially small, being of the form $\exp(-\text{const}/v)$, where v is the speed of the bobsleigh.

A first application of these ideas to solitons is to the kink. This is the minimal energy static solution of some Lorentz invariant scalar field theory defined on a line, stabilized by its topological charge. It is unique apart from its location. Let us write the solution as $\phi(x - a)$, where a is the location. The manifold of minima of the energy is the line itself, parametrized by a. This is called the one-kink moduli space, $\mathcal{M}_1 = \mathbb{R}$. Now suppose the kink is slowly moving. We make the "rigid motion" ansatz

$$\phi(t, x) = \phi(x - a(t)). \tag{4.82}$$

The effective Lagrangian of the rigid kink has a kinetic term given by the field kinetic energy

$$T = \frac{1}{2} \int_{-\infty}^{\infty} (\partial_0 \phi)^2 \, dx \tag{4.83}$$

which reduces to $T = \frac{1}{2} M \dot{a}^2$, where M is the kink mass. The potential energy is constant and can be neglected. The reduced equation of motion is $\ddot{a} = 0$, whose solution is $a(t) = vt + \text{const}$, i.e. motion at constant velocity v. This example sounds even more trivial than that of the rigid body, since the moduli space has no intrinsic curvature. However, it is not completely trivial. The ansatz (4.82) is not an exact solution for any varying function $a(t)$. An exact solution is the Lorentz boosted kink $\phi(\gamma(x - vt))$ where $\gamma = 1/\sqrt{1 - v^2}$. The interpretation is that a quasi-rigid motion is possible, but the shape of the kink is adiabatically deformed and the potential energy increased as a result of the motion. The deformation is small for small v, as γ is close to 1. (An analogy is with a bobsleigh going steadily round a circular track, slightly above the bottom.) If the static kink were set in motion in its undeformed shape, then its shape would vibrate and it would radiate – a relativistic phenomenon verified in numerical simulations. The conclusion is that there is an effective one-dimensional dynamics on moduli space which is valid for low speed motion, but one needs to be a little careful. The speed should be non-relativistic, but it doesn't have to be negligibly slow.

A slightly more complicated example is the dynamics of one Skyrmion. There is a unique shape for the static soliton, but it has six degrees of freedom associated with the symmetries of the underlying Lagrangian, three for translations in $\mathbb{R}^3$ and three for rotations. The Skyrmion is very like a spherical rigid body free to move and rotate. There is an effective dynamics on $\mathcal{M}_1 = \mathbb{R}^3 \times SO(3)$, which depends on the mass and moment of inertia of the Skyrmion. However, the rotational motion and also the centre of mass motion adiabatically deform the Skyrmion. There is also a new phenomenon. A slowly rotating Skyrmion can lose energy by exciting the asymptotic pion radiation field. This effect can not be avoided, as it is by the Lorentz boosted kink, but it is algebraically small if the pion field is massless. The effect is probably exponentially small or possibly absent if the pion field is massive and the rotational frequency is much less than the lowest vibrational frequency of the pion field, which is proportional to the pion mass parameter. The "rigid" dynamics of a Skyrmion is therefore accurate at slow speeds, but it has its limitations.

The most interesting extension of the ideas here occurs in a Lagrangian field theory with solitons satisfying a Bogomolny equation, of which we shall later discuss a number of examples, including critically coupled vortices and BPS monopoles. Here there is an integer topological charge N, and the minimal energy static fields have energy $E = c|N|$ for some positive constant c. The minimal energy is attained by fields satisfying a PDE which is first order in spatial derivatives – this is the Bogomolny equation. Furthermore, there is a surprisingly large moduli space of solutions of the Bogomolny equation. The moduli space $\mathcal{M}_N$ of solutions of topological charge N has dimension $k|N|$ for some integer k ($k = 2$ for vortices, and $k = 4$ for monopoles). For most values of N, this is much bigger than the dimension of the symmetry group of the theory. The interpretation is that there are N-soliton solutions which are nonlinear superpositions of N individual solitons, where each constituent has its own k degrees of freedom. At least, this is the interpretation when the solitons have a moderate or large separation. When they are close, the solitons can merge and lose their identities, and this is reflected in the global geometry of moduli space. Because the potential energy is constant throughout $\mathcal{M}_N$, whether the solitons are close together or far apart, there is no interaction energy between the solitons, provided they are at rest.

The dynamics of solitons can be modelled by a Lagrangian dynamics on $\mathcal{M}_N$ with a kinetic energy obtained as the restriction of the kinetic energy of the full field theory [279]. One assumes the field configuration is exactly a solution of the Bogomolny equation for all time, with the parameters, or moduli of the solution, varying slowly with time. The kinetic energy is a quadratic expression in the time derivatives of the moduli, whose coefficient matrix can be interpreted as the metric on $\mathcal{M}_N$. It is not easy to calculate this metric explicitly in most cases, but we shall explain later how progress can be made. One interesting example is the metric on the two-monopole moduli space, calculated by Atiyah and Hitchin using a remarkable indirect method. The second ingredient in the reduced Lagrangian is the potential energy, but this is constant and has no direct effect. Since the Lagrangian is purely kinetic, as for a rigid body pivoted at its centre of mass, the reduced dynamics on moduli space is geodesic

motion at constant speed. The geodesics are interpreted as motion of the
N solitons, either soliton scattering or a bounded motion. In general,
some of the moduli correspond to the soliton positions and some to their
internal orientations, which means that part of the motion of each soliton
is internal, and may be interpreted as an unquantized charge or spin. The
solitons do not move in straight lines, unless they are infinitely separated.
In a naive sense, they experience forces. These forces are not due to the
potential energy, but to the intrinsic curvature of the moduli space. It
is a geometrical effect. It is quite often possible to calculate these forces
in detail for well separated solitons, using physical reasoning. Inevitably,
these forces are proportional to the square of the speed of the solitons,
where speed means speed of the spatial motion and/or internal motion.
An explicit metric can be calculated for N well separated monopoles using
this approach, even though the exact metric on $\mathcal{M}_N$ is only known for
$|N| = 1$ or 2.

One general feature of the reduction to moduli space dynamics is that it
is compatible with the spatial symmetries and global internal symmetries
of the original Lagrangian. The symmetry group acts on solutions of the
Bogomolny equation, preserving the minimal energy property, and hence
acts on $\mathcal{M}_N$. There are two consequences. First, there are conserved
Noether charges for the reduced dynamics, which are the reductions of
the corresponding expressions in field theory. Second, one may apply the
principle of symmetric criticality to find symmetric motions in moduli
space, and these correspond to similar symmetric motions in the field
theory.

There must be limitations to the accuracy of the reduced dynamics,
just as there are for rigid body dynamics. First of all, one expects some
adiabatic deformation of the solitons, due to their motion. Also one must
anticipate some transfer of energy from the moduli space motion into vibrational modes of the transverse field. Physically, soliton motion couples
to radiation. An estimate has been made of this for two-monopole scattering. The total energy radiated is algebraically small at non-relativistic
monopole speeds, $v \ll 1$, the total energy radiated being of order v^5 compared to the kinetic energy of order v^2, and this gives confidence that
for this simple type of soliton scattering, the moduli space dynamics is
reliable. Generally speaking, it appears that the moduli space dynamics
is exact, if it is regarded as the formal, non-relativistic limit of soliton
dynamics in field theory.

A problem with geodesic motion is that it can be incomplete. This
happens for lumps, though not for vortices and monopoles. Geodesic
motion can reach the boundary of moduli space in a finite time. Then
the field becomes singular and the moduli space dynamics breaks down.
However, the moduli space dynamics may not be misleading, because

similar singularities may form even if the full field equation is solved.

There are some rigorous mathematical studies of the reduction of N-soliton dynamics in field theory to an effective dynamics on $\mathcal{M}_N$, both for vortices and monopoles, by Stuart [385, 386]. Stuart has proved that the field dynamics for fields close to moduli space can be uniquely decomposed, by orthogonally projecting onto moduli space. A field configuration is characterized, at any instant, by a point in $\mathcal{M}_N$, and a residual field in an orthogonal direction. The field equation is shown to split in such a way that the equation for geodesic motion on $\mathcal{M}_N$ is formally the leading part of the equation for the projected motion. There are corrections to this which are carefully estimated. Provided the initial field is close to moduli space, and the initial time derivative of the field has its dominant component parallel to moduli space, and is small, then the moduli space dynamics is reliable, at least for a finite time. The estimate of this time is of order $1/v$, where v is a characteristic initial soliton speed. This means the geodesic motion is reliable for a distance of order 1, and it is plausible that it is reliable for a simple soliton scattering process. As the solitons approach, and later as they separate, they move along approximately straight line trajectories at constant speed, reflecting the exact solutions for well separated solitons. The main scattering process takes place in a finite distance, of the order of the soliton length scale, which is 1. Therefore, the result for the scattering angle as a function of impact parameter, assuming geodesic motion, should be reliable for slow speeds. However, for bounded soliton motion, Stuart's results suggest one can have less confidence in the moduli space dynamics, for large time intervals.

Finally, we mention some generalizations. Suppose a rigid body is pivoted at a point other than the centre of mass. Then the reduced Lagrangian on $SO(3)$ has a kinetic term and a gravitational potential term which depends on the orientation of the body. The reduced dynamics is geodesic motion modified by a potential, the equation of motion being of the type (2.5) with $f_{ij} = 0$. This is still reliable provided the typical frequencies are small compared with the elastic vibration frequencies. The frequencies of the rigid body motion are no longer determined so much by the initial speed, but rather by the strength of the gravitational force.

Similarly, for solitons, there are situations where a moduli space can be defined, with both a metric and potential on it. There is then a reduced dynamics on moduli space governing the soliton motion, where the forces are partly static (the gradient of the potential), and partly geometrical. For example, if a Lagrangian field theory has couplings close to the critical values where a Bogomolny equation occurs, then one may define the moduli space as in the critically coupled case, and restrict the Lagrangian to motion in this space. This usually gives an unmodified metric, and a small potential, working to first approximation. One is often able to understand the static forces between solitons when they are well separated, using some physical intuition.

Our discussion so far has focussed exclusively on second order dynamical field theories where the kinetic terms are quadratic in time derivatives, and the reduced dynamics has kinetic terms quadratic in velocities. We are also interested in Lagrangian field theories with solitons where the kinetic terms are linear in time derivatives, but the moduli space of static solutions is as before. Here again we make the ansatz that the reduced dynamics on moduli space can be obtained by restricting the Lagrangian to solutions of the Bogomolny equation with time varying moduli. The reduced equation of motion is of type (2.22). There is usually no motion on moduli space unless a small non-constant potential is present. In contrast to examples of second order dynamics, there is less mathematical work, or physical intuition, demonstrating the validity of this approach.

We shall make a similar ansatz when dealing with first order, dissipative field dynamics, that is, gradient flow. The gradient flow in field configuration space can be restricted to a gradient flow on moduli space, but the precise or optimal way to define the space is still debatable. At critical coupling, there is no potential gradient and hence no motion on moduli space. Here, Demoulini and Stuart have studied rigorously the flow from a general point in field configuration space down to the moduli space. Less clear is how to deal with the close-to-critical case. The unmodified moduli space is one possible stage for a non-trivial gradient flow, with an equation of motion of type (2.24), but this is only an approximation. Close by, an exact attractor for the field theory gradient flow appears to exist. This has been investigated numerically for vortices in the charge 2 sector, but rigorous results are lacking. We discuss this further in Section 7.7.2. Gradient flow on an attractor may also be the best way to obtain an effective moduli space in systems like the Skyrme model, where there is no Bogomolny equation; see Section 9.8.

5

Kinks

5.1 Bogomolny bounds and vacuum structure

The most elementary topological solitons occur in one space dimension and involve a single scalar field [343]. Consider the Lagrangian density

$$\mathcal{L} = \frac{1}{2}\partial_\mu\phi\partial^\mu\phi - U(\phi),\tag{5.1}$$

where ϕ is a real scalar field and $U(\phi)$ is a real non-negative function of ϕ. The Euler-Lagrange field equation which follows from (5.1) is the nonlinear wave equation

$$\partial_\mu\partial^\mu\phi + \frac{dU}{d\phi} = 0.\tag{5.2}$$

Let $U_{\min}$ be the global minimum of the potential $U(\phi)$. By the addition of a suitable constant to U, which of course does not alter the field equation, it is always possible to arrange that $U_{\min} = 0$, which we will assume to be the case from now on.

The potential energy is

$$V = \int_{-\infty}^{\infty}\left(\frac{1}{2}\phi'^2 + U(\phi)\right)dx,\tag{5.3}$$

with contributions from $U(\phi)$ and the gradient of the field $\phi' = \frac{\partial\phi}{\partial x}$. The total energy is $T + V$, where T is the kinetic energy

$$T = \frac{1}{2}\int_{-\infty}^{\infty}\dot\phi^2\,dx\tag{5.4}$$

and $\dot\phi = \frac{\partial\phi}{\partial t}$.

Let $\mathcal{V}$ denote the set of vacuum fields (which we assume are isolated)

$$\mathcal{V} = \{\phi_0,\text{ such that }\phi_0' = \dot\phi_0 = 0,\text{ and }U(\phi_0) = U_{\min}\}.\tag{5.5}$$

As described in Chapter 4, the existence of topological solitons depends on there being multiple vacua, so that $\mathcal{V}$ contains more than one component. In other words, $\pi_0(\mathcal{V})$ needs to be non-trivial. A finite energy field configuration is then classified topologically by an element (ϕ_-, ϕ_+) of $\pi_0(\mathcal{V}) \times \pi_0(\mathcal{V})$, where $\phi_\pm = \lim_{x\to\pm\infty} \phi(x)$. Solutions which interpolate between different vacua, that is $\phi_+ \neq \phi_-$, are known generically as kinks, a name suggested by the shape of the scalar field when plotted as a function of x.

If $\phi_+ = \phi_-$ then by a continuous deformation, the field can be transformed into the constant vacuum solution $\phi(x) = \phi_+$, which has zero energy. If, on the other hand, $\phi_+ \neq \phi_-$, then the field can not be continuously deformed to a constant zero energy solution by deformations which keep the energy finite, since any field for which $\phi(\pm\infty) \notin \mathcal{V}$ has infinite energy. This is the fundamental reason for the stability of a kink solution, since time evolution is an example of a continuous deformation for which the energy remains finite.

Recall from Chapter 4 that in one space dimension, the combination in V of a potential term and a term quadratic in the field gradient is sufficient to evade Derrick's theorem, and allow static soliton solutions. Under a spatial dilation the two contributions to the potential energy scale in opposite ways, producing a balancing effect, and the minimal energy is attained at the finite and non-zero scale where the virial theorem

$$\int_{-\infty}^{\infty} \frac{1}{2}\phi'^2 \, dx = \int_{-\infty}^{\infty} U(\phi) \, dx \qquad (5.6)$$

holds.

By a series of simple manipulations, it is possible to derive a lower bound on the energy E of any field configuration in terms of topological data, the bound only depending on the field values at spatial infinity. The key inequality is simply

$$\left(\frac{1}{\sqrt{2}}\phi' \pm \sqrt{U(\phi)} \right)^2 \geq 0. \qquad (5.7)$$

Expanding this inequality and integrating over space, we obtain

$$\int_{-\infty}^{\infty} \left(\frac{1}{2}\phi'^2 + U(\phi) \right) dx \geq \pm \int_{-\infty}^{\infty} \sqrt{2U(\phi)} \, \phi' \, dx. \qquad (5.8)$$

Therefore, for static fields

$$E \geq \left| \int_{-\infty}^{\infty} \sqrt{2U(\phi)} \, \phi' \, dx \right| = \left| \int_{\phi_-}^{\phi_+} \sqrt{2U(\phi)} \, d\phi \right|. \qquad (5.9)$$

The same bound also holds for time dependent fields, as T is positive. Since $U(\phi) \geq 0$, we may introduce a superpotential $W(\phi)$ such that

$U(\phi) = \frac{1}{2}(\frac{dW}{d\phi})^2$, and then the right-hand side of (5.9) can be integrated to give the bound in the form

$$E \geq |W(\phi_+) - W(\phi_-)|. \tag{5.10}$$

This observation is due to Bogomolny [56], and energy bounds of this general type, where the energy is bounded from below in terms of solely topological data, are known as Bogomolny bounds.

Clearly, to attain equality in the Bogomolny bound the field must be static, $\dot{\phi} = 0$, and satisfy one of the first order Bogomolny equations

$$\phi' = \pm\sqrt{2U(\phi)}, \tag{5.11}$$

where solutions of the equation with the $+$ sign (if they exist) are called kinks and those with the $-$ sign antikinks. For these solutions, the two contributions to the energy density, $\frac{1}{2}\phi'^2$ and $U(\phi)$, are pointwise the same, a stronger statement than the virial theorem (5.6).

Solutions of the Bogomolny equations (5.11) are global minima of the energy within a given topological class of fields, so they are critical points of the energy function and hence automatically static solutions of the second order field equation (5.2). It is easy to confirm this explicitly by differentiating (5.11) to give

$$\phi'' = \pm\frac{1}{\sqrt{2U}}\frac{dU}{d\phi}\phi' = \frac{dU}{d\phi}. \tag{5.12}$$

Given a static kink solution it is, of course, a trivial task to Lorentz boost it and obtain a solution in which the kink moves with any speed less than the speed of light (which is 1 in our units).

5.2 ϕ^4 kinks

In this section we discuss in detail the simplest model with kinks, where there are just two vacua, that is, $\pi_0(\mathcal{V}) = \mathbb{Z}_2$. To obtain two vacua with a potential which is polynomial in ϕ^2 requires at least quartic terms. To be specific, consider a potential of the form

$$U(\phi) = \mu + \nu\phi^2 + \lambda\phi^4 \tag{5.13}$$

where μ, ν, λ are real constants and $\lambda > 0$ in order that the energy is bounded from below. If $\nu \geq 0$ then the potential (5.13) has a unique global minimum at $\phi = 0$, so $\mathcal{V}$ has only one component and there are no kinks. Therefore, from now on we assume that $\nu < 0$, and for convenience write $\nu = -2m^2\lambda$, where m is a positive real constant. In order to set $U_{\min} = 0$ we choose $\mu = \lambda m^4$. The potential (5.13) is now

$$U(\phi) = \lambda(m^2 - \phi^2)^2, \tag{5.14}$$

and it is clear that degenerate global minima occur at $\phi = m$ and $\phi = -m$, so there are two vacua, which we denote by $\mathcal{V}_+$ and $\mathcal{V}_-$.

This example is known as the ϕ^4 model [102, 336] and the full Lagrangian density is

$$\mathcal{L} = \frac{1}{2}\partial_\mu\phi\partial^\mu\phi - \lambda(m^2 - \phi^2)^2 \,, \tag{5.15}$$

with the corresponding field equation

$$\partial_\mu\partial^\mu\phi - 4\lambda(m^2 - \phi^2)\phi = 0 \,. \tag{5.16}$$

The topological content of a given field configuration is captured by the topological charge

$$N = \frac{\phi_+ - \phi_-}{2m} \,, \tag{5.17}$$

where $\phi_\pm$ are the field values at $x = \pm\infty$. This takes the possible values $N \in \{0, 1, -1\}$. Although it is rather trivial in this example, note that N may be written as the integral over space of a topological charge density,

$$N = \frac{1}{2m} \int_{-\infty}^{\infty} \phi' \, dx. \tag{5.18}$$

The first possibility, $N = 0$, means that the field interpolates between the same vacuum values, so it lies in the same topological sector as one of the vacuum solutions, $\phi(x) = \pm m$, to which it may be continuously deformed.

The minimal energy solution with $N = 1$ is the kink, which interpolates between the vacua $\mathcal{V}_-$ and $\mathcal{V}_+$ as x increases from $-\infty$ to ∞. The related solution with $N = -1$ is called the antikink and is obtained by making the replacement $\phi \mapsto -\phi$ in the kink solution. Note that there are no multi-kink solutions with $N > 1$, since fields of this kind are not compatible with the finite energy boundary conditions. However, a field configuration containing a finite mixture of kinks and antikinks alternating along the line can be constructed, but there are no static solutions of this type. We discuss the interaction between a kink and an antikink at the end of this section.

The Bogomolny energy bound for the ϕ^4 model is

$$E \geq \left| \int_{\phi_-}^{\phi_+} \sqrt{2\lambda}(m^2 - \phi^2) \, d\phi \right| = \left| \sqrt{2\lambda} \left[m^2\phi - \frac{1}{3}\phi^3 \right]_{\phi_-}^{\phi_+} \right| = \frac{4}{3}m^3\sqrt{2\lambda}|N| \,. \tag{5.19}$$

Both for the kink and antikink, $|N| = 1$, so the bound in these sectors is $E \geq \frac{4}{3}m^3\sqrt{2\lambda}$. Equality is attained if the Bogomolny equation (5.11) is satisfied, which in the ϕ^4 model reads

$$\phi' = \sqrt{2\lambda}(m^2 - \phi^2) \,. \tag{5.20}$$

The choice of the + sign gives a kink, rather than an antikink. This equation is easily integrated to yield the kink solution

$$\phi(x) = m \tanh\left(\sqrt{2\lambda}m(x-a)\right), \tag{5.21}$$

where a is an arbitrary constant of integration. The energy density of the kink is

$$\mathcal{E} = \frac{1}{2}\phi'^2 + \lambda(m^2 - \phi^2)^2 = 2\lambda m^4 \text{sech}^4\left(\sqrt{2\lambda}m(x-a)\right), \tag{5.22}$$

from which it is easy to confirm that $E = \int_{-\infty}^{\infty} \mathcal{E} \, dx = \frac{4}{3}m^3\sqrt{2\lambda}$. The energy E is also the rest mass, M, of the kink. Note that ϕ has value 0 (the value mid-way between the two vacua $\pm m$) at the point $x = a$, which is also the point at which the energy density is maximal, and equal to $2\lambda m^4$. The point a is therefore naturally interpreted as the position of the kink, and is a free parameter, corresponding to the translation invariance of the Lagrangian. This is the only free parameter in the kink solution, so the moduli space for a kink is simply $\mathcal{M}_1 = \mathbb{R}$.

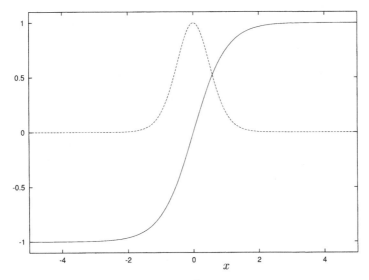

Fig. 5.1.   The field $\phi(x)$ of the ϕ^4 kink (solid curve) and its energy density (dashed curve).

In Fig. 5.1 we plot the kink solution (5.21) and the energy density (5.22) for the choice of parameters $\lambda = \frac{1}{2}$, $m = 1$ and $a = 0$. The characteristic kink shape is clear, as is the localized lump-like nature of the energy density. By a redefinition of the field and length units, the constants λ

and m in the ϕ^4 model can always be scaled to equal any given positive
values, so the qualitative features of the kink do not depend on them.

By applying a Lorentz boost we obtain the moving kink solution

$$\phi(t, x) = m \tanh\left(\sqrt{2\lambda}m\gamma(x - vt - a)\right), \qquad (5.23)$$

where $-1 < v < 1$ is the velocity of the kink and $\gamma = 1/\sqrt{1 - v^2}$ is the
Lorentz factor. This solution has energy $E = \frac{4}{3}\gamma m^3\sqrt{2\lambda}$. In the non-
relativistic limit, where $|v| \ll 1$, the moving kink simplifies to $\phi(t, x) = m \tanh\left(\sqrt{2\lambda}m(x - vt - a)\right)$ and since $\dot\phi = -v\phi'$ its kinetic energy is

$$T = \frac{1}{2}v^2 \int_{-\infty}^{\infty} \phi'^2 \, dx. \qquad (5.24)$$

By the virial theorem, this equals $\frac{1}{2}Mv^2$.

To conclude this section we compute the interaction energy of a well
separated kink-antikink pair, and show that there is an attractive force
between the two, as one might expect. To simplify the presentation, and
given the above comments, we set $\lambda = \frac{1}{2}$ and $m = 1$, it being possible to
reintroduce arbitrary values of these constants by a simple rescaling.

There are a number of ways to derive the interaction energy. One way
is to compute the energy of a static solution of the field equation in which
delta-function sources are introduced to pin the kink and antikink at a
given separation [342]. Here we follow the procedure of ref. [278], and
compute the interaction energy by identifying the force produced on one
soliton by the other with the rate of change of momentum.

For a general theory of the form (5.1), the momentum on the semi-
infinite interval $(-\infty, b]$ is

$$P = -\int_{-\infty}^{b} \dot\phi\phi' \, dx. \qquad (5.25)$$

The force on this interval, F, is given by the time derivative of the momentum

$$F = \dot P = -\int_{-\infty}^{b} (\ddot\phi\phi' + \dot\phi\dot\phi') \, dx = \left[-\frac{1}{2}(\dot\phi^2 + \phi'^2) + U(\phi)\right]_{-\infty}^{b}, \qquad (5.26)$$

where to obtain the final expression we have used the field equation (5.2)
and integrated the total derivative terms. This shows that the force on the
interval can be identified with the difference of pressure at the endpoints.

Now consider a kink-antikink pair, with the antikink at position $-a$
and the kink at position a, where $a \gg 1$. A field configuration of this
form can be obtained by the superposition

$$\phi(x) = \phi_1(x) + \phi_2(x) + 1, \qquad (5.27)$$

where $\phi_1(x)$ is the antikink and $\phi_2(x)$ is the kink, given explicitly by

$$\phi_1(x) = -\tanh(x+a), \quad \phi_2(x) = \tanh(x-a). \tag{5.28}$$

Let the endpoint of the interval, b, lie between the kink and antikink, and far from each, that is, $-a \ll b \ll a$. Then throughout the interval the sum $\phi_2 + 1$ is close to zero, so we can linearize in this combination. To leading order this produces the result

$$
\begin{aligned}
F &= \left[-\frac{1}{2}\phi_1'^2 + U(\phi_1) - \phi_1'\phi_2' + (1+\phi_2)\frac{dU}{d\phi}(\phi_1) \right]_{-\infty}^{b} \\
&= \left[-\phi_1'\phi_2' + (1+\phi_2)\phi_1'' \right]_{-\infty}^{b} \tag{5.29}
\end{aligned}
$$

where, to obtain the second expression, we have used the fact that the antikink solves the Bogomolny equation (5.11) to cancel the first two terms in the first expression, and used the static version of the field equation (5.2) to replace the last term. Since we are dealing with a field configuration whose spatial derivatives fall off exponentially fast at infinity there is clearly no contribution from the lower limit in the expression (5.29). To evaluate the contribution from the upper limit we recall that the point b is far from both the kink and antikink, so we may use the asymptotic forms

$$\phi_1(x) \sim -1 + 2e^{-2(x+a)}, \quad \phi_2(x) \sim -1 + 2e^{2(x-a)}. \tag{5.30}$$

This leads to the expression for the force

$$F = 32e^{-2R} = \frac{dE_{\text{int}}}{dR}, \tag{5.31}$$

where we have defined the kink-antikink separation $R = 2a$, and equated the force with the derivative of the interaction energy E_{int}. Note that F is independent of b, as it should be since b was just a parameter introduced to perform the calculation and has no physical significance. Therefore we may identify F with the force on the antikink, produced by the kink. The asymptotic interaction energy is, finally,

$$E_{\text{int}} = -16e^{-2R}, \tag{5.32}$$

which is negative and decreases as the separation decreases, indicating an attractive force between the kink and antikink.

This picture of the kink-antikink interaction is confirmed by numerical simulations of the full time dependent field equation, starting with a well separated kink-antikink pair at rest. The kink and antikink move toward each other and annihilate into radiation. This is also true if they are sent toward each other with a speed which is much less than the speed of light, but if the speed is great enough the situation is more complicated and depends sensitively on the speed [307, 72].

5.3 Sine-Gordon kinks

The ϕ^4 model is not a very rich system from the point of view of multi-solitons, since there are no topological sectors which may be thought of as describing multi-kink fields, except those including antikinks as well. In this section we turn to the sine-Gordon model [378], in which the vacua are labelled by an arbitrary integer, so that field configurations corresponding to any number of solitons are allowed. This is the general situation in the more complicated, higher-dimensional theories with solitons that we discuss later.

The Lagrangian density defining the sine-Gordon model is

$$\mathcal{L} = \frac{1}{2}\partial_\mu\phi\partial^\mu\phi - (1 - \cos\phi)\,, \tag{5.33}$$

where we have chosen appropriate length and energy units to scale away any possible parameters of the model. The sine-Gordon field equation which follows from this is

$$\partial_\mu\partial^\mu\phi + \sin\phi = 0\,. \tag{5.34}$$

It is self-evident that the zero energy vacua of this model are given by the constant solutions $\phi = 2\pi n$, where $n \in \mathbb{Z}$ is any integer, so

$$\pi_0(\mathcal{V}) = \mathbb{Z}\,. \tag{5.35}$$

Let (ϕ_-, ϕ_+) denote the vacuum values attained by the field at $x = \pm\infty$. The Lagrangian density (5.33) is invariant under 2π shifts of the field, $\phi \mapsto \phi \pm 2\pi$, so, without loss of generality, we can choose to set $\phi_- = 0$, though we will not insist on this. The topological sectors of the model are indexed by the integer $N = (\phi_+ - \phi_-)/2\pi$. Once again, this topological charge may be trivially written as the integral of a charge density

$$N = \frac{1}{2\pi}\int_{-\infty}^{\infty} \phi'\, dx\,. \tag{5.36}$$

N counts the net number of solitons.

The Bogomolny bound in this case is

$$E \geq \int_0^{2\pi N} 2\left|\sin\frac{\phi}{2}\right|\, d\phi = 4|N|\left[-\cos\frac{\phi}{2}\right]_0^{2\pi} = 8|N| \tag{5.37}$$

where, in evaluating the integral, we have used the periodicity of the integrand and the fact that the range of integration is an N-fold cover of the interval $[0, 2\pi]$. This bound is attained by solutions of one of the Bogomolny equations

$$\phi' = \pm 2\sin\frac{\phi}{2}\,. \tag{5.38}$$

Restricting to kink solutions, by choosing the $+$ sign, we can integrate directly to give

$$\phi(x) = 4 \tan^{-1} e^{x-a} \tag{5.39}$$

where a is the arbitrary constant of integration. Taking the solution branch for which $\phi_- = 0$, we see that $\phi_+ = 2\pi$, so this solution has topological charge $N = 1$, and therefore it describes a single kink. The antikink solution, which solves the Bogomolny equation with the $-$ sign, is obtained by the replacement $\phi \mapsto -\phi$.

For the kink solution (5.39), $\phi(a) = \pi$, and since π is the field value half-way between the vacuum values 0 and 2π, one should interpret a as the position of the kink. This is confirmed by examining the energy density

$$\mathcal{E} = 4 \operatorname{sech}^2(x - a), \tag{5.40}$$

which is maximal at $x = a$. From this expression it is simple to check that $E = \int_{-\infty}^{\infty} \mathcal{E} \, dx = 8$.

Since the general solution of the Bogomolny equation is a kink of unit topological charge, there are no multi-kink solutions of the Bogomolny equation. It follows that there is a repulsive force between two kinks. This is because any field configuration with $N = 2$ must obey the strict Bogomolny bound $E > 16$, but in the limit in which two kinks are infinitely separated, the energy will approach $E = 16$, the sum of the energies of the two individual kinks. Thus, the potential energy of two kinks decreases as they separate and there is a corresponding repulsive force. By a similar calculation as that done in the previous section to compute the asymptotic kink-antikink interaction energy in the ϕ^4 model, the asymptotic interaction energy of two sine-Gordon kinks is found to be

$$E_{\text{int}} = 32e^{-R}, \tag{5.41}$$

where R is the separation between the kinks. This result was obtained by Perring and Skyrme [329], who used the sine-Gordon model as a toy model for a more realistic three-dimensional field theory, the one now known as the Skyrme model. The Skyrme model and its soliton solutions are the subject of Chapter 9.

Although there are no static multi-soliton solutions in the sine-Gordon model, there are time dependent solutions, which describe the scattering of two or more kinks. Rather unusually, such solutions can be written down explicitly in closed form. The reason is that the sine-Gordon model in one dimension is an integrable system. Integrable soliton equations are not the topic of this book and it would involve a lengthy digression to introduce even the main concepts. Here we will content ourselves with a few comments, and as a simple example, explicitly construct a two-soliton solution of the sine-Gordon equation using some of the integrable systems machinery [2].

Although there is no generally accepted universal definition of an integrable system, there are a number of common features which arise in most PDEs considered to belong to this category. They include the existence of an infinite number of conserved quantities, the property that the given equation can be written as the compatibility condition of an overdetermined linear system, known as a Lax Pair, and the applicability of solution generating techniques such as the inverse scattering method and Bäcklund transformations. It is this final feature which we now use.

It is convenient here to introduce lightcone coordinates, $x_\pm = \frac{1}{2}(x \pm t)$, and the corresponding derivatives, $\partial_\pm = \partial/\partial x_\pm$, in terms of which the sine-Gordon field equation (5.34) becomes

$$\partial_-\partial_+\phi = \sin\phi. \tag{5.42}$$

Now consider the following pair of equations

$$\partial_+\psi = \partial_+\phi - 2\beta\sin\left(\frac{\phi+\psi}{2}\right), \quad \partial_-\psi = -\partial_-\phi + \frac{2}{\beta}\sin\left(\frac{\phi-\psi}{2}\right), \tag{5.43}$$

which is known as a Bäcklund transformation, and may be thought of as determining the field ψ, given the field ϕ. β is a non-zero real constant, called the Bäcklund parameter. This pair of equations is subject to the compatibility condition $\partial_-\partial_+\psi = \partial_+\partial_-\psi$, because of the symmetry of second partial derivatives, which implies that

$$\partial_-\partial_+\phi - \beta\cos\left(\frac{\phi+\psi}{2}\right)(\partial_-\phi + \partial_-\psi)$$
$$= -\partial_+\partial_-\phi + \frac{1}{\beta}\cos\left(\frac{\phi-\psi}{2}\right)(\partial_+\phi - \partial_+\psi). \tag{5.44}$$

After using Eqs. (5.43) to eliminate the single derivative terms, this simplifies to the sine-Gordon equation, $\partial_-\partial_+\phi = \sin\phi$. Similarly, subjecting the Bäcklund transformation (5.43) to the compatibility condition $\partial_-\partial_+\phi = \partial_+\partial_-\phi$ yields the sine-Gordon equation for ψ, that is, $\partial_-\partial_+\psi = \sin\psi$. Thus, the Bäcklund transformation is a mapping between solutions of the sine-Gordon equation and can be used to generate new solutions from known solutions. Since the Bäcklund transformation contains a free parameter β, this extra parameter is introduced into the new solution, in addition to a constant of integration.

As an example, if we start with the trivial vacuum solution $\phi = 0$, then Eqs. (5.43) take the simplified form

$$\partial_+\psi = -2\beta\sin\frac{\psi}{2}, \quad \partial_-\psi = -\frac{2}{\beta}\sin\frac{\psi}{2}. \tag{5.45}$$

These equations are easily integrated to give the solution

$$\psi(x_+, x_-) = 4 \tan^{-1} e^{-\beta x_+ - x_-/\beta + \alpha} \qquad (5.46)$$

where α is a constant of integration. Making the identifications

$$v = \frac{1 - \beta^2}{1 + \beta^2}, \quad \gamma = \frac{1}{\sqrt{1 - v^2}} = -\frac{1 + \beta^2}{2\beta}, \quad a = \frac{2\beta\alpha}{1 + \beta^2} \qquad (5.47)$$

where $\beta < 0$, this solution can be written as

$$\psi(t, x) = 4 \tan^{-1} e^{\gamma(x - vt - a)}, \qquad (5.48)$$

which we recognize as the Lorentz boosted version of the one-kink solution (5.39).

The real power of the Bäcklund transformation is that it leads to a purely algebraic method of constructing multi-kink solutions, evading the task of having to explicitly integrate Eqs. (5.43), which may be tricky for a complicated seed solution ϕ. This algebraic construction arises by considering two solutions ψ_1, ψ_2, obtained from Eqs. (5.43) by starting with the same seed solution $\phi = \psi_0$ but using two different values, β_1, β_2, of the Bäcklund parameter. By manipulating the equations, it can be shown that a theorem of permutability holds, so that the solution ψ_{12}, obtained by applying the Bäcklund transformation with parameter β_2 to the seed solution ψ_1, is (with an appropriate choice of integration constants) equal to the solution ψ_{21}, obtained by applying the Bäcklund transformation with parameter β_1 to the seed solution ψ_2. The consistency condition $\psi_{12} = \psi_{21}$ leads to the explicit relation

$$\psi_{12} = \psi_{21} = 4 \tan^{-1} \left[\left(\frac{\beta_1 + \beta_2}{\beta_2 - \beta_1} \right) \tan \left(\frac{\psi_1 - \psi_2}{4} \right) \right] - \psi_0, \qquad (5.49)$$

giving a new solution ψ from the triplet of known solutions ψ_0, ψ_1, ψ_2.

We have already seen that starting with the vacuum solution $\psi_0 = 0$, the Bäcklund transformation produces the one-kink solutions $\psi_j = 4 \tan^{-1} e^{\theta_j}$ $(j = 1, 2)$, where $\theta_j = -\beta_j x_+ - x_-/\beta_j + \alpha_j$. Substituting these into Eq. (5.49) yields the further solution

$$\psi(x_+, x_-) = 4 \tan^{-1} \left[\left(\frac{\beta_1 + \beta_2}{\beta_2 - \beta_1} \right) \frac{\sinh \frac{\theta_1 - \theta_2}{2}}{\cosh \frac{\theta_1 + \theta_2}{2}} \right]. \qquad (5.50)$$

For simplicity, set $\beta_1 = -1/\beta_2 \equiv \beta$, and $\alpha_1 = \alpha_2 = 0$. Then (5.50) becomes

$$\psi(t, x) = 4 \tan^{-1} \left[\frac{v \sinh(\gamma x)}{\cosh(\gamma v t)} \right] \qquad (5.51)$$

where v and γ are related to β as before by Eqs. (5.47). Since this solution interpolates between the vacua -2π and 2π as x increases from $-\infty$ to ∞, it is in the $N = 2$ sector, and therefore describes a time dependent two-kink field. As we have already noted, there are no static two-kink solutions and this is consistent with the fact that (5.51) degenerates in the limit $v \to 0$.

In order to interpret this solution it is useful to rewrite it in the form

$$\tan \frac{\psi}{4} = e^{\gamma(x-a)} - e^{-\gamma(x+a)} \tag{5.52}$$

where $a > 0$ is the time dependent function

$$a(t) = \frac{1}{\gamma} \log \left(\frac{2}{v} \cosh(\gamma v t) \right). \tag{5.53}$$

If $|vt| \gg 1$ then $a \sim |vt| + \delta$ is also large, where $\delta = -(\log v)/\gamma$. Considering the expression (5.52) in this limit we see that near the point $x = a$ the second term on the right-hand side is exponentially small, and may be neglected, and the remaining term describes a single kink moving with speed v and located at $x = a \sim |vt| + \delta$. Similarly, near $x = -a$ the first term may be neglected, leaving the second term, which describes a kink moving with speed v and having position $x = -a \sim -(|vt|+\delta)$. We therefore see that this solution describes two kinks which are well separated for $|t|$ large, with both approaching the origin at speed v for t negative, and separating at the same speed for t positive. They feel the repulsive kink-kink force, and smoothly bounce back off each other, the time of closest approach being $t = 0$. The solution is an even function of t so the motion is symmetric about $t = 0$. Note that the interpretation of the coordinate a as half the separation of two individual kinks is only valid when a is large, so it should not generally be used near $t = 0$ to estimate the distance of closest approach. In Fig. 5.2 we plot the energy density at various times for the two-kink solution (5.51) with $v = 0.2$. The total energy is 16γ. It is worth noting that the two-kink solution was found by Perring and Skyrme by performing numerical simulations and examining plots, such as those in Fig. 5.2, from which they were able to guess the exact solution.

Usually, this scattering solution is given a different interpretation. It is claimed that the two kinks pass through each other, rather than bouncing back, with the kinks accelerated through the collision process. The only effect of the collision is then that each kink position is shifted forward by an amount 2δ, as compared to the position the kink would have had if there had been no interaction at all. Since we are dealing with two identical solitons, there is no way to distinguish between forward and backward scattering, so both interpretations appear to be equally valid.

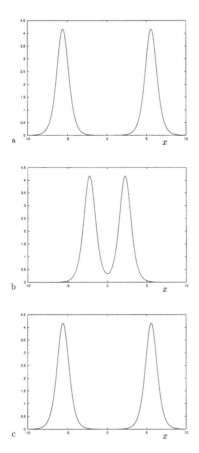

Fig. 5.2. The energy density of a sine-Gordon two-kink solution at times (a) $t = -20$, (b) $t = 0$, (c) $t = 20$. Each kink has an initial speed $v = 0.2$.

However, we have already seen that there is a repulsive force between two kinks, so, at least at low speeds, it clearly only makes physical sense if the two solitons scatter backwards, since they do not have enough kinetic energy to overcome the repulsive potential.

It is interesting to compare the exact two-kink solution with the approximate motion one would predict using the asymptotic interaction energy (5.41). For two kinks with positions $\pm a(t)$, the approximate equation of motion is

$$\ddot{a} = 4e^{-2a}, \tag{5.54}$$

where we have equated the force $F = 32e^{-2a}$ with the product of the kink acceleration $\ddot{a}$ and its mass, which is 8. If we impose $\dot{a}(-\infty) = -v$, so that the kinks each have an initial speed v, and $\dot{a}(0) = 0$, to fix the time of closest approach at $t = 0$, then the solution is

$$a(t) = \log \left(\frac{2}{v} \cosh(vt) \right) . \tag{5.55}$$

This is the low velocity limit of the exact expression (5.53), obtained by replacing the Lorentz factor γ by 1. Hence the approximate dynamics accurately models the true motion for low speeds $v \ll 1$, validating the asymptotic force law. The closest approach of the kinks is approximately $2 \log(2/v)$. One source of the error for relativistic speeds is that the two kinks do not remain far apart for all t, so that the terms neglected in the asymptotic expression for the force become significant.

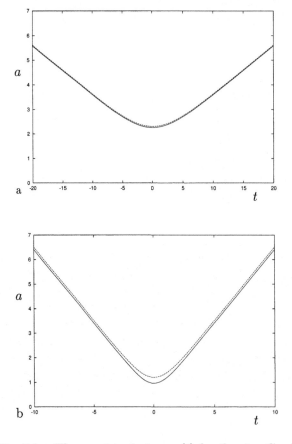

Fig. 5.3. The exact trajectory $a(t)$ for the sine-Gordon two-kink solution (solid curve) and the approximate trajectory (dashed curve) derived from the asymptotic force law. The speeds are (a) $v = 0.2$, (b) $v = 0.6$.

In Fig. 5.3 we display the exact kink trajectories given by expression

(5.53) and the approximate trajectories obtained from (5.55), for speeds $v = 0.2$ and $v = 0.6$.

The integrability of the sine-Gordon model also allows the construction, for example by the use of Bäcklund transformations, of explicit solutions describing kink-antikink scattering. Unlike most topological solitons, which at low speeds annihilate with antisolitons into radiation, the kink and antikink scatter elastically. This is a consequence of the infinite number of conserved quantities which prevent total annihilation. In fact, the solution describing the symmetric collision of a kink and antikink gives, at a particular time, a constant vacuum field, so one may think in terms of the instantaneous annihilation of the kink and antikink, but the two reappear at a later time with precisely the same form and speeds as they had initially. There is also an exact time periodic solution in the charge zero sector, known as a breather, which may be interpreted as a kink-antikink bound state, with the kink and antikink oscillating around their centre of mass.

In Chapter 4 we explained that in d space dimensions there are two main types of topological soliton – those in which the topology arises due to non-trivial vacuum values of a linear field at spatial infinity and the topological charge is an element of the homotopy group $\pi_{d-1}(\mathcal{V})$, and those in which there is a nonlinear field which is constant at infinity and the topological charge is associated with the mapping of the whole of space into a target manifold Y, which gives an element of the homotopy group $\pi_d(Y)$. So far in this chapter on kinks we have only encountered the first type. However, as we now describe, the sine-Gordon model has an alternative formulation as a nonlinear model with target space a circle, in which the kink becomes an example of the second type of soliton.

To formulate the sine-Gordon model as a nonlinear scalar field model, we introduce the two-component unit vector

$$\boldsymbol{\phi} = (\phi_1, \phi_2) = (\sin \phi, \cos \phi).$$ (5.56)

In terms of this field the sine-Gordon Lagrangian density (5.33) becomes

$$\mathcal{L} = \frac{1}{2}\partial_\mu \boldsymbol{\phi} \cdot \partial^\mu \boldsymbol{\phi} - (1 - \phi_2) + \nu(1 - \boldsymbol{\phi} \cdot \boldsymbol{\phi}),$$ (5.57)

where the Lagrange multiplier ν is introduced to constrain $\boldsymbol{\phi}$ to lie on the circle $\boldsymbol{\phi} \cdot \boldsymbol{\phi} = 1$. Finite energy now requires that the field takes the (unique) vacuum value $\boldsymbol{\phi} = (0, 1)$ at spatial infinity, which corresponds in the previous formulation to the field ϕ being an integer multiple of 2π at infinity.

Since $\boldsymbol{\phi}(-\infty)$ and $\boldsymbol{\phi}(\infty)$ must have the same value, the points $x = -\infty$ and $x = +\infty$ can be identified, thereby compactifying space from $\mathbb{R}$ to S^1.

At any given time, the field is therefore a map $\phi : S^1 \mapsto S^1$, where the domain is compactified space and the target is the unit circle. The map has an associated degree, or winding number, N, which also determines its class in $\pi_1(S^1) = \mathbb{Z}$. This may be computed as

$$N = \frac{1}{2\pi} \int_{-\infty}^{\infty} \varepsilon_{ab} \phi'_a \phi_b \; dx \qquad (5.58)$$

and is easily seen to be equal to the topological charge (5.36) defined previously. In the above expression ε_{ab} is the alternating tensor in two dimensions, with $\varepsilon_{12} = -\varepsilon_{21} = 1$, and all other components zero. In this formulation, the sine-Gordon model is a lower-dimensional analogue, with the addition of a potential term, of the sigma model we discuss in Chapter 6, and of the Skyrme model.

5.4 Generalizations

There are a number of generalizations of the kink models we have discussed so far. The most interesting are, of course, the higher-dimensional systems admitting localized topological solitons which are our main concern in the rest of this book. However, the kink solutions themselves can be trivially embedded into a higher-dimensional theory as solutions which are independent of all but one spatial direction. Thus, if we consider a general kink model of the form (5.1), but in three space dimensions, and require the field to be a function only of t and x^1, and independent of x^2 and x^3, then the field equation is the same as for one-dimensional kinks. The kink-like solutions now have infinite energy, because they have infinite extent in two spatial directions, but they have finite energy per unit area. These types of solution are known as domain walls and are of importance in condensed matter applications, since they can be formed in phase transitions. There are also possible cosmological applications, if phase transitions occurred in the early universe [407].

For a theory in which the potential U has more than two isolated degenerate minima an interesting phenomenon can occur for domain walls, namely, there can exist static solutions in which three or more domain walls meet at a junction. The simplest family of theories with solutions of this type [151, 73] has a single complex scalar field ϕ and Lagrangian density

$$\mathcal{L} = \frac{1}{4} \partial_\mu \bar{\phi} \partial^\mu \phi - |W'(\phi)|^2 \,, \qquad (5.59)$$

where $W(\phi)$ is a holomorphic function of ϕ, known as the superpotential.

For the present discussion we restrict to static configurations and consider fields which are independent of x^3, so that effectively we have a model in the (x^1, x^2) plane. Here, a single domain wall lies along a line and the field ϕ has a non-trivial dependence only on the transverse spatial coordinate, but a domain wall junction involves several domain walls embedded in different directions, and meeting near a point, so ϕ has non-trivial dependence on x^1 and x^2.

Let z be the complex coordinate in the plane, $z = x^1 + ix^2$, and denote the partial derivatives with respect to z and $\bar{z}$ by ∂_z and $\partial_{\bar{z}}$ respectively, i.e.

$$\partial_z = \frac{1}{2}(\partial_1 - i\partial_2), \quad \partial_{\bar{z}} = \frac{1}{2}(\partial_1 + i\partial_2). \tag{5.60}$$

Then the static energy associated with (5.59) may be written as

$$E = \int \left(\frac{1}{2}(|\partial_z\phi|^2 + |\partial_{\bar{z}}\phi|^2) + |W'(\phi)|^2 \right) d^2x. \tag{5.61}$$

Variation of this energy produces the second order field equation

$$\partial_{\bar{z}}\partial_z\phi = \overline{W''(\phi)}W'(\phi). \tag{5.62}$$

There is a 1-parameter family of Bogomolny equations associated with this system, given by

$$\partial_z\phi = e^{i\alpha}\overline{W'(\phi)} \tag{5.63}$$

where α is a constant phase. It is simple to verify that solutions of this first order Bogomolny equation satisfy the second order field equation. Explicitly,

$$\partial_{\bar{z}}\partial_z\phi = e^{i\alpha}\partial_{\bar{z}}\,\overline{W'(\phi)} = e^{i\alpha}\overline{W''(\phi)}\,\partial_{\bar{z}}\bar{\phi} = \overline{W''(\phi)}W'(\phi), \tag{5.64}$$

where the penultimate expression is obtained using the holomorphic property of the superpotential, $\partial\overline{W'(\phi)}/\partial\phi = 0$, and the final expression makes use of the complex conjugate of the Bogomolny equation. Note that the phase factor $e^{i\alpha}$ in (5.63) can be removed by a phase rotation of the coordinate z, which corresponds to a spatial rotation in the plane.

Fields satisfying the anti-Bogomolny equations $\partial_{\bar{z}}\phi = e^{i\beta}\overline{W'(\phi)}$ also satisfy the second order field equation.

Consider now a single domain wall in this type of model, lying along the x^2-axis, and satisfying the Bogomolny equation (5.63). Since the field is independent of x^2, we have $\partial_z = \frac{1}{2}\partial_1$ and (5.63) reduces to a one-dimensional kink equation. (Notice that for a field independent of x^2, $\partial_z = \partial_{\bar{z}}$, so the domain wall simultaneously satisfies the anti-Bogomolny equation, with $\beta = \alpha$. A similar remark applies to domain walls in any direction, but $\beta \neq \alpha$ in general.) From the general discussion earlier in this chapter, we anticipate that there will be kink solutions which interpolate between pairs of distinct vacua of the potential $|W'|^2$ as x^1

covers the real line. The superpotential form of the model allows a simple, but important, observation regarding these kink solutions, which follows from the equation

$$|\partial_1 \phi|^2 = \partial_1 \phi \partial_1 \bar{\phi} = 2e^{i\alpha} \overline{W'(\phi)} \partial_1 \bar{\phi} = 2e^{i\alpha} \partial_1 \overline{W(\phi)}, \qquad (5.65)$$

where again we have made use of the Bogomolny equation and the fact that $W(\phi)$ is holomorphic. The upshot of this formula is that the combination $e^{i\alpha} \partial_1 \overline{W}$ is real, implying that the imaginary part of $e^{i\alpha} \overline{W}$ is constant as x^1 varies. This means that although the kink traces out a non-trivial curved path in the ϕ-plane connecting two distinct vacuum values, when viewed in the W-plane the path is simply a straight line.

To obtain a solution of the Bogomolny equation which is a domain wall junction we need a potential with at least three distinct vacua. To be specific, we choose the quartic superpotential

$$W(\phi) = \phi - \frac{1}{4}\phi^4 \qquad (5.66)$$

with vacuum values of the potential $|W'|^2$ occurring at the cube roots of unity $\phi = 1, \omega, \omega^2$, where $\omega = e^{2\pi i/3}$. There are therefore three types of domain wall separating pairs of vacua. It is now easy to imagine dividing the plane symmetrically into three sectors, as shown in Fig. 5.4, with a different vacuum value occurring in the interior of each sector and the boundary between any two sectors being the domain wall associated with the kink solution connecting the two corresponding vacua. This is a domain wall junction. There is no rigorous proof that such a solution of the Bogomolny equation (5.63) exists, but it has been constructed numerically [360] (and explicit solutions are known in a related model [320]). When pictured in W-space, the field configuration maps $\mathbb{R}^2$ into the interior of the equilateral triangle with vertices $\phi = 1, \omega, \omega^2$, whose sides are the straight lines associated with the three domain walls which form the junction, as described above. A domain wall which interpolates between the vacuum values ϕ_A and ϕ_B has tension (energy per unit length) $\mu_{AB} = |W(\phi_A) - W(\phi_B)|$. A junction in which three or more domain walls meet can be associated with a polygon in the W-plane – an equilateral triangle in the example above – and the angles between the domain walls are precisely those for which there is a balance of tensions. This means that the directions of the walls and the directions of the associated sides of the polygon are the same, up to a rigid rotation.

A network of junctions can be formed, connected by domain wall segments, and leading to tilings of the plane [360], but it is easily seen that such a network can not be a global solution of the first order Bogomolny equation. It is a solution of the second order field equation such that

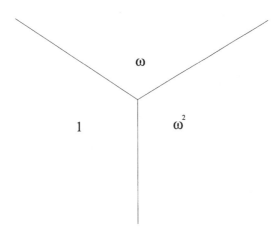

Fig. 5.4. A sketch of a domain wall junction. The lines represent the domain walls and the symbol in the interior of each sector denotes the vacuum value asymptotically attained there.

each junction is locally close to a solution of the Bogomolny equation $\partial_z \phi = e^{i\alpha} \overline{W'(\phi)}$ or the anti-Bogomolny equation $\partial_{\bar{z}} \phi = e^{i\beta} \overline{W'(\phi)}$, with these two possibilities alternating around the network, and the phases also differing from junction to junction. Note that because infinitely long domain walls satisfy both the Bogomolny and anti-Bogomolny equation, there is no contradiction here along each domain wall segment, provided it is long.

In the above, we effectively introduced a second real scalar field, by using a single complex scalar. A related generalization of scalar field theory is the extension of this idea, to include even more fields, with potentials such that the vacuum structure becomes more complicated and allows for the possibility of many different species of soliton, interpolating between the different vacua. Perhaps the mathematically richest class of such models is affine Toda field theory with imaginary coupling. There is an example of affine Toda field theory based on each compact semi-simple Lie algebra, and the number of fields is equal to the rank of this algebra. The potential is a sum of exponentials of the fields involving coefficients which depend on the algebra through data such as the simple roots. With real fields, this model has a unique vacuum, and hence no soliton solutions, but if the fields are allowed to be complex then the Toda potential has multiple stationary points and there are a number of topological charges, associated with the weights of the fundamental representations of the Lie algebra. Although allowing the fields to be complex means that the energy density is no longer real it turns out that the total energy and momentum of the soliton solutions are real, so these solitons are worth studying [189]. The affine Toda theory based on the simplest algebra A_1 has just a single field, and it reproduces the sine-Gordon model if the field is taken to be purely imaginary. Thus the general imaginary affine Toda field equation may be considered as a multi-field version of the sine-Gordon equation, and indeed it shares many of its properties, such as being integrable, which allows the construction of explicit soliton solutions. However, even at the classical level there are still a number of open puzzles concerning these theories, since the number of solitons that are expected, corresponding to the possible topological charges, is much greater than the number of solutions currently known; see ref. [92] for a review.

As a final generalization of kinks we consider a modification in which the spatial domain $\mathbb{R}$ is replaced by a circle, S^1, of finite radius. Solitons on compact and periodic domains are of physical interest because they model regions of high soliton density, where the solitons are expected to form a crystal. In the case of kinks, the crystal will be a one-dimensional kink chain.

An example of a kink chain occurs in the sine-Gordon model [300] if the field $\phi(t, x)$ is taken to be periodic in x with period L, modulo a 2π shift, that is,

$$\phi(t, x + L) = \phi(t, x) + 2\pi. \tag{5.67}$$

In the nonlinear interpretation of the sine-Gordon model, the field is strictly periodic. Let us now consider static fields, and suppress the time variable t. We can restrict to a unit cell, $x \in [0, L]$, with the boundary conditions $\phi(0) = 0$ and $\phi(L) = 2\pi$, so that this cell contains precisely one kink. The energy per kink is given by integrating the usual energy density over the unit cell,

$$E = \int_0^L \left(\frac{1}{2}\phi'^2 + 1 - \cos\phi \right) dx, \tag{5.68}$$

and the energy-minimizing field configuration will satisfy the static sine-Gordon equation

$$\phi'' = \sin\phi. \tag{5.69}$$

Integrating this equation once (and making a choice of sign) we arrive at

$$\phi' = 2\sqrt{\sin^2\frac{\phi}{2} + \frac{1 - k^2}{k^2}} \tag{5.70}$$

where $k \in (0, 1]$ is a constant of integration which is related to the period L, as we describe below. Comparing equations (5.70) and (5.38) we see that the Bogomolny bound is attained only if $k = 1$, which corresponds to the limit $L \to \infty$.

To solve Eq. (5.70) we make the changes of variable

$$\psi = \sin \frac{\phi}{2}, \quad X = \frac{x}{k}, \tag{5.71}$$

which transform the equation into

$$\left(\frac{d\psi}{dX} \right)^2 = (1 - \psi^2)(k^2 \psi^2 + 1 - k^2). \tag{5.72}$$

This is (see for example ref. [5]) the standard form of the equation satisfied by the Jacobi elliptic function $\mathrm{cn}_k(X)$, with modulus k. The solution of Eq. (5.70) is therefore

$$\phi(x) = 2 \sin^{-1} \mathrm{cn}_k \left(\frac{x - L/2}{k} \right), \tag{5.73}$$

where we have set the constant of integration equal to $-L/2$ in order to position the kink at the centre of the cell.

The period of the solution (5.73) must be equal to L, which gives the relation

$$L = 2k K_k \tag{5.74}$$

where K_k is the complete elliptic integral of the first kind with modulus k,

$$K_k = \int_0^{\frac{1}{2}\pi} \frac{d\theta}{\sqrt{1 - k^2 \sin^2 \theta}}. \tag{5.75}$$

Substituting the solution (5.73) into the expression (5.68) gives the energy per kink

$$E = \frac{8E_k - 4(1 - k^2)K_k}{k}, \tag{5.76}$$

where E_k is the complete elliptic integral of the second kind with modulus k,

$$E_k = \int_0^{\frac{1}{2}\pi} \sqrt{1 - k^2 \sin^2 \theta} \, d\theta. \tag{5.77}$$

In Fig. 5.5 we plot the energy (5.76) as a function of the period L. It tends to the Bogomolny bound, $E = 8$, in the limit $L \to \infty$ ($k \to 1$), and is strictly monotonically increasing as the period (and hence k) decreases, in accordance with the fact that there are repulsive forces between kinks.

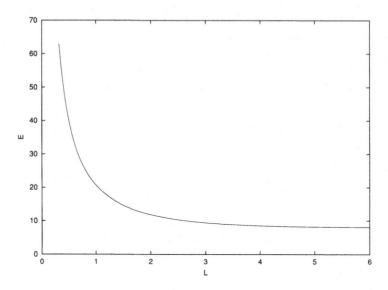

Fig. 5.5.   The energy per kink of the sine-Gordon kink chain as a function of the period L.

There are two limits in which the kink chain simplifies. The first is the small period limit $L \ll 1$, when the gradient term dominates the energy, leading to the asymptotic linear solution

$$\phi(x) = \frac{2\pi x}{L}. \tag{5.78}$$

The second is the infinite period limit, $L \to \infty$, when the solution (after a shift by $L/2$) reduces to the earlier expression (5.39) for a kink with position $a = 0$.

If the period is large, which corresponds to $0 \leq 1 - k \ll 1$, standard asymptotic expressions [5] for the elliptic integrals K_k and E_k can be used to obtain from (5.74) and (5.76) the asymptotic relations

$$k \sim 1 - 8e^{-L}, \quad E \sim 8 + 4(1 - k) \tag{5.79}$$

from which we can recover the asymptotic interaction energy of a kink pair

$$E_{\text{int}} = 32e^{-L} \tag{5.80}$$

by subtracting off the free kink energy, 8. This agrees with the earlier relation (5.41) after recognizing that in the periodic case L is also the separation between kinks.

6

Lumps and rational maps

6.1 Lumps in the $O(3)$ sigma model

One of the simplest systems admitting static topological soliton solutions is the $O(3)$ sigma model in the plane [436]. Strictly speaking, it is perhaps incorrect to use the term soliton for these solutions since, as we describe in detail below, they have an instability associated with changes in their scale. To reflect this lack of stability these structures are sometimes referred to as lumps, rather than solitons, and we will adopt this nomenclature here. Despite this shortcoming, it is still worthwhile studying these lumps, particularly because of the simplicity in constructing exact solutions. Lump solutions are given explicitly by rational maps between Riemann spheres, and since rational maps will play a vital role in later chapters on three-dimensional topological solitons, it is useful to familiarize ourselves with their properties in this concrete setting.

A sigma model is a nonlinear scalar field theory, where the field takes values in a target space which is a curved Riemannian manifold, usually with a large symmetry. The simplest example is the $O(3)$ sigma model, in which the target space is the unit 2-sphere, S^2. To formulate the model we parametrize the field as a three-component unit vector, $\boldsymbol{\phi} = (\phi_1, \phi_2, \phi_3)$. The Lagrangian density, for the model in $(d+1)$-dimensional Minkowski space-time, is simply that of a massless free theory

$$\mathcal{L} = \frac{1}{4}\partial_\mu \boldsymbol{\phi} \cdot \partial^\mu \boldsymbol{\phi} + \nu(1 - \boldsymbol{\phi} \cdot \boldsymbol{\phi}) \tag{6.1}$$

with the constraint $\boldsymbol{\phi} \cdot \boldsymbol{\phi} = 1$ enforced by using the Lagrange multiplier ν. The resulting nonlinear Euler-Lagrange equation, after elimination of ν, is

$$\partial_\mu \partial^\mu \boldsymbol{\phi} + (\partial_\mu \boldsymbol{\phi} \cdot \partial^\mu \boldsymbol{\phi})\boldsymbol{\phi} = 0. \tag{6.2}$$

The dot product in $\partial_\mu \boldsymbol{\phi} \cdot \partial^\mu \boldsymbol{\phi}$ means that the Euclidean metric of $\mathbb{R}^3$ is being used, and this becomes the standard metric on the target S^2 when

131

the constraint $\phi \cdot \phi = 1$ is imposed. The $O(3)$ in the model's name refers to the global symmetry of the target S^2, corresponding to rotations

$$\phi \mapsto M\phi \tag{6.3}$$

where $M \in O(3)$ is a constant matrix. The "sigma" refers to the fact that the model is sometimes formulated in terms of fields (ϕ_1, ϕ_2, σ), where ϕ_1 and ϕ_2 are locally unconstrained and $\sigma = \sqrt{1 - \phi_1^2 - \phi_2^2}$ is dependent on these.

From now on we deal only with the situation in which space-time is (2+1)-dimensional, since there is then the following topological classification of finite energy fields. The energy of a static field configuration is

$$E = \frac{1}{4} \int \partial_i \phi \cdot \partial_i \phi \, d^2x, \tag{6.4}$$

where $i = 1, 2$ runs over the spatial indices only. For this to be finite ϕ must tend to a constant vector at spatial infinity, which without loss of generality we may take to be $\phi^\infty = (0, 0, 1)$. In the vacuum, ϕ takes this value everywhere. This boundary condition spontaneously breaks the $O(3)$ symmetry (6.3) to an $O(2)$ symmetry rotating the components ϕ_1, ϕ_2. Moreover, there is a compactification of space to $\mathbb{R}^2 \cup \{\infty\} \cong S^2$, so that at a fixed time the field ϕ is a based map $\phi : S^2 \mapsto S^2$. The relevant homotopy group is $\pi_2(S^2) = \mathbb{Z}$, which implies that each field configuration is characterized by an integer topological charge N, the topological degree of the map ϕ. This has an explicit representation as the integral

$$N = \frac{1}{4\pi} \int \phi \cdot (\partial_1 \phi \times \partial_2 \phi) \, d^2x \tag{6.5}$$

where the integrand is the pull-back of the normalized, standard area form on S^2. The charge N is interpreted as the number of lumps in the field configuration, since generically there are N well separated, localized regions where the energy density is concentrated, each supporting one unit of charge, though this simplified picture breaks down as lumps approach each other.

Since the static energy density is quadratic in spatial derivatives, and space is two-dimensional, a spatial rescaling does not change the energy. The model is in fact conformally invariant. This does not rule out static solutions, but it means that each solution lies in a 1-parameter family of solutions related by rescalings.

By integrating the inequality

$$(\partial_i \phi \pm \varepsilon_{ij} \phi \times \partial_j \phi) \cdot (\partial_i \phi \pm \varepsilon_{ik} \phi \times \partial_k \phi) \geq 0 \tag{6.6}$$

over the plane, and using the expressions (6.4) and (6.5) for the energy and topological charge, it is a simple exercise to derive the Bogomolny bound

$$E \geq 2\pi|N|\,, \tag{6.7}$$

which is a lower bound on the energy in terms of the number of lumps. Equality occurs if and only if the field is a solution of one of the first order Bogomolny equations

$$\partial_i \phi \pm \varepsilon_{ij} \phi \times \partial_j \phi = 0\,. \tag{6.8}$$

As pointed out by Belavin and Polyakov [46], the Bogomolny equations (6.8) are best analysed by making the following changes of variable. Let R denote the Riemann sphere coordinate on the target space, that is, $R = (\phi_1 + i\phi_2)/(1 + \phi_3)$, and let $z = x^1 + ix^2$ be the complex coordinate in the spatial plane. Generally, R is a function of z and $\bar{z}$. The Lagrangian density (6.1) now takes the form

$$\mathcal{L} = \frac{\partial_\mu R \partial^\mu \bar{R}}{(1 + |R|^2)^2} \tag{6.9}$$

and there is no constraint. The denominator factor means that the target space is still the unit 2-sphere with its standard metric. The Lagrangian density (6.9) is referred to as that of the $\mathbb{CP}^1$ sigma model, though it is equivalent to the $O(3)$ sigma model as far as classical solutions are concerned. In terms of $R(z, \bar{z})$, the above expressions for the energy and topological charge take the form

$$E = 2 \int \frac{|\partial_z R|^2 + |\partial_{\bar{z}} R|^2}{(1 + |R|^2)^2}\, d^2x\,, \quad N = \frac{1}{\pi} \int \frac{|\partial_z R|^2 - |\partial_{\bar{z}} R|^2}{(1 + |R|^2)^2}\, d^2x \tag{6.10}$$

where $\partial_z = \frac{\partial}{\partial z} = \frac{1}{2}(\partial_1 - i\partial_2)$ and $\partial_{\bar{z}} = \frac{\partial}{\partial \bar{z}} = \frac{1}{2}(\partial_1 + i\partial_2)$. Clearly, $E \geq 2\pi|N|$. The Bogomolny equation (6.8) (with the $+$ sign) is equivalent to the Cauchy-Riemann equation

$$\partial_{\bar{z}} R = 0 \tag{6.11}$$

whose solutions are holomorphic functions $R(z)$, that is, R is independent of $\bar{z}$. Choosing the $-$ sign in (6.8) gives the equation $\partial_z R = 0$ satisfied by antiholomorphic functions $R(\bar{z})$. It is immediately obvious from the expressions (6.10) that holomorphic or antiholomorphic functions saturate the Bogomolny bound, $E = \pm 2\pi N$, with the positive sign for the holomorphic case, which we restrict to from now on. Notice that for holomorphic functions the energy density is 2π times the topological charge density.

Because the function $R(z)$ is a map to the Riemann sphere, it is acceptable for R to have poles. If R has a pole at z_0, the image is simply the point $(0, 0, -1)$ on the target S^2, and neither the energy density nor the topological charge density has a singularity there. The requirements that $R(z)$ has a definite limit as $z \to \infty$, and that the total energy is finite, force $R(z)$ to be a rational function of z. Recall from Section 3.2 that a rational map is given by the ratio of two polynomials in the variable z,

$$R(z) = \frac{p(z)}{q(z)}, \tag{6.12}$$

where p and q have no common factors, that is, no common roots. Also, by counting preimages, the topological degree of the rational map is $k_{\mathrm{alg}} = \max\{\deg(p), \deg(q)\}$, the highest power of z in either the numerator or denominator. For a function $R(z)$, the expression for N in (6.10) can be rewritten as

$$N = \frac{1}{4\pi} \int \left(\frac{1 + |z|^2}{1 + |R|^2} |\partial_z R| \right)^2 \frac{2i \, dz \, d\bar{z}}{(1 + |z|^2)^2}, \tag{6.13}$$

which we recognize as the pull-back under the map R of the normalized area form $2i \, dR \, d\bar{R}/4\pi(1 + |R|^2)^2$ on the target sphere. Of course this definition of the degree agrees with the degree defined by counting preimages, so that $N = k_{\mathrm{alg}}$. (In the area element in (6.13) there is an implied wedge product, so $2i \, dz \, d\bar{z} = 2i(dx^1 + i \, dx^2)(dx^1 - i \, dx^2) = 4 \, dx^1 \, dx^2 = 4 \, d^2 x$.)

In summary, a rational map of degree N is a solution of the Bogomolny equation of the $O(3)$ sigma model with topological charge N and energy $2\pi N$. It is referred to as an N-lump solution. More general meromorphic functions satisfy the Bogomolny equation, but have infinite energy.

Lumps are also required to satisfy the boundary condition at infinity, $\phi(\infty) = \phi^\infty$, or equivalently, $R(\infty) = 0$. This base point condition means that N-lump solutions are in one-to-one correspondence with the space of based rational maps, satisfying $\deg(p) < \deg(q)$. For such maps, the denominator q can be normalized to be a monic polynomial of degree N, a polynomial $z^N + q_1 z^{N-1} + \cdots + q_N$ with leading coefficient 1, and the numerator is a polynomial $p_1 z^{N-1} + \cdots + p_N$ whose coefficients are not all zero. The condition that the numerator and denominator have no common roots is a single polynomial inequality in the coefficients

$$\mathrm{Res}(p_1, \ldots, p_N, q_1, \ldots, q_N) \neq 0, \tag{6.14}$$

where Res is called the resultant (formerly, eliminant) of p and q, and is given by the $(2N - 1) \times (2N - 1)$ determinant [362]

$$
\text{Res} = \begin{vmatrix}
p_N & \cdots & \cdots & p_1 & & & & \\
& p_N & \cdots & \cdots & p_1 & & & \\
& & \cdots & \cdots & \cdots & \cdots & & \\
& & & p_N & \cdots & \cdots & \cdots & p_1 \\
q_N & \cdots & \cdots & q_1 & 1 & & & \\
& q_N & \cdots & \cdots & q_1 & 1 & & \\
& & \cdots & \cdots & \cdots & \cdots & & \\
& & q_N & \cdots & \cdots & \cdots & q_1 & 1
\end{vmatrix}.
$$

(6.15)

The moduli space $\mathcal{M}_N$ of based rational maps of degree N is therefore naturally a complex manifold of complex dimension $2N$, and hence real dimension $4N$. It is the complement, in $\mathbb{C}^{2N}$, of the hypersurface defined by the equation $\text{Res} = 0$. Multiplying any rational map by a complex number of unit magnitude preserves the base point condition and has no effect on the energy density of the map. There is therefore an internal $U(1)$ global symmetry group, which acts on moduli space.

For $N = 1$, the four-dimensional moduli space is $\mathcal{M}_1 = \mathbb{C}^* \times \mathbb{C}$, where $\mathbb{C}^*$ is the set of non-zero complex numbers (which is also a multiplicative group). A point $(\lambda e^{i\chi}, a)$ in $\mathcal{M}_1$, with λ real and positive, χ real, and a complex, corresponds to the lump solution

$$
R(z) = \frac{\lambda e^{i\chi}}{z - a}.
$$

(6.16)

The energy density of this solution has the form of a rotationally symmetric lump, with a maximum at $z = a$. The constant a is therefore the position of the lump in the plane. Note that at the point $z = a$, $R(a) = \infty$ or equivalently $\phi = (0, 0, -1)$, so the position of the lump may also be defined as the point in space where the field takes the value on the target sphere antipodal to the vacuum value. The constant λ determines the radius of the lump. More precisely, the integral of the topological charge density over the disc $|z - a| \le \lambda$ is exactly $\frac{1}{2}$, so λ is a reasonable definition of the radius. The angle χ is the internal phase of the lump.

The lump solution (6.16) is rotationally symmetric about a in the following sense. For convenience, position the lump at the origin by setting $a = 0$. Then, under a spatial rotation $z \mapsto e^{i\theta}z$, the solution $R(z)$ is mapped to $e^{-i\theta}R(z)$. But this change can be removed by acting with an element of the internal $U(1)$ symmetry group of the model. In particular, this implies that the energy density is strictly invariant under rotations. The full symmetry group of the solution is $O(2)$, because in addition to the rotational symmetry, the solution also satisfies $R(\bar{z}) = \overline{R(z)}e^{2i\chi}$, so a reflection in space is equivalent to a reflection in the target space together with an internal phase rotation.

By definition, all points in $\mathcal{M}_1$ correspond to field configurations with the same energy, $E = 2\pi$, so we see that the energy of a single lump is independent of its radius. This is the source of the instability of a lump solution in dynamical processes, since collisions of lumps (or the interaction of a lump with radiation modes of the field) can lead to the radius of a lump tending either to zero or to infinity. Although this has not been rigorously proved analytically, there have been a number of different numerical studies [257, 262, 332, 203] and all the results support this conjectured behaviour, with the radius evolving essentially linearly with time. Thus lump collapse, leading to a singular field, can occur in a finite time.

A generic point in the $4N$-dimensional space $\mathcal{M}_N$ describes N well separated charge 1 lumps, with ϕ close to its vacuum value ϕ^∞ in between. The $4N$ parameters may be interpreted as a position, radius and phase for each of the N lumps. In terms of the degree N rational map $R(z)$, the positions of the lumps are the poles of the map, and the radius and phase of the lump associated with a particular pole are given by the modulus and phase of the residue of the pole. The condition that the lumps are well separated, which is required for this interpretation to be valid, is that the distance between any two poles is large in comparison with the modulus of any residue.

From this discussion it is natural to imagine that coincident lumps correspond to rational maps with higher order poles, so that a pole of order n may be interpreted as n lumps at the same position. This is indeed correct, although the lumps highly distort each other as they come close together, and the picture of individual radii and phases is not applicable. In particular, a solution of the form

$$R(z) = \frac{\lambda^N}{z^N} \tag{6.17}$$

describes N lumps coincident at the origin, but for $N > 1$ the energy density is zero at the origin and maximal on a circle of radius $\lambda\left((N-1)/(N+1)\right)^{1/2N}$, so the solution is a ring rather than a lump.

The dynamics of N lumps can be approximated by geodesic motion in the moduli space $\mathcal{M}_N$, with the metric determined by the restriction to the moduli space of the kinetic energy

$$T = \int \frac{|\partial_0 R|^2}{(1 + |R|^2)^2} \, d^2x. \tag{6.18}$$

Unlike for vortices or monopoles, as described in the following chapters, this procedure does not lead to a well defined metric in all directions in $\mathcal{M}_N$. The reason is that some tangent vectors have infinite length, that is, there are moduli space coordinates for which the kinetic energy associated with changing their values is infinite. One says that these coordinates have infinite inertia. The simplest example is on $\mathcal{M}_1$, whose four real coordinates appear in the explicit solution (6.16). If the radius λ is allowed to be time dependent then T is infinite, since the integral multiplying the term $\dot{\lambda}^2$ is divergent. So a charge 1 lump can not collapse or expand in the geodesic approximation. Similarly, a time dependent phase leads to infinite kinetic energy, so the phase can not change. Only changes in the position of the lump, a, lead to finite values of the kinetic energy, so within the geodesic approximation a single lump moves with constant velocity, with a fixed radius and phase. This is the slow motion approximation to the exact solution obtained by Lorentz boosting the static lump. The general situation, for $N \geq 1$, can be understood by expanding the rational map $R(z) = p(z)/q(z)$ in a series in $1/z$, that is, about the point at infinity. This gives

$$R(z) = \frac{c_1}{z} + \frac{c_2}{z^2} + \frac{c_3}{z^3} + \cdots \tag{6.19}$$

recalling that $R(\infty) = 0$. If c_1 is time dependent, then the leading contribution to the kinetic energy density for large $|z|$ is $|\dot{c}_1|^2/|z|^2$. Integrating this with the measure $d^2x \sim |z|\,d|z|$ gives a logarithmic divergence. Therefore c_1 has infinite inertia and must be constant in time. Equivalently, p_1, the leading coefficient of $p(z)$, must be constant in time.

Note that for a circularly symmetric multi-lump solution, such as (6.17) with $N > 1$, the kinetic energy associated with a time dependent λ (which is the radius of the ring up to a constant factor) is finite. For example, if $N = 2$ then the kinetic energy is simply

$$T = \pi^2 \dot{\lambda}^2 \tag{6.20}$$

so that the geodesic approximation leads to a radius which evolves linearly in time, either expanding indefinitely or collapsing to a point in finite time. The second possibility, in which the geodesic hits the boundary of moduli space in a finite time, shows that the moduli space $\mathcal{M}_2$ is geodesically incomplete. Generally, for $N > 1$, $\mathcal{M}_N$ is geodesically incomplete.

Although lumps have the tendency to shrink to zero radius (or to expand indefinitely), the scattering of two lumps can still be investigated within the geodesic approximation, provided the scattering takes place before the lumps shrink to a point. This aspect was first investigated by Ward [416] and later by Leese [258], who made a thorough investigation of the problem using numerical methods to compute a variety of geodesics. Here, we briefly recount the main features of the simplest head-on collision of two lumps.

The eight-dimensional moduli space $\mathcal{M}_2$ is parametrized by the complex constants $\beta, \gamma, \delta, \varepsilon$ in the general charge 2 lump solution

$$R(z) = \frac{\beta z + \gamma}{z^2 + \delta z + \varepsilon}. \qquad (6.21)$$

By fixing the centre of mass at the origin, the constant δ can be set to zero. At least one of the lumps shrinks to zero radius at points on the boundary of $\mathcal{M}_2$, where the numerator and denominator in the rational map (6.21) have a common root. This is given by the equation

$$\gamma^2 + \beta^2 \varepsilon = 0 \qquad (6.22)$$

which is a special case of the equation Res $= 0$, involving the resultant of the numerator and denominator. The inertia term for the parameter β is infinite, because β is the leading coefficient of the numerator, so β must take a fixed value. The two remaining parameters, γ and ε, are complex coordinates on a family of four-dimensional manifolds, Σ_β, labelled by the constant β. Let us now set $\beta = 0$. The metric on Σ_0 was computed explicitly in ref. [416] in terms of complete elliptic integrals. There is a two-dimensional totally geodesic submanifold $\widetilde{\Sigma}_0 \subset \Sigma_0$ obtained by imposing the reflection symmetry $x^2 \mapsto -x^2$, realized for the rational maps as the condition $\overline{R(\bar{z})} = R(z)$, which forces both γ and ε to be real. The interpretation of points in $\widetilde{\Sigma}_0$ as two well separated lumps is valid if $|\gamma| \ll |\varepsilon|$. The lumps are then positioned at the points $z = \pm i\sqrt{\varepsilon}$, with equal radii $|\gamma/2\sqrt{\varepsilon}|$. Thus if ε is negative the two lumps are located on the x^1-axis, whereas if ε is positive the lumps are on the x^2-axis. If $\varepsilon = 0$ then the solution is circularly symmetric, and the two lumps are coincident at the origin and form a ring.

In Fig. 6.1 we plot the energy density of the static two-lump solution with $\gamma = 1$ and three values of ε.

If we consider the head-on collision of two lumps approaching along the x^1-axis, then initially we have $\varepsilon < 0$ and $\dot{\varepsilon} > 0$. Provided that $\dot{\varepsilon}$ remains positive, ε passes through the value zero and changes sign. Thus in a head-on collision of this type one expects right-angle scattering through the ring solution; a generic feature of topological soliton dynamics which we will encounter again when we discuss vortices, monopoles and Skyrmions. A caveat to the above right-angle scattering result is that the lumps could shrink to a point before scattering takes place. The initial values of $\varepsilon, \dot{\varepsilon}, \gamma, \dot{\gamma}$ determine when (and if) the geodesic flow hits the boundary of $\mathcal{M}_2$, given by (6.22), whose intersection with $\widetilde{\Sigma}_0$ is the

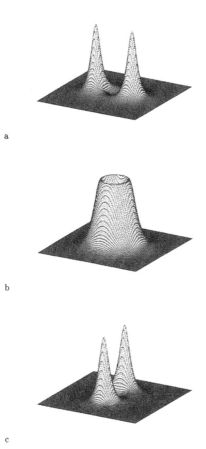

Fig. 6.1. The energy density of a static two-lump solution with parameters $\gamma = 1$ and (a) $\varepsilon = -1$, (b) $\varepsilon = 0$, (c) $\varepsilon = 1$.

line $\gamma = 0$. The three possibilities, each of which may be realized, are that the lumps shrink to zero size before scattering, after right-angle scattering, or not at all, in which case they scatter at right angles and expand indefinitely. A numerical simulation of the time dependent field equation [437] confirms both the dynamical behaviour found from the geodesic approximation and the fact that radiative effects are small.

More general geodesics in Σ_β lead to more complicated dynamics, which demonstrate that all the parameters of the lumps, in particular their phases, have a marked influence on the motion. For a detailed exposition we refer the reader to refs. [258, 416].

The metric on each $(4N-2)$-dimensional submanifold of $\mathcal{M}_N$, for which all inertia terms are finite, is Kähler [108]. To see this, introduce coordinates a_i and b_i which are the zeros and poles of the rational map, so that

$$R(z) = \frac{p(z)}{q(z)} = \frac{(z - a_1) \cdots (z - a_N)}{(z - b_1) \cdots (z - b_N)}. \tag{6.23}$$

The choice of base point condition is here $R(\infty) = 1$, rather than $R(\infty) = 0$ as earlier. The kinetic energy (6.18) is finite if the condition $\sum_{j=1}^{N} (\dot{a}_j - \dot{b}_j) = 0$ is imposed. It may be expressed as

$$
\begin{aligned}
T &= \int \frac{|\dot{p}(z)q(z) - \dot{q}(z)p(z)|^2}{(|p(z)|^2 + |q(z)|^2)^2} \, d^2x \\
&= \int \frac{1}{(|p(z)|^2 + |q(z)|^2)^2} \left| \dot{a}_i \frac{\partial}{\partial a_i} - \dot{b}_i \frac{\partial}{\partial b_i} \right|^2 |p(z)|^2 |q(z)|^2 \, d^2x \\
&= \dot{u}_\alpha \dot{\bar{u}}_\beta \int \frac{\partial}{\partial u_\alpha} \frac{\partial}{\partial \bar{u}_\beta} \log(|p(z)|^2 + |q(z)|^2) \, d^2x \\
&= g_{\alpha\beta} \dot{u}_\alpha \dot{\bar{u}}_\beta
\end{aligned}
\tag{6.24}
$$

where $u_\alpha = a_\alpha$ for $\alpha = 1, \ldots, N$, and $u_\alpha = b_{\alpha-N}$ for $\alpha = N+1, \ldots, 2N$. The metric therefore has the Kähler representation

$$g_{\alpha\beta} = \frac{\partial}{\partial u_\alpha} \frac{\partial}{\partial \bar{u}_\beta} \mathcal{K} \tag{6.25}$$

where $\mathcal{K}$ is the Kähler potential

$$\mathcal{K} = \int \log(|p(z)|^2 + |q(z)|^2) \, d^2x. \tag{6.26}$$

Note that this Kähler potential is divergent, but the divergent term is independent of the coordinates u_α, so does not contribute to the metric.

Ruback [351] was able to generalize the above result and show that the metric on the moduli space of holomorphic maps, defined by the sigma model kinetic energy, is Kähler whenever both the domain and target manifolds are Kähler.

In the following sections we discuss modifications of the $O(3)$ sigma model in which either the plane is replaced by a different domain or additional terms are included in the Lagrangian. Before this, we end this section by briefly describing the generalization in which the target space $\mathbb{CP}^1$ $(= S^2)$ is replaced by the complex projective space $\mathbb{CP}^m$.

$\mathbb{CP}^m$ has real dimension $2m$ and may be defined as the space of complex lines in $\mathbb{C}^{m+1}$. To introduce explicit coordinates one may use the equivalence class $[\mathbf{f}]$ of complex $(m + 1)$-component (column) vectors $\mathbf{f} = (f_1, \ldots, f_{m+1})$, with two vectors equivalent if one is a (non-zero) constant multiple of the other. One approach to define the $\mathbb{CP}^m$ sigma model is to introduce the $(m+1) \times (m+1)$ hermitian projector

$$P = \frac{\mathbf{f}\,\mathbf{f}^\dagger}{|\mathbf{f}|^2} \tag{6.27}$$

which clearly satisfies $P^2 = P = P^\dagger$. Note that all vectors $\mathbf{f}$ within a given equivalence class produce the same projector P, so that P is well defined on $\mathbb{CP}^m$. The $\mathbb{CP}^m$ sigma model has Lagrangian density

$$\mathcal{L} = \mathrm{Tr}(\partial_\mu P \partial^\mu P)\,, \tag{6.28}$$

which leads to the field equation

$$[\partial_\mu \partial^\mu P, P] = 0\,. \tag{6.29}$$

Our previous formulation of the $\mathbb{CP}^1$ model (6.9) can be recovered from this formulation by setting $\mathbf{f} = (1, R)$.

As with the $\mathbb{CP}^1$ model, the minimal energy static solutions are given by holomorphic (or antiholomorphic) maps, that is, vectors $\mathbf{f}$ in which each component is a meromorphic function of z. The requirement of finite energy, and the resulting boundary conditions, mean that these functions are rational. Using the projective equivalence of $\mathbf{f}$ we can multiply by a common denominator to make all the components of $\mathbf{f}$ into polynomials in z. For example, in the $\mathbb{CP}^1$ case, $\mathbf{f} = (1, R)$ is replaced by $\mathbf{f} = (q, p)$. Then the degree of the map, N, is the highest degree of these polynomials. The energy of these $\mathbb{CP}^m$ lump solutions is $2\pi N$ and they satisfy a first order Bogomolny equation. However, there are some differences between the $\mathbb{CP}^m$ model with $m > 1$ and the $\mathbb{CP}^1$ model. Perhaps the most important is that for $m > 1$ there are static solutions of the second order field equation (6.29) that are not solutions of the Bogomolny equation, in other words, not holomorphic. We shall discuss this aspect a little more in Chapter 11.

6.2 Lumps on a sphere and symmetric maps

The static $O(3)$ (or $\mathbb{CP}^1$) sigma model is conformally invariant, so replacing the domain $\mathbb{R}^2$ by the unit 2-sphere S^2 leads to the same static energy and Euler-Lagrange equation, although, as we shall see, the dynamics is different. To be explicit we will again use the Riemann sphere coordinate z on the domain, related to the standard angular coordinates θ, φ by the relation $z = \tan\frac{\theta}{2}\, e^{i\varphi}$. For holomorphic fields the static energy may then be written as

$$E = \frac{1}{2} \int \frac{|\partial_z R|^2 (1 + |z|^2)^2}{(1 + |R|^2)^2}\, \frac{2i\, dz\, d\bar{z}}{(1 + |z|^2)^2} \tag{6.30}$$

where we have isolated the term $2idzd\bar{z}/(1+|z|^2)^2$, which is the standard area 2-form $\sin\theta\, d\theta d\varphi$ on the unit sphere.

The compactification of the plane in the previous section meant that topologically we were already considering maps from a 2-sphere, but of infinite radius rather than unit radius. However, there is a subtle difference between the model defined on the plane and on the unit 2-sphere, since in the first case there is an arbitrary, but fixed, field value at spatial infinity which breaks the global $O(3)$ symmetry to an $O(2)$ symmetry. For maps from the unit sphere, there is no such symmetry breaking and the whole $O(3)$ symmetry group acts on field configurations, and on solutions. This means that on the sphere all rational maps correspond to lump solutions and there is no base point condition. The moduli space is therefore $(4N+2)$-dimensional. Moreover, the parameters of the rational maps can vary with time in an arbitrary way, and the kinetic energy remains finite.

In later applications to monopoles and Skyrmions we will be dealing with rational maps between Riemann spheres which are highly symmetric, so it is useful to introduce the main ideas here, where certain lump solutions on a sphere are given explicitly by such maps. Let us recall what it means for a rational map to be symmetric under a group $K \subset SO(3)$. This analysis was first presented in ref. [193], in the context of studying monopoles and Skyrmions.

Consider a spatial rotation $k \in SO(3)$, which acts on the Riemann sphere as an $SU(2)$ Möbius transformation

$$z \mapsto k(z) = \frac{\gamma z + \delta}{-\bar{\delta}z + \bar{\gamma}} \qquad \text{where} \qquad |\gamma|^2 + |\delta|^2 = 1. \qquad (6.31)$$

Similarly, a rotation $M \in SO(3)$ of the target 2-sphere acts as

$$R \mapsto M(R) = \frac{\Gamma R + \Delta}{-\bar{\Delta}R + \bar{\Gamma}} \qquad \text{where} \qquad |\Gamma|^2 + |\Delta|^2 = 1. \qquad (6.32)$$

A map is K-symmetric if, for each $k \in K$, there exists a target space rotation M_k which counteracts the effect of the spatial rotation, that is,

$$R(k(z)) = M_k(R(z)). \qquad (6.33)$$

Note that, in general, the rotations on the domain and target spheres will not be the same, so that $(\gamma, \delta) \neq (\Gamma, \Delta)$. However, for consistency, we require that the pairs (k, M_k) have the same composition rule as in K, so $M_{k_1}M_{k_2} = M_{k_1 k_2}$ $\forall k_1, k_2 \in K$. The map $k \mapsto M_k$ is therefore a homomorphism.

Since the realization of the $SO(3)$ action on the domain and target is by $SU(2)$ transformations, the group K should really be replaced by its double group in $SU(2)$, which we still call K. This is the group with twice as many elements, obtained by including both elements of $SU(2)$ which correspond to each element of $SO(3)$. In particular, it includes both ± 1 in $SU(2)$. We will then take the map $k \mapsto M_k$ to be a homomorphism of K into $SU(2)$. The fact that we are dealing with the double group is important since it has representations which are not representations of the original group. From now on it is to be understood that when we refer to a group K we actually mean the double group.

Determining the existence of symmetric rational maps, and computing particular ones, is a matter of classical representation theory. We are concerned with degree N polynomials in z, which form the carrier module for $\underline{N+1}$, the $(N+1)$-dimensional irreducible representation of $SU(2)$. To see this explicitly, introduce the homogeneous coordinates (z_0, z_1) on $\mathbb{CP}^1$, with $z = z_1/z_0$. Then $X_\pm$ and H, defined by

$$X_+ = z_1 \frac{\partial}{\partial z_0}, \quad X_- = z_0 \frac{\partial}{\partial z_1}, \quad H = -z_0 \frac{\partial}{\partial z_0} + z_1 \frac{\partial}{\partial z_1} \tag{6.34}$$

act on degree N homogeneous polynomials in z_0, z_1 and are a basis for $su(2)$ satisfying

$$[H, X_\pm] = \pm 2X_\pm, \quad [X_+, X_-] = H. \tag{6.35}$$

As a representation of $SU(2)$, $\underline{N+1}$ is irreducible, but if we consider the restriction to a subgroup $K \subset SU(2)$, $\underline{N+1}|_K$, this will in general be reducible. What we are interested in are its irreducible components, and tables of these subductions can be found, for example, in ref. [10].

The simplest case in which a K-symmetric degree N rational map exists is if

$$\underline{N+1}|_K = E \oplus \cdots \tag{6.36}$$

where E denotes any two-dimensional irreducible representation of K. In this case a basis for E consists of two degree N polynomials which can be taken to be the numerator p and denominator q of the rational map. Because E is a two-dimensional representation of K, each element k, through its action on (z_0, z_1), acts on this pair by a linear transformation

$$(p, q) \mapsto (ap + bq, cp + dq), \tag{6.37}$$

but this is precisely what can be compensated by M_k. (p here denotes the homogenized version of the polynomial $p(z)$, that is, $p_0 z_1^N + p_1 z_0 z_1^{N-1} + \cdots + p_N z_0^N$, and similarly for q.) A subtle point which needs to be addressed is that the two basis polynomials may have a common root, in which case the resulting rational map is degenerate and does not correspond to a genuine degree N map, but rather one of lower degree.

More complicated situations can arise. For example, if

$$\underline{N+1}|_K = A_1 \oplus A_2 \oplus \cdots \tag{6.38}$$

where A_1 and A_2 denote one-dimensional representations, then a whole 1-parameter family of maps can be obtained by taking an arbitrary, constant multiple of the ratio of the two polynomials which are the basis elements for A_1 and A_2. An n-parameter family of K-symmetric maps can be constructed if the decomposition contains $n+1$ copies of a two-dimensional representation, that is,

$$\underline{N+1}|_K = (n+1)E \oplus \cdots \tag{6.39}$$

where the n (complex) parameters correspond to the freedom in the choice of one copy of E from $(n+1)E$.

For a detailed explanation of how to calculate symmetric maps by computing appropriate projectors, using the characters of the relevant representations, see ref. [193].

The simplest example of a symmetric map is the spherically symmetric one-lump solution $R(z) = z$. Any rotation of the domain can be counteracted by performing the same rotation on the target. It is immediately clear from the energy formula (6.30) that this solution has an energy density distributed uniformly over the sphere. The solution $R(z) = \lambda z$, where λ is a real constant, is axially symmetric. The energy density at the North pole ($z = 0$) is $\lambda^2/2$ whereas at the South pole ($z = \infty$) it is $1/2\lambda^2$. Therefore if $\lambda^2 \gg 1$ the energy density is localized around the North pole and if $\lambda^2 \ll 1$ it is localized around the South pole.

For a degree 1 map the energy density is always positive over the whole sphere, but from the expression (6.30) it is clear that for a general map $R(z) = p(z)/q(z)$, the energy density will vanish at the zeros of the Wronskian

$$W(z) = p'(z)q(z) - q'(z)p(z) . \tag{6.40}$$

Generically, the Wronskian is a polynomial in z of degree $2N - 2$, where N is the degree of R, but it can have lower degree, in which case one says that it has roots at infinity. In this way, $W(z)$ always has $2N - 2$ zeros, counted with multiplicity. These zeros are the locations at which the multi-valued inverse of the map $R : S^2 \mapsto S^2$ has branch points. For a K-symmetric map the Wronskian changes only by a (non-zero) constant factor under a spatial rotation $k \in K$, because the linear transformation (6.37) just replaces W by $(ad - bc)W$, so the locations of its $2N - 2$ zeros are invariant under K. The zeros of the Wronskian are therefore an important characteristic of the map and will play a key role in later chapters of this book.

Table 6.1. Irreducible representations of T.

irreps of T	A	A_1	A_2	E'	E'_1	E'_2	F
dimension	1	1	1	2	2	2	3

Table 6.2. Irreducible representations of O.

irreps of O	A	A_1	E	E'_1	E'_2	F_1	F_2	G'
dimension	1	1	2	2	2	3	3	4

Table 6.3. Irreducible representations of Y.

irreps of Y	A	E'_1	E'_2	F_1	F_2	G	G'	H	I'
dimension	1	2	2	3	3	4	4	5	6

Rational maps with Platonic symmetries are particularly interesting, so we will briefly discuss the simplest examples here, using the formalism introduced above. For this purpose we need to recall some basic facts about the irreducible representations of the rotation groups of the Platonic solids. T, O and Y denote, respectively, the groups of rotational symmetries of the tetrahedron, the octahedron/cube, and the icosahedron/dodecahedron. In Tables 6.1, 6.2 and 6.3 we list our notation for the irreducible representations of T, O and Y, giving the dimension of each representation and denoting by a $'$ those which are only representations of the double group. For example, Table 6.1 summarizes the following information: T has three one-dimensional representations, which are the trivial representation A, and two conjugate representations A_1 and A_2. There is also a three-dimensional representation F, which is obtained as $\underline{3}|_T$, the restriction of the representation $\underline{3}$ of $SO(3)$ to the tetrahedral subgroup. In addition to these representations there are three two-dimensional representations of the tetrahedral double group, which we denote by E', E'_1 and E'_2. E' is obtained as $\underline{2}|_T$, the restriction of the fundamental representation of $SU(2)$ to T, and E'_1 and E'_2 are conjugate representations.

There are certain important polynomials, known as Klein polynomials [237], which form one-dimensional representations of the Platonic symmetry groups and are constructed as follows. Take the example of the tetrahedron. Scale a regular tetrahedron so that its vertices lie on the unit 2-sphere. Using the Riemann sphere coordinate z, the positions of the vertices correspond to four values $z_1, \ldots, z_4$. Now construct the unique monic polynomial of degree 4 which has these roots. In an appropriate orientation this procedure yields the Klein polynomial

$$T_v = z^4 + 2\sqrt{3}iz^2 + 1 \tag{6.41}$$

associated with the vertices of a tetrahedron. It is invariant under the action of any element t of T, possibly up to a constant factor, because t just permutes the roots. Applying the same procedure to the centres of the faces and to the mid-points of the edges of the tetrahedron (in the same orientation) produces the two further Klein polynomials

$$\begin{align} T_f &= z^4 - 2\sqrt{3}iz^2 + 1 \tag{6.42} \\ T_e &= z^5 - z. \tag{6.43} \end{align}$$

Note that a tetrahedron has six edges, but the polynomial T_e is only of degree 5. This is because in the orientation we have chosen, the mid-point of one of the edges is at the South pole, where $z = \infty$. So T_e should really be regarded as a degree 6 polynomial with one root at infinity. All three polynomials T_v, T_f, T_e transform as one-dimensional representations under the Möbius transformations of the tetrahedral group T, for the reason we just gave in the case of T_v. In fact they transform, respectively, as the representations A_1, A_2 and A.

Applying the above construction to the vertices, face centres and edge mid-points of the octahedron and icosahedron produces the Klein polynomials

$$\begin{align} \mathcal{O}_v &= z^5 - z \tag{6.44} \\ \mathcal{O}_f &= z^8 + 14z^4 + 1 \tag{6.45} \\ \mathcal{O}_e &= z^{12} - 33z^8 - 33z^4 + 1 \tag{6.46} \\ \mathcal{Y}_v &= z^{11} + 11z^6 - z \tag{6.47} \\ \mathcal{Y}_f &= z^{20} - 228z^{15} + 494z^{10} + 228z^5 + 1 \tag{6.48} \\ \mathcal{Y}_e &= z^{30} + 522z^{25} - 10005z^{20} - 10005z^{10} - 522z^5 + 1 \tag{6.49} \end{align}$$

where the notation is self-explanatory. Recall that the cube and dodecahedron are dual to the octahedron and icosahedron, respectively. Their Klein polynomials are just as above, but vertices are exchanged with face centres. For a tetrahedron, exchanging vertices and face centres gives another tetrahedron, rotated relative to the first one by 90°. These dual tetrahedra have the same edge mid-points, and together, their vertices are those of a cube, since $T_v T_f = \mathcal{O}_f$.

Let us now turn to a concrete example, and construct the rational map of lowest degree which is tetrahedrally symmetric. Of course, the spherically symmetric map $R = z$ is automatically K-symmetric for any $K \subset SO(3)$, but we ignore this degree 1 map. We have already seen that for a degree N map the $2N - 2$ zeros of the Wronskian must be strictly invariant, and for n points on a sphere to be invariant under the tetrahedral group requires that $n \geq 4$, with the lower limit $n = 4$ corresponding to placing the four points on the vertices of a tetrahedron. From this we see that T-symmetric maps must satisfy the condition that $N \geq 3$. We can check the first possibility, $N = 3$, by applying the group theory formalism developed above. The relevant decomposition is

$$\underline{4}|_T = E_1' \oplus E_2' \tag{6.50}$$

so there is a unique (up to orientations of the domain and target spheres) T-symmetric degree 3 map corresponding to the first component in the above decomposition. The map associated with the second component can be obtained from that of the first by a rotation of the domain, reflecting the fact that the representations E_1' and E_2' are conjugate by an $SO(3)$ element. The T-symmetric map associated with the E_1' representation must be a genuine degree 3 map, since the only T-symmetric map of lower degree is the $N = 1$ spherical map. To explicitly calculate the map, $R(z)$, one can apply the algorithm described in detail in ref. [193], given the characters of the representation E_1'. However, for a map of low degree it is simpler and more instructive to compute it by applying the group generators directly to a general map. For this example we begin by requiring $R(z)$ to be symmetric under two independent 180° rotations contained in the tetrahedral group. In terms of the Riemann sphere coordinates these two symmetries are realized by

$$R(-z) = -R(z) \qquad \text{and} \qquad R\left(\frac{1}{z}\right) = \frac{1}{R(z)}. \tag{6.51}$$

The first condition implies that the numerator of R is even in z and the denominator is odd, or vice versa. These two possibilities are related by a Möbius transformation, so we choose the former and ignore the latter. Imposing the second condition as well gives us maps of the form

$$R(z) = \frac{\sqrt{3}az^2 - 1}{z(z^2 - \sqrt{3}a)} \tag{6.52}$$

with a complex. The inclusion of the $\sqrt{3}$ factor is a convenience. Tetrahedral symmetry is obtained by imposing the further condition

$$R\left(\frac{iz + 1}{-iz + 1}\right) = \frac{iR(z) + 1}{-iR(z) + 1} \tag{6.53}$$

which is satisfied by (6.52) if $a = \pm i$, the two choices being related by the $90°$ spatial rotation $z \mapsto iz$, followed by a rotation of the target sphere. Note that $z \mapsto (iz + 1)/(-iz + 1)$ sends $0 \mapsto 1 \mapsto i \mapsto 0$ and hence generates the $120°$ rotation cyclically permuting three Cartesian axes.

The T-symmetric map

$$R(z) = \frac{\sqrt{3}iz^2 - 1}{z^3 - \sqrt{3}iz} \tag{6.54}$$

also has a reflection symmetry, represented by the relation $R(i\bar{z}) = i\overline{R(z)}$. This reflection extends the symmetry group T to T_d, where the subscript d denotes that the plane of the reflection symmetry contains a C_2-axis, which is the case for a tetrahedron. Alternatively, the group T could be extended by inversion $z \mapsto -1/\bar{z}$, which produces the group T_h, though from the above discussion it is clear that there are no T_h-symmetric maps of degree 3. The rotation groups O and Y can also be extended by inversion to produce the groups O_h and Y_h, which are the full symmetry groups of a cube and icosahedron, respectively.

It is interesting to look at the Wronskian of maps of the form (6.52),

$$W(z) = -\sqrt{3}a(z^4 + \sqrt{3}(a - a^{-1})z^2 + 1). \tag{6.55}$$

For $a = i$, W is proportional to the tetrahedral Klein polynomial $\mathcal{T}_v$, and for $a = -i$ it is proportional to $\mathcal{T}_f$. In both cases the zeros of the Wronskian are tetrahedrally invariant, as anticipated. Thus an examination of the Wronskian is an alternative, for fixing the coefficient a in the family of maps (6.52), to the slightly more complicated computation of imposing the $120°$ rotation symmetry (6.53) directly. From the Wronskian, we know that for $a = \pm i$ the energy density is minimal, and in fact zero, on the face centres of a tetrahedron. A calculation of the energy density reveals that it is maximal on the vertices of the same tetrahedron.

Turning our attention to octahedrally symmetric maps, the $2N - 2$ zeros of the Wronskian must be placed on the sphere with octahedral symmetry, which requires at least six points, when they can be located at the vertices of an octahedron. Thus the lowest possible degree for the map is $N = 4$. Then the decomposition required is

$$\underline{5}|_O = E \oplus F_2, \tag{6.56}$$

which demonstrates the existence of an O-symmetric degree 4 map associated with the two-dimensional representation E. A computation along the lines illustrated above produces the map

$$R(z) = \frac{z^4 + 2\sqrt{3}iz^2 + 1}{z^4 - 2\sqrt{3}iz^2 + 1} \tag{6.57}$$

which is in fact O_h-symmetric, due to the additional inversion symmetry $R(-1/\bar{z}) = 1/\overline{R(z)}$. The Wronskian of this map is proportional to the Klein polynomial $\mathcal{O}_v$, so the energy density is zero on the six face centres of a cube, and in fact it is maximal on the eight vertices of the cube.

Finally, for the icosahedral symmetry group the $2N - 2$ zeros of the Wronskian can be placed on the twelve vertices of an icosahedron if $N = 7$. The decomposition in this case is

$$\underline{8}|_Y = E'_2 \oplus I', \tag{6.58}$$

again demonstrating the existence of a unique Y-symmetric degree 7 map corresponding to the two-dimensional representation E'_2. This map can be written as

$$R(z) = \frac{z^7 - 7z^5 - 7z^2 - 1}{z^7 + 7z^5 - 7z^2 + 1} \tag{6.59}$$

and is Y_h-symmetric. Its Wronskian is proportional to the Klein polynomial $\mathcal{Y}_v$, with the energy density being zero on the twelve face centres of a dodecahedron, and maximal on its twenty vertices.

All these symmetric rational maps are examples of lump solutions on the 2-sphere. In addition to being symmetric, they illustrate that the energy density of a degree N lump is not necessarily localized around N points. Other examples of symmetric maps will be computed in later chapters, for their application to monopoles and Skyrmions.

So far we have only discussed static lumps on the 2-sphere, but to conclude this section we briefly mention the dynamics of lumps. Since the domain is compact, the metric induced on the moduli space $\mathcal{M}_N$ from the sigma model kinetic energy is well defined everywhere. The moduli space for a single lump, $\mathcal{M}_1$, is six-dimensional, and the whole $SO(3)$ symmetry group of the target sphere acts on it, not just the unbroken $SO(2)$ symmetry as in the plane. (We ignore the further symmetry combining spatial and target sphere reflections.) In most of this moduli space, the six coordinates may be interpreted as a position and size for the lump, plus the three Euler angles of the target space $SO(3)$ action. The metric on $\mathcal{M}_1$ has been computed explicitly and some geodesics identified [381, 31]. The metric is geodesically incomplete, reflecting the fact that a lump can shrink to a point in finite time, in agreement with what appears to happen in the plane, although in the latter case the geodesic approximation could not be applied to study this issue for a single lump. The lump dynamics is surprisingly rich, including not only motion at constant speed on a great circle but also more complicated dynamics in which a spinning lump bounces indefinitely between antipodal points on the sphere with a time dependent size.

The fixed point set of a symmetry group action is a totally geodesic submanifold, this being an example of the principle of symmetric criticality, so applying symmetries is a useful way to simplify lump dynamics in the geodesic approximation. In particular, if the fixed point set of a symmetry group is a real one-dimensional submanifold then this is automatically a geodesic and the metric need not even be computed. For example, a geodesic in the moduli space $\mathcal{M}_4$ can be obtained by considering tetrahedrally symmetric degree 4 maps (with fixed orientations in both the domain and target space). The appropriate decomposition is

$$\underline{5}|_T = A_1 \oplus A_2 \oplus F \,, \tag{6.60}$$

so there is a 1-parameter family of degree 4 maps obtained as constant multiples of the ratio of the basis polynomials for the one-dimensional representations A_1 and A_2. Since the basis polynomials for A_1 and A_2 are the tetrahedral Klein polynomials $\mathcal{T}_v$ and $\mathcal{T}_f$, the family of maps is

$$R(z) = c \, \frac{z^4 + 2\sqrt{3} i z^2 + 1}{z^4 - 2\sqrt{3} i z^2 + 1} \tag{6.61}$$

where c is a complex parameter. By imposing a reflection symmetry, c can be restricted to be real, and the maps are T_d-symmetric. This gives the desired geodesic. The map is degenerate for $c = 0$ and $c = \infty$, and a geodesic corresponds to c monotonically increasing (or decreasing) in the interval $c \in (0, \infty)$. The transformation $c \mapsto 1/c$ is equivalent to the $90°$ rotation $z \mapsto iz$, and when $c = 1$ the map (6.61) becomes the octahedrally symmetric map (6.57). An examination of the energy density shows that this geodesic describes a motion in which four highly localized lumps on the vertices of a tetrahedron (for $0 < c \ll 1$) spread out until the energy density is localized on the edges and especially the vertices of a cube ($c = 1$), and finally becomes highly localized once more, but this time around the vertices of the tetrahedron dual to the initial one ($c \gg 1$).

We can display this behaviour by plotting a surface whose height above the unit sphere is proportional to the energy density at that point on the sphere. Five such surfaces are shown in Fig. 6.2 corresponding to the values $c = \frac{4}{5}, \frac{9}{10}, 1, \frac{10}{9}, \frac{5}{4}$. The whole motion, $c \in (0, \infty)$, takes place in a finite time, confirming again the geodesic incompleteness of the moduli space. We will encounter this motion again in Chapter 8, where it will have a different interpretation in terms of monopole scattering, and takes an infinite time. Other geodesics can be obtained by a similar application of symmetries to maps of a particular degree, and we will see some of these later, again in the monopole context, which is where they were first considered. More complicated geodesics in the moduli space of higher charge lumps on a sphere could be investigated, but this has not been done.

Finally, it has been shown [359] that the metric on the moduli space $\mathcal{M}_N$ of the $\mathbb{CP}^1$ model whose domain is any compact Riemann surface

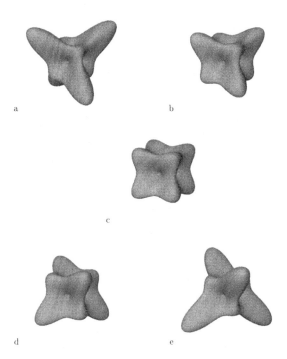

Fig. 6.2. Surfaces displaying the energy density for five different configurations of tetrahedrally symmetric 4-lumps.

is well defined but geodesically incomplete, so that (within the geodesic approximation) a lump can always shrink to a point in finite time. Also, the metric is well defined on the whole moduli space if gravitational self-interactions are included for the model in the plane [383], but again the metric is geodesically incomplete.

In the following section we discuss several modifications of the $\mathbb{CP}^1$ model in the plane, which remove the size instability of a lump, so that the lumps may properly be termed topological solitons.

6.3 Stabilizing the lump

The first, and perhaps most obvious, way in which the $O(3)$ sigma model can be modified to remove the size instability of a lump is by breaking the conformal invariance of the static energy. This is done by the introduction of extra terms into the Lagrangian which scale both as negative and positive powers of a spatial dilation factor. The example we consider here is known as the Baby Skyrme model [330] and is given by the Lagrangian density

$$\mathcal{L} = \frac{1}{4}\partial_\mu\phi \cdot \partial^\mu\phi - \frac{1}{8}(\partial_\mu\phi \times \partial_\nu\phi) \cdot (\partial^\mu\phi \times \partial^\nu\phi) - \frac{m^2}{2}(1 - \phi_3) \quad (6.62)$$

with the constraint $\phi \cdot \phi = 1$ implied. The name derives from the fact that it is a planar analogue of the three-dimensional Skyrme model, which we discuss in Chapter 9, and it has a smaller target space.

The first term in (6.62) is that of the $O(3)$ sigma model, the second term, known as a Skyrme term, is higher order in first derivatives, and the final contribution with no derivatives becomes a mass term for the fields ϕ_1, ϕ_2, when these are considered as small fluctuations around the vacuum $\phi = (0, 0, 1)$.

Although any term which is more than quadratic in first derivatives would suffice from the point of view of providing a preferred scale, the Skyrme term above is uniquely selected out as the lowest order Lorentz invariant expression that leads to a field equation involving time derivatives of no more than second order. The mass term is far from unique, and indeed any term which contains no field derivatives would do equally well. The particular mass term in (6.62) is motivated by an analogy with the one traditionally used in the three-dimensional Skyrme model, but other possibilities have been investigated [263, 388, 421]. The term in (6.62) has its minimum at just one point on the target, $\phi = (0, 0, 1)$; the alternatives have minima at the two points $(0, 0, \pm 1)$, or possibly on the whole circle $\phi_3 = 0$. These mass terms all depend just on ϕ_3, and therefore maintain the $O(2)$ symmetry between ϕ_1 and ϕ_2. Nevertheless, these different choices lead to very different qualitative results, such as whether multi-lump bound states exist or not, and from that point of view the model is not so elegant since there is no motivation to prefer one particular choice over another. In the three-dimensional Skyrme model a mass term is not essential, so this complication does not assume the same importance.

The static energy associated with the Lagrangian density (6.62) is

$$E = \int \left(\frac{1}{4}\partial_i\phi \cdot \partial_i\phi + \frac{1}{4}(\partial_1\phi \times \partial_2\phi) \cdot (\partial_1\phi \times \partial_2\phi) + \frac{m^2}{2}(1 - \phi_3) \right) d^2x \, .$$

$$(6.63)$$

Let us apply Derrick's scaling argument to this expression, denoting by E_j the contribution to the energy from the term which is of order j in derivatives. We see that under a rescaling, $\mathbf{x} \mapsto \mu\mathbf{x}$, the energy transforms to $e(\mu)$, where

$$e(\mu) = E_2 + \mu^2 E_4 + \frac{E_0}{\mu^2} \, . \quad (6.64)$$

For any map of non-zero degree, E_2, E_4 and E_0 are all positive. The minimum of the energy therefore occurs at a finite non-zero value of μ. A soliton solution will have a preferred size, at which the energy contributions from the Skyrme and mass terms, E_4 and E_0, are equal.

Of course, the topological classification of field configurations is the same as in the $O(3)$ sigma model, with the topological charge N given by (6.5), and since the additional terms can only increase the energy, the Bogomolny bound (6.7) is still valid, though it can no longer be attained.

The charge 1 soliton, known as a Baby Skyrmion, has the circularly symmetric form (see Eq. (4.46))

$$\phi = (\sin f(\rho) \cos \theta, \sin f(\rho) \sin \theta, \cos f(\rho)) \tag{6.65}$$

where ρ, θ are polar coordinates in the plane and $f(\rho)$ is a real profile function satisfying the boundary conditions $f(0) = \pi$ and $f(\infty) = 0$. Note that the single lump solution in the $O(3)$ sigma model can be written in this form, with $f(\rho) = \cos^{-1}((\rho^2 - \lambda^2)/(\rho^2 + \lambda^2))$, where λ is the radius of the lump. Substituting the ansatz (6.65) into the energy of the Baby Skyrme model (6.63) gives

$$E = \pi \int_0^\infty \left(\frac{1}{2} f'^2 + \frac{\sin^2 f}{2\rho^2}(1 + f'^2) + m^2(1 - \cos f) \right) \rho \, d\rho. \tag{6.66}$$

The profile function f must satisfy the variational equation

$$\left(1 + \frac{\sin^2 f}{\rho^2}\right) f'' + \left(1 - \frac{\sin^2 f}{\rho^2}\right) \frac{f'}{\rho} + \frac{\sin 2f}{2\rho^2}(f'^2 - 1) - m^2 \sin f = 0. \tag{6.67}$$

Linearizing this equation reveals the large ρ asymptotic behaviour

$$f(\rho) \sim \frac{A}{\sqrt{\rho}} e^{-m\rho} \tag{6.68}$$

so that the soliton has an exponential decay, unlike the algebraic decay of a lump. The profile function can only be obtained by solving equation (6.67) numerically, but this is a simple task using a shooting method, and leads to an energy of $E = 1.57 \times 2\pi$ for the choice $m^2 = 0.1$ [330]. The energy of a Baby Skyrmion therefore exceeds the Bogomolny bound by around 50%, although this value is highly dependent on the choice of m, with the energy approaching the Bogomolny bound 2π as $m \to 0$. However, the limit can not be reached, because the size of the Baby Skyrmion becomes infinite in the limit.

As in the pure sigma model, each soliton in the Baby Skyrme model has an internal phase corresponding to the freedom to rotate the components ϕ_1, ϕ_2. The interaction of two well separated solitons depends upon their relative phase, and by a suitable choice of this the asymptotic forces can be made attractive. This leads to multi-soliton bound states, with the minimal energy charge 2 soliton being circularly symmetric (the analogue of the lump solution $R(z) = \lambda^2/z^2$), but higher charge solutions are less symmetric. For more details we refer the reader to the papers [330, 331], in which soliton dynamics is also investigated, and it is verified that in a head-on collision of two solitons in the attractive channel there is the ubiquitous right-angle scattering.

Structures very similar to Baby Skyrmions are of importance in condensed matter physics where, amongst other applications, they play an important role in quantum Hall ferromagnets [380]. In a classical field theory approach to quantum Hall ferromagnets the static energy function is given by

$$E = \nu^2 \int \partial_i \phi \cdot \partial_i \phi \, d^2x + \eta^2 \int \int \frac{\phi^*(\mathbf{x})\phi^*(\mathbf{x}')}{|\mathbf{x} - \mathbf{x}'|} \, d^2x \, d^2x' + m^2 \int (1 - \phi_3) \, d^2x$$

(6.69)

where ν, η, m are constants and $\phi^* = \frac{1}{4\pi} \phi \cdot (\partial_1 \phi \times \partial_2 \phi)$ is the topological charge density. The first term in (6.69) is that of the pure sigma model, representing the exchange interaction between Heisenberg spins. The second term is a non-local Coulomb energy, with the electric charge density expressed in terms of the topological charge density since the solitons are electric charge carriers. The final term is the standard Zeeman energy for spins in a background magnetic field. The energy is very similar to that of the Baby Skyrme model, but the Skyrme term has been replaced by the non-local Coulomb term. The Coulomb term provides the same stabilizing effect as the Skyrme term, since it also scales as a positive power of the rescaling factor μ (but this time linearly, rather than quadratically). The circularly symmetric soliton solution of unit charge has been computed numerically [3] and recently multi-soliton solutions have also been investigated [410].

There is an alternative approach to stabilizing the sigma model lump due to Leese [259], which is mathematically elegant, and involves introducing a (classical) internal spin which prevents the lump from collapsing to a point. In terms of the $\mathbb{CP}^1$ formulation the model is defined by the Lagrangian density

$$\mathcal{L} = \frac{\partial_\mu R \partial^\mu \bar{R} - \alpha^2 |R|^2}{(1 + |R|^2)^2}$$

(6.70)

where α is a positive constant. Note that if $\alpha = 0$ then we recover the $\mathbb{CP}^1$ sigma model. If we were to consider static fields then the additional term would not stabilize a lump, since there is no higher order derivative term to provide a scaling balance. Instead, we consider time dependent fields of a very special form. The additional term respects the global $U(1)$

symmetry, $R \mapsto e^{i\chi}R$. The associated Noether charge, Q, is given by

$$Q = i \int \frac{\bar{R}\partial_0 R - R\partial_0 \bar{R}}{(1 + |R|^2)^2} \, d^2x \, . \tag{6.71}$$

If we recall the expression for the topological charge

$$N = \frac{i}{2\pi} \int \frac{\varepsilon_{ij}\partial_i R\partial_j \bar{R}}{(1 + |R|^2)^2} \, d^2x \, , \tag{6.72}$$

then the simple inequality

$$\int \frac{(\partial_i R \pm i\varepsilon_{ij}\partial_j R)(\partial_i \bar{R} \mp i\varepsilon_{ik}\partial_k \bar{R}) + |\partial_0 R \pm i\alpha R|^2}{(1 + |R|^2)^2} \, d^2x \geq 0 \tag{6.73}$$

implies a lower bound on the total energy

$$E \geq 2\pi|N| + |\alpha Q| \tag{6.74}$$

in terms of the topological and Noether charges. The bound (6.74) is clearly attained when

$$\partial_i R \pm i\varepsilon_{ij}\partial_j R = 0 \quad \text{and} \quad \partial_0 R \pm i\alpha R = 0 \, . \tag{6.75}$$

The first equation in (6.75) is the same as in the pure sigma model and requires that in its spatial dependence, R is a holomorphic (or antiholomorphic) function of $z = x^1 + ix^2$. The second equation is new and implies that the field has an internal spin, a constant motion in the global $U(1)$ phase direction with frequency $\pm\alpha$. Choosing both N and Q to be positive we have the solution

$$R(t, z) = e^{-i\alpha t} R_0(z) \tag{6.76}$$

where $R_0(z)$ is a degree N rational map in z, with the base point condition $R_0(\infty) = 0$.

These solutions were termed Q-lumps in ref. [259], in analogy with a type of non-topological soliton known as a Q-ball [349, 256, 86], which is stabilized by a Noether charge alone and similarly spins in internal space. Despite these similarities, the properties and dynamics of Q-lumps [259] are very different from those of Q-balls [44] (even in $(2 + 1)$ dimensions), due to the topological aspects of Q-lumps.

Although Q-lumps are not static solutions their time dependence resides only in the internal phase, which means that all physical quantities, such as the energy density, are static. These solutions may therefore be termed stationary.

As the Q-lump solutions (6.76) minimize the energy for fixed values of Q and N, they are automatically critical points of the constrained variational problem and hence solve the second order field equation. Moreover, because they represent the global energy minima (for given Q and N) there can be no unstable modes. The zero modes associated with scale invariance of lumps in the pure sigma model are absent for Q-lumps, due to the fact that the internal spin lifts the degeneracy between solitons of different radii. This is illustrated by the family of circularly symmetric Q-lumps

$$R(t, z) = \frac{e^{-i\alpha t}\lambda^N}{z^N}.$$
(6.77)

Using formula (6.71) we find this solution carries the λ-dependent Noether charge

$$Q = \frac{2\pi^2\alpha\lambda^2}{N^2}\operatorname{cosec}\frac{\pi}{N},$$
(6.78)

and the energy bound (6.74) is attained, with E depending non-trivially on λ.

Q-lumps are therefore stable against a change in radius, as are Baby Skyrmions, but in contrast to the latter the radius is not fixed but is determined by the value of the Noether charge Q, which is a free parameter. Note that equation (6.78) shows that a unit charge Q-lump ($N = 1$) is not a finite energy configuration, so we must restrict to $N > 1$. Unit charge Q-lumps can occur, well separated from other Q-lumps, but they must be part of a larger configuration with total topological charge $N > 1$ and a rational map $R_0(z)$ which has no $1/z$ term in its expansion around $z = \infty$.

The scattering of Q-lumps can be investigated using the geodesic approximation, which reveals that even in a head-on collision the scattering can be exotic, due to the internal spin of the Q-lumps. For more details we refer the reader to ref. [259].

The $\mathbb{CP}^1$ Q-lumps discussed above can be generalized to a whole class of Kähler sigma models with potential terms, provided the target manifold has a Killing vector field with at least one fixed point [4].

A final method of stabilizing the $O(3)$ sigma model lump is by gauging the unbroken $U(1)$ symmetry and including a Maxwell term for the gauge field [366]. The static energy in this model is given by

$$E = \frac{1}{4}\int\left(B^2 + D_i\phi \cdot D_i\phi + (1 - \phi_3)^2\right)d^2x,$$
(6.79)

where a_i is the $U(1)$ gauge potential with magnetic field $B = \partial_1 a_2 - \partial_2 a_1$, and again the constraint $\phi \cdot \phi = 1$ is implied. The covariant derivative $D_i\phi$ is given by

$$D_i\phi = \partial_i\phi + a_i\,\phi^\infty \times \phi$$
(6.80)

where, as earlier, $\phi^\infty = (0, 0, 1)$ is the vacuum vector.

As for Q-lumps, there is no finite energy solution for a single gauged lump, but for $N > 1$ there is a whole moduli space of gauged lumps, with energy $E = 2\pi N$, satisfying Bogomolny equations. The Bogomolny equations for positive N are

$$D_1\phi = -\phi \times D_2\phi \tag{6.81}$$
$$B = 1 - \phi_3 \tag{6.82}$$

but unfortunately, unlike in the pure sigma model or its Q-lump version, these Bogomolny equations are not integrable, so the solutions must be constructed numerically. Thus gauged lumps are similar to vortices in this respect. Another similarity with vortices is that the solitons carry magnetic flux, though unlike for vortices, the flux is related to the size of the soliton, and is not quantized. The energy is degenerate with respect to the size, or equivalently the flux, but there is no zero mode instability, since it requires infinite energy to change the flux.

7

Vortices

7.1 Ginzburg-Landau energy functions

Vortices are solitons in two dimensions, with a finite core size. When considered in a three-dimensional theory vortices become extended objects whose core is a straight line or curve, but we shall present the two-dimensional theory, in which the vortices are particle-like. This theory also describes the three-dimensional situation where all the fields are constant in one spatial direction, so that one has one or more straight, parallel vortices. For the theory and applications of genuinely curved, interacting vortices in three dimensions (a spaghetti of vortices), see the book of Vilenkin and Shellard [407].

For most of this chapter, we suppose space is the two-dimensional plane $\mathbb{R}^2$. A general space-time point is denoted by x or $(t, \mathbf{x})$, and in Cartesian coordinates $\mathbf{x} = (x^1, x^2)$. Sometimes, the coordinates are combined as x^μ ($\mu = 0, 1, 2$) where $x^0 = t$. We shall sometimes identify $\mathbb{R}^2$ with the complex plane $\mathbb{C}$, and denote a spatial point by z, where $z = x^1 + ix^2$. Where there is the possibility of fields with circular symmetry, we shall use polar coordinates ρ, θ.

The basic field theory with vortices is one having a scalar field with two real components $(\phi_1(x), \phi_2(x))$, and the internal symmetry

$$\phi_a(x) \mapsto R_{ab}\phi_b(x) \tag{7.1}$$

where $R \in SO(2)$. This symmetry ensures that an individual vortex can be circularly symmetric.

It is very often convenient to combine the two field components into a single complex field

$$\phi(x) = \phi_1(x) + i\phi_2(x). \tag{7.2}$$

The $SO(2)$ rotation

$$R_{ab}(\alpha) = \begin{pmatrix} \cos\alpha & -\sin\alpha \\ \sin\alpha & \cos\alpha \end{pmatrix} \tag{7.3}$$

is then replaced by the $U(1)$ phase rotation

$$\phi(x) \mapsto e^{i\alpha}\phi(x). \tag{7.4}$$

Heuristically, a field configuration contains a vortex centred at a point $\mathbf{X}$ if $\mathbf{X}$ is an isolated point where ϕ vanishes, and if along a small circle enclosing $\mathbf{X}$ anticlockwise, the phase of ϕ increases by $2\pi n$, with n a non-zero integer called the multiplicity of the vortex. n is generically ± 1, since higher multiplicity vortices break up under small perturbations of the field.

Field theories with vortices are of two types, global and gauged, and their solutions are called, correspondingly, global vortices and gauged vortices. In a global theory there is only the complex scalar field $\phi(x)$. In a gauged theory this is coupled to an electromagnetic field with gauge group $U(1)$. The fields are now $\phi(x)$ and the electromagnetic gauge potential $a_\mu(x)$, with time and space components $a_0(x)$ and $\mathbf{a}(x) = (a_1(x), a_2(x))$.

Several types of vortex dynamics may be considered, depending on the physical application. There are correspondingly several types of Lagrangian and field equations, some Lorentz invariant, and others not. There is also dissipative vortex dynamics, which has no Lagrangian formulation.

We start by discussing static fields. The expressions defining their energy are known as Ginzburg-Landau (GL) energy functions [152]. In a global theory, the GL energy is of the form

$$V = \int \left(\frac{1}{2}\boldsymbol{\nabla}\bar{\phi} \cdot \boldsymbol{\nabla}\phi + U(\bar{\phi}\phi) \right) d^2x, \tag{7.5}$$

which is invariant under the global internal symmetry (7.4). The single complex field equation, obtained by varying V with respect to $\bar{\phi}$, is

$$\nabla^2\phi - 2U'(\bar{\phi}\phi)\phi = 0, \tag{7.6}$$

where ∇^2 is the two-dimensional Laplacian. Its real and imaginary parts are what one obtains by working directly with ϕ_1 and ϕ_2. The equation obtained by varying V with respect to ϕ is the complex conjugate of (7.6),

$$\nabla^2\bar{\phi} - 2U'(\bar{\phi}\phi)\bar{\phi} = 0, \tag{7.7}$$

and is automatically satisfied if (7.6) is satisfied.

U depends only on $|\phi|$, the magnitude of ϕ, and is usually assumed to be a polynomial of at most quartic or sextic order in $|\phi|$, which means it is quadratic or cubic in $|\phi|^2 = \bar{\phi}\phi$. In the former case,

$$U = \mu + \nu\bar{\phi}\phi + \frac{\lambda}{8}(\bar{\phi}\phi)^2 \tag{7.8}$$

where μ, ν, λ are real, constant coefficients. (The factor $\frac{1}{8}$ will be convenient later.) For a stable theory λ must be positive. We adjust μ so that $U_{\min}$, the minimal value of U, is zero. Thus, for ν negative, we can rewrite U in the form

$$U = \frac{\lambda}{8}(m^2 - \bar{\phi}\phi)^2 \tag{7.9}$$

where m is positive. The vacuum manifold $\mathcal{V}$ is the circle $|\phi| = m$, with $\pi_1(\mathcal{V}) = \mathbb{Z}$. This is the interesting case, with the possibility of vortices. If ν is positive or zero, then $\mathcal{V}$ is the one point $\phi = 0$, with trivial first homotopy group. There is no possibility of topological solitons in the plane in this case, so we do not discuss it further.

For U of the form (7.9) the global GL energy is

$$V = \frac{1}{2}\int \left(\boldsymbol{\nabla}\bar{\phi}\cdot\boldsymbol{\nabla}\phi + \frac{\lambda}{4}(m^2 - \bar{\phi}\phi)^2\right) d^2x \tag{7.10}$$

and the field equation simplifies to

$$\nabla^2\phi + \frac{\lambda}{2}(m^2 - \bar{\phi}\phi)\phi = 0, \tag{7.11}$$

which is the classic, complex Ginzburg-Landau equation in two dimensions. For the remainder of this chapter we shall assume that in both the global and gauged theory U has the symmetry-breaking, quartic form (7.9), unless explicitly stated otherwise.

The vacuum solutions, which minimize V, are of the form $\phi = me^{i\chi}$, where χ is an arbitrary phase which must be constant for the gradient energy to vanish. The choice of χ spontaneously breaks the global $U(1)$ symmetry. Derrick's theorem, discussed in Section 4.2, actually rules out any other finite energy, static solutions in the global theory. By the scaling argument, such a solution has to satisfy

$$E_0 = \frac{\lambda}{8}\int(m^2 - \bar{\phi}\phi)^2 d^2x = 0, \tag{7.12}$$

so $|\phi| = m$ everywhere. If we write $\phi = me^{i\chi}$, and substitute into (7.11), we find that $\nabla^2\chi = 0$ and $\boldsymbol{\nabla}\chi \cdot \boldsymbol{\nabla}\chi = 0$. Therefore χ is constant, so ϕ is a vacuum solution. We shall return to the global theory later, and see that interesting vortex solutions of the field equation do exist, but they have logarithmically divergent energies.

The Ginzburg-Landau energy in the gauged theory is

$$V = \frac{1}{2} \int \left(B^2 + \overline{D_i \phi} D_i \phi + \frac{\lambda}{4} (m^2 - \bar{\phi}\phi)^2 \right) d^2 x, \tag{7.13}$$

the two-dimensional version of (2.102). This is invariant under a gauge transformation

$$\phi(\mathbf{x}) \quad \mapsto \quad e^{i\alpha(\mathbf{x})} \phi(\mathbf{x}) \tag{7.14}$$

$$a_i(\mathbf{x}) \quad \mapsto \quad a_i(\mathbf{x}) + \partial_i \alpha(\mathbf{x}), \tag{7.15}$$

where $e^{i\alpha(\mathbf{x})}$ is a spatially varying phase rotation. The ingredients of V are the gauge invariant quantity $\bar{\phi}\phi$, the covariant gradient of the scalar field $D_i \phi = \partial_i \phi - i a_i \phi$, and the magnetic field

$$B = f_{12} = \partial_1 a_2 - \partial_2 a_1. \tag{7.16}$$

In two space dimensions, the Maxwell field tensor has only three independent components. There is the single magnetic component B, which from a three-dimensional viewpoint is the magnetic field in the $-x^3$ direction. The spatial part of the field tensor can be expressed in terms of B as $f_{ij} = \varepsilon_{ij} B$. The two components of the electric field are $e_1 = \partial_0 a_1 - \partial_1 a_0$ and $e_2 = \partial_0 a_2 - \partial_2 a_0$, but they do not contribute to the static GL energy.

The field equations associated with the energy (7.13) are obtained by varying with respect to $\bar{\phi}$, a_1 and a_2 as the independent fields. They are

$$D_i D_i \phi + \frac{\lambda}{2} (m^2 - \bar{\phi}\phi)\phi \;\; = \;\; 0 \tag{7.17}$$

$$\varepsilon_{ij} \partial_j B + \frac{i}{2}(\bar{\phi} D_i \phi - \phi \overline{D_i \phi}) \;\; = \;\; 0. \tag{7.18}$$

Equation (7.18) is a two-dimensional version of Ampère's equation $\nabla \times \mathbf{b} = \mathbf{J}$. Therefore

$$J_i = \frac{i}{2}(\bar{\phi} D_i \phi - \phi \overline{D_i \phi}) \tag{7.19}$$

can be interpreted as the electric current in the plane. (The sign is consistent with (2.106) as a spatial index has been lowered.)

The vacuum is unique in the gauged GL theory. The energy is minimized if $|\phi| = m$, $D_i \phi = 0$ and $B = 0$, everywhere. The first condition requires that $\phi(\mathbf{x}) = m e^{i\chi(\mathbf{x})}$, and the last condition requires a_i to be pure gauge, i.e.

$$a_i(\mathbf{x}) = \partial_i \alpha(\mathbf{x}). \tag{7.20}$$

$D_i \phi$ then vanishes if

$$im(\partial_i \chi - \partial_i \alpha)e^{i\chi} = 0, \tag{7.21}$$

so $\partial_i(\chi - \alpha) = 0$ and therefore $\alpha = \chi + $ const. Thus a vacuum field is of the form

$$\phi = me^{i\chi}, \qquad a_i = \partial_i \chi . \qquad (7.22)$$

By the gauge transformation $e^{-i\chi}$, it becomes the simple vacuum

$$\phi = m, \qquad a_i = 0 . \qquad (7.23)$$

The field $\phi = me^{i\chi}$, $a_i = 0$, with χ a constant, is a vacuum too, but it is gauge equivalent to (7.23). So the vacuum is unique. Nevertheless, the fact that the minima of $(m^2 - \bar{\phi}\phi)^2$ lie on a circle is still significant – it leads to vortices.

Note that the condition $D_i\phi = 0$ is by itself quite strong. It implies that $\bar{\phi}D_i\phi + \phi\overline{D_i\phi} = \bar{\phi}\partial_i\phi + \phi\partial_i\bar{\phi} = \partial_i(\bar{\phi}\phi) = 0$, so $|\phi|$ is constant. It also implies that

$$[D_1 , D_2]\phi = -if_{12} \phi = 0 , \qquad (7.24)$$

so $B = f_{12}$ must vanish wherever $\phi \neq 0$. Since $|\phi|$ is constant, $B = 0$ everywhere if $\phi \neq 0$ at just one point, and in particular if ϕ is required to be non-zero at spatial infinity.

Derrick's theorem does not rule out non-vacuum, finite energy solutions in the gauged GL theory. The scaling argument just shows that for such solutions, the two contributions to the energy

$$E_4 = \frac{1}{2} \int B^2 \, d^2x , \qquad E_0 = \frac{\lambda}{8} \int (m^2 - \bar{\phi}\phi)^2 \, d^2x \qquad (7.25)$$

are equal. Indeed, the gauged GL theory with $U = \frac{\lambda}{8}(m^2 - \bar{\phi}\phi)^2$ is the paradigm for a theory possessing topological solitons in two dimensions.

It will be helpful, in the following sections, to have an expression for the gauged and global GL energies in polar coordinates. Cartesian and polar coordinates and their differentials are related by

$$x^1 = \rho\cos\theta , \qquad x^2 = \rho\sin\theta \qquad (7.26)$$
$$dx^1 = \cos\theta \, d\rho - \rho\sin\theta \, d\theta , \qquad dx^2 = \sin\theta \, d\rho + \rho\cos\theta \, d\theta , \quad (7.27)$$

so $dx^1 \wedge dx^2 = \rho \, d\rho \wedge d\theta$. The 1-form gauge potential a is coordinate invariant, so

$$a = a_1 \, dx^1 + a_2 \, dx^2 = a_\rho \, d\rho + a_\theta \, d\theta , \qquad (7.28)$$

and therefore

$$a_\rho = a_1 \cos\theta + a_2 \sin\theta , \qquad a_\theta = -a_1\rho\sin\theta + a_2\rho\cos\theta . \qquad (7.29)$$

The field tensor has the one component $f_{\rho\theta} = \partial_\rho a_\theta - \partial_\theta a_\rho = \rho B$. The general energy expression for a gauge theory on a Riemannian manifold

X is the integral of the spatial part of the Lagrangian density (2.103). Here, X is $\mathbb{R}^2$, with the metric $d\rho^2 + \rho^2 d\theta^2$, so $h_{\rho\rho} = 1, h_{\theta\theta} = \rho^2, h_{\rho\theta} = 0$. The gauged GL energy in polars is therefore

$$V = \frac{1}{2} \int_0^\infty \int_0^{2\pi} \left(\frac{1}{\rho^2} f_{\rho\theta}^2 + \overline{D_\rho\phi} D_\rho\phi + \frac{1}{\rho^2} \overline{D_\theta\phi} D_\theta\phi + \frac{\lambda}{4} (m^2 - \bar\phi\phi)^2 \right) \rho d\rho\, d\theta \tag{7.30}$$

where the covariant derivatives are $D_\rho\phi = \partial_\rho\phi - ia_\rho\phi$ and $D_\theta\phi = \partial_\theta\phi - ia_\theta\phi$. The analogous expression in the global theory is

$$V = \frac{1}{2} \int_0^\infty \int_0^{2\pi} \left(\partial_\rho\bar\phi\partial_\rho\phi + \frac{1}{\rho^2}\partial_\theta\bar\phi\partial_\theta\phi + \frac{\lambda}{4}(m^2 - \bar\phi\phi)^2 \right) \rho\, d\rho\, d\theta. \tag{7.31}$$

7.2 Topology in the global theory

Let us consider a field configuration $\phi(\mathbf{x})$ in the global GL theory whose energy density approaches zero rapidly as $|\mathbf{x}| \to \infty$. From (7.31), the energy in polars, we see that $|\phi| \to m$ and $\partial_\rho\phi \to 0$ as $\rho \to \infty$. Let us assume that $\lim_{\rho\to\infty} \phi(\rho, \theta)$ exists. Denote the limiting form $\phi^\infty(\theta) = me^{i\chi^\infty(\theta)}$, and call this the value of ϕ on the circle at infinity. For such a field, ϕ^∞ is a map from the circle at infinity S^1_∞ to the vacuum manifold $\mathcal{V} = S^1$,

$$\phi^\infty : S^1_\infty \mapsto S^1. \tag{7.32}$$

ϕ^∞ is single-valued, so $\chi^\infty(\theta)$ must have the property $\chi^\infty(2\pi) = \chi^\infty(0) + 2\pi N$, for some integer N. N is the winding number, or degree, of the map (7.32), and is the topological charge of the field configuration.

Despite the non-trivial topology of the vacuum manifold, there are no *finite energy* field configurations with non-zero topological charge in the global theory. This is because the contribution of the angular gradient of ϕ to the energy density is $O\left(\frac{1}{\rho^2}\right)$ provided $\chi^\infty(\theta)$ is differentiable. Its contribution to the total energy, outside a circle of sufficiently large radius ρ_0, is

$$\frac{1}{2}m^2 \int_{\rho_0}^\infty \int_0^{2\pi} \frac{1}{\rho} (\partial_\theta\chi^\infty)^2 \, d\rho\, d\theta. \tag{7.33}$$

The angular and radial integrals separate, and the radial one is logarithmically divergent unless $\int_0^{2\pi} (\partial_\theta\chi^\infty)^2 \, d\theta$ vanishes. Thus finite energy requires that $\lim_{\mathbf{x}\to\infty} \phi = me^{i\chi^\infty}$ for some constant phase χ^∞, and hence $N = 0$. The vacuum manifold therefore plays no significant role for finite energy fields in the global theory.

7.3 Topology in the gauged theory

Suppose that $\{\phi(\mathbf{x}), a_i(\mathbf{x})\}$ is a finite energy field configuration in the gauged GL theory. Finite energy imposes the boundary condition $|\phi| \to m$ as $|\mathbf{x}| \to \infty$. We would like to deduce that ϕ has a limiting form on the circle at infinity. Because of gauge invariance this is not immediately possible. However, from the energy expression in polars, (7.30), we see that the radial covariant derivative $D_\rho \phi$ tends to zero as $|\mathbf{x}| \to \infty$. Consider any radial line $0 \le \rho < \infty$, with θ fixed. For large ρ, $\phi \sim m e^{i\chi}$, so asymptotically

$$D_\rho \phi = im(\partial_\rho \chi - a_\rho) e^{i\chi} = 0 \tag{7.34}$$

and therefore $a_\rho = \partial_\rho \chi$. Let us now transform the field to the radial gauge $a_\rho = 0$. If $a_\rho \ne 0$ initially, then the gauge transformation

$$g(\rho, \theta) = \exp\left(-i \int_0^\rho a_\rho(\rho', \theta) \, d\rho'\right) \tag{7.35}$$

does this. g is a smooth function everywhere, and $\lim_{\mathbf{x}\to 0} g(\mathbf{x}) = 1$. The gauge transformation also changes ϕ and a_θ, but these are still smooth functions in the plane if a_ρ was initially smooth. In the new gauge, $\partial_\rho \phi \to 0$ as $|\mathbf{x}| \to \infty$, so ϕ has a limiting value along each radial line

$$\lim_{\rho\to\infty} \phi(\rho, \theta) = \phi^\infty(\theta) = m e^{i\chi^\infty(\theta)}, \tag{7.36}$$

and this defines ϕ on the circle at infinity. The only remaining gauge freedom is to multiply ϕ by a constant phase factor $e^{i\alpha}$. One might imagine that gauge transformations of the type $e^{i\alpha(\theta)}$ are still allowed, but these are ill defined at $\mathbf{x} = \mathbf{0}$ if $\alpha(\theta)$ is a non-constant function. In this sense the radial gauge is different from an axial gauge, e.g. $a_1 = 0$, where more gauge freedom remains.

The finiteness of the integrals of $\frac{1}{\rho^2}\overline{D_\theta \phi} D_\theta \phi$ and $\frac{1}{\rho^2} f_{\rho\theta}^2$ implies that $D_\theta \phi \to 0$ and $f_{\rho\theta} \to 0$ as $|\mathbf{x}| \to \infty$. In the radial gauge, $f_{\rho\theta} = 0$ implies that $\partial_\rho a_\theta = 0$, so a_θ has a limit

$$\lim_{\rho\to\infty} a_\theta(\rho, \theta) = a_\theta^\infty(\theta) \tag{7.37}$$

which defines the gauge potential on the circle at infinity. The vanishing of $D_\theta \phi$ now implies that

$$\partial_\theta \chi^\infty - a_\theta^\infty = 0, \tag{7.38}$$

so the gauge potential equals the derivative of the phase of ϕ on the circle at infinity. Note the important difference from the global theory – it is not necessary for χ^∞ to be constant.

ϕ^∞ is again a map from the circle at infinity S^1_∞ to the vacuum manifold $\mathcal{V} = S^1$,

$$\phi^\infty : S^1_\infty \mapsto S^1 , \tag{7.39}$$

and has the integer winding number

$$N = \frac{1}{2\pi} \int_0^{2\pi} \partial_\theta \chi^\infty(\theta) \, d\theta = \frac{1}{2\pi} \left(\chi^\infty(2\pi) - \chi^\infty(0) \right) . \tag{7.40}$$

N is the topological charge of the field configuration $\{\phi(\mathbf{x}), a_i(\mathbf{x})\}$. The actual value of χ^∞ is not quite fixed, even in the radial gauge, because the constant gauge transformation $e^{i\alpha}$ shifts $\chi^\infty(\theta)$ to $\chi^\infty(\theta) + \alpha$.

N does not in fact depend on the gauge choice we have made. To see this, consider a smooth gauge transformation $g(\mathbf{x})$. For each $\rho \geq 0$, g is a map from the circle of radius ρ to $U(1)$, with winding number $N_g(\rho)$, say. This gauge transformation would change N to $N + N_g(\infty)$. However, by continuity, N_g is independent of ρ, and smoothness of g at the origin implies that $N_g(0) = 0$. So $N_g(\infty) = 0$, and hence the topological charge N is gauge invariant.

We have defined N in terms of the winding of the scalar field ϕ at infinity. However, because of the correlation (7.38) between the scalar field and gauge potential, N equals c_1, the first Chern number of the magnetic field. Recall from Section 3.4 that (by Stokes' theorem)

$$c_1 = \frac{1}{2\pi} \int_{\mathbb{R}^2} B \, d^2 x = \frac{1}{2\pi} \int_0^{2\pi} a_\theta^\infty(\theta) \, d\theta . \tag{7.41}$$

From (7.38) and (7.40) it follows that the last expression is equal to N. So magnetic flux is quantized in the gauged GL theory, in units of 2π, and the total flux Φ is $2\pi N$.

There is a third topological characterization of the winding number N. It is the total vortex number, that is, the number of points in the plane, with multiplicity taken into account, where $\phi = 0$. For this, we need to assume that the zeros of ϕ are a finite set of isolated points $\{A, B, C, \ldots\}$ with multiplicities $\{n_A, n_B, n_C, \ldots\}$ (see Fig. 7.1). The winding number of ϕ along the circle at infinity, N, is just the sum of these multiplicities

$$N = n_A + n_B + n_C + \cdots . \tag{7.42}$$

To see this, consider the deformation of the circle C_A into the curve C'_A. By continuity, the increase of the phase of ϕ around C_A, which is $2\pi n_A$, is the same as that around C'_A, as no new zero is enclosed. Then the increase of phase around C_{AB} is the sum of the increases around C'_A and C_B, and hence C_A and C_B. By extending this construction to enclose more zeros, and eventually all of them, we arrive at (7.42).

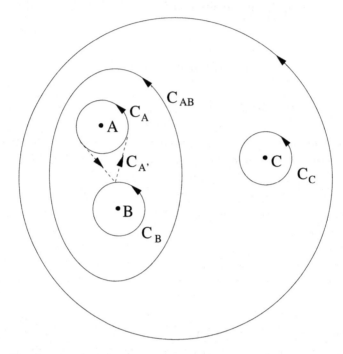

Fig. 7.1.   The points A, B, C denote isolated zeros of ϕ. The circle C_A is deformed into the curve C'_A. The increase of phase of ϕ around the curve C_{AB} is equal to the sum of the increases around the curves $C_{A'}$ and C_B, which is equal to the sum of the increases around the curves C_A and C_B.

A is a simple zero of ϕ if $n_A = \pm 1$. If ϕ has winding number $N > 0$, and only simple zeros, then there are at least N of them. If there are $N + N'$ zeros of multiplicity 1, there must be N' zeros of multiplicity -1. A sufficiently small, smooth deformation of a field with isolated simple zeros simply shifts those zeros. They therefore have a certain stability. However, by a continuous deformation of the field, it is possible to have simple zeros of multiplicity 1 coalesce into higher multiplicity zeros. Also zeros of opposite multiplicities can coalesce into a zero of ϕ with zero multiplicity, and then annihilate, leaving ϕ no longer vanishing at all in the neighbourhood of the annihilation point. Both these processes can also be reversed.

We shall see later that static and time dependent solutions of the GL equations generally have their energy density and magnetic field peaked around the zeros of ϕ. Hence the zeros of ϕ give not just global topological data, but also local dynamical information about the fields. We shall identify the zeros of ϕ with the locations of vortices. Note that the energy density at a zero of ϕ is at least $\frac{\lambda}{8} m^4$.

7.4 Vortex solutions

In the gauged GL theory, for all values of the coupling constant λ except $\lambda = 1$, all known finite energy static solutions of the field equations have circular symmetry about some point, and a reflection symmetry. By translational symmetry we can choose that point to be the origin. Some consequences of circular symmetry were presented in Section 4.3, but we shall go into further detail here. In the global GL theory too, there are circularly symmetric solutions, although with a logarithmically divergent energy. The basic solution in each case, with winding number $N = 1$, we shall call a vortex [6]. The discrete transformation $\phi \mapsto \bar{\phi}$, together with $\mathbf{a} \mapsto -\mathbf{a}$ in the gauged theory, converts a vortex into an antivortex, with the same energy and $N = -1$. Solutions with higher winding number, $|N| > 1$, are multi-vortices.

It is natural to discuss these circularly symmetric solutions in polar coordinates. Thus the fields are $\phi(\rho, \theta)$ in the global theory, supplemented by $a_\rho(\rho, \theta)$ and $a_\theta(\rho, \theta)$ in the gauged theory. We shall use the circular and reflection symmetries to obtain a reduced GL energy function, an integral just over the radial coordinate ρ. Its variational equations are the reduced field equations. By the principle of symmetric criticality, described in Section 4.4, solutions of these reduced equations give solutions of the full field equations in the plane.

The action of an element $R(\beta)$ of the spatial rotation group $SO(2)$ is $\{\rho \mapsto \rho,\ \theta \mapsto \theta + \beta\}$, and the operator generating rotations is $\frac{\partial}{\partial \theta}$. A scalar field configuration $\phi(\rho, \theta)$ is rotationally invariant in the naive sense if $\phi(\rho, \theta + \beta) = \phi(\rho, \theta)$ for all β, that is, ϕ depends only on ρ. Equivalently, $\frac{\partial \phi}{\partial \theta} = 0$. Such a ϕ has zero winding number. If ϕ is also assumed to have the reflection symmetry $\phi(\rho, \theta) = \bar{\phi}(\rho, -\theta)$, then it is real. The global GL energy for a field of this type is

$$V = \pi \int_0^\infty \left(\left(\frac{d\phi}{d\rho} \right)^2 + \frac{\lambda}{4}(m^2 - \phi^2)^2 \right) \rho\, d\rho, \tag{7.43}$$

and the corresponding variational equation is

$$\frac{d^2\phi}{d\rho^2} + \frac{1}{\rho}\frac{d\phi}{d\rho} + \frac{\lambda}{2}(m^2 - \phi^2)\phi = 0. \tag{7.44}$$

The energy is minimized by the vacuum solution $\phi = m$ (and the equivalent solution $\phi = -m$). Derrick's theorem, which is applicable for circularly symmetric fields with zero winding number, rules out any other non-singular solution with $|\phi| \to m$ as $\rho \to \infty$.

In the gauged theory, naive rotational invariance means that

$$\frac{\partial \phi}{\partial \theta} = \frac{\partial a_\rho}{\partial \theta} = \frac{\partial a_\theta}{\partial \theta} = 0. \tag{7.45}$$

ϕ, a_ρ and a_θ depend only on ρ, so $N = 0$. It is still possible to perform a ρ-dependent gauge transformation, without introducing any θ-dependence, and transform to the radial gauge $a_\rho = 0$. Reflection symmetry makes ϕ real. The GL energy reduces to the expression

$$V = \pi \int_0^\infty \left(\frac{1}{\rho^2} \left(\frac{da_\theta}{d\rho} \right)^2 + \left(\frac{d\phi}{d\rho} \right)^2 + \frac{1}{\rho^2} a_\theta^2 \phi^2 + \frac{\lambda}{4} (m^2 - \phi^2)^2 \right) \rho \, d\rho,$$

$$(7.46)$$

whose variational equations are

$$\frac{d^2\phi}{d\rho^2} + \frac{1}{\rho} \frac{d\phi}{d\rho} - \frac{1}{\rho^2} a_\theta^2 \phi + \frac{\lambda}{2} (m^2 - \phi^2) \phi = 0 \qquad (7.47)$$

$$\frac{d^2 a_\theta}{d\rho^2} - \frac{1}{\rho} \frac{da_\theta}{d\rho} - a_\theta \phi^2 = 0. \qquad (7.48)$$

Again, the solution of minimal energy is the vacuum, $\phi = m$, $a_\theta = 0$. There are no other finite energy solutions, satisfying the boundary conditions $\phi(\infty) = m$, $a_\theta(\infty) = 0$ and $\phi(0)$ finite, $a_\theta(0) = 0$. This is because Eq. (7.48) excludes the possibility of a_θ having a positive maximum or negative minimum. So a_θ has to vanish, and then Eq. (7.47) reduces to the equation (7.44) of the global GL theory, with only the vacuum as a solution.

A more interesting realization of circular symmetry is possible, which exploits the internal symmetry of the GL theory under global phase rotations $\tilde{R}(\alpha) : \phi \mapsto e^{i\alpha}\phi$. Consider the $SO(2)$ action by the combined rotations and phase rotations $(R(\beta), \tilde{R}(\kappa\beta))$, with κ a constant. This is a lift of the naive $SO(2)$ action, because

$$(R(\beta_1), \tilde{R}(\kappa\beta_1)) \cdot (R(\beta_2), \tilde{R}(\kappa\beta_2)) = (R(\beta_1 + \beta_2), \tilde{R}(\kappa(\beta_1 + \beta_2))). \quad (7.49)$$

κ must be an integer, so that $(R(2\pi), \tilde{R}(2\pi\kappa))$ is the identity. (If not, ϕ would need to be invariant under a 2π rotation, which does nothing, together with a non-trivial phase rotation by $2\pi\kappa$. This would force ϕ to vanish, and is incompatible with the boundary conditions.)

Invariance of ϕ under this combined $SO(2)$ action requires

$$\phi(\rho, \theta + \beta) = e^{i\kappa\beta} \phi(\rho, \theta). \qquad (7.50)$$

Let $\phi(\rho)$ denote $\phi(\rho, 0)$. Then (7.50) is satisfied if and only if

$$\phi(\rho, \theta) = e^{i\kappa\theta} \phi(\rho). \qquad (7.51)$$

It is clear that the winding number of such a field (at $\rho = \infty$) is $N = \kappa$.

We may assume that $N \neq 0$, since $N = 0$ is the case of naive rotational symmetry. Since the winding number is such an important quantity, we rewrite the formula for ϕ as

$$\phi(\rho, \theta) = e^{iN\theta}\phi(\rho) . \tag{7.52}$$

Infinitesimally, the generator of the combined $SO(2)$ action is $(\frac{\partial}{\partial\theta}, iN)$, and ϕ is invariant if it is annihilated by the operator $\frac{\partial}{\partial\theta} - iN$. For a field of the form (7.52), with $\phi(\rho)$ real, the global GL energy is

$$V = \pi \int_0^\infty \left(\left(\frac{d\phi}{d\rho}\right)^2 + \frac{N^2}{\rho^2}\phi^2 + \frac{\lambda}{4}(m^2 - \phi^2)^2 \right) \rho \, d\rho , \tag{7.53}$$

and the field equation reduces to

$$\frac{d^2\phi}{d\rho^2} + \frac{1}{\rho}\frac{d\phi}{d\rho} - \frac{N^2}{\rho^2}\phi + \frac{\lambda}{2}(m^2 - \phi^2)\phi = 0 \tag{7.54}$$

with boundary conditions $\phi(\infty) = m$ and $\phi(0) = 0$.

From now on we make the choice of parameters $m = 1$ and $\lambda = 2$ in the global GL theory. Other values of m and λ correspond to a rescaling of the field ϕ, and of the length scale, and hence also of the energy. The equation (7.54) becomes

$$\frac{d^2\phi}{d\rho^2} + \frac{1}{\rho}\frac{d\phi}{d\rho} - \frac{N^2}{\rho^2}\phi + (1 - \phi^2)\phi = 0 . \tag{7.55}$$

Solutions exist for any $N \neq 0$ [177], and can be found numerically [317]. Near $\rho = 0$, $\phi(\rho) \sim \rho^N$, so $\phi(\rho, \theta) \sim \rho^N e^{iN\theta}$. The solution with winding number N is therefore a vortex of multiplicity N. The asymptotic form of ϕ as $\rho \to \infty$ is $\phi(\rho) \sim 1 - \frac{N^2}{2\rho^2} - \frac{N^2(N^2+8)}{8\rho^4}$ with corrections involving higher even powers of $\frac{1}{\rho}$. The only difficulty with these solutions is their logarithmically divergent energy, but one may regularize the energy, for example by the method discussed in Section 7.13.

Even without this regularization, there is an interesting variant of the virial theorem, which gives some information about the energy. Let us multiply Eq. (7.55) by $2\rho^2 \frac{d\phi}{d\rho}$, obtaining

$$\frac{d}{d\rho}\left(\rho^2\left(\frac{d\phi}{d\rho}\right)^2 - N^2\phi^2 - \frac{1}{2}\rho^2(1 - \phi^2)^2 \right) + \rho(1 - \phi^2)^2 = 0 . \tag{7.56}$$

Integrating, and using the boundary conditions, we conclude that

$$\int_0^\infty (1 - \phi(\rho)^2)\rho \, d\rho = N^2 , \tag{7.57}$$

or equivalently

$$E_0 = \frac{1}{4} \int (1 - \bar{\phi}\phi)^2 \, d^2x = \frac{1}{2}\pi N^2 \,, \tag{7.58}$$

where the integral is over the whole plane. This is an example of a Derrick-Pohozaev identity [124].

Let us now turn to the gauged GL theory. Here also, we shall fix $m = 1$. λ is left as a free parameter, as the vortices depend on it in a non-trivial way. The phase rotation $e^{iN\beta}$ accompanying a rotation by β is a global one, independent of ρ and θ. It has no action on the gauge potential (a_ρ, a_θ). Therefore, the combined $SO(2)$ action leaves the fields invariant provided

$$
\begin{aligned}
\phi(\rho, \theta) &= e^{iN\theta}\phi(\rho) \\
a_\rho(\rho, \theta) &= a_\rho(\rho) \\
a_\theta(\rho, \theta) &= a_\theta(\rho) \,.
\end{aligned} \tag{7.59}
$$

Again, a ρ-dependent gauge transformation is possible, which changes $\phi(\rho)$ and can be used to set $a_\rho = 0$. The reflection symmetry $\phi(\rho, \theta) = \bar{\phi}(\rho, -\theta)$ again makes $\phi(\rho)$ real. The boundary conditions are $\phi(\infty) = 1$, $a_\theta(\infty) = N$ and $\phi(0) = 0$, $a_\theta(0) = 0$. The conditions at $\rho = \infty$ ensure that

$$D_\theta\phi = \partial_\theta\phi - ia_\theta\phi = (iN - iN)e^{iN\theta} = 0 \,. \tag{7.60}$$

The conditions at $\rho = 0$ ensure that ϕ is single-valued and the gauge potential is non-singular there.

So we see the three different meanings of N appearing. N is the winding number of ϕ at infinity. Since $\int_0^{2\pi} a_\theta(\infty) \, d\theta = 2\pi N$, the total magnetic flux is $2\pi N$. Since ϕ vanishes at $\rho = 0$, there is a vortex or multi-vortex centred at the origin. The increase of the phase of ϕ around the origin is $2\pi N$, so the multiplicity of the vortex is N. There can be no further zeros of ϕ if all zeros are isolated.

For fields of the form (7.59), with $a_\rho = 0$ and $\phi(\rho)$ real, the gauged GL energy is

$$V = \pi \int_0^\infty \left(\frac{1}{\rho^2}\left(\frac{da_\theta}{d\rho}\right)^2 + \left(\frac{d\phi}{d\rho}\right)^2 + \frac{1}{\rho^2}(N - a_\theta)^2\phi^2 + \frac{\lambda}{4}(1 - \phi^2)^2 \right) \rho \, d\rho \,, \tag{7.61}$$

and the field equations reduce to

$$\frac{d^2\phi}{d\rho^2} + \frac{1}{\rho}\frac{d\phi}{d\rho} - \frac{1}{\rho^2}(N - a_\theta)^2\phi + \frac{\lambda}{2}(1 - \phi^2)\phi = 0 \tag{7.62}$$

$$\frac{d^2a_\theta}{d\rho^2} - \frac{1}{\rho}\frac{da_\theta}{d\rho} + (N - a_\theta)\phi^2 = 0 \,. \tag{7.63}$$

Harden and Arp calculated the basic solution, with $N = 1$ [171]. Plohr [333] and Berger and Chen [50] have established rigorously that solutions satisfying the boundary conditions exist for all $N \neq 0$ and all $\lambda > 0$. They are minima of the energy in the class of fields with circular symmetry. The fields are smooth, including at the origin. By the principle of symmetric criticality, they give solutions of the full field equations in the plane, but not necessarily absolute minima of the energy. If $N > 0$, both $\phi(\rho)$ and $a_\theta(\rho)$ are strictly increasing as ρ increases from 0 to ∞.

Solutions have been obtained numerically, and Fig. 7.2 shows the profile functions $\phi(\rho), a_\theta(\rho)$ of the basic $N = 1$ vortex, for $\lambda = \frac{1}{2}$, 1 and 2. The vortex has a core size of order 1, beyond which the fields approach the vacuum values exponentially fast. It is of interest to plot some further physical quantities. Figure 7.3 shows, for the $\lambda = 1$ solution, the energy density and the magnetic field $B = \frac{1}{\rho}\partial_\rho a_\theta$, both as functions of ρ. Ampère's equation and the circular symmetry of the magnetic field imply that the current is

$$\mathbf{J} = \partial_\rho \left(\frac{1}{\rho} \partial_\rho a_\theta \right) \hat{\mathbf{t}}, \tag{7.64}$$

where $\hat{\mathbf{t}}$ is the unit vector in the positive θ direction.

Figure 7.4 shows profiles of multi-vortices with $N = 2, 3, 4$ and $\lambda = 1$. Figure 7.5 displays the associated energy densities and magnetic fields. Finally, Fig. 7.6 plots the energies of vortices with $1 \leq N \leq 4$ and $\lambda = \frac{1}{2}$, 1 and 2. Notice that for $\lambda = 1$, the energy grows linearly with N, for $\lambda > 1$ the growth is faster than linear, and for $\lambda < 1$ it is slower than linear.

We can understand the asymptotic form of the profile functions $\phi(\rho)$ and $a_\theta(\rho)$, both for $\rho \sim 0$ and $\rho \to \infty$. Near the origin, Eqs. (7.62) and (7.63) imply that there are expansions for ϕ and a_θ of the form $\phi(\rho) = \rho^N F(\rho^2)$ and $a_\theta(\rho) = \rho^2 G(\rho^2)$, where F and G are series in ρ^2 with non-zero constants as leading terms. This behaviour is observed in Fig. 7.4. If we convert to Cartesian coordinates, which are better than polars near the origin, we find that

$$\begin{aligned} \phi &= e^{iN\theta}\rho^N F(\rho^2) \\ a_1 &= -x^2\, G(\rho^2) \\ a_2 &= x^1\, G(\rho^2), \end{aligned} \tag{7.65}$$

since $a_1 = -\frac{1}{\rho^2}x^2 a_\theta$ and $a_2 = \frac{1}{\rho^2}x^1 a_\theta$ if $a_\rho = 0$. Since $\rho^N e^{iN\theta} = (x^1 + ix^2)^N$ and $\rho^2 = (x^1)^2 + (x^2)^2$, it follows that the vortex solutions are real analytic functions of x^1 and x^2 in a neighbourhood of the origin. A similar analyticity result would hold even for solutions which are not circularly symmetric. In particular, $|\phi| \sim \rho^n$ near a vortex of multiplicity n, where ρ is the distance from the vortex. We shall use this later.

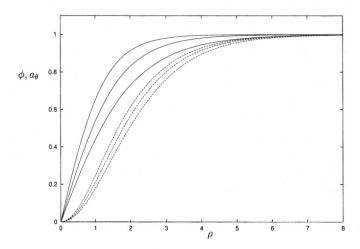

Fig. 7.2. The profile functions $\phi(\rho)$ (solid curves) and $a_\theta(\rho)$ (dashed curves) for the $N = 1$ vortex with $\lambda = 0.5, 1.0, 2.0$. The curves move to the left with increasing λ.

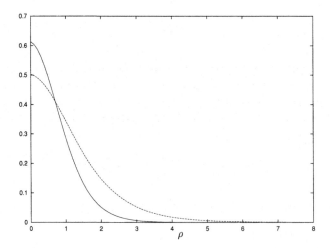

Fig. 7.3. The energy density (solid curve) and magnetic field B (dashed curve) for the $N = 1$ vortex with $\lambda = 1$.

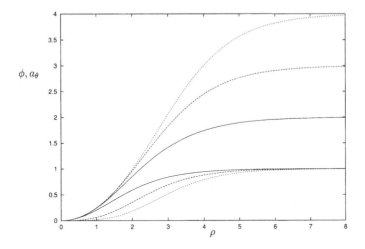

Fig. 7.4. The profile functions $\phi(\rho)$ and $a_\theta(\rho)$ for the vortex with $N = 2$ (solid curves), $N = 3$ (dashed curves) and $N = 4$ (dotted curves). Here $\lambda = 1$.

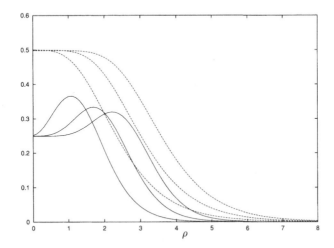

Fig. 7.5. Energy densities (solid curves) and magnetic fields B (dashed curves) for vortices with $N = 2, 3, 4$. The curves move to the right with increasing N, and $\lambda = 1$.

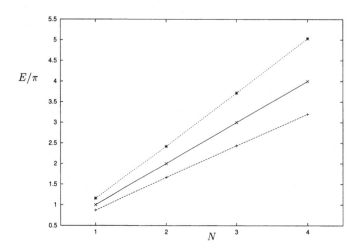

Fig. 7.6.   The energy (in units of π) for vortices with $N = 1, 2, 3, 4$ and $\lambda = 0.5$ (dashed curve), $\lambda = 1.0$ (solid line) and $\lambda = 2.0$ (dotted curve).

To understand the behaviour as $\rho \to \infty$, let

$$\phi(\rho) \quad = \quad 1 - \sigma(\rho) \tag{7.66}$$

$$a_\theta(\rho) \quad = \quad N - \psi(\rho). \tag{7.67}$$

The linearized forms of Eqs. (7.62) and (7.63) are the modified Bessel equations

$$\frac{d^2\sigma}{d\rho^2} + \frac{1}{\rho}\frac{d\sigma}{d\rho} - \lambda\sigma \quad = \quad 0 \tag{7.68}$$

$$\frac{d^2\psi}{d\rho^2} - \frac{1}{\rho}\frac{d\psi}{d\rho} - \psi \quad = \quad 0, \tag{7.69}$$

whose decaying solutions give the asymptotic expressions

$$\phi(\rho) \quad \sim \quad 1 - \frac{A_{\rm s}}{2\pi}K_0(\sqrt{\lambda}\rho) \tag{7.70}$$

$$a_\theta(\rho) \quad \sim \quad N - \frac{A_{\rm m}}{2\pi}\rho K_1(\rho). \tag{7.71}$$

The leading exponential term in the modified Bessel functions $K_0(\rho)$ and $K_1(\rho)$ for large ρ is $\sqrt{\frac{\pi}{2\rho}}e^{-\rho}$ [5]. The coefficients $A_{\rm s}$ and $A_{\rm m}$, associated with the decay of the scalar and magnetic fields, need to be determined numerically, by solving the full nonlinear equations with appropriate boundary conditions at $\rho = 0$.

In fact, the asymptotic expressions (7.70) and (7.71) have somewhat limited validity. For $\lambda \ll 1$ they are valid only for very large ρ, much greater than $1/\sqrt{\lambda}$. For $\lambda > 4$ they are not valid, because of forcing terms neglected in the linearization leading to (7.68) and (7.69) [333, 328]. Instead, for $\lambda \gg 1$, the solution of the nonlinear equations (7.62) and (7.63), except in a very small neighbourhood of $\rho = 0$, is approximately

$$\phi(\rho) \;=\; 1 - \frac{1}{\lambda}\left(\frac{A_{\mathrm{m}}^{\infty}}{2\pi}K_1(\rho)\right)^2 \tag{7.72}$$

$$a_\theta(\rho) \;=\; N - \frac{A_{\mathrm{m}}^{\infty}}{2\pi}\rho K_1(\rho)\,, \tag{7.73}$$

so ϕ is very close to 1 for almost all ρ. The constant A_{m}^{∞} is $2\pi N$ [50].

The asymptotic forms of the scalar field and gauge potential inform us of the forces to expect between well separated vortices. This is discussed in Section 7.6.

At the critical value of the coupling constant, $\lambda = 1$, there are more solutions of the static equations than just the circularly symmetric ones. There is a whole moduli space $\mathcal{M}_N$ of N-vortex solutions, which has dimension $\dim \mathcal{M}_N = 2N$, because the vortices can be at arbitrary locations. All these solutions satisfy first order Bogomolny equations, and they minimize the potential energy in the charge N sector of the theory. They are discussed further in Section 7.8.

7.5 Forces between gauged vortices

This section is concerned with the energetics of two or more gauged vortices as a function of their separation. If the energy decreases as the vortices separate, we shall say that there is a repulsive force between them; if it increases then they attract. We will need to specify a complete dynamical version of the GL theory before we can deduce how the vortices actually move relative to one another. The discussion of forces between global vortices is postponed until Section 7.13.

We shall assume that λ is of order 1, being neither very close to zero nor very large. There are several ways to investigate the energy of multi-vortex configurations, some more rigorous than others. From all of these, a coherent picture emerges. For $\lambda < 1$ vortices attract, and for $\lambda > 1$ vortices repel. For $\lambda = 1$ the vortices are in neutral equilibrium, which allows static multi-vortex solutions, with the vortex locations at arbitrary points of the plane. Vortices and antivortices always attract.

Consider first the energies E_N of the circularly symmetric N-vortex solutions discussed in the last section. Since $E_N < NE_1$ for $\lambda < 1$, N coincident vortices have an energy less than N vortices "at infinity".

This implies that vortices attract, provided an N-vortex configuration with N well separated vortices has energy arbitrarily close to NE_1. Such a configuration can be constructed as follows.

Take a set of N circular discs in the plane, each with radius R and with minimal separation L, where $1 \ll R \ll L$. Let $\widetilde{\Sigma}$ denote the complement of all the discs. In $\widetilde{\Sigma}$ we may construct a field configuration with $|\phi| = 1$, $D_i\phi = 0$ and $B = 0$, and with the phase of ϕ having winding 2π around each disc. To get the windings right, take the complex polynomial

$$p(z) = \prod_{r=1}^{N}(z - Z_r) \qquad (7.74)$$

where Z_r is the centre of the rth disc (we are here identifying $\mathbb{R}^2$ with $\mathbb{C}$). Then set

$$\phi = \frac{p(z)}{|p(z)|} \qquad (7.75)$$

in $\widetilde{\Sigma}$. $\arg\phi$ increases by 2π around each point Z_r. We define the gauge potential in $\widetilde{\Sigma}$ by

$$a_i = \partial_i(\arg\phi). \qquad (7.76)$$

This ensures that $D_i\phi = 0$, and also $B = 0$. As $|z| \to \infty$, $\phi \sim z^N/|z|^N$, so in polar coordinates $\phi = e^{iN\theta}$ along the circle at infinity. Therefore, the winding number of ϕ is N.

To complete the construction, we need to glue a one-vortex solution into each disc. For the rth disc, take the exact, circularly symmetric solution, and deform it slightly – an exponentially small change – so that $|\phi| = 1$ and $D_i\phi = 0$ exactly for $|\mathbf{x}| \geq R$. Then, automatically, $B = 0$ for $|\mathbf{x}| \geq R$. Next, translate the vortex so that its centre is at Z_r, and perform a gauge transformation so that the phase of ϕ, which is $\arg(z - Z_r)$, matches the phase of $p(z)/|p(z)|$ along the boundary of the disc. This is possible, because the windings are the same. What is required is approximately a constant phase rotation, because

$$\arg p(z) = \arg(z - Z_r) + \sum_{q \neq r}\arg(z - Z_q) \qquad (7.77)$$

and the final sum is approximately constant on the boundary of the rth disc, since $R \ll L$.

The resulting field configuration is continuous, and can be smoothed out by smoothing the transition across the boundaries of the discs. Its energy differs by an amount of order $e^{-2\sqrt{\lambda}R}$ from NE_1. This is because the field in $\widetilde{\Sigma}$ has zero energy, and the vortex inside each disc differs from a true solution of the field equations only in that its tail ($|\mathbf{x}| > O(R)$) has been adjusted. The energy difference becomes arbitrarily small as R, and L, become sufficiently large.

A more refined discussion shows that smooth field configurations with winding number N can be constructed with ϕ having simple zeros at any N given points. It is plausible, although we do not here offer a proof, that the solution with N coincident vortices (and energy E_N) can be continuously deformed into a configuration of well separated vortices (and asymptotic energy NE_1), with a monotonic increase of energy. That would show that the vortices everywhere attract.

For $\lambda > 1$, a very similar argument can be given, showing that vortices repel. The only change is that the energy difference is of order e^{-2R} rather than $e^{-2\sqrt{\lambda}R}$. Since $E_N > NE_1$ in this case, the vortices can probably be separated from coalescence to infinite separation with a monotonically decreasing energy.

Further insight comes from the stability or otherwise of the circularly symmetric N-vortex solution. For $N = 1$, the solution is stable for all λ. For $N > 1$, the stability depends on λ. If $\lambda < 1$, the solution is stable, as one expects if vortices attract. If $\lambda > 1$ it is unstable, and the number of unstable modes corresponds to the number of ways the multiple zero can split up into simple zeros, as one expects if vortices repel. These stability results were conjectured by Jaffe and Taubes [223], and an argument for the instability in the $\lambda > 1$ case was given by Bogomolny [56]. They have been rigorously established by Gustafson and Sigal [167].

A numerical investigation of the vortex interaction energy was carried out by Jacobs and Rebbi [222]. They considered two-vortex configurations, with simple zeros of ϕ separated (along the x^1-axis) by s. They calculated numerically the minimal energy field configuration $\{\phi(\mathbf{x}), a_i(\mathbf{x})\}$ with these given zeros. This is an example of constrained minimization. The energy, for $0 \leq s \leq 12$ and for $\lambda = \frac{1}{2}$, 1 and 2, is plotted in Fig. 7.7. The graphs show clearly that two vortices attract if $\lambda < 1$ and repel if $\lambda > 1$, and that the energy is independent of s if $\lambda = 1$. The fields satisfying the constrained minimization problem need to have discontinuous derivatives at the zeros, but otherwise they satisfy the field equations.

7.6 Forces between vortices at large separation

In this section we shall discuss more precisely how the interaction energy of two well separated gauged GL vortices depends on distance. It is not easy to give a completely rigorous analysis of this. However, a number of different approaches all yield the same answer.

The interaction energy $E_{\text{int}}(s)$ of two unit winding vortices at a separation $s \gg 1$ is the total energy E minus the constant $2E_1$, representing the energy of infinitely separated vortices. E_{int} is calculated for a field

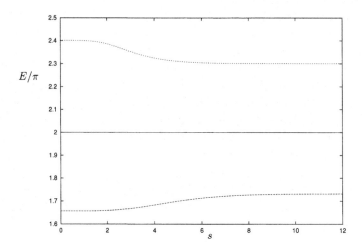

E/π

Fig. 7.7.   The energy (in units of π) of two vortices as a function of their separation s for $\lambda = 0.5$ (dashed curve), $\lambda = 1.0$ (solid line), $\lambda = 2.0$ (dotted
curve).

where each vortex minimally distorts the other vortex. This is possible
for $s \gg 1$. The field equations are linearized in the region well away from
the vortex cores. One has to be more careful than in Section 7.5 where
we rather crudely glued together an N-vortex field. There are two contributions to E_{int}, coming physically from the scalar field and from the
magnetic field. These both decay exponentially with s, but at different
rates. The result is

$$E_{\text{int}}(s) = -\frac{A_s^2}{2\pi}K_0(\sqrt{\lambda}s) + \frac{A_m^2}{2\pi}K_0(s) \qquad (7.78)$$

where A_s and A_m are the coefficients in the asymptotic expressions (7.70)
and (7.71).

A result of this kind for E_{int} was first obtained by Bettencourt and
Rivers [52] by a method that is outlined below (with rather crude approximations for the values of A_s and A_m). Speight rederived (7.78) from
a different point of view, explained below, and calculated the constants
A_s and A_m numerically for a range of values of λ [382]. For $\lambda = 1$,
$A_s = A_m = 10.6$. As λ increases, A_s increases whereas A_m decreases, so,
for example, $A_s = 14.5$ and $A_m = 8.9$ when $\lambda = 2$. Using a remarkable
indirect approach, involving dualities in string theory, Tong has obtained
the analytic value $A_s = A_m = 2\pi\,8^{1/4} \simeq 10.57$, for $\lambda = 1$ [404], but this
value has not yet been understood directly from the vortex equations.

The important point is that in (7.78) the scalar A_s^2 term is negative and the magnetic A_m^2 term is positive, with both decaying exponentially as s increases. Thus the scalar field produces an attraction and the magnetic field produces a repulsion. For $\lambda < 1$, the scalar term dominates and vortices attract. For $\lambda > 1$, the magnetic term dominates and vortices repel. For $\lambda = 1$, the interaction energy is exactly zero, to this level of approximation.

Subleading exponential corrections to E_{int} might be expected. However, there is no sense in actually calculating these, because the separation between extended objects has a degree of uncertainty, albeit exponentially small in the vortex case. So further corrections to the formula (7.78) would not be physically meaningful. Although the separation of the zeros of the scalar field may be regarded as a precise quantity, there is nothing that says that this is physically the precise separation of the vortices.

If λ is close to unity, then the two terms in (7.78) are both significant, even though one dominates the other, as both are larger than the uncertainties just mentioned. If $\lambda > 4$, the asymptotic form of ϕ is not given by (7.70), and one should just retain the second term in (7.78), which in any case completely dominates the first.

Let us now turn to the calculations of E_{int}. Let $\{\phi^{(1)}, a_i^{(1)}\}$ be the fields of vortex 1 (in the absence of vortex 2) and $\{\phi^{(2)}, a_i^{(2)}\}$ the fields of vortex 2 (in the absence of vortex 1). A neat way to construct the desired superposition of the two vortices is Abrikosov's product ansatz

$$\begin{aligned} \phi &= \phi^{(1)}\phi^{(2)} \\ a_i &= a_i^{(1)} + a_i^{(2)} . \end{aligned} \tag{7.79}$$

ϕ is topologically correct. It has simple zeros at the zeros of $\phi^{(1)}$ and $\phi^{(2)}$, and its winding at infinity is the sum of the windings of $\phi^{(1)}$ and $\phi^{(2)}$. The ansatz is gauge invariant in the sense that if $\{\phi^{(1)}, a_i^{(1)}\}$ is gauge transformed by $e^{i\alpha^{(1)}}$ (not necessarily constant), and $\{\phi^{(2)}, a_i^{(2)}\}$ gauge transformed by $e^{i\alpha^{(2)}}$, then $\{\phi, a_i\}$ is (consistently) gauge transformed by $e^{i(\alpha^{(1)}+\alpha^{(2)})}$.

One has $\phi^{(1)} = (1 - \sigma^{(1)})e^{i\chi^{(1)}}$, $\phi^{(2)} = (1 - \sigma^{(2)})e^{i\chi^{(2)}}$ in the region far from both vortex cores, where $\sigma^{(1)}$ and $\sigma^{(2)}$ are exponentially small. Therefore

$$\phi = (1 - \sigma^{(1)} - \sigma^{(2)})e^{i(\chi^{(1)}+\chi^{(2)})} , \tag{7.80}$$

with a negligible correction, so the physically important, gauge invariant magnitudes $\sigma^{(1)}$ and $\sigma^{(2)}$ are linearly superposed. Also, the magnetic fields are exactly superposed, since

$$B = (\partial_1 a_2^{(1)} - \partial_2 a_1^{(1)}) + (\partial_1 a_2^{(2)} - \partial_2 a_1^{(2)}) = B^{(1)} + B^{(2)} . \tag{7.81}$$

Further, the covariant derivative $D_j\phi$ is a linear superposition of $D_j^{(1)}\phi^{(1)}$ and $D_j^{(2)}\phi^{(2)}$, since

$$
\begin{aligned}
D_j\phi &= (\partial_j - i(a_j^{(1)} + a_j^{(2)}))\phi^{(1)}\phi^{(2)} \\
&= \partial_j\phi^{(1)}\phi^{(2)} + \phi^{(1)}\partial_j\phi^{(2)} - ia_j^{(1)}\phi^{(1)}\phi^{(2)} - ia_j^{(2)}\phi^{(1)}\phi^{(2)} \\
&= (D_j^{(1)}\phi^{(1)})\phi^{(2)} + \phi^{(1)}(D_j^{(2)}\phi^{(2)}).
\end{aligned}
\tag{7.82}
$$

In the region far from the vortex cores, $\phi^{(1)}$ and $\phi^{(2)}$ can be approximated by $e^{i\chi^{(1)}}$ and $e^{i\chi^{(2)}}$ respectively. The sum in (7.82) is therefore a superposition of $D_j^{(1)}\phi^{(1)}$ and $D_j^{(2)}\phi^{(2)}$ in a gauge coherent sense.

Now suppose one vortex is at a distance s from the other, with $s \gg 1$. Bettencourt and Rivers showed that the dominant contribution to the interaction energy comes from the overlap between the tails of the fields of the two vortices, which are given by the asymptotic expressions (7.70) and (7.71). The vortex cores can be treated as regularized delta-function sources for these asymptotic fields. The energy depends only on the gauge invariant quantities, and can be simplified using the formulae (7.80)–(7.82), and reduced to an integration over products of Bessel functions. The integration can be carried out, leading to the formula (7.78).

Speight has calculated the interaction energy by giving the vortices the following physical interpretation [382]. It is as if each vortex, when viewed from far away, behaves as a point-like object in the plane, carrying both a scalar charge A_s and a magnetic dipole moment $A_m\hat{n}$, where $\hat{n}$ is the unit vector perpendicular to the plane. The scalar charge A_s is interpreted as a source for a (new) scalar field $\tilde{\phi}$ obeying the linear equation

$$
(-\nabla^2 + \lambda)\tilde{\phi} = A_s\delta^2(\mathbf{x} - \mathbf{X})
\tag{7.83}
$$

where $\mathbf{X}$ is the vortex centre and δ^2 denotes the two-dimensional delta-function. The scalar interaction energy of the vortex with a second vortex at $\tilde{\mathbf{X}}$ of charge A_s is $-A_s\tilde{\phi}(\tilde{\mathbf{X}})$. The solution of (7.83) is

$$
\tilde{\phi} = \frac{A_s}{2\pi}K_0(\sqrt{\lambda}|\mathbf{x} - \mathbf{X}|),
\tag{7.84}
$$

so the scalar interaction is $-\frac{A_s^2}{2\pi}K_0(\sqrt{\lambda}s)$ where $s = |\tilde{\mathbf{X}} - \mathbf{X}|$, as in the first term of (7.78).

The dipole moment $A_m\hat{n}$ can be thought of as due to a small current loop in the plane. It is interpreted as a source for a (new) gauge potential $\tilde{\mathbf{a}}$. The equation satisfied by $\tilde{\mathbf{a}}$, analogous to (7.83), is

$$
(-\nabla^2 + 1)\tilde{\mathbf{a}} = -A_m\hat{n} \times \nabla\delta^2(\mathbf{x} - \mathbf{X}).
\tag{7.85}
$$

The solution is

$$\widetilde{\mathbf{a}} = -\frac{A_{\mathrm{m}}}{2\pi}\hat{\mathbf{n}} \times \nabla K_0(\mathbf{x} - \mathbf{X})\,. \tag{7.86}$$

In terms of polar coordinates centred at $\mathbf{X}$, $\widetilde{a}_\rho = 0$ and

$$\widetilde{a}_\theta = \frac{A_{\mathrm{m}}}{2\pi}\rho K_1(\rho) \tag{7.87}$$

since $K_1 = -K_0'$. Now we can calculate the magnetic interaction between a dipole at $\widetilde{\mathbf{X}}$ and the potential $\widetilde{\mathbf{a}}$ due to the dipole at $\mathbf{X}$. The magnetic field associated with $\widetilde{\mathbf{a}}$ is

$$\begin{aligned}
\widetilde{B} = \frac{1}{\rho}\frac{d\widetilde{a}_\theta}{d\rho} &= \frac{A_{\mathrm{m}}}{2\pi}\left(\frac{1}{\rho}K_1(\rho) + K_1'(\rho)\right) \\
&= -\frac{A_{\mathrm{m}}}{2\pi}\left(\frac{1}{\rho}K_0'(\rho) + K_0''(\rho)\right) \\
&= -\frac{A_{\mathrm{m}}}{2\pi}K_0(\rho)\,, \tag{7.88}
\end{aligned}$$

using the equation satisfied by K_0. Therefore, the dipole-dipole interaction between two vortices at separation s is $\frac{A_{\mathrm{m}}^2}{2\pi}K_0(s)$, as in the second term of (7.78).

Together, the scalar interaction and the magnetic interaction give the interaction energy between the vortices.

7.7 Dynamics of gauged vortices

The Ginzburg-Landau energy function determines how the energy of a configuration with several vortices depends on the vortex separations. However, this does not by itself determine how the vortices move. We shall discuss three types of dynamical field equations, which lead to three types of vortex motion. These are not equally well understood theoretically.

7.7.1 Second order dynamics

This type of dynamics comes from the Lorentz invariant extension of GL theory, which is called scalar electrodynamics, or the abelian Higgs model, and was introduced in Section 2.6. The spatial and time derivatives of both the scalar field and gauge potential appear quadratically in the Lagrangian density $\mathcal{L}$. With the standard symmetry-breaking quartic potential, the Lagrangian is

$$L = \int\left(-\frac{1}{4}f_{\mu\nu}f^{\mu\nu} + \frac{1}{2}\overline{D_\mu\phi}D^\mu\phi - \frac{\lambda}{8}(1 - \bar{\phi}\phi)^2\right)d^2x\,. \tag{7.89}$$

Its kinetic part is

$$T = \frac{1}{2} \int \left(e_1^2 + e_2^2 + \overline{D_0\phi}D_0\phi \right) d^2x \tag{7.90}$$

with e_1 and e_2 the components of the electric field. The potential part, V, is the GL energy (7.13), with $m = 1$. The Euler-Lagrange field equations are

$$D_\mu D^\mu \phi - \frac{\lambda}{2}(1 - \bar\phi\phi)\phi = 0 \tag{7.91}$$

$$\partial_\mu f^{\mu\nu} + \frac{i}{2}(\bar\phi D^\nu\phi - \phi\overline{D^\nu\phi}) = 0. \tag{7.92}$$

The general, finite energy solution of these equations is complicated. It can involve vortices and antivortices colliding and annihilating, with a complicated radiation pattern.

We collect here the set of conserved quantities. First of all there is the conserved topological charge, N. N is the winding number of ϕ on the circle at infinity, and the total magnetic flux is $2\pi N$. One also has the conservation laws for the geometrical Noether charges, the energy, momentum and angular momentum. The conserved energy is

$$E = T + V = \frac{1}{2} \int \Big(e_i e_i + B^2 + \overline{D_0\phi}D_0\phi$$

$$+ \ \overline{D_i\phi}D_i\phi + \frac{\lambda}{4}(1 - \bar\phi\phi)^2 \Big) d^2x. \tag{7.93}$$

To find the conserved momentum, it helps to use the improvement method first mentioned at the end of Section 2.6. Momentum is associated with translation invariance of the Lagrangian. Naively, an infinitesimal translation in the x^i direction gives variations

$$\begin{aligned} \Delta\phi &= \partial_i\phi \\ \Delta\bar\phi &= \partial_i\bar\phi \\ \Delta a_j &= \partial_i a_j. \end{aligned} \tag{7.94}$$

To improve these, we include the effect of an infinitesimal gauge transformation with parameter $-a_i$. This gives gauge covariant variations

$$\begin{aligned} \tilde\Delta\phi &= \partial_i\phi - ia_i\phi = D_i\phi \\ \tilde\Delta\bar\phi &= \partial_i\bar\phi + ia_i\bar\phi = \overline{D_i\phi} \\ \tilde\Delta a_j &= \partial_i a_j - \partial_j a_i = \varepsilon_{ij}B. \end{aligned} \tag{7.95}$$

The conserved momentum is then

$$P_i = -\int \left(\frac{\partial\mathcal{L}}{\partial(\partial_0\phi)}\tilde\Delta\phi + \frac{\partial\mathcal{L}}{\partial(\partial_0\bar\phi)}\tilde\Delta\bar\phi + \frac{\partial\mathcal{L}}{\partial(\partial_0 a_j)}\tilde\Delta a_j \right) d^2x$$

$$= -\int \left(\frac{1}{2} \overline{D_0\phi} D_i\phi + \frac{1}{2} D_0\phi \overline{D_i\phi} + \varepsilon_{ij} e_j B \right) d^2x. \tag{7.96}$$

Angular momentum is associated with rotational invariance. In two dimensions, the vector field generating rotations is $\boldsymbol{\xi} = (-x^2, x^1)$ and the naive variation of a dynamical field is its Lie derivative in the direction $\boldsymbol{\xi}$. For the fields we are considering here, the variations are

$$\begin{aligned}
\Delta\phi &= x^1\partial_2\phi - x^2\partial_1\phi \\
\Delta\bar\phi &= x^1\partial_2\bar\phi - x^2\partial_1\bar\phi \\
\Delta a_1 &= x^1\partial_2 a_1 - x^2\partial_1 a_1 + a_2 \\
\Delta a_2 &= x^1\partial_2 a_2 - x^2\partial_1 a_2 - a_1,
\end{aligned} \tag{7.97}$$

the final terms in Δa_1 and Δa_2 arising because a_1 and a_2 are components of a 1-form. Improvement is achieved by including the effect of an infinitesimal gauge transformation with parameter $-(x^1 a_2 - x^2 a_1)$. This gives covariant variations

$$\begin{aligned}
\tilde\Delta\phi &= x^1\partial_2\phi - x^2\partial_1\phi - i(x^1 a_2 - x^2 a_1)\phi = x^1 D_2\phi - x^2 D_1\phi \\
\tilde\Delta\bar\phi &= x^1\partial_2\bar\phi - x^2\partial_1\bar\phi + i(x^1 a_2 - x^2 a_1)\bar\phi = x^1 \overline{D_2\phi} - x^2 \overline{D_1\phi} \\
\tilde\Delta a_1 &= x^1\partial_2 a_1 - x^2\partial_1 a_1 + a_2 - \partial_1(x^1 a_2 - x^2 a_1) = -x^1 B \quad (7.98) \\
\tilde\Delta a_2 &= x^1\partial_2 a_2 - x^2\partial_1 a_2 - a_1 - \partial_2(x^1 a_2 - x^2 a_1) = -x^2 B,
\end{aligned}$$

recalling that $B = \partial_1 a_2 - \partial_2 a_1$. The conserved angular momentum is

$$\begin{aligned}
\ell &= \int \left(\frac{\partial\mathcal{L}}{\partial(\partial_0\phi)}\tilde\Delta\phi + \frac{\partial\mathcal{L}}{\partial(\partial_0\bar\phi)}\tilde\Delta\bar\phi + \frac{\partial\mathcal{L}}{\partial(\partial_0 a_1)}\tilde\Delta a_1 + \frac{\partial\mathcal{L}}{\partial(\partial_0 a_2)}\tilde\Delta a_2 \right) d^2x \\
&= \int x^1 \left(\frac{1}{2}\overline{D_0\phi}D_2\phi + \frac{1}{2}D_0\phi\overline{D_2\phi} - e_1 B \right) \\
&\quad - x^2 \left(\frac{1}{2}\overline{D_0\phi}D_1\phi + \frac{1}{2}D_0\phi\overline{D_1\phi} + e_2 B \right) d^2x. \tag{7.99}
\end{aligned}$$

Finally, there is the conserved electric charge

$$Q = -\frac{i}{2}\int (\bar\phi D_0\phi - \phi\overline{D_0\phi}) \, d^2x. \tag{7.100}$$

Within this theory, an $N = 1$ vortex behaves like a particle [318] whose rest mass M is the static GL energy, E_1. The static solution can be Lorentz boosted to give an exact solution representing a vortex moving at an arbitrary speed up to the speed of light. A superposition of such solutions is possible, representing well separated vortices released from rest, or given arbitrary individual velocities, and set to collide.

The geometry of a two-vortex collision is indicated in Fig. 7.8.

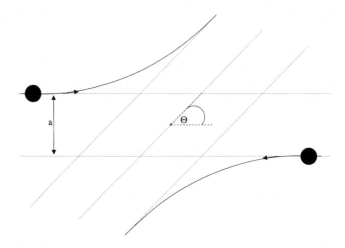

Fig. 7.8.   The geometry of a two-vortex collision. The impact parameter is a
and the scattering angle is Θ.

In the frame of reference where the centre of mass remains at rest at
the origin, the initial trajectories are parallel straight lines, separated by
a distance a, known as the impact parameter. A collision at slow speed
is almost adiabatic, and the outgoing vortices carry approximately the
same energy as the ingoing ones.   There are always just two zeros of
the scalar field, symmetrically located. If a is small, or zero, the field
can instantaneously pass very close to the circularly symmetric static
solution, where the zeros coincide.   The asymptotic trajectories of the
outgoing vortices are also parallel lines with separation a (to conserve
angular momentum). The interesting quantity is the scattering angle Θ
and its dependence on a.

While the vortices are separated by a distance much greater than 1, and
the speeds are non-relativistic, the motion of each vortex is determined
by a Newtonian equation of motion

$$M\ddot{\mathbf{x}} = \mathbf{F} \tag{7.101}$$

where M is the rest mass of the vortex, and $\mathbf{F}$ is minus the gradient of the
interaction energy E_{int}, given by Eq. (7.78). Because the force is central,
the angular momentum $\ell = 2M\varepsilon_{ij}x^i\dot{x}^j$ is conserved. (The factor of 2
occurs because there are two vortices.) From the asymptotic trajectories,
one sees that $\ell = \frac{1}{2}Mva$, where v is the initial relative speed and a the
impact parameter.

The Newton equation is not valid if the vortices approach to a distance of order unity, and here numerical simulations are usually needed. Not only is the force not well defined, but neither is the rest mass. Numerical studies of two-vortex scattering have been carried out by Matzner [299], by Shellard and Ruback [374], and by Moriarty, Myers and Rebbi [306]. The result for $\Theta(a)$, when $\lambda = 1$, is shown in Fig. 7.9.

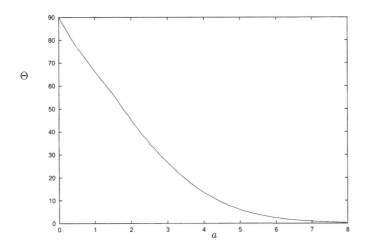

Fig. 7.9. The scattering angle Θ as a function of the impact parameter a for two-vortex scattering.

In relativistic collisions, radiation is produced. Wave-like oscillations are emitted from the collision region, and carry away energy. The amount of radiated energy is negligible (of order 1% of the kinetic energy of the vortices, or less) until the vortex speeds reach about half the speed of light [306]. Numerically, it is quite hard to see the radiation directly. However, the vortices emerging from a collision are observed to have slower speeds than the incoming ones, and this is interpreted as evidence for energy loss via radiation. The radiation is easily seen numerically in higher speed collisions, for example, at 0.9 times the speed of light.

Probably the most interesting phenomenon in two-vortex dynamics is the right-angle scattering that occurs in a head-on collision ($a = 0$) [352]. For $\lambda < 1$, vortices attract, and in a head-on collision the field can easily pass through a configuration in which the zeros of ϕ coincide at the origin. Following this, the vortices always emerge at right angles, and usually they escape to infinity. The same phenomenon occurs when $\lambda = 1$. When $\lambda > 1$, the vortices repel, so for slow initial speeds the kinetic energy is not sufficient for them to reach a coincident two-vortex configuration. The vortices bounce back off each other, as in a collision of billiard balls. The scattering angle is 180°. However, if the kinetic energy is sufficiently large, the vortices will scatter at right angles. The energy threshold is just slightly larger than that needed to produce the circularly symmetric two-vortex solution, as the dynamical field must pass close to this. Figure 7.10 displays the energy density during the right-angle scattering of two vortices with $\lambda = 1$.

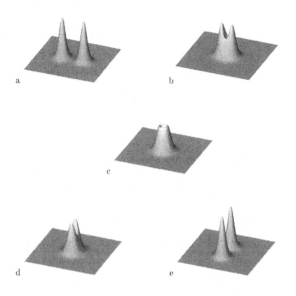

Fig. 7.10. Energy density plots at increasing times during the right-angle scattering of two vortices in a head-on collision.

The phenomenon of right-angle scattering shows in an essential way that vortices are not just Newtonian point particles, or extended Newtonian particles (hard discs). Unlike in Newtonian dynamics, the vortices are *classically indistinguishable* and can not be labelled. This is because the fields are the fundamental objects, and the vortex locations are simply the zeros of the scalar field. Two zeros can be followed until coincidence, and treated as particle trajectories, but when the zeros emerge at right angles, one can no longer say which outgoing zero corresponds to which incoming one. One interpretation of what happens in the collision is that some of the matter making up each vortex is exchanged. Each outgoing vortex is made up of one half of each ingoing vortex. In contrast, point particles or rigid extended ones maintain their identities, and have precise trajectories which satisfy Newton's laws. Sudden changes of their velocity, as in ideal billiard ball collisions, are interpreted as due to impulses (infinite forces acting instantaneously). The right-angle scattering of vortex zeros does not involve hard objects and an impulse. Quite the contrary; the zeros are completely soft, and they move superluminally (faster than light) close to their collision, so they can not carry any energy or momentum at the moment of collision. Also the fields deform quite smoothly as the zeros scatter.

A good analogy is with the evolution of an ellipse as the parameters change. Consider the equation

$$\frac{(x^1)^2}{(1-\varepsilon)^2} + \frac{(x^2)^2}{(1+\varepsilon)^2} = 1 \tag{7.102}$$

as ε moves through zero. This defines an ellipse whose shape for $\varepsilon < 0$, $\varepsilon = 0$, and $\varepsilon > 0$ is shown in Fig. 7.11.

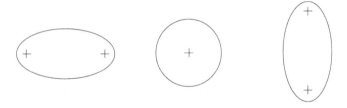

Fig. 7.11.   The smooth evolution of an ellipse and the right-angle scattering of its foci.

The foci are at $(\pm 2\sqrt{|\varepsilon|}, 0)$ for $\varepsilon \leq 0$, and at $(0, \pm 2\sqrt{\varepsilon})$ for $\varepsilon \geq 0$. They scatter through a right angle while the ellipse smoothly changes shape. For another analogy, consider the complex polynomial $z^2 + c$. The polynomial evolves smoothly if c moves at constant speed along the real axis through $c = 0$, but its roots scatter at right angles in the complex plane, and just before and after scattering, the roots have arbitrarily large speeds. Note that the foci of an ellipse, and the roots of a quadratic polynomial, are a pair of unlabelled, and hence unordered points. This seems to preclude a Newtonian equation of motion for them, accommodating collisions.

The right-angle scattering of vortices has been investigated in other ways too. An argument based purely on topology and symmetry has been given by Rosenzweig and Srivastava [350]. Abdelwahid and Burzlaff have given a more analytic explanation in terms of an initial value problem

[1]. They consider a dynamical field which at $t = 0$ is the exact static solution of two coincident vortices, and whose first time derivative splits the zeros. There is a symmetry under the operation

$$(t, x^1, x^2) \mapsto (-t, -x^2, x^1), \tag{7.103}$$

which combines a $90°$ rotation with time reversal. The leading part of the solution for the scalar field, at small $|\mathbf{x}|$ and small t, is

$$\phi = \alpha(x^1 + ix^2)^2 + \beta t \tag{7.104}$$

with α and β real and positive. This just changes sign (a global gauge transformation) under the operation (7.103). The zeros of ϕ are at $x^1 + ix^2 = \pm\sqrt{-\frac{\beta}{\alpha}t}$, and scatter through a right angle at $t = 0$. The solution represents vortices approaching coincidence along the x^1-axis, and separating along the x^2-axis. The solution is not quite physical. It represents incoming vortices together with radiation, turning into outgoing vortices with radiation. Nevertheless, this is close to the physical situation of colliding vortices with no incoming radiation, since at slow speeds the radiation field is small.

The right-angle scattering of two vortices in a head-on collision has a generalization for N vortices [272, 14]. If N vortices approach non-relativistically in a symmetric star-shaped formation, along radial lines separated by an angle $2\pi/N$, and simultaneously collide, then N vortices emerge from the collision in a similar star, but rotated relative to the incoming one by π/N. Again a reasonable interpretation is that the matter making up each outgoing vortex is acquired equally from the two vortices which are on the closest lines of the incoming star. See Fig. 7.12 for the example of four colliding vortices.

In a high speed N-vortex collision, the final configuration may consist of $N + N'$ vortices and N' antivortices (of unit winding). There is a threshold kinetic energy $2M$ for a vortex-antivortex pair to be produced. The mechanism of pair production is not well understood, and surprisingly, numerical simulations show that in the gauged GL theory such pair production seems hardly ever to occur, even at very high energy [305].

At critical coupling ($\lambda = 1$), one can say considerably more about non-relativistic vortex motion, because one can model the dynamics by geodesic motion on the moduli space of exact static solutions. This will be discussed from Section 7.8 onwards.

7.7.2 Gradient flow

Here, we consider the dissipative dynamics of gauged vortices, which arises from the gradient flow equations for the static GL energy function [83].

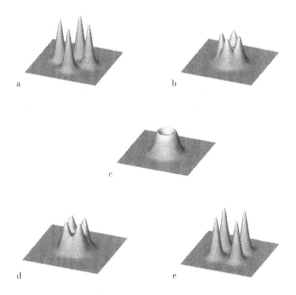

Fig. 7.12.   Energy density plots at increasing times during the $\pi/4$ scattering of four vortices with C_4 symmetry.

Following the discussion in Section 2.8, we take the field equations of the Lorentz invariant theory, and in a gauge covariant way, replace terms with double time derivatives by terms with a single time derivative. The resulting gradient flow equations are

$$\kappa D_0 \phi \;=\; D_i D_i \phi + \frac{\lambda}{2}(1 - \bar{\phi}\phi)\phi \tag{7.105}$$

$$\kappa e_i \;=\; -\varepsilon_{ij}\partial_j B - \frac{i}{2}(\bar{\phi}D_i\phi - \phi\overline{D_i\phi}) \tag{7.106}$$

with κ positive, a special case of Eqs. (2.167) and (2.168). These are known as the Gorkov-Eliashberg, or TDGL (Time Dependent Ginzburg-Landau) equations [164].

Recall that these equations should be supplemented by Gauss' law

$$\partial_i e_i = -\frac{i}{2}(\bar{\phi}D_0\phi - \phi\overline{D_0\phi}), \tag{7.107}$$

but this follows automatically for any choice of a_0 by taking the divergence of (7.106) and using the expression for $D_i D_i \phi$ in (7.105). In particular, one can safely choose the gauge $a_0 = 0$, and then the gradient flow equations are

$$\kappa\partial_0\phi = D_iD_i\phi + \frac{\lambda}{2}(1 - \bar{\phi}\phi)\phi \qquad (7.108)$$

$$\kappa\partial_0 a_i = -\varepsilon_{ij}\partial_j B - \frac{i}{2}(\bar{\phi}D_i\phi - \phi\overline{D_i\phi}), \qquad (7.109)$$

and Gauss' law simplifies too. The geometrical reason for these simplifications was discussed in Section 2.8.

Note that (7.106) is a two-dimensional Ampère equation, $-\varepsilon_{ij}\partial_j B = J_i^{\text{total}}$, where $\mathbf{J}^{\text{total}}$ is the sum of the supercurrent $\mathbf{J}^{\text{S}} = \frac{i}{2}(\bar{\phi}\mathbf{D}\phi - \phi\overline{\mathbf{D}\phi})$ and a normal current $\mathbf{J}^{\text{N}} = \kappa\mathbf{e}$. So κ is the Ohmic conductivity.

The key property of the gradient flow is that it is in the direction in which the GL energy V decreases. In addition, the momentum and angular momentum, given by the formulae (7.96) and (7.99), vanish. This is because these quantities measure the rate of flow in field configuration space in the directions defined by spatial translations and rotations. But these directions are tangent to the hypersurfaces $V = $ const, whereas the gradient flow is orthogonal to them. The electric charge Q is also zero.

What solutions to the gradient flow equation does one expect if the initial data are a non-coincident multi-vortex configuration? For $\lambda < 1$, vortices attract, so vortices approach and coalesce. For $\lambda > 1$, vortices repel each other and separate, eventually moving out to infinity. When the vortices are far apart, the velocity of each vortex is proportional to the force acting on it, but the forces are exponentially small. The motion is therefore very slow.

The circularly symmetric, coincident N-vortex solution is static, but it is unstable, and if perturbed, will generally split up into separated vortices. Let us consider in more detail the two-vortex case. We may define a submanifold M_2 of the field configuration space $\mathcal{C}_2$, as follows. The circularly symmetric solution has two linearly independent unstable modes. One of these separates the vortex zeros in the x^1 direction, and the orthogonal mode separates them in the direction at 45° to this (since separation in the x^2 direction is the negative of separation in the x^1 direction). Having perturbed the vortices by the first mode, the gradient flow evolves the field to two well separated vortices. The zeros of ϕ move along the x^1-axis in opposite directions at the same speeds, by reflection symmetry. This gives one gradient flow curve in $\mathcal{C}_2$. Acting with the rotation group $SO(2)$ we obtain all the gradient flow curves which descend from the circularly symmetric solution. The union of these curves forms a smooth two-dimensional surface embedded in $\mathcal{C}_2$. If we allow for all possible translations of these fields, we get a four-dimensional manifold $M_2 \subset \mathcal{C}_2$, invariant under the gradient flow.

M_2 is the unstable manifold of the circularly symmetric solution and its translates, and is probably an attractor of the gradient flow in C_2. Starting with almost any finite energy field configuration in the two-vortex sector, the flow descends rapidly towards the two-vortex configurations lying on M_2, and then evolves as in M_2. Rather special initial data of codimension two in C_2 will flow to one of the circularly symmetric solutions, and form the stable manifold of these solutions. The stable manifold includes all fields with the same circular symmetry as the solutions, but different radial dependence.

Moore [305] has solved the gradient flow equations and determined the geometry of the manifold M_2 numerically, showing that it has a similar structure to $\mathcal{M}_2$, the moduli space of static two-vortex solutions at $\lambda = 1$. However, it is difficult to accurately calculate the intrinsic geometry of M_2, whereas we shall be able to determine the geometry of $\mathcal{M}_2$ in considerable detail.

Rather more precise properties of the gradient flow have been proved in the case $\lambda = 1$. In this critical case, in the sector of vortex number N, Jaffe and Taubes have shown that there are *no* static solutions of the gauged GL equation except those with minimal energy [223], and the solutions of minimal energy form the moduli space $\mathcal{M}_N$. The absence of higher stationary points means that a gradient flow curve in C_N can not end anywhere except on $\mathcal{M}_N$. Demoulini and Stuart have proved that for arbitrary finite energy initial data in C_N, the gradient flow equation has a solution well defined at all later times, and the fields remain smooth [106]. $\mathcal{M}_N$ is a global attractor. Moreover, the fields approach a definite point in $\mathcal{M}_N$ (after suitable gauge fixing). This is not quite obvious. Although the potential energy function V is constant on $\mathcal{M}_N$, there are gradients of V in the neighbourhood of $\mathcal{M}_N$. However, these are not sufficient to push the vortices out to infinite separation. The vortices approach a particular configuration with zeros at finite points of the plane.

It may appear that the gradient flow equations and their solutions are of rather limited interest, as they simply describe relaxation to a stable configuration. However, they rather accurately model the field evolution in real thin superconductors. This is perhaps a surprise, as superconductors are associated with currents that persist indefinitely, and not with the dissipative Ohmic currents that occur in ordinary conductors. However, vortex motion is a dissipative process in superconductors. This is simply illustrated by what happens if a steady current passes through a superconductor which contains a vortex, as sketched in Fig. 7.13. The detailed gradient flow dynamics is quite complicated to calculate, but the conclusion is that the vortex moves at right angles to the current, and this is confirmed experimentally [326]. (If one thinks of the applied current as circulating around a distant multi-vortex, then the motion of the vortex is towards or away from the multi-vortex.) Part of the current

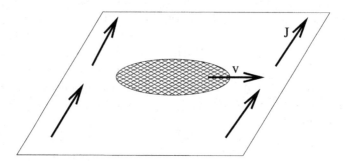

Fig. 7.13.   The motion of a vortex when a steady current **J** is passed through a superconductor.

passes through the normal core of the vortex, where the superconducting condensate, represented by ϕ, is absent, and here there is Ohmic dissipation. The physical reason is that the moving vortex generates an electric field perpendicular to both the vortex velocity and the magnetic field in the vortex core (thought of as perpendicular to the plane of the superconductor). The electric field is therefore in the direction opposite to the current, and hence there is dissipation. In real superconductors, carrying large currents, it is inevitable that some vortices penetrate the superconductor, because the current produces large magnetic fields. Unless the vortices can be pinned, there is dissipation, and this limits the size of the currents that can be set up. Thus the technology of superconductors requires a careful choice of geometry for the current-carrying wires, plus the introduction of vortex pinning centres into the superconducting material.

   Chapman has studied the gradient flow equations for superconductors of finite extent. The main interest is that the external conditions of a superconductor can be suddenly changed, and one can ask how the fields then evolve, and how the vortices move. The external changes can include change of temperature, affecting the sign of $\lambda - 1$, change of externally imposed current, or change of externally imposed magnetic field [81, 82]. Suppose, for example, that a thin superconductor of finite size is bathed in a perpendicular magnetic field. If $\lambda > 1$, which corresponds to a Type II superconductor, there will be vortices penetrating the superconductor. If the field is switched off, the vortices will drift towards the boundary under gradient flow and eventually there will be no more magnetic flux. The details of this are somewhat more complicated than what is described by Eqs. (7.105) and (7.106) because of the boundary conditions and the three-dimensional structure of the field. Analogously, flux flows in from the boundary if the external field is switched on. The flux enters the superconductor at the point on the boundary where the curvature is greatest. The end point of this process is the establishment of the Abrikosov lattice of vortices, whose lattice spacing is determined by the strength of the external field. The Abrikosov lattice is discussed further in Section 7.14.2.

A further interesting process involving relaxation to equilibrium is the annihilation of vortex-antivortex pairs. It is theoretically straightforward to set up an initial field with a well separated vortex and antivortex. There is an attraction between vortex and antivortex (for all λ) because the magnetic and scalar forces act in the same direction. The gradient flow leads to annihilation of the pair, and the fields evolve to the vacuum. This is rigorously proved in the case $\lambda = 1$ [106]. In principle, this process can be observed experimentally by setting up a large magnetic field gradient, so there are vortices and antivortices in different parts of the superconductor. Vortices will tend to drift towards and annihilate antivortices, being replaced by new vortices entering at the boundary.

7.7.3 First order dynamics

In the gauged GL theory, defined in two space and one time dimension, it is possible to add to the Lagrangian density a Chern-Simons term $\frac{1}{2} f \wedge a$ (or a multiple of this) [216, 364]. Its component form is

$$\mathcal{L}_{\text{CS}} = \frac{1}{4} \varepsilon^{\mu\nu\sigma} f_{\mu\nu} a_\sigma = \frac{1}{2} (Ba_0 + e_1 a_2 - e_2 a_1), \qquad (7.110)$$

which in no way depends on the Lorentz metric of space-time. In particular, it is invariant under the Euclidean group of translations and rotations, but it is parity odd.

We showed in Chapter 3 that the Chern-Simons action

$$S_{\text{CS}} = \int_{t_1}^{t_2} \mathcal{L}_{\text{CS}} \, d^2x \, dt \qquad (7.111)$$

is gauge invariant, provided suitable boundary conditions are satisfied. Also the formal variation of S_{CS} with respect to a_σ is $\frac{1}{2} \varepsilon^{\mu\nu\sigma} f_{\mu\nu}$, which is gauge invariant. Thus the field equations get this additional term if S_{CS} is added to the GL action.

One effect of the Chern-Simons term is to modify Gauss' law. It now has the form

$$\nabla \cdot \mathbf{e} + B = j^0 \qquad (7.112)$$

where j^0 is the electric charge density. Generally, the electric field decays rapidly as $|\mathbf{x}| \to \infty$, so integration of (7.112) over the plane gives the relation

$$Q = \Phi \tag{7.113}$$

between the total magnetic flux Φ and the total electric charge Q. Therefore vortices become charged [217].

Another effect of the Chern-Simons term on vortices is to deflect their motion away from the direction in which the static force acts. So $\lambda > 1$ vortices, which would tend to move directly away from each other, will now move along spiral trajectories, consistent with the parity violation.

Many Chern-Simons variants of GL theory have been studied, and we discuss two of them. The first is called pure Chern-Simons theory [169]. Here, the Maxwell term in the Lagrangian of scalar electrodynamics (2.99) is removed, and replaced by a Chern-Simons term. The Lagrangian is the Lorentz invariant expression

$$L = \int \left(\frac{1}{2}\overline{D_\mu\phi}D^\mu\phi + \frac{1}{4}\varepsilon^{\mu\nu\sigma} f_{\mu\nu} a_\sigma - U(\bar{\phi}\phi) \right) d^2x, \tag{7.114}$$

where U is not necessarily quartic in $|\phi|$. The field equations are

$$D_\mu D^\mu \phi + 2U'(\bar{\phi}\phi)\phi = 0 \tag{7.115}$$

$$\frac{1}{2}\varepsilon^{\mu\nu\sigma} f_{\mu\nu} + \frac{i}{2}(\bar{\phi}D^\sigma\phi - \phi\overline{D^\sigma\phi}) = 0, \tag{7.116}$$

a mixture of second order dynamics for the scalar field and first order dynamics for the gauge field. The $\sigma = 0$ component of (7.116) is the Gauss law

$$B + \frac{i}{2}(\bar{\phi}D_0\phi - \phi\overline{D_0\phi}) = 0, \tag{7.117}$$

which has no $\nabla \cdot \mathbf{e}$ term any more. The magnetic field is therefore precisely equal to the electric charge density at every point.

One may seek stationary solutions for this system, with a_0 non-zero, satisfying the field equations

$$a_0^2\phi + D_i D_i\phi - 2U'(\bar{\phi}\phi)\phi = 0 \tag{7.118}$$

$$\varepsilon_{ij}\partial_j a_0 - \frac{i}{2}(\bar{\phi}D_i\phi - \phi\overline{D_i\phi}) = 0 \tag{7.119}$$

$$B + a_0\bar{\phi}\phi = 0. \tag{7.120}$$

For the special sextic potential

$$U = \frac{1}{8}(1 - \bar{\phi}\phi)^2\bar{\phi}\phi \tag{7.121}$$

these equations can be reduced to a first order system of Bogomolny equations with static vortex solutions [190, 217]. We shall discuss this further in Section 7.8.

We turn now to the second variant. Since the Chern-Simons term is linear in the time derivative of the gauge potential, it is natural in the GL Lagrangian to replace the quadratic term in the time derivative of the scalar field by the linear term $\frac{i}{2}(\bar{\phi}\,D_0\phi - \phi\overline{D_0\phi})$, which is gauge invariant and real. This breaks Lorentz invariance, although it does not violate the two-dimensional Euclidean symmetry. Particularly interesting is the Schrödinger-Chern-Simons model with Lagrangian [287]

$$L_{\text{SCS}} \;=\; \int \Big(\frac{i\gamma}{2}(\bar{\phi}D_0\phi - \phi\overline{D_0\phi}) + \mu(Ba_0 + e_1a_2 - e_2a_1) - \gamma a_0$$
$$-\frac{1}{2}B^2 - \frac{1}{2}\overline{D_i\phi}D_i\phi - \frac{\lambda}{8}(1 - \bar{\phi}\phi)^2 + a_i J_i^{\text{T}}\Big)\,d^2x \quad (7.122)$$

where γ, μ, λ are real constants, and J_i^{T} a constant vector. This contains the static GL potential energy, including the B^2 term, but no e_ie_i term. The contribution γa_0 is rather unusual [32]. Under a gauge transformation $e^{i\alpha(t,\mathbf{x})}$ it varies by the time derivative $\gamma\partial_0\alpha$, so the action is invariant provided α is the same at the initial and final time.

The field equations obtained from this Lagrangian are

$$i\gamma D_0\phi \;=\; -\frac{1}{2}D_iD_i\phi - \frac{\lambda}{4}(1 - \bar{\phi}\phi)\phi \qquad (7.123)$$
$$-\varepsilon_{ij}\partial_j B \;=\; J_i^{\text{S}} - J_i^{\text{T}} + 2\mu\varepsilon_{ij}e_j \qquad (7.124)$$
$$2\mu B \;=\; \gamma(1 - \bar{\phi}\phi)\,. \qquad (7.125)$$

The first equation is a gauge covariant nonlinear Schrödinger equation. The second is an Ampère equation. The total current is the combination of J_i^{S}, the usual supercurrent (7.19), J_i^{T}, the constant external "transport" current, and a Hall current orthogonal to the electric field [114]. 2μ is the Hall conductivity. The third equation is the Gauss law for this system, and involves no time derivatives. It is a constraint on initial data, and is preserved by the time evolution, since one can show using the first two equations that

$$\frac{\partial}{\partial t}\left(2\mu B - \gamma(1 - \bar{\phi}\phi)\right) = 0\,. \qquad (7.126)$$

The term γa_0 in the Lagrangian contributes the γ in Gauss' law, and allows the possibility that asymptotically, $|\phi| = 1$ and $B = 0$. This means the model can accommodate vortices of the kind we have been discussing.

A noteworthy feature of this model is its Galilean invariance. This generalizes the Galilean invariance of the nonlinear Schrödinger equation. A Galilean transformation involves a boost in velocity by $\mathbf{v}$, so $\mathbf{x} \mapsto \mathbf{x}' = \mathbf{x} - \mathbf{v}t$. It is the low velocity limit of a Lorentz boost. The fields $\{\phi, a_\mu\}$ transform to

$$\phi'(t, \mathbf{x}) \;=\; \phi(t, \mathbf{x}')e^{i\gamma(\mathbf{v}\cdot\mathbf{x} - \frac{1}{2}|\mathbf{v}|^2 t)} \qquad (7.127)$$

$$\mathbf{a}'(t,\mathbf{x}) = \mathbf{a}(t,\mathbf{x}') \tag{7.128}$$

$$a_0'(t,\mathbf{x}) = a_0(t,\mathbf{x}') - \mathbf{v}\cdot\mathbf{a}(t,\mathbf{x}'). \tag{7.129}$$

These satisfy the field equations if the original fields do, provided we transform the transport current to $(\mathbf{J}^{\mathrm{T}})' = \mathbf{J}^{\mathrm{T}} - \gamma\mathbf{v}$. If the original fields describe some motion of vortices, then the transformed fields describe a similar motion but with a superimposed drift velocity $\mathbf{v}$.

Physically, the Galilean transformation changes the transport current because it changes the asymptotic fields. If B and e_i are asymptotically zero, then $\mathbf{J}^{\mathrm{T}}$ is the asymptotic value of $\mathbf{J}^{\mathrm{S}}$. Suppose that $|\phi| \to 1$ and $\mathbf{D}\phi \to \mathbf{0}$ as $|\mathbf{x}| \to \infty$, so $\mathbf{J}^{\mathrm{T}} = \mathbf{0}$. After the transformation, $|\phi| \to 1$ and $\mathbf{D}\phi \to i\gamma\mathbf{v}\phi$, so $(\mathbf{J}^{\mathrm{T}})' = -\gamma\mathbf{v}$. The drift of vortices, at velocity $\mathbf{v}$, is the response.

The motion of a vortex parallel to a superimposed external current is non-dissipative, and characteristic of Chern-Simons dynamics [104, 319]. It is the analogue of the Magnus drift of fluid vortices. Having understood the role of the transport current, we from now on only consider the closed system where $\mathbf{J}^{\mathrm{T}} = \mathbf{0}$.

Another feature of this Schrödinger-Chern-Simons model is seen most easily in the gauge $a_0 = 0$. Assuming that Gauss' law is satisfied, the field equations reduce to

$$i\gamma\partial_0\phi = -\frac{1}{2}D_iD_i\phi - \frac{\lambda}{4}(1-\bar{\phi}\phi)\phi \tag{7.130}$$

$$2\mu\varepsilon_{ij}\partial_0 a_j = -\varepsilon_{ij}\partial_j B - J_i^{\mathrm{S}}. \tag{7.131}$$

Notice how similar these are to the gradient flow equations of GL theory (7.108) and (7.109). Schrödinger-Chern-Simons dynamics and gradient flow dynamics are in fact orthogonal in configuration space, in the sense discussed in Section 2.6, because at each point $\mathbf{x}$, $\{i\partial_0\phi, \varepsilon_{ij}\partial_0 a_j\}$ is orthogonal to $\{\partial_0\phi, \partial_0 a_i\}$. This suggests that vortex motion in this model will be at right angles to the vortex motion occurring in gradient flows. In particular, two vortices should circle around one another, like fluid vortices of equal strength. We shall establish this later, for a rather restricted range of the parameters γ, μ and λ, using the moduli space approach to vortex dynamics.

We conclude by noting some of the conservation laws for finite energy fields satisfying the boundary conditions $|\phi| \to 1$, $D_i\phi \to 0$ and $B \to 0$ as $|\mathbf{x}| \to \infty$ [288, 174]. Since the Lagrangian density is linear in the time derivatives of the fields, no time derivatives appear in the conserved quantities. There is, as usual, a conserved winding number N, which defines the net vortex number. From (7.125) it follows that, at all times,

$$\int B\,d^2x = \frac{\gamma}{2\mu}\int(1-\bar{\phi}\phi)\,d^2x = 2\pi N. \tag{7.132}$$

The conserved energy is the standard GL expression

$$E = \frac{1}{2} \int \left(B^2 + \overline{D_i \phi} D_i \phi + \frac{\lambda}{4} (1 - \bar{\phi}\phi)^2 \right) d^2 x .$$

(7.133)

The conserved momentum has the two components

$$P_i = -\gamma \int (J_i^S - \varepsilon_{ij} x_j B) \, d^2 x .$$

(7.134)

Because the Lagrangian density is not gauge invariant, it is necessary to use the improvement method and to add a further total derivative term to obtain this result. The conserved angular momentum is

$$\ell = -\gamma \int \left(\varepsilon_{ij} x_i J_j^S + \frac{1}{2} |\mathbf{x}|^2 B \right) d^2 x .$$

(7.135)

The preceeding conservation laws can be reexpressed in terms of a field vorticity density $\mathcal{W} = -\varepsilon_{ij} \partial_i J_j^S + B$. This was noted in a slightly different model by Papanicolaou and Tomaras [325]. The integral of $\mathcal{W}$ is $2\pi N$, and

$$P_i = \gamma \varepsilon_{ij} \int x_j \mathcal{W} \, d^2 x , \quad \ell = -\frac{\gamma}{2} \int |\mathbf{x}|^2 \mathcal{W} \, d^2 x ,$$

(7.136)

as is easily verified by integrating by parts. The Schrödinger-Chern-Simons model therefore has analogies with ideal fluid dynamics in two dimensions [37], and also with certain ferromagnetic models [324], where the conserved momentum and angular momentum are moments of a vorticity density.

7.8 Vortices at critical coupling

It has been recognized for a long time that the gauged GL model has physically and mathematically special properties when $\lambda = 1$, the critical value of the coupling [152]. This value separates Type I ($\lambda < 1$) and Type II ($\lambda > 1$) superconductivity. Some real materials are close to this critical value. Pure niobium has $\lambda \simeq 2.4$, and certain alloys (for example, lead with 1%–2% thallium) have λ arbitrarily close to critical coupling [326].

Interest in the $\lambda = 1$ case really started with the famous paper of Bogomolny [56]. At critical coupling, the static GL energy function

$$E = V = \frac{1}{2} \int \left(B^2 + \overline{D_i \phi} D_i \phi + \frac{1}{4} (1 - \bar{\phi}\phi)^2 \right) d^2 x$$

(7.137)

can be rewritten as

$$E = \frac{1}{2} \int \Bigg\{ \left(B - \frac{1}{2}(1 - \bar{\phi}\phi) \right)^2 + \left(\overline{D_1 \phi} - i\overline{D_2 \phi} \right)\left(D_1 \phi + i D_2 \phi \right)$$
$$+ B - i\left(\partial_1 (\bar{\phi} D_2 \phi) - \partial_2 (\bar{\phi} D_1 \phi) \right) \Bigg\} d^2 x .$$

(7.138)

The last expression is a two-dimensional curl, so its integral can be re-expressed as a line integral around the circle at infinity. If the energy is finite, so that $|\phi| \to 1$ and $D_i\phi \to 0$ as $|\mathbf{x}| \to \infty$, then this line integral vanishes. Also, the integral of B is 2π times the winding number of the field, N. Thus

$$E = \frac{1}{2} \int \left\{ \left(B - \frac{1}{2}(1 - \bar\phi\phi) \right)^2 + \left(\overline{D_1\phi} - i\overline{D_2\phi} \right) \left(D_1\phi + iD_2\phi \right) \right\} d^2x + \pi N .$$
$$(7.139)$$

Since the integrand in (7.139) is non-negative, we deduce that $E \geq \pi N$, which is non-trivial if $N > 0$ (since it is clear from (7.137) that $E \geq 0$). Similarly, by reversing some signs, we can show (for any N) that $E \geq \pi(-N)$. Thus, generally, the energy satisfies the Bogomolny bound

$$E \geq \pi |N| . \qquad (7.140)$$

The Bogomolny bound is said to be saturated if there is equality, and $E = \pi |N|$. For $N = 0$ this only occurs for the vacuum field. For $N > 0$, the bound is saturated if and only if the fields satisfy

$$D_1\phi + iD_2\phi = 0 \qquad (7.141)$$
$$B - \frac{1}{2}(1 - \bar\phi\phi) = 0 , \qquad (7.142)$$

which we shall call the first and second Bogomolny equations. For $N < 0$, one has $E = \pi |N|$ if $D_1\phi - iD_2\phi = 0$ and $B + \frac{1}{2}(1 - \bar\phi\phi) = 0$. The sign of N is reversed by a reflection, e.g. $(x^1, x^2) \mapsto (x^1, -x^2)$, together with the sign reversal of a_2. From now on we consider only the case of $N > 0$.

Fields satisfying the Bogomolny equations (7.141) and (7.142) (assuming there are some) are automatically minima of the energy within their topological class, and hence guaranteed to be stable. Let us check that they also satisfy the GL field equations (7.17) and (7.18), which represent the weaker condition for a stationary point of the energy. The Bogomolny equations imply that

$$(D_1 - iD_2)(D_1 + iD_2)\phi = 0 , \qquad (7.143)$$

so

$$(D_1 D_1 + D_2 D_2)\phi = -i[D_1, D_2]\phi = -B\phi = -\frac{1}{2}(1 - \bar\phi\phi)\phi , \qquad (7.144)$$

which is one of the field equations. Also

$$\partial_1 B = \frac{1}{2}\partial_1(1 - \bar\phi\phi) = -\frac{1}{2}(\bar\phi D_1\phi + \overline{D_1\phi}\phi) = \frac{i}{2}(\bar\phi D_2\phi - \phi\overline{D_2\phi}) , \qquad (7.145)$$

and similarly $\partial_2 B = -\frac{i}{2}(\bar{\phi}D_1\phi - \phi\overline{D_1\phi})$, which are the remaining equations.

The two equations (7.141) and (7.142) are of a different character. The first is a covariant generalization of the Cauchy-Riemann equations, and occurs in other variants of the GL theory. The second equation is characteristic of the specific GL energy (7.137).

A deep study of the Bogomolny equations was made by Taubes [396, 397], and presented in the book by Jaffe and Taubes [223]. One result is that for a field obeying the Bogomolny equations, the multiplicities of the zeros of ϕ are all positive. Thus solutions of the Bogomolny equations with winding number N have a finite number of positive multiplicity zeros, with the sum of the multiplicities being N. The solutions therefore represent N vortices, located at these points, with no antivortices present.

Another observation of Taubes is that (7.141) can be used to eliminate the gauge potential a_i from the pair of Bogomolny equations. Let us write $\phi = |\phi|e^{i\chi}$ and then define

$$h = \log|\phi|^2\,, \tag{7.146}$$

so $\phi = e^{\frac{1}{2}h+i\chi}$. h is gauge invariant and finite, except at the points where $\phi = 0$. h vanishes on the circle at infinity, where $|\phi| = 1$. Equation (7.141) becomes

$$\partial_1\left(\frac{1}{2}h+i\chi\right) - ia_1 + i\partial_2\left(\frac{1}{2}h+i\chi\right) + a_2 = 0 \tag{7.147}$$

so

$$a_1 = \frac{1}{2}\partial_2 h + \partial_1\chi\,, \quad a_2 = -\frac{1}{2}\partial_1 h + \partial_2\chi\,. \tag{7.148}$$

The magnetic field B is now

$$B = \partial_1 a_2 - \partial_2 a_1 = -\frac{1}{2}\nabla^2 h\,. \tag{7.149}$$

Hence the equation (7.142) becomes

$$\nabla^2 h + 1 - e^h = 0\,, \tag{7.150}$$

and χ is eliminated.

This equation is valid except at the zeros of ϕ, where h has logarithmic singularities, and becomes infinitely negative. We can allow for these singularities by including delta-function sources, giving the final equation for h

$$\nabla^2 h + 1 - e^h = 4\pi\sum_{r=1}^{N}\delta^2(\mathbf{x} - \mathbf{X}_r) \tag{7.151}$$

where $\{\mathbf{X}_r\}$ are the positions of the (simple) zeros of ϕ. To verify that Eq. (7.151) has the correct delta-functions, recall that near a simple zero $\mathbf{X}$, $|\phi| \sim |\mathbf{x} - \mathbf{X}|$, so $h \sim 2\log|\mathbf{x} - \mathbf{X}|$ and $\nabla^2 h = 4\pi\delta^2(\mathbf{x} - \mathbf{X})$. Since $e^h \sim |\mathbf{x} - \mathbf{X}|^2$, e^h has no singularity. If the zero has multiplicity n, then $|\phi| \sim |\mathbf{x} - \mathbf{X}|^n$, so $h \sim 2n\log|\mathbf{x} - \mathbf{X}|$ and the delta-function source needs a coefficient $4\pi n$. But this is implied by (7.151), because a zero of multiplicity n contributes n times to the sum. We also observe that since a_1 and a_2 are not singular at $\mathbf{X}$, Eqs. (7.148) require $\nabla\chi$ to have a singularity to cancel the singularity in ∇h. In terms of polar coordinates ρ, θ centred at $\mathbf{X}$, $h \sim 2n\log\rho$ and therefore $\partial_\rho h \sim 2n/\rho$, $\partial_\theta h \sim 0$. Hence $\partial_\rho\chi \sim 0$, $\partial_\theta\chi \sim n$, so the increase of χ around $\mathbf{X}$ is $2\pi n$. This confirms Taubes' more rigorous argument that only zeros of positive multiplicity can occur for solutions of the Bogomolny equations.

The strategy for solving the Bogomolny equations with winding number N is to first fix N points in the plane $\{\mathbf{X}_1, \mathbf{X}_2, \ldots, \mathbf{X}_N\}$. These points are unordered, and not necessarily distinct. Then solve (7.151) subject to the boundary condition $h \to 0$ as $|\mathbf{x}| \to \infty$. For $N = 1$ there is a unique circularly symmetric solution, centred at the origin. This is the basic Bogomolny vortex, with energy, and hence mass, equal to π. Taubes has proved that there exists a unique solution for any N-tuple of points. This is what one expects at critical coupling, where there are no forces between vortices. An important property of this N-vortex solution is that h is everywhere negative, so $|\phi| < 1$ and $B > 0$ everywhere [223]. This is because the boundary condition and absence of positive singularities of h imply that if h were anywhere positive it would attain a maximal positive value at some point; and there, $\nabla^2 h \leq 0$ and $1 - e^h < 0$, contradicting (7.151). h carries all the gauge invariant information. Thus $|\phi|^2 = e^h$, $B = -\frac{1}{2}\nabla^2 h$, and $J_i = \frac{1}{2}\varepsilon_{ij}\partial_j h \, e^h$. The electric current flows along the contours of h, and $\frac{1}{2}e^h$ plays the role of a stream function.

Solutions of (7.151) can be constructed numerically. After fixing the points $\{\mathbf{X}_r\}$, small discs of radius ε centred at those points are inserted, on the boundary of which h is set equal to $2\log\varepsilon$ (or $2n\log\varepsilon$ if the point is of multiplicity n). On a large circle or square (representing infinity) h is set equal to zero. The equation (7.150) is then solved in the region between. Examples of solutions are shown in Fig. 7.14.

$\mathcal{M}_N$, the moduli space of solutions of the Bogomolny equations with winding number N, is the space of unordered N-tuples of points in the plane. As a manifold this is $(\mathbb{R}^2)^N/S_N$ where S_N is the permutation group on N objects. Normally, a "manifold" of the type X^N/S_N – the Nth symmetrized power of a manifold X – is not smooth. There are singularities where two or more points of X coincide, because the orbit under S_N of such a point in X^N is of smaller size than the generic orbit

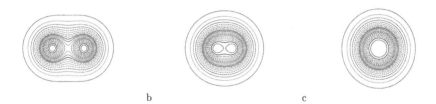

a   b   c

Fig. 7.14. Contour plots of $|\phi|^2$ for two vortices with separation s, for (a) $s = 4$, (b) $s = 2$, (c) $s = 0$.

of size $N!$. If X is $\mathbb{R}^2$ (or any other two-dimensional manifold) this argument doesn't apply. Identify $\mathbb{R}^2$ with $\mathbb{C}$, and use the standard complex coordinate z. Let the complex vortex positions be $\{Z_1, \ldots, Z_N\}$, where $Z_r = X_r^1 + iX_r^2$, and define

$$p(z) = \prod_{r=1}^{N}(z - Z_r)\,. \tag{7.152}$$

p is a polynomial of degree N and monic (the leading coefficient is unity), and has an expansion

$$p(z) = z^N + p_1 z^{N-1} + \cdots + p_N\,. \tag{7.153}$$

There is a one-to-one correspondence between the set of complex coefficients $\{p_1, \ldots, p_N\}$ and the set of unordered points $\{Z_1, \ldots, Z_N\}$. The points determine the coefficients, since there are explicit formulae for $\{p_r\}$ as elementary symmetric polynomials in the roots

$$p_1 = -(Z_1 + \cdots + Z_N)\,, \quad \ldots\,, \quad p_N = (-1)^N Z_1 Z_2 \cdots Z_N\,, \tag{7.154}$$

and conversely, the coefficients determine $p(z)$ and hence its N unordered roots. Now the set of coefficients $\{p_1, \ldots, p_N\}$ is an *ordered* set of arbitrary complex numbers. Therefore $\mathcal{M}_N = \mathbb{C}^N$, with these coefficients as global coordinates. It is perhaps rather paradoxical that $\mathbb{C}^N/S_N$ is diffeomorphic to $\mathbb{C}^N$, but that is what this argument shows. The moduli space can be regarded as having the usual smooth manifold structure of $\mathbb{C}^N$.

This moduli space structure can immediately be seen to be consistent with the right-angle scattering of two vortices, or the π/N scattering of N vortices. For suppose we consider the smooth trajectory in $\mathcal{M}_N$

$$p_1 = 0\,, \quad p_2 = 0\,, \quad \ldots\,, \quad p_N = t \tag{7.155}$$

where t (real) can be thought of as time. The corresponding polynomial is

$$p(z; t) = z^N + t \tag{7.156}$$

whose roots for $t \leq 0$ are $Z_r = |t|^{1/N} e^{2\pi i r/N}$, $r = 1, \ldots, N$, and whose roots for $t \geq 0$ are $Z_r = |t|^{1/N} e^{\pi i(2r-1)/N}$, $r = 1, \ldots, N$. This implies that the star of outgoing vortex trajectories is rotated by π/N with respect to the star of incoming trajectories.

We conclude this section with a brief discussion of the Bogomolny equations for pure Chern-Simons vortices [190, 217], and their solutions [213]. The static equations to be solved are (7.118)–(7.120), where $U(\bar{\phi}\phi) = \frac{1}{8}(1 - \bar{\phi}\phi)^2 \bar{\phi}\phi$, so $2U'(\bar{\phi}\phi) = \frac{1}{4} - \bar{\phi}\phi + \frac{3}{4}(\bar{\phi}\phi)^2$. They are satisfied provided the first order Bogomolny equations

$$D_1\phi + iD_2\phi = 0 \tag{7.157}$$

$$B - \frac{1}{2}(1 - \bar{\phi}\phi)\bar{\phi}\phi = 0 \tag{7.158}$$

$$a_0 + \frac{1}{2}(1 - \bar{\phi}\phi) = 0 \tag{7.159}$$

hold. Equation (7.120) is obviously satisfied. Equation (7.118) follows by a similar calculation to that leading to (7.144), because here $-B = 2U'(\bar{\phi}\phi) - a_0^2$. Differentiating (7.159) and using (7.157), one shows that (7.119) is satisfied.

As for the GL vortices, one can make the ansatz $\phi = e^{\frac{1}{2}h + i\chi}$. The first Bogomolny equation determines the spatial part of the gauge potential as before, and the second equation becomes the gauge invariant scalar equation

$$\nabla^2 h + e^h - e^{2h} = 0. \tag{7.160}$$

Finally, $a_0 = -\frac{1}{2}(1 - e^h)$.

Two types of boundary condition and vortex are possible. Topological vortices have $|\phi| \to 1$ as $|\mathbf{x}| \to \infty$, so $h \to 0$. There is an integer winding number N, and quantized flux $2\pi N$. Taubes' methods, and index theory calculations, show that there is a $2N$-dimensional moduli space of N-vortex solutions. The vortex centres, where ϕ vanishes, are again N arbitrary points in the plane. The main difference between the Chern-Simons vortices and GL vortices is that the magnetic field is peaked on an annulus around each vortex centre, rather than having a maximum at the centre. The non-topological vortices have boundary condition $|\phi| \to 0$ as $|\mathbf{x}| \to \infty$, so h diverges. For more on these solutions, see [213, 232].

7.9 Moduli space dynamics

Here, we begin to explore the idea that at or near critical coupling, we may approximate the field theory dynamics of GL vortices by a reduced dynamics on the moduli space, provided the vortices are slowly moving. We refer back to Section 4.5 for the basic formalism and its justification.

As we explained in the last section, the moduli space $\mathcal{M}_N$ for N critically coupled GL vortices is the $2N$-dimensional manifold $\mathbb{C}^N$, whose complex Cartesian coordinates are the coefficients $\{p_1, \ldots, p_N\}$ of the polynomial $p(z)$ whose roots $\{Z_1, \ldots, Z_N\}$ are the vortex positions. However, we usually find it most convenient to use as coordinates $\{Z_1, \ldots, Z_N\}$ themselves. These are locally well defined (with some arbitrary choice of order) when the vortices are separated. We can check, after certain calculations, that the structures on $\mathcal{M}_N$ are smooth as some or all of the vortices coincide. Occasionally we will use real coordinates $\{y^i : 1 \le i \le 2N\}$ on $\mathcal{M}_N$.

At critical coupling, the second order dynamics reduces to free, geodesic motion on the moduli space. The reduced Lagrangian for N-vortex motion is purely kinetic, and defines a metric on $\mathcal{M}_N$, which we need to study and calculate. Using the coordinates $\{Z_1, \ldots, Z_N\}$, one may write the metric as

$$ds^2 = \sum_{r,s=1}^{N} \left(g_{rs} dZ_r dZ_s + g_{r\bar{s}} dZ_r d\bar{Z}_s + g_{\bar{r}\bar{s}} d\bar{Z}_r d\bar{Z}_s \right), \tag{7.161}$$

where $g_{rs}, g_{r\bar{s}}, g_{\bar{r}\bar{s}}$ are functions of the coordinates and their complex conjugates. Reality of the metric requires that $g_{\bar{r}\bar{s}} = \bar{g}_{rs}$ and $g_{r\bar{s}} = \bar{g}_{s\bar{r}}$, so that as an $N \times N$ matrix, $g_{r\bar{s}}$ is Hermitian. We shall see in the next section that $g_{rs} = g_{\bar{r}\bar{s}} = 0$, and that the metric is actually Kähler. The reduced Lagrangian is then

$$L_{\text{red}} = \frac{1}{2} \sum_{r,s=1}^{N} g_{r\bar{s}} \dot{Z}_r \dot{\bar{Z}}_s, \tag{7.162}$$

leading to the equation for geodesic motion at constant speed on $\mathcal{M}_N$. (The subscript $\bar{s}$ is just a convenient shorthand; $\bar{s}$, like s, runs from 1 to N, and the bar simply indicates the pairing with $d\bar{Z}_s$ or $\dot{\bar{Z}}_s$.)

If one is slightly away from critical coupling, that is, $\lambda \simeq 1$, then N-vortex motion is still well approximated by a motion on the moduli space $\mathcal{M}_N$. The metric need not be changed, but one should include the perturbation of the potential energy

$$V_{\text{red}} = \frac{\lambda - 1}{8} \int (1 - \bar{\phi}\phi)^2 \, d^2x \tag{7.163}$$

as a potential on $\mathcal{M}_N$. In principle, this can be evaluated. Given a set of vortex positions $\{Z_1, \ldots, Z_N\}$, one must solve for h, and then

$$V_{\text{red}} = \frac{\lambda - 1}{8} \int (1 - e^h)^2 \, d^2x. \tag{7.164}$$

The total reduced Lagrangian is

$$L_{\text{red}} = \frac{1}{2} \sum_{r,s=1}^{N} g_{r\bar{s}} \dot{Z}_r \dot{\bar{Z}}_s - V_{\text{red}} \,. \tag{7.165}$$

Varying the action with respect to $\bar{Z}_s$ leads to the equation for geodesic motion modified by the effect of small forces

$$g_{r\bar{s}}\ddot{Z}_r + \frac{\partial g_{r\bar{s}}}{\partial Z_u} \dot{Z}_r \dot{Z}_u + \frac{\partial g_{r\bar{s}}}{\partial \bar{Z}_u} \dot{Z}_r \dot{\bar{Z}}_u - \frac{\partial g_{r\bar{u}}}{\partial \bar{Z}_s} \dot{Z}_r \dot{\bar{Z}}_u + 2\frac{\partial V_{\text{red}}}{\partial \bar{Z}_s} = 0 \,. \tag{7.166}$$

Because of the Kähler property, the third and fourth terms cancel, and the equation of motion simplifies to

$$g_{r\bar{s}}\ddot{Z}_r + \frac{\partial g_{r\bar{s}}}{\partial Z_u} \dot{Z}_r \dot{Z}_u + 2\frac{\partial V_{\text{red}}}{\partial \bar{Z}_s} = 0 \,. \tag{7.167}$$

At critical coupling, $\lambda = 1$, the field theory gradient flow has the moduli space $\mathcal{M}_N$ as an attractor, but the motion on $\mathcal{M}_N$ itself is trivial. Close to critical coupling, the attractor appears to persist. For $\lambda > 1$ it is the unstable manifold of the circularly symmetric, coincident N-vortex solution (and its translates). We assume that this can be approximated by $\mathcal{M}_N$ with its unchanged metric, on which V_{red} is the potential. The reduced N-vortex gradient flow dynamics is then given by the equation of motion on $\mathcal{M}_N$ [387]

$$\kappa g_{r\bar{s}}\dot{Z}_r + 2\frac{\partial V_{\text{red}}}{\partial \bar{Z}_s} = 0 \,, \tag{7.168}$$

which is a complex version of Eq. (2.24). The potential energy V_{red} produces a force, and hence motion in the direction of the force.

The first order Schrödinger-Chern-Simons dynamics of vortices leads to a reduced Lagrangian [287]

$$L_{\text{red}} = -\sum_{i=1}^{2N} \mathcal{A}_i(\mathbf{y})\dot{y}^i - V_{\text{red}}(\mathbf{y}) \,, \tag{7.169}$$

analogous to Eq. (2.21). This will be shown in Section 7.12. L_{red} is again defined on the moduli space $\mathcal{M}_N$, but is expressed in terms of real coordinates $\{y^1, \ldots, y^{2N}\}$. $\mathcal{A}_i$ is an abelian gauge potential, or connection, on moduli space. The equation of motion is

$$\mathcal{F}_{ij}\dot{y}^j + \frac{\partial V_{\text{red}}}{\partial y^i} = 0 \tag{7.170}$$

where

$$\mathcal{F}_{ij} = \frac{\partial \mathcal{A}_j}{\partial y^i} - \frac{\partial \mathcal{A}_i}{\partial y^j} \,. \tag{7.171}$$

It turns out that $\mathcal{F}_{ij}$ is the symplectic, Kähler 2-form associated with the metric on $\mathcal{M}_N$. At critical coupling, V_{red} vanishes, and there is no motion. Away from critical coupling, the motion is along a curve in one of the hypersurfaces $V_{\text{red}} = \text{const}$, determined by the initial data. It remains unclear whether the equation of motion on moduli space (7.170) actually provides a good approximation to the Schrödinger-Chern-Simons field theory dynamics of the vortices, even for initial data close to the moduli space.

7.10 The metric on $\mathcal{M}_N$

In this section, we explain the method of Samols for determining the form of the metric $g_{r\bar{s}}$ on moduli space [363]. This developed from Strachan's study [384] of the moduli space metric for an integrable model of vortices on the hyperbolic plane, which we consider in Section 7.14.3. Although the metric is defined initially as an integral over the plane, and can not be given explicitly, it is a remarkable feature of the Bogomolny equations that the metric can be expressed in terms of local data in the neighbourhood of each vortex.

Suppose that $\{\phi(t), a_i(t)\}$ is a family of N-vortex solutions of the Bogomolny equations (7.141) and (7.142), with distinct vortex locations $\{Z_r(t)\}$ slowly varying with time. The time derivatives of the fields are $\{\partial_0\phi, \partial_0 a_i\}$ and they satisfy the time derivatives of the Bogomolny equations. Suppose they also satisfy Gauss' law, with $a_0 = 0$,

$$\partial_i\partial_0 a_i + \frac{i}{2}(\bar{\phi}\partial_0\phi - \phi\partial_0\bar{\phi}) = 0. \tag{7.172}$$

The kinetic energy is then

$$T = \frac{1}{2}\int (\partial_0 a_i\partial_0 a_i + \partial_0\bar{\phi}\partial_0\phi)\,d^2x. \tag{7.173}$$

Our task is to express T in terms of $\{\dot{Z}_r\}$ and $\{\dot{\bar{Z}}_r\}$, and from this to extract the metric on $\mathcal{M}_N$.

By analogy with (5.60), we define

$$a_z = \frac{1}{2}(a_1 - ia_2), \quad a_{\bar{z}} = \frac{1}{2}(a_1 + ia_2), \tag{7.174}$$

and reexpress the kinetic energy as

$$T = \frac{1}{2}\int (4\partial_0 a_{\bar{z}}\partial_0 a_z + \partial_0\bar{\phi}\partial_0\phi)\,d^2x. \tag{7.175}$$

As earlier, let $\phi = e^{\frac{1}{2}h+i\chi}$, and define

$$\eta = \partial_0(\log\phi) = \frac{1}{2}\partial_0 h + i\partial_0\chi, \tag{7.176}$$

so

$$\partial_0 \phi = \phi \eta. \tag{7.177}$$

The first Bogomolny equation can be written as

$$\partial_{\bar{z}} \phi - i a_{\bar{z}} \phi = 0. \tag{7.178}$$

Therefore $a_{\bar{z}} = -i \partial_{\bar{z}} \log \phi$, and hence

$$\partial_0 a_{\bar{z}} = -i \partial_{\bar{z}} \eta. \tag{7.179}$$

Combining (7.177) and (7.179), and their complex conjugates, we can rewrite the kinetic energy (7.175) in terms of η as

$$T = \frac{1}{2} \int (4 \partial_z \bar{\eta} \partial_{\bar{z}} \eta + e^h \bar{\eta} \eta) \, d^2 x. \tag{7.180}$$

From (7.175), it follows that the integrand here is finite.

We establish next a differential equation obeyed by η. Recall that the Laplacian is

$$\nabla^2 = 4 \partial_{\bar{z}} \partial_z. \tag{7.181}$$

The second Bogomolny equation reduces, as we have seen, to $\nabla^2 h + 1 - e^h = 0$ away from the zeros of ϕ. Taking the time derivative, we obtain

$$(\nabla^2 - e^h) \partial_0 h = 0. \tag{7.182}$$

Also, since

$$\partial_i \partial_0 a_i = 2(\partial_z \partial_0 a_{\bar{z}} + \partial_{\bar{z}} \partial_0 a_z) = -2i \partial_z \partial_{\bar{z}} \eta + 2i \partial_{\bar{z}} \partial_z \bar{\eta} = \nabla^2 \partial_0 \chi \tag{7.183}$$

and

$$\frac{i}{2} (\bar{\phi} \partial_0 \phi - \phi \partial_0 \bar{\phi}) = \frac{i}{2} (\bar{\phi} \phi \eta - \phi \bar{\phi} \bar{\eta}) = -e^h \partial_0 \chi, \tag{7.184}$$

Gauss' law (7.172) takes the form

$$(\nabla^2 - e^h) \partial_0 \chi = 0. \tag{7.185}$$

Combining Eqs. (7.182) and (7.185), we deduce that

$$(\nabla^2 - e^h) \eta = 0. \tag{7.186}$$

The non-singular real operator $\nabla^2 - e^h$ plays an important role in what follows.

Equation (7.186) is only valid away from the zeros of ϕ, where η has singularities. Near the moving zero Z_r,

$$\phi = (z - Z_r) e^k \tag{7.187}$$

where k is smooth and finite, so

$$\log \phi = \log(z - Z_r) + k. \tag{7.188}$$

Taking the time derivative gives

$$\eta = \frac{-\dot{Z}_r}{z - Z_r} + O(1). \tag{7.189}$$

Thus η has a pole singularity at each vortex location Z_r, with residue $-\dot{Z}_r$.

Now recall the basic result $\nabla^2 \log |z - Z_r|^2 = 4\pi \delta^2(z - Z_r)$, or equivalently

$$\nabla^2(\log(z - Z_r) + \log(\bar{z} - \bar{Z}_r)) = 4\pi \delta^2(z - Z_r). \tag{7.190}$$

Differentiating with respect to Z_r we obtain

$$\nabla^2 \left(\frac{-1}{z - Z_r} \right) = 4\pi \frac{\partial}{\partial Z_r} \delta^2(z - Z_r) = -4\pi \partial_z \delta^2(z - Z_r). \tag{7.191}$$

The completed version of equation (7.186) is therefore

$$(\nabla^2 - e^h)\eta = -4\pi \sum_{r=1}^N \dot{Z}_r \, \partial_z \, \delta^2(z - Z_r), \tag{7.192}$$

where we have noted that $e^h \eta$ has no singularity, since near Z_r, $e^h \sim |z - Z_r|^2$, which cancels the pole in η.

The solution of this equation is found as follows. Differentiating the equation for h,

$$\nabla^2 h + 1 - e^h = 4\pi \sum_{r=1}^N \delta^2(z - Z_r), \tag{7.193}$$

gives

$$(\nabla^2 - e^h)\frac{\partial h}{\partial Z_r} = -4\pi \, \partial_z \, \delta^2(z - Z_r). \tag{7.194}$$

By comparing (7.194) and (7.192), and noting that $\frac{\partial h}{\partial Z_r}$ has a pole at Z_r, we deduce that

$$\eta = \sum_{r=1}^N \dot{Z}_r \frac{\partial h}{\partial Z_r}. \tag{7.195}$$

There is no ambiguity here, since the operator $\nabla^2 - e^h$ is negative definite and has no zero modes (non-singular, bounded eigenfunctions with zero eigenvalue).

Let us now divide up the integral (7.180) as

$$T = \frac{1}{2} \int_{\mathbb{R}^2 - D} (4\partial_z \bar{\eta} \partial_{\bar{z}} \eta + e^h \bar{\eta} \eta) \, d^2 x + \frac{1}{2} \int_D (4\partial_z \bar{\eta} \partial_{\bar{z}} \eta + e^h \bar{\eta} \eta) \, d^2 x, \quad (7.196)$$

where D is a union of non-overlapping discs D_r centred at Z_r, of radius ε. We can neglect the second integral in the limit $\varepsilon \to 0$. The first integral we can reexpress as

$$
\begin{aligned}
T &= 2 \int_{\mathbb{R}^2 - D} \partial_z (\bar{\eta} \partial_{\bar{z}} \eta) \, d^2 x - \frac{1}{2} \int_{\mathbb{R}^2 - D} \bar{\eta} (\nabla^2 - e^h) \eta \, d^2 x \\
&= 2 \int_{\mathbb{R}^2 - D} \partial_z (\bar{\eta} \partial_{\bar{z}} \eta) \, d^2 x \quad\quad\quad\quad\quad\quad\quad\quad (7.197)
\end{aligned}
$$

using (7.186). This can be turned into a sum of line integrals along the boundary of D using the following integral identity. Let S be a domain in the plane with boundary ∂S, and $f(z, \bar{z})$ a differentiable function. Then

$$\int_S \partial_z f \, d^2 x = \frac{i}{2} \int_{\partial S} f \, d\bar{z}. \quad (7.198)$$

This is a consequence of the more familiar identity

$$\int_S \frac{\partial f}{\partial x^j} \, d^2 x = \int_{\partial S} f \, n_j \, dl \quad (7.199)$$

where $\mathbf{n}$ is the outward normal on ∂S, and the observation that $(n_1 - in_2)dl = id\bar{z}$ on ∂S. Using (7.198), we obtain

$$T = -i \sum_{r=1}^{N} \int_{C_r} \bar{\eta} \partial_{\bar{z}} \eta \, d\bar{z} \quad (7.200)$$

where C_r is the boundary of the disc D_r (taken anticlockwise). There is no contribution from the circle at infinity as η decays exponentially fast as $|\mathbf{x}| \to \infty$. This is a localized expression for T, but not our final one.

Let us now expand h around the point Z_r, say. The expansion has the form

$$
\begin{aligned}
h(z, \bar{z}) &= 2 \log |z - Z_r| + a_r + \frac{1}{2} \bar{b}_r (z - Z_r) + \frac{1}{2} b_r (\bar{z} - \bar{Z}_r) + \bar{c}_r (z - Z_r)^2 \\
&\quad + d_r (z - Z_r)(\bar{z} - \bar{Z}_r) + c_r (\bar{z} - \bar{Z}_r)^2 + O(|z - Z_r|^3), \quad (7.201)
\end{aligned}
$$

combining a log term and a convergent Taylor series, with a_r and d_r real [223]. It is straightforward to verify that (7.193) is satisfied to the order shown provided $d_r = -\frac{1}{4}$. There are no local constraints on a_r, b_r or c_r. These coefficients are completely determined by the positions of the other vortices, but not in an explicitly known way.

The coefficient b_r (and its complex conjugate $\bar{b}_r$) is the most important for us. It is twice the value of the derivative with respect to $\bar{z}$ of the regularized function $h(z, \bar{z}) - 2 \sum_r \log |z - Z_r|$, at Z_r. Geometrically, it measures the extent to which contours of h close to Z_r differ from circles centred at Z_r. In fact, the contours remain circular to leading order in their size, but the centre shifts, because of the effect of the other vortices. More precisely, one can show starting with (7.201) that the contour of radius ε has centre at $Z_r - \frac{1}{2}\varepsilon^2 b_r$ for small ε (see Fig. 7.15). We shall show, shortly, that the kinetic energy T can be expressed in terms of derivatives of b_r.

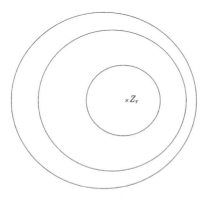

Fig. 7.15. Sketch showing the contours of h close to Z_r. The contours remain circular but the centre drifts from Z_r due to the effect of the other vortices.

We digress briefly to derive some symmetry properties of derivatives of $\frac{\partial h}{\partial Z_r}$, and hence of b_r. Equation (7.194) shows that $\frac{\partial h}{\partial Z_r}$ is a type of Green's function for the operator $\nabla^2 - e^h$, decaying as $|z| \to \infty$, and with a derivative of a delta-function as source. We temporarily use the notation $G(z; Z_r)$ for $\frac{\partial h}{\partial Z_r}$. G has a pole-type singularity at Z_r, but is smooth elsewhere. (h itself has singularities at all points Z_s, but the singular terms are independent of Z_r if $r \neq s$.) For $r \neq s$ we have

$$\int G(z; Z_s)(\nabla^2 - e^h)G(z; Z_r)\, d^2x \;=\; -4\pi \int G(z; Z_s)\, \partial_z \delta^2(z - Z_r)\, d^2x$$

$$=\; 4\pi \int \partial_z G(z; Z_s)\, \delta^2(z - Z_r)\, d^2x$$

$$=\; 4\pi\, \partial_z G(Z_r; Z_s)\,, \qquad (7.202)$$

where ∂_z acts on the first argument. If we integrate by parts, to make ∇^2 act on $G(z; Z_s)$ we get the same result with Z_r and Z_s exchanged. Thus, in the earlier notation, we have the symmetry property

$$\partial_z \frac{\partial h}{\partial Z_s}\bigg|_{z=Z_r} = \partial_z \frac{\partial h}{\partial Z_r}\bigg|_{z=Z_s}. \tag{7.203}$$

The calculation is not valid for $r = s$ because in (7.202) we would be multiplying a delta-function by a function with a singularity at the same location.

Since the operator $\nabla^2 - e^h$ and the basic delta-function are both real, the complex conjugate of (7.194) is

$$(\nabla^2 - e^h)\frac{\partial h}{\partial \bar{Z}_r} = -4\pi\partial_{\bar{z}}\delta^2(z - Z_r), \tag{7.204}$$

and a similar calculation as above leads to the second symmetry property

$$\partial_z \frac{\partial h}{\partial \bar{Z}_s}\bigg|_{z=Z_r} = \partial_{\bar{z}} \frac{\partial h}{\partial Z_r}\bigg|_{z=Z_s}. \tag{7.205}$$

This is valid for all r and s, because neither quantity here is singular if $r = s$.

From (7.201), we evaluate that for $r \neq s$

$$\partial_z \frac{\partial h}{\partial Z_s}\bigg|_{z=Z_r} = \partial_z \left(\frac{\partial a_r}{\partial Z_s} + \frac{1}{2}\frac{\partial \bar{b}_r}{\partial Z_s}(z - Z_r) + \frac{1}{2}\frac{\partial b_r}{\partial Z_s}(\bar{z} - \bar{Z}_r) + \cdots \right)\bigg|_{z=Z_r}$$

$$= \frac{1}{2}\frac{\partial \bar{b}_r}{\partial Z_s}, \tag{7.206}$$

and similarly

$$\partial_z \frac{\partial h}{\partial Z_r}\bigg|_{z=Z_s} = \frac{1}{2}\frac{\partial \bar{b}_s}{\partial Z_r}. \tag{7.207}$$

Hence the symmetry property (7.203) implies that

$$\frac{\partial \bar{b}_r}{\partial Z_s} = \frac{\partial \bar{b}_s}{\partial Z_r} \tag{7.208}$$

together with the complex conjugate of this

$$\frac{\partial b_r}{\partial \bar{Z}_s} = \frac{\partial b_s}{\partial \bar{Z}_r}. \tag{7.209}$$

We have demonstrated these results only for $r \neq s$, but they are clearly trivially valid if $r = s$ too. The second symmetry property (7.205) implies that

$$\frac{\partial \bar{b}_r}{\partial \bar{Z}_s} = \frac{\partial b_s}{\partial Z_r} \tag{7.210}$$

and this is valid for all r, s. This completes our digression.

Now we shall obtain formulae for $\bar{\eta}$ and $\partial_{\bar{z}}\eta$ on C_s, and evaluate the integral (7.200). Near Z_s, we have

$$
\begin{aligned}
h &= \log(z - Z_s) + \log(\bar{z} - \bar{Z}_s) + a_s + \frac{1}{2}\bar{b}_s(z - Z_s) + \frac{1}{2}b_s(\bar{z} - \bar{Z}_s) \\
&+ \bar{c}_s(z - Z_s)^2 - \frac{1}{4}(z - Z_s)(\bar{z} - \bar{Z}_s) + c_s(\bar{z} - \bar{Z}_s)^2 \\
&+ O(|z - Z_r|^3),
\end{aligned}
\tag{7.211}
$$

so

$$
\begin{aligned}
\frac{\partial h}{\partial Z_r} &= -\frac{\delta_{rs}}{z - Z_s} + \frac{\partial a_s}{\partial Z_r} + \frac{1}{2}\frac{\partial \bar{b}_s}{\partial Z_r}(z - Z_s) - \frac{1}{2}\bar{b}_s\delta_{rs} \\
&+ \frac{1}{2}\frac{\partial b_s}{\partial Z_r}(\bar{z} - \bar{Z}_s) - 2\bar{c}_s(z - Z_s)\delta_{rs} + \frac{1}{4}(\bar{z} - \bar{Z}_s)\delta_{rs} \\
&+ O(|z - Z_s|^2)
\end{aligned}
\tag{7.212}
$$

(no sum over s). Also

$$
\partial_{\bar{z}}\left(\frac{\partial h}{\partial Z_r}\right) = \frac{1}{2}\frac{\partial b_s}{\partial Z_r} + \frac{1}{4}\delta_{rs} + O(|z - Z_s|). \tag{7.213}
$$

Therefore, using (7.195),

$$
\bar{\eta} = \sum_{r=1}^{N} \dot{Z}_r\left(\frac{-\delta_{rs}}{\bar{z} - \bar{Z}_s}\right) + O(1) = \frac{-\dot{Z}_s}{\bar{z} - \bar{Z}_s} + O(1), \tag{7.214}
$$

and similarly,

$$
\partial_{\bar{z}}\eta = \sum_{r=1}^{N} \dot{Z}_r\left(\frac{1}{2}\frac{\partial b_s}{\partial Z_r} + \frac{1}{4}\delta_{rs}\right) + O(|z - Z_s|). \tag{7.215}
$$

The pole term in $\bar{\eta}$ will give a finite result when we integrate around C_s. Combining (7.214) and (7.215) we obtain

$$
\int_{C_s} \bar{\eta}\partial_{\bar{z}}\eta \, d\bar{z} = 2\pi i \dot{\bar{Z}}_s \sum_{r=1}^{N} \dot{Z}_r\left(\frac{1}{2}\frac{\partial b_s}{\partial Z_r} + \frac{1}{4}\delta_{rs}\right), \tag{7.216}
$$

using the antiholomorphic version of the residue theorem. By summing over all the circles C_s, we find that the kinetic energy expression (7.200) reduces to

$$
T = \frac{1}{2}\pi \sum_{r,s=1}^{N} \left(\delta_{rs} + 2\frac{\partial b_s}{\partial Z_r}\right) \dot{Z}_r \dot{\bar{Z}}_s. \tag{7.217}
$$

This is our fundamental result. The symmetry relation (7.210) implies that T is real. (In the work of Strachan and Samols, the physical reality of T was used to argue that (7.210) had to hold. The Green's function argument above gives a more direct understanding.*) From (7.217) we read off that the metric on the N-vortex moduli space $\mathcal{M}_N$ is

$$ds^2 = \pi \sum_{r,s=1}^{N} \left(\delta_{rs} + 2\frac{\partial b_s}{\partial Z_r} \right) dZ_r \, d\bar{Z}_s \,, \tag{7.218}$$

so

$$g_{r\bar{s}} = \pi \left(\delta_{rs} + 2\frac{\partial b_s}{\partial Z_r} \right) \,. \tag{7.219}$$

As promised, there are no $dZ_r \, dZ_s$ or $d\bar{Z}_r \, d\bar{Z}_s$ terms.

For well separated vortices, the contours of $|\phi|$ close to each zero of ϕ are hardly affected by the other vortices – the effect is exponentially small in the separation. Therefore b_s and its derivatives are exponentially small, and if we ignore their contribution completely, the kinetic energy is

$$T = \frac{1}{2}\pi \sum_{r=1}^{N} \dot{Z}_r \dot{\bar{Z}}_r \,, \tag{7.220}$$

and the corresponding asymptotic metric is

$$ds^2 = \pi \sum_{r=1}^{N} dZ_r \, d\bar{Z}_r \,, \tag{7.221}$$

the standard flat metric on $\mathbb{C}^N$ multiplied by π, the mass of a vortex. More precisely, because the vortices are indistinguishable, the asymptotic form of the moduli space is $\mathbb{C}^N/S_N$, with this flat metric. Asymptotically, the geodesic motion on moduli space corresponds to each vortex having an independent straight line motion in $\mathbb{C}$ at constant velocity.

An important consequence of the symmetry property (7.210) is that the metric on $\mathcal{M}_N$ is Kähler (a property first noted by Ruback [353]). A Hermitian metric tensor $g_{r\bar{s}}$ is Kähler if the associated 2-form

$$\omega = \frac{i}{2} \sum_{r,s=1}^{N} g_{r\bar{s}} dZ_r \wedge d\bar{Z}_s \tag{7.222}$$

is closed, that is, $d\omega = 0$. For the metric tensor (7.219), the Kähler 2-form is

$$\omega = \frac{i\pi}{2} \left(\sum_{r=1}^{N} dZ_r \wedge d\bar{Z}_r + 2\sum_{r,s=1}^{N} \frac{\partial b_s}{\partial Z_r} dZ_r \wedge d\bar{Z}_s \right) \,, \tag{7.223}$$

---

* N. S. M. thanks H. Brezis for this wizard idea.

so

$$
\begin{aligned}
d\omega &= i\pi \sum_{r,s,t} \left(\frac{\partial^2 b_s}{\partial \bar{Z}_t \partial Z_r} d\bar{Z}_t \wedge dZ_r \wedge d\bar{Z}_s + \frac{\partial^2 b_s}{\partial Z_t \partial Z_r} dZ_t \wedge dZ_r \wedge d\bar{Z}_s \right) \\
&= i\pi \sum_{r,s,t} \left(\frac{\partial^2 \bar{b}_r}{\partial \bar{Z}_t \partial \bar{Z}_s} d\bar{Z}_t \wedge dZ_r \wedge d\bar{Z}_s + \frac{\partial^2 b_s}{\partial Z_t \partial Z_r} dZ_t \wedge dZ_r \wedge d\bar{Z}_s \right) \\
&= 0,
\end{aligned}
\tag{7.224}
$$

where we have used (7.210) to rewrite the first term, and used the symmetry property of second partial derivatives.

The Kähler property is intrinsically interesting. It shows that the complex coordinates we are using to parametrize moduli space are natural. It will allow certain calculations of a global nature using methods of cohomology when we generalize to the situation of vortices on a compact manifold.

One consequence is that there is locally a real function $\mathcal{K}$ on $\mathcal{M}_N$, such that

$$
b_r = \frac{\partial \mathcal{K}}{\partial \bar{Z}_r},
\tag{7.225}
$$

and hence

$$
ds^2 = \pi \sum_{r,s=1}^N \left(\delta_{rs} + 2\frac{\partial^2 \mathcal{K}}{\partial Z_r \partial \bar{Z}_s} \right) dZ_r \, d\bar{Z}_s.
\tag{7.226}
$$

The Kähler potential is

$$
\pi \sum_{r=1}^N Z_r \bar{Z}_r + 2\pi \mathcal{K}.
\tag{7.227}
$$

Both the symmetry relations (7.208) and (7.210) follow from (7.225). Since $\mathcal{M}_N$ is topologically trivial, $\mathcal{K}$ should be globally well defined. Regrettably, we have no formula for $\mathcal{K}$ in terms of other quantities that we have been considering.

The Kähler property is also at the root of the following results. If all the vortices are translated by the same amount, then the contours of h translate rigidly, so the coefficients b_s are unaltered. Thus

$$
\left(\sum_{r=1}^N \frac{\partial}{\partial Z_r} \right) b_s = \left(\sum_{r=1}^N \frac{\partial}{\partial Z_r} \right) \bar{b}_s = 0.
\tag{7.228}
$$

Using (7.208) and (7.210), we deduce that

$$
\frac{\partial}{\partial \bar{Z}_s} \left(\sum_{r=1}^N \bar{b}_r \right) = \frac{\partial}{\partial Z_s} \left(\sum_{r=1}^N \bar{b}_r \right) = 0,
\tag{7.229}
$$

so $\sum_r \bar{b}_r$, and hence $\sum_r b_r$, is constant on moduli space. Since $b_r \to 0$ as the vortices separate, this constant is zero, so [363]

$$\sum_{r=1}^{N} b_r = 0. \tag{7.230}$$

Similarly, by considering a rotation we find that $\sum_r \bar{Z}_r b_r$ is real [347].

The result (7.230) can be presented in the following way. If we define the centre of mass coordinate $Z = \frac{1}{N}(\sum_r Z_r)$, and define a set of relative coordinates $W_r = Z_r - Z$, $1 \le r \le N - 1$, which are unaffected by an overall translation, then the metric separates in the form

$$ds^2 = N\pi \, dZ \, d\bar{Z} + \sum_{r,s=1}^{N-1} \tilde{g}_{r\bar{s}} dW_r \, d\bar{W}_s, \tag{7.231}$$

with the coefficient of $dZ \, d\bar{Z}$ independent of the relative positions of the vortices.

The functions b_r on moduli space have singularities as vortices coincide. To calculate the nature of the singularity, suppose just two vortices are very close, at $Z_1 = Z + \varepsilon$ and $Z_2 = Z - \varepsilon$. Then the equation for h near these two vortices has a solution of the form

$$h = 2\log|z - (Z+\varepsilon)| + 2\log|z - (Z-\varepsilon)| + a + \frac{1}{2}\bar{b}(z-Z) + \frac{1}{2}b(\bar{z}-\bar{Z}) + \cdots \tag{7.232}$$

where a and b remain finite as $\varepsilon \to 0$. We now expand around $Z + \varepsilon$, writing $z - (Z - \varepsilon)$ as $2\varepsilon + (z - (Z+\varepsilon))$. Provided $|z - (Z+\varepsilon)| < |2\varepsilon|$, we can reorganize the expansion (7.232) as

$$\begin{aligned}
h &= 2\log|z - (Z+\varepsilon)| + a + 2\log|2\varepsilon| + \frac{1}{2}\bar{b}\varepsilon + \frac{1}{2}b\bar{\varepsilon} \\
&+ \frac{1}{2\varepsilon}(z - (Z+\varepsilon)) + \frac{1}{2\bar{\varepsilon}}(\bar{z} - (\bar{Z}+\bar{\varepsilon})) \\
&+ \frac{1}{2}\bar{b}(z - (Z+\varepsilon)) + \frac{1}{2}b(\bar{z} - (\bar{Z}+\bar{\varepsilon})) + \cdots. \tag{7.233}
\end{aligned}$$

We see that

$$b_1 = \frac{1}{\bar{\varepsilon}} + b \tag{7.234}$$

as ε gets small. A similar expansion around $Z - \varepsilon$ implies that $b_2 = -\frac{1}{\bar{\varepsilon}} + b$.

Thus the coefficients b_1 and b_2 develop pole-type singularities as vortices 1 and 2 approach each other. $b_1 + b_2$ has no pole, consistent with the general property $\sum_r b_r = 0$. If a number of vortices cluster together, then the corresponding coefficients b_r all develop poles, whose strengths depend on the relative configuration of the cluster.

7.11 Two-vortex scattering

The metric on $\mathcal{M}_2$ is rather simple, as a result of translational and rotational symmetry. Let the two vortices have locations $Z + W$ and $Z - W$. The centre of mass motion decouples, and the metric has the form

$$ds^2 = 2\pi dZ \, d\bar{Z} + f^2(W, \bar{W}) \, dW \, d\bar{W} \,, \tag{7.235}$$

where, by rotational invariance, f depends only on $|W|$. In polar coordinates ρ, θ, with $W = \rho e^{i\theta}$, the metric is

$$ds^2 = 2\pi dZ \, d\bar{Z} + f^2(\rho)(d\rho^2 + \rho^2 d\theta^2) \,. \tag{7.236}$$

For the rest of this section we shall only discuss centred vortices, whose reduced moduli space $\mathcal{M}_2^0$ is the circularly symmetric surface with metric

$$ds^2 = f^2(\rho)(d\rho^2 + \rho^2 d\theta^2) \,. \tag{7.237}$$

We can express f in terms of the coefficients b_1 and b_2 that occur in the expansion of h around the vortex locations. Rotation and reflection symmetry imply that $b_1 = b(\rho)e^{i\theta}$ and $b_2 = -b(\rho)e^{i\theta}$, where $b(\rho)$ is real. The general formula (7.218) for the metric then implies that

$$f^2(\rho) = 2\pi \left(1 + \frac{1}{\rho} \frac{d}{d\rho} \left(\rho b(\rho) \right) \right) \,. \tag{7.238}$$

By solving the equation for h numerically, Samols [363] calculated $b(\rho)$, and hence $f^2(\rho)$. The result is shown in Fig. 7.16.

As expected, $b(\rho)$ decays exponentially as $\rho \to \infty$, so $f^2 \to 2\pi$. For small ρ, $b(\rho)$ has the form $b(\rho) \sim \frac{1}{\rho} - \frac{1}{2}\rho + \alpha\rho^3 + \cdots$, so $f^2(\rho) \sim 8\pi\alpha\rho^2$. Recall that the vortices are unordered, and that replacing $W = \rho e^{i\theta}$ by $W = \rho e^{i(\theta+\pi)}$ just exchanges them. The range of θ is therefore $0 \le \theta \le \pi$, with $\theta = 0$ and $\theta = \pi$ identified. Recall also that the polynomial $p(z)$ associated with a pair of vortices at W and $-W$ is

$$p(z) = (z - W)(z + W) = z^2 - W^2 \,, \tag{7.239}$$

so $w = W^2$ is a good complex coordinate on the reduced moduli space. Since $w = \rho^2 e^{2i\theta}$ we expect ρ^2 and 2θ to be "good" polar coordinates on the moduli space (with ranges $\rho^2 \ge 0$, $0 \le 2\theta \le 2\pi$). A smooth metric should therefore have the structure, for small ρ,

$$ds^2 = \gamma(d(\rho^2)^2 + (\rho^2)^2 \, d(2\theta)^2) = 4\gamma\rho^2(d\rho^2 + \rho^2 d\theta^2) \,. \tag{7.240}$$

But this is exactly what we have found, with $\gamma = 2\pi\alpha$.

$\mathcal{M}_2^0$ can be isometrically embedded in $\mathbb{R}^3$ as a surface of revolution (since $d(\log f)/d(\log \rho) \le 1$, and therefore the circumference of the circles

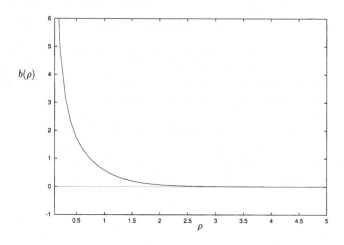

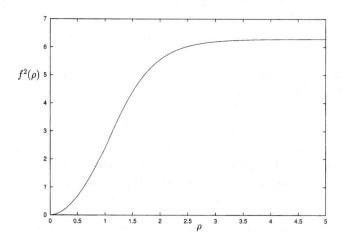

Fig. 7.16.   The function $b(\rho)$, used in the computation of the two-vortex metric, and the metric coefficient $f^2(\rho)$.

$\rho = $ const does not grow too rapidly with ρ). A metrically correct sketch of the surface is shown in Fig. 7.17. As $\rho \to \infty$, the metric becomes flat, up to exponentially small corrections. $\mathcal{M}_2^0$ is asymptotically a plane with opposite points (ρ, θ) and $(\rho, \theta + \pi)$ identified, and is therefore a smoothed cone of half-opening angle $30°$.

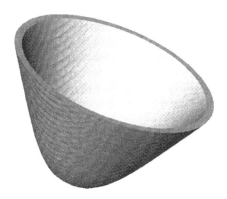

Fig. 7.17. A representation of (part of) the manifold $\mathcal{M}_2^0$ as a surface of revolution isometrically embedded in $\mathbb{R}^3$.

The leading exponentially small correction for large ρ has been calculated, by directly studying the equation (7.151) for h with two well separated sources [295]. One finds that

$$b(\rho) \sim \frac{A_s^2}{2\pi^2} K_1(2\rho) \tag{7.241}$$

where A_s is the constant introduced in Section 7.6, whose numerical value is 10.6, and K_1 is the modified Bessel function. It follows that

$$f^2(\rho) \sim 2\pi \left(1 - \frac{A_s^2}{\pi^2} K_0(2\rho)\right) . \tag{7.242}$$

The result can also be understood using the model mentioned in Section 7.6, where the vortices are point-like and carry a scalar charge and a magnetic dipole moment. Two such objects in relative motion interact, because the scalar and magnetic interactions have a different dependence on velocity and do not exactly cancel. From the effective Lagrangian for the vortices, which is purely kinetic, one can extract the metric on the moduli space for two well separated vortices. We shall explain in detail in Section 8.12 how a similar analysis for BPS monopoles allows one to calculate the asymptotic form of the metric on the N-monopole moduli space.

Samols made an interesting observation about the global geometry of the moduli space $\mathcal{M}_2^0$. Clearly, the rounded cone has smaller area than the flat cone which it approaches asymptotically. The two metrics are $ds^2 = f^2(\rho)(d\rho^2 + \rho^2 d\theta^2)$ and $ds^2 = 2\pi(d\rho^2 + \rho^2 d\theta^2)$, with $\rho \geq 0$, $0 \leq \theta \leq \pi$. The difference in area is [363]

$$\pi \int_0^\infty (2\pi - f^2(\rho))\rho \, d\rho = -2\pi^2 \int_0^\infty \frac{d}{d\rho}\left(\rho b(\rho)\right) \, d\rho = -2\pi^2 [\rho b(\rho)]_0^\infty = 2\pi^2.$$
(7.243)

The possibility of giving a precise value for this stems from the Kähler property of the metric.

Geodesics on the moduli space can easily be found. Geodesic motion conserves energy

$$E = \frac{1}{2}f^2(\rho)(\dot{\rho}^2 + \rho^2\dot{\theta}^2)$$
(7.244)

and angular momentum

$$\ell = f^2(\rho)\rho^2\dot{\theta}.$$
(7.245)

For given ℓ, the radial motion is determined by solving

$$\frac{d\rho}{dt} = \frac{1}{f(\rho)}\sqrt{2E - \frac{\ell^2}{f(\rho)^2\rho^2}}.$$
(7.246)

ρ decreases from infinity to the finite value given by $f(\rho)\rho = \ell/\sqrt{2E}$ and then increases to infinity again. Knowing the radial motion, the angular motion follows by solving

$$\frac{d\theta}{dt} = \frac{\ell}{f^2(\rho)\rho^2}.$$
(7.247)

Some geodesics are shown using Cartesian coordinates ($x^1 = \rho\cos\theta$, $x^2 = \rho\sin\theta$) in Fig. 7.18. The time dependence is not indicated here, and the motion is not at a constant speed in these coordinates, except asymptotically. Each geodesic ($\rho(t), \theta(t)$) gives the time dependent position of one vortex, and the second vortex is at ($\rho(t), \theta(t) + \pi$). Therefore Fig. 7.18 shows the actual trajectories of the vortices in the plane. Because this is geodesic motion, we can change the relative speed of the vortices, but the trajectories are unaffected while the geodesic approximation remains valid. Notice that the vortices always repel each other. This is a purely geometrical effect related to their motion – there is no repulsive potential at critical coupling.

There is right-angle scattering in head-on collisions, as anticipated earlier. It is a clear consequence of the rounded cone structure of the moduli

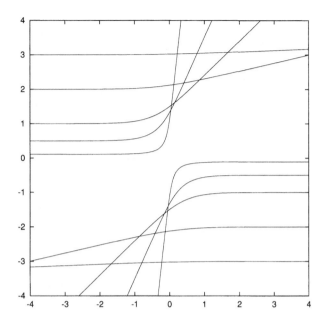

Fig. 7.18. The trajectories of the zeros of the Higgs field for several two-vortex scatterings with non-zero impact parameter.

space [374]. One type of geodesic goes straight through the apex of the cone. The incoming part of the geodesic is at a fixed angle $\theta = \theta_0$, and the outgoing part is at $\theta = \theta_0 + \frac{\pi}{2}$, because going right round the cone corresponds to increasing θ by π. This jump in angle on moduli space corresponds to right-angle scattering of the vortices in the plane.

The detailed geometrical structure of the moduli spaces $\mathcal{M}_N$, for $N > 2$, is little understood. However, we know that the centre of mass decouples, and that the moduli space is asymptotically $\mathbb{C}^N/S_N$, where S_N is the permutation group, and $\mathbb{C}^N$ has its flat metric. The leading exponentially small corrections to the flat metric are given in ref. [295]. It is likely that the geodesic motion always makes the vortex trajectories repel each other. A consequence would be that there are no closed, or even bounded, geodesics on $\mathcal{M}_N$. This is definitely the case on $\mathcal{M}_2$, because a bounded geodesic can only occur if $f(\rho)\rho$ has the same value for two different values of ρ, and the numerical evidence rules this out.

In $\mathcal{M}_N$ there is a geodesic submanifold consisting of centred N-vortex configurations with C_N symmetry, where C_N is the cyclic group of rotations by multiples of $2\pi/N$. This is a surface of revolution, which is asymptotically a cone of half-opening angle $\sin^{-1}\left(\frac{1}{N}\right)$. The simplest geodesic, which passes through the apex, represents the π/N scattering of N vortices that we discussed in Section 7.7.1.

7.12 First order dynamics near critical coupling

We turn now to the first order, Schrödinger-Chern-Simons model, introduced in Section 7.7.3, show that it has vortices, and discuss how, close to critical coupling, the vortex motion can be modelled by motion on moduli space [287]. Let us set the transport current J_i^{T} to zero, and first seek static solutions of the field equations. Setting time derivatives to zero, and $a_0 = 0$, we see that the field equations (7.123) and (7.124) reduce to the static equations of GL theory, which have circularly symmetric vortex solutions for all λ. However, these solutions do not generally satisfy the third field equation (7.125), and the resolution of this paradox is that a_0 is non-zero, and there is a radial electric field accompanying the vortex.

The electric field is absent if we make the special choice of parameters $\mu = \gamma$ and $\lambda = 1$. Then, static N-vortex solutions of the Bogomolny equations

$$(D_1 + iD_2)\phi = 0 \tag{7.248}$$

$$B - \frac{1}{2}(1 - \bar{\phi}\phi) = 0 \tag{7.249}$$

satisfy (7.123) and (7.124), and remarkably, the second Bogomolny equation coincides with Gauss' law (7.125), so we have a solution of the complete system of equations, for vortices at arbitrary locations. (Note that this works only because of the term γa_0 in the Lagrangian which leads to the term γ in (7.125).)

Horváthy and collaborators noted a neat extension of this, for $\lambda = (2\gamma\mu - \gamma^2)/\mu^2$ and γ/μ in the range where λ is positive [175]. There is a modified set of Bogomolny equations, including a non-zero a_0, and stationary N-vortex solutions.

From now on, we assume that $\mu = \gamma$ and that λ is close to, but not necessarily equal to 1. We assume that the field is an N-vortex solution of the Bogomolny equations, with time dependent vortex locations $\{Z_1(t), \ldots, Z_N(t)\}$. Gauss' law is satisfied. There is a reduced Lagrangian on the moduli space $\mathcal{M}_N$, obtained from the kinetic and potential terms of the Schrödinger-Chern-Simons Lagrangian, and this determines the motion of the vortices.

The residual potential term is, as earlier,

$$V_{\mathrm{red}} = \frac{\lambda - 1}{8} \int (1 - e^h)^2 \, d^2x \,, \tag{7.250}$$

where $e^h = |\phi|^2$. This is a well defined function on moduli space, smoothly varying with the vortex locations. There is no singularity as vortices coalesce. Unfortunately, we have little explicit information about this potential. However, we know numerically that for two vortices, the integral (without the factor $\frac{\lambda-1}{8}$) decreases as the vortex separation increases.

More can be said about the kinetic term [287]

$$T = \gamma \int \left(\frac{i}{2}(\bar{\phi}D_0\phi - \phi\overline{D_0\phi}) + Ba_0 + e_1a_2 - e_2a_1 - a_0 \right) d^2x . \quad (7.251)$$

If we set $\phi = \phi_1 + i\phi_2$, and integrate by parts, we find

$$\begin{aligned}
T &= -\gamma \int (\phi_1\partial_0\phi_2 - \phi_2\partial_0\phi_1 + a_1\partial_0a_2 - a_2\partial_0a_1) \, d^2x \\
&\quad + \gamma \int (2B - 1 + \phi_1^2 + \phi_2^2) \, a_0 \, d^2x .
\end{aligned} \quad (7.252)$$

For fields obeying the Bogomolny equations, the second integral vanishes, so we need not worry about the value of a_0. Remarkably, the first integral can be calculated by a similar method as for the metric on the moduli space. The integrand is a total derivative and the integral can be reduced to line integrals along small circles around each vortex location Z_r. The result is a sum of terms linear in $\dot{Z}_r$ and $\dot{\bar{Z}}_r$, whose coefficients depend only on Z_r and on the quantity b_r that occurs in the expansion of h around Z_r. In addition there is a contribution from the time derivative of a phase χ_r which can be associated to each vortex. It is convenient to use real notation here, writing $Z_r = X_r^1 + iX_r^2$ and $b_r = b_r^1 + ib_r^2$. Then the final expression for the kinetic energy is

$$T_{\text{red}} = \pi\gamma \sum_{r=1}^{N} \left((X_r^2 + 2b_r^2)\dot{X}_r^1 - (X_r^1 + 2b_r^1)\dot{X}_r^2 \right) + 2\pi\gamma \sum_{r=1}^{N} \dot{\chi}_r . \quad (7.253)$$

The last term, involving the phases, is a total time derivative, and has no effect on the reduced equations of motion. However, its presence ensures that T_{red} has no singularity as the vortices coincide, despite the pole-type singularity of b_r.

The complete Lagrangian on moduli space is

$$L_{\text{red}} = T_{\text{red}} - V_{\text{red}} \quad (7.254)$$

with T_{red} and V_{red} given by (7.253) and (7.250). This is of the general form (7.169), where $\{y^1, \ldots, y^{2N}\} = \{X_1^1, X_1^2, \ldots, X_N^1, X_N^2\}$. The equation of motion for the rth vortex,

$$\frac{d}{dt}\left(\frac{\partial L_{\text{red}}}{\partial \dot{X}_r^i} \right) - \frac{\partial L_{\text{red}}}{\partial X_r^i} = 0 , \quad (7.255)$$

has components

$$2\pi\gamma\left[\dot{X}_r^1 + \left(\frac{\partial b_r^1}{\partial X_s^1} + \frac{\partial b_s^2}{\partial X_r^2}\right)\dot{X}_s^1 + \left(\frac{\partial b_r^1}{\partial X_s^2} - \frac{\partial b_s^1}{\partial X_r^2}\right)\dot{X}_s^2\right] - \frac{\partial V_{\text{red}}}{\partial X_r^2} = 0 \quad (7.256)$$

$$2\pi\gamma\left[\dot{X}_r^2 + \left(\frac{\partial b_r^2}{\partial X_s^2} + \frac{\partial b_s^1}{\partial X_r^1}\right)\dot{X}_s^2 + \left(\frac{\partial b_r^2}{\partial X_s^1} - \frac{\partial b_s^2}{\partial X_r^1}\right)\dot{X}_s^1\right] + \frac{\partial V_{\text{red}}}{\partial X_r^1} = 0, \quad (7.257)$$

and is of the form (7.170).

Translational symmetry leads to a conserved momentum with components

$$P_1 = 2\pi\gamma\sum_{r=1}^{N}(X_r^2 + b_r^2), \quad P_2 = -2\pi\gamma\sum_{r=1}^{N}(X_r^1 + b_r^1). \quad (7.258)$$

We saw earlier that $\sum_r b_r^1 = \sum_r b_r^2 = 0$, so the conservation of momentum means that the naive centre of the N-vortex system

$$\mathbf{X} = \frac{1}{N}\sum_{r=1}^{N}\mathbf{X}_r \quad (7.259)$$

does not move. Rotational symmetry implies a conserved angular momentum

$$\ell = -\pi\gamma\sum_{r=1}^{N}\mathbf{X}_r \cdot (\mathbf{X}_r + 2\mathbf{b}_r). \quad (7.260)$$

Since $\sum_r \mathbf{X}_r \cdot \mathbf{b}_r$ does not become singular as vortices coalesce, conservation of ℓ implies that $|\mathbf{X}_r|^2$ can not become infinite. Therefore vortex motion is bounded for this type of dynamics. It was verified in [288] that the same conserved quantities are obtained from the field theoretic expressions (7.136), when the vortices satisfy the Bogomolny equations and have slowly moving locations.

We have yet to show that the symplectic form underlying this first order vortex dynamics is actually the Kähler form of the Samols metric on moduli space. To see this, note that

$$\pi\gamma\sum_{r=1}^{N}\left((X_r^2 + 2b_r^2)\dot{X}_r^1 - (X_r^1 + 2b_r^1)\dot{X}_r^2\right) \quad (7.261)$$

can be expressed in complex form as

$$\sum_{r=1}^{N}(A_r\dot{Z}_r + A_{\bar{r}}\dot{\bar{Z}}_r), \quad (7.262)$$

where

$$A_r = \frac{i\pi\gamma}{2}\left(\bar{Z}_r + 2\bar{b}_r\right), \quad A_{\bar{r}} = -\frac{i\pi\gamma}{2}\left(Z_r + 2b_r\right).$$ (7.263)

These are the components of a connection on moduli space whose field strength is the 2-form

$$
\begin{aligned}
\mathcal{F} &= \sum_{r,s}\left(\frac{\partial A_{\bar{s}}}{\partial Z_r} - \frac{\partial A_r}{\partial \bar{Z}_s}\right) dZ_r \wedge d\bar{Z}_s \\
&= -i\pi\gamma \sum_{r,s}\left(\delta_{rs} + 2\frac{\partial b_s}{\partial Z_r}\right) dZ_r \wedge d\bar{Z}_s,
\end{aligned}
$$ (7.264)

which is -2γ times the Kähler 2-form (7.223).

It is easy to understand the motion of two vortices. Suppose their positions are $\mathbf{X} + \mathbf{Y}$ and $\mathbf{X} - \mathbf{Y}$. The centre, $\mathbf{X}$, is a constant of motion. Let $\mathbf{Y} = \rho(\cos\theta, \sin\theta)$. Then $\mathbf{b}_1 = b(\rho)(\cos\theta, \sin\theta)$, and $\mathbf{b}_2 = -\mathbf{b}_1$. Also, $\dot{\chi}_1 + \dot{\chi}_2 = 2\dot{\theta}$. The potential depends only on ρ. The Lagrangian therefore simplifies to

$$L_{\text{red}} = -2\pi\gamma\left(\rho^2 + 2\rho b(\rho)\right)\dot{\theta} + 4\pi\gamma\dot{\theta} - V_{\text{red}}(\rho).$$ (7.265)

The term $4\pi\gamma\dot{\theta}$ does not affect the equations of motion, but it cancels the singularity produced by the term $b(\rho) \sim \frac{1}{\rho}$ for small ρ.

The equations of motion arising from (7.265) are

$$\dot{\rho} = 0$$ (7.266)

$$2\pi\gamma\frac{d}{d\rho}\left(\rho^2 + 2\rho b(\rho)\right)\dot{\theta} = -\frac{dV_{\text{red}}}{d\rho}.$$ (7.267)

ρ is constant, and so is the angular velocity $\dot{\theta}$. The vortices circulate around one another at constant separation and constant speed. If ρ is large, $\dot{\theta}$ is exponentially small. If ρ is close to zero, $dV_{\text{red}}/d\rho = O(\rho^3)$, because V_{red} has a stationary point at $\rho = 0$, and is a smooth function of the "good" radial coordinate ρ^2. The coefficient of $\dot{\theta}$ is $O(\rho^3)$ too. Therefore the angular velocity approaches a finite limit as $\rho \to 0$. If $\gamma > 0$, and also $\lambda > 1$, so $dV_{\text{red}}/d\rho$ is negative, then the two vortices circle each other anticlockwise.

Symmetry implies that if there are p vortices at the vertices of a regular p-gon and q coincident vortices at the centre, possibly with $q = 0$, then the p-gon will rigidly rotate about the centre.

7.13 Global vortex dynamics

In this section we briefly discuss the dynamics of global vortices. We consider first the Lorentz invariant dynamics, with the field equation second order in time derivatives, although the vortices do not necesarily move at relativistic speeds. The first important issue is that of the static forces between vortices. For gauged vortices we could determine these by examining the potential energy of two vortices as a function of their separation. However, for global vortices even the potential energy of a single vortex is infinite, as we showed earlier, so this issue is clearly more subtle. A regularized energy must first be found, so that one can deal with finite quantities. The problem has been addressed by Ovchinnikov and Sigal in a series of papers [321], together with other aspects of global vortex dynamics which we shall mention below. Fortunately, the divergent part of the energy depends only on N and not on the vortex positions, so the regularization just removes a divergent constant. Hence the forces between the vortices are unaffected by the regularization.

As before, we fix the parameter values $m = 1$, $\lambda = 2$. The regularized potential energy is defined by

$$\tilde{V} = \frac{1}{2} \int \left(\boldsymbol{\nabla}\bar{\phi} \cdot \boldsymbol{\nabla}\phi + \frac{1}{2}(1 - \bar{\phi}\phi)^2 - \frac{N^2}{\rho^2}\eta(\rho) \right) d^2x , \qquad (7.268)$$

where N is the topological charge of the configuration and η is a smooth cutoff function with the properties that $\eta(\rho) = 0$ for $\rho \leq 1$ and $\eta(\rho) = 1$ for $\rho \geq 2$, with a smooth interpolation in between, the details of which are not important for the well separated vortices that we are interested in. Clearly the purpose of the additional negative contribution to the energy is to subtract off the divergent part. The next step is to define a configuration, C, by specifying k points in the complex plane, $\{Z_1, \ldots, Z_k\}$, and k non-zero integers, $\{n_1, \ldots, n_k\}$. These points give the positions of the vortices (or antivortices) and the integers are the associated multiplicities (which can be of either sign). A field ϕ has configuration C, and one writes config$(\phi) = C$, if ϕ has a zero of multiplicity n_r at $z = Z_r$, for $1 \leq r \leq k$, and has no other zeros. $N = n_1 + \cdots + n_k$ is the total topological charge. The potential energy of a configuration, $V(C)$, is defined as

$$V(C) = \inf\{\tilde{V} \,|\, \text{config}(\phi) = C\} . \qquad (7.269)$$

Physically, the energy of a configuration is the minimal regularized energy compatible with the given positions and multiplicities of the vortices. If the dependence of the energy $V(C)$ on the vortex positions Z_r can be found, then we have determined the static vortex forces. It can be shown that, for $k > 1$ and providing all inter-vortex separations are much greater than 1, there are no stationary points of $V(C)$ with respect to variations of all the vortex positions. In other words, as expected, there are no static well separated multi-vortex configurations. Moreover, when all the vortices are well separated the leading order contributions to $V(C)$ are given by [321]

$$V(C) = \sum_{r=1}^{k} V_{n_r} - \pi \sum_{\substack{r,s=1 \\ r \neq s}}^{k} n_r n_s \log |Z_r - Z_s| + \cdots , \qquad (7.270)$$

where V_{n_r} denotes the regularized energy of a single charge n_r vortex.

If we consider two unit charge vortices with separation $L \gg 1$ then the above formula yields

$$V(L) = 2V_1 - 2\pi \log L , \qquad (7.271)$$

so that there is a repulsive force between two global vortices which is inversely proportional to their separation. In the case of second order dynamics, the vortices accelerate away from each other. For a similar vortex-antivortex configuration the potential is

$$V(L) = 2V_1 + 2\pi \log L , \qquad (7.272)$$

so there is an attraction. The effective inertial mass of these vortices has been calculated by Moore [305]. It is not constant but has a logarithmic dependence on the distance to the other vortex.

These qualitative features have been confirmed by numerical simulations of the field equation

$$\partial_0 \partial_0 \phi - \nabla^2 \phi - (1 - \bar{\phi}\phi)\phi = 0 , \qquad (7.273)$$

which indeed show a repulsion between vortices and an attraction between a vortex and an antivortex. There are static axially symmetric multi-vortex solutions, but these are all unstable and decay into well separated single vortices if perturbed. Because the dynamics is second order in time derivatives, two vortices can be set in motion toward each other with any given speed. If this speed is low then there is not enough kinetic energy to overcome the potential repulsion. The vortices do not collide but instead scatter back-to-back. If the speed is increased then there is a critical value at which the repulsion is overcome, and the two vortices scatter at right angles in a head-on collision, as for gauged vortices. If the initial speed is very close to the speed of light then many vortex-antivortex pairs are produced during the collision, although these subsequently annihilate [407, 39].

Let us now turn to the first order dynamics given by the equation

$$i\partial_0 \phi = -\nabla^2 \phi - (1 - \bar{\phi}\phi)\phi . \qquad (7.274)$$

This is the Gross-Pitaevski, or nonlinear Schrödinger equation, modelling the dynamics of a superfluid. The motion of well separated superfluid vortices has been investigated by Fetter [130] and many others. The following remarks are based on more recent work of Ovchinnikov and Sigal [321], and Colliander and Jerrard [90]. The potential $V(C)$ is again relevant. By substituting well separated configurations into an effective action the dynamics can be approximated by a Hamiltonian system with the Hamiltonian proportional to $V(C)$. Explicitly, the leading order contribution to the dynamics of the vortices is given by

$$\dot{Z}_r = -\frac{2i}{\pi n_r} \frac{\partial V}{\partial \bar{Z}_r} \tag{7.275}$$

where, of course, there is no sum on the repeated index. For a well separated vortex-vortex configuration ($n_1 = n_2 = 1$) the equations become

$$\dot{Z}_1 = -\dot{Z}_2 = 2i \frac{(Z_1 - Z_2)}{|Z_1 - Z_2|^2}, \tag{7.276}$$

so the centre of vorticity $\frac{1}{2}(Z_1 + Z_2)$ is conserved during the dynamics. Without loss of generality we can set this to be the origin, and then the solution is

$$Z_1 = -Z_2 = \frac{L}{2} e^{4it/L^2}, \tag{7.277}$$

so the two vortices rotate around each other anticlockwise with period $\frac{1}{2}\pi L^2$ while their separation is $L \gg 1$. The above Hamiltonian description neglects the effects of radiation in the system, but this can also be studied (though it is a much more difficult problem) and leads to the result that the rotating pair radiate and move apart with a growth law $L \sim t^{1/6}$ [321].

For a vortex-antivortex pair ($n_1 = -n_2 = 1$) the equations are

$$\dot{Z}_1 = \dot{Z}_2 = -2i \frac{(Z_1 - Z_2)}{|Z_1 - Z_2|^2} \tag{7.278}$$

so this time $Z_1 - Z_2 = L$ is the conserved quantity, which we may take to be real without loss of generality. The relevant solution in this case is

$$Z_1 = \frac{L}{2} - \frac{2it}{L}, \quad Z_2 = -\frac{L}{2} - \frac{2it}{L} \tag{7.279}$$

so the vortex-antivortex pair simply translate at constant speed $2/L$, determined by their separation $L \gg 1$, and in the direction perpendicular to the line joining them. Clearly this description must break down for small L since the speed has a maximal value. In fact there is a critical separation L_* such that for $L > L_*$ there is a travelling wave solution which describes a vortex-antivortex pair moving parallel to each other, whereas for $L \leq L_*$ an initialized vortex-antivortex pair emits a shock wave and eventually annihilates [228, 321].

Similar results are found for the gradient flow, controlled by the equation

$$\partial_0\phi = \nabla^2\phi + (1 - \bar{\phi}\phi)\phi\,. \tag{7.280}$$

This has been studied by Neu [317] and E [117]; and more recently by Lin [266] and Jerrard and Soner [227]. Vortices now repel, and a vortex-antivortex pair attract. The equations of motion for well separated vortices or antivortices are the gradient flow equations for the regularized potential $V(C)$.

There are many other interesting mathematical results on global vortices and their dynamics, particularly in cases of extreme parameter values. We refer the interested reader to the book by Bethuel, Brezis and Hélein [51] and the papers by Jerrard and collaborators.

7.14 Varying the geometry

We have so far considered several aspects of gauged vortex dynamics in the plane. However, there are a number of reasons for being interested in non-planar geometries. A thin superconductor may be formed in a non-planar shape. Alternatively, we may be interested in vortices in the plane at a certain density. This can be achieved by imposing periodic boundary conditions in the plane – giving a torus – and specifying a finite number of vortices on the torus. There are some novel results for critically coupled vortices on a general, compact Riemann surface, X, without boundary. First, the number of vortices that may satisfy the Bogomolny equations is constrained by the area of the surface. Second, the moduli space is compact, and its total volume can be calculated. Making certain approximations, we can use the result to consider the statistical mechanics of a fluid of vortices by taking the limit, as $N, A \to \infty$, of the dynamics of N vortices on a surface of area A. Finally, it is mathematically interesting to consider vortices on the hyperbolic plane of curvature $-\frac{1}{2}$. Here the Bogomolny equations for critically coupled vortices reduce to Liouville's equation, which is solvable in terms of rational functions.

To discuss vortices on a compact surface X, we need to use the formalism introduced in Section 3.4, where the scalar field is a section and the gauge field a connection on a $U(1)$ bundle over the surface X. The scalar field ϕ is locally a complex function on X, and the gauge potential components a_i are combined into a connection 1-form $a = a_1 dx^1 + a_2 dx^2$. The spatial part of the field tensor, the magnetic field, is the 2-form $f = da = (\partial_1 a_2 - \partial_2 a_1) dx^1 \wedge dx^2$. The first Chern number

$$c_1 = \frac{1}{2\pi} \int_X f \tag{7.281}$$

takes an integer value N, and is the only topological invariant of the bundle. It can be demonstrated that the number of zeros of ϕ, counted with multiplicity, equals N. So the first Chern number is the net number of vortices minus antivortices.

We also need a Riemannian metric on X. Locally, we can choose coordinates (x^1, x^2) which are "isothermal", meaning that the metric is the flat metric times a conformal factor $\Omega(x^1, x^2)$. Space-time is $\mathbb{R} \times X$, with metric

$$ds^2 = dt^2 - \Omega(x^1, x^2)((dx^1)^2 + (dx^2)^2). \tag{7.282}$$

The Lagrangian for scalar electrodynamics on X is

$$L = \int_X \left(-\frac{1}{4} f_{\mu\nu} f^{\mu\nu} + \frac{1}{2} \overline{D_\mu \phi} D^\mu \phi - \frac{\lambda}{8}(1 - \bar\phi\phi)^2 \right) \Omega \, d^2x. \tag{7.283}$$

The same patches $\{U_p\}$ and transition functions $\{e^{-i\alpha^{(qp)}}\}$ on X can be used as in the static theory. ϕ and a_i are defined on these patches as before, and a_0 is a global function. The Lagrangian is then globally well defined and gauge invariant. As usual, it can be split into kinetic and potential terms, $L = T - V$. For the metric (7.282),

$$T = \frac{1}{2} \int_X \left(e_1^2 + e_2^2 + \Omega \overline{D_0 \phi} D_0 \phi \right) d^2x \tag{7.284}$$

$$V = \frac{1}{2} \int_X \left(\Omega^{-1} B^2 + \overline{D_1 \phi} D_1 \phi + \overline{D_2 \phi} D_2 \phi + \frac{\lambda \Omega}{4}(1 - \bar\phi\phi)^2 \right) d^2x, \tag{7.285}$$

where B, as usual, denotes f_{12}.

We may apply the Bogomolny argument to V at critical coupling, $\lambda = 1$, obtaining

$$\begin{aligned} V = \ &\frac{1}{2} \int_X \left\{ \Omega^{-1} \left(B - \frac{\Omega}{2}(1 - \bar\phi\phi) \right)^2 \right. \\ &\left. + \left(\overline{D_1 \phi} - i \overline{D_2 \phi} \right) \left(D_1 \phi + i D_2 \phi \right) + B \right\} d^2x. \end{aligned} \tag{7.286}$$

Thus, if $N > 0$, we have the usual bound

$$V \geq \pi N, \tag{7.287}$$

with equality for fields satisfying the Bogomolny equations [363]

$$D_1 \phi + i D_2 \phi = 0 \tag{7.288}$$

$$B - \frac{\Omega}{2}(1 - \bar\phi\phi) = 0. \tag{7.289}$$

As before, one may reduce these to a single gauge invariant equation by setting $\phi = e^{\frac{1}{2}h + i\chi}$. Here χ depends on the choice of gauge within each patch, and will vary discontinuously from patch to patch. Nevertheless, after eliminating a_1 and a_2 using (7.288), one obtains

$$\nabla^2 h + \Omega - \Omega e^h = 0 \qquad (7.290)$$

where ∇^2 is the standard Laplacian. As in the plane, this equation is only valid away from the zeros of ϕ. The zeros of ϕ, which all have positive multiplicity, are delta-function sources for h, and the full equation for h is

$$\nabla^2 h + \Omega - \Omega e^h = 4\pi \sum_{r=1}^{N} \delta^2(\mathbf{x} - \mathbf{X}_r). \qquad (7.291)$$

A simple, but astute observation was made by Bradlow [67] concerning these equations. Integrating (7.289) over X, one obtains

$$2 \int_X B \, d^2x + \int_X |\phi|^2 \Omega \, d^2x = \int_X \Omega \, d^2x. \qquad (7.292)$$

Therefore

$$4\pi N + \int_X |\phi|^2 \Omega \, d^2x = A, \qquad (7.293)$$

where A is the total area of X, since $\Omega \, d^2x$ is the area element. From this follows Bradlow's inequality

$$A \geq 4\pi N \qquad (7.294)$$

since $|\phi|^2$ is non-negative. For a given number of vortices N, there can be no solutions of the Bogomolny equations unless the area of X is at least $4\pi N$. Equivalently, for a surface of given area A, the number of Bogomolny vortices can not exceed the integer part of $A/4\pi$. The same bound emerges from (7.291) by integrating over X.

If $A > 4\pi N$, there is no simple way to solve (7.291) for a general surface. However, at the Bradlow bound $A = 4\pi N$, one may solve both Bogomolny equations. Note from (7.293) that in this case $\phi = 0$ on all of X (so it doesn't make sense to introduce h). Equation (7.288) is trivially solved, and from (7.289), $B = \frac{\Omega}{2}$, implying that the magnetic flux per unit area has the constant value $\frac{1}{2}$. Implicitly, the gauge potential is determined by this, up to a choice of gauge. (There is some choice for the holonomy around non-contractible loops if $\pi_1(X)$ is non-trivial.)

Close to the Bradlow bound, with A slightly larger than $4\pi N$, the magnetic field is approximately $B = \frac{\Omega}{2}$, and this again determines the gauge potential. Equation (7.288) is then a linear equation for ϕ, the same as that which defines the lowest Landau level of electron states in a background magnetic field. The normalization of ϕ is determined by

(7.293). So we see that in this case the moduli of the vortices are very closely related to the parameters of Landau level states.

In general, the solutions of (7.291) on a compact surface X have rather similar properties to those on the plane. It is convenient to use the local complex coordinate $z = x^1 + ix^2$. One may specify any N points $\{Z_1, \ldots, Z_N\}$ on the surface as the vortex locations, and then (for $A > 4\pi N$) there is a unique solution to (7.291). This was established by Bradlow [67] and also by García-Prada [142]. h has an expansion of the form (7.201) in the neighbourhood of the point Z_r, with leading term $h \sim 2\log|z - Z_r|$. The modification from the flat space case is that $d_r = -\frac{1}{4}\Omega(Z_r)$. $\mathcal{M}_N$, the N-vortex moduli space, is X^N/S_N, the symmetrized Nth power of X. As in the planar case, this has a smooth manifold structure, even though the orbits of S_N are not all of the same type. Samols calculated that the metric on $\mathcal{M}_N$ is

$$ds^2 = \pi \sum_{r,s=1}^{N} \left(\Omega(Z_r)\delta_{rs} + 2\frac{\partial b_s}{\partial Z_r} \right) dZ_r \, d\bar{Z}_s \, . \qquad (7.295)$$

This is again Kähler.

The simplest example is where X is a 2-sphere. The 2-sphere may be identified with the complex projective line $\mathbb{CP}^1$, the complex plane $\mathbb{C}$ together with a point at infinity. The points $\{Z_1, \ldots, Z_N\}$ may then be identified with the roots of a polynomial

$$p(z) = p_0 z^N + p_1 z^{N-1} + \cdots + p_N \, , \qquad (7.296)$$

where the only constraint on the coefficients is that they are not all zero. Generically $p_0 \neq 0$, and this polynomial has N finite roots. However, if the leading non-zero coefficient is p_n, then there are $N - n$ finite roots, and we interpret the polynomial as having a root of multiplicity n at infinity. So in all cases there are N roots, giving N unordered points on $\mathbb{CP}^1$. Note that multiplying the entire polynomial by a non-zero complex constant does not change the roots. So the moduli are the $N+1$ coefficients $\{p_0, \ldots, p_N\}$ modulo multiplication by such a constant. The moduli space is therefore $\mathbb{CP}^N$, and we have shown explicitly that

$$(\mathbb{CP}^1)^N/S_N = \mathbb{CP}^N \, . \qquad (7.297)$$

For a surface X of higher genus g, there is also a geometrical description of X^N/S_N [270]. This is simplest if $N > g$. Then X^N/S_N is a $\mathbb{CP}^{N-g}$ bundle over the Jacobian of X, which is a complex g-torus $\mathbb{T}^g$. The projection from X^N/S_N to $\mathbb{T}^g$ is the standard Abel-Jacobi map obtained by integrating the g independent, holomorphic 1-forms on X from a base point Z_0 to each of the points Z_r and summing over r. For $N \leq g$, X^N/S_N is a complex submanifold of $\mathbb{T}^g$.

7.14.1 Volume of moduli space

Here, we show that for Bogomolny vortices on a compact surface X, the total volume of the moduli space $\mathcal{M}_N = X^N/S_N$ can be calculated, even though the metric is not known explicitly [289].

The metric on $\mathcal{M}_N$ is as given by the formula (7.295). If the second term were absent, it would be

$$ds^2 = \pi \sum_{r=1}^{N} \Omega(Z_r)dZ_r\, d\bar{Z}_r \qquad (7.298)$$

which is π times the naive metric on X^N/S_N, determined by the given metric on X. The total volume would be $\pi^N A^N/N!$, where A is the area of X. However, this naive metric has conical singularities when two or more vortex positions coincide. Recall now the area deficit of the moduli space of two centred vortices in the plane, given by (7.243). Compared to the naive conical metric, the true metric has area $2\pi^2$ less. This calculation suggests that one may be able to calculate the exact volume of $\mathcal{M}_N$, a volume likely to be less than $\pi^N A^N/N!$.

The starting point of the volume calculation is the Kähler 2-form associated with the metric (7.295),

$$\omega = \frac{i\pi}{2} \sum_{r,s=1}^{N} \left(\Omega(Z_r)\delta_{rs} + 2\frac{\partial b_s}{\partial Z_r} \right) dZ_r \wedge d\bar{Z}_s\,. \qquad (7.299)$$

The volume form on moduli space is $\omega^N/N!$, so the total volume is

$$\mathrm{Vol}(\mathcal{M}_N) = \int_{\mathcal{M}_N} \frac{\omega^N}{N!}\,. \qquad (7.300)$$

This is because, on any Kähler manifold of complex dimension N, there are local, complex, normal coordinates $w_1, \ldots, w_N$ such that the metric is $\sum_r dw_r d\bar{w}_r$ and the Kähler form is $\omega = \frac{i}{2}\sum_r dw_r \wedge d\bar{w}_r$. The local volume element is then $\prod_r(\frac{i}{2}dw_r \wedge d\bar{w}_r) = \omega^N/N!$. Similarly, the integral of ω over any complex curve in $\mathcal{M}_N$ (but not an arbitrary real 2-surface) is its area.

Since ω is a closed 2-form, we can use the homology ring structure of $\mathcal{M}_N$ to express $\mathrm{Vol}(\mathcal{M}_N)$ in terms of $[\omega]$, where $[\omega]$ denotes the integral of ω over the one or more generating 2-cycles of the homology ring. If such a cycle is represented by a complex curve, then $[\omega]$ is its area.

The easiest case where a calculation of the volume is possible is for the moduli space of N vortices on a standard 2-sphere of radius R, with conformal factor $\Omega = 4R^2/(1 + |z|^2)^2$ and area $A = 4\pi R^2$ [285]. When $X = S^2$, then $\mathcal{M}_N = \mathbb{CP}^N$, and the homology of $\mathbb{CP}^N$ is generated by a single 2-cycle, which may be taken to be any complex line in $\mathbb{CP}^N$. A complex line arises if the coordinates $\{p_0, \ldots, p_N\}$ depend linearly on a $\mathbb{CP}^1$ parameter t, for example, $p_0, \ldots, p_{N-1}$ fixed, and $p_N = t$.

Now consider the 2-cycle in $\mathbb{CP}^N$ corresponding to all vortices being coincident at a variable point t. The associated polynomial is

$$p(z) = (z - t)^N = z^N - Ntz^{N-1} + \cdots + (-1)^N t^N. \tag{7.301}$$

The coefficients depend on t as $p_0 = 1, p_1 = -Nt, \ldots, p_N = (-1)^N t^N$, so this cycle is homologically N times the generating line. To see this, consider its intersection with an $(N-1)$-dimensional hyperplane of the form

$$c_0 p_0 + \cdots + c_N p_N = 0 \tag{7.302}$$

(for some constants $c_0, \ldots, c_N$). The intersection gives the equation for t

$$c_0 - c_1(Nt) + \cdots + c_N(-1)^N t^N = 0 \tag{7.303}$$

which has N solutions. This number of intersection points is the same as for N copies of a generating line, each of which would intersect the hyperplane once.

The restriction of the Kähler metric to this 2-cycle where all vortices are coincident is of the form

$$ds^2 = N\pi \left(\Omega(t) + 2\frac{\partial b}{\partial t} \right) dt\, d\bar{t}. \tag{7.304}$$

b is the coefficient in the expansion of h,

$$h = 2N \log |z - t| + a + \frac{1}{2}\bar{b}(z - t) + \frac{1}{2}b(\bar{z} - \bar{t}) + \cdots, \tag{7.305}$$

where it is assumed that all zeros of ϕ are at $z = t$. We can now calculate the area of this 2-cycle because we can evaluate the dependence of b on t explicitly.

Recall the interpretation of b in terms of the contours of h close to t. In general, these are approximately circles, and the circle in the z-plane of radius ε is centred at $t - \frac{1}{2N}\varepsilon^2 b$, to this order in ε. Now the symmetry helps. On a sphere of radius R, the square of the chordal distance between z and t is

$$\frac{4R^2|z - t|^2}{(1 + |z|^2)(1 + |t|^2)}, \tag{7.306}$$

and for given t, h can only depend on this. Therefore, the contours of h in the z-plane are given by the equation

$$\frac{|z - t|^2}{(1 + |z|^2)} = \text{const}, \tag{7.307}$$

so are all exactly circular. The circle of radius ε has centre $t + \varepsilon^2 t/(1+|t|^2)$ so $b = -2Nt/(1+|t|^2)$. It follows that

$$\frac{\partial b}{\partial t} = -\frac{2N}{(1+|t|^2)^2}. \tag{7.308}$$

Therefore the metric (7.304) on the 2-cycle of N coincident vortices is

$$ds^2 = N\left(1 - \frac{N}{R^2}\right)\frac{4\pi R^2}{(1+|t|^2)^2}\, dt\, d\bar{t}. \tag{7.309}$$

This is a simple multiple of the metric of the underlying 2-sphere, as it must be, by symmetry. The area of this 2-cycle in $\mathcal{M}_N$, obtained by integrating over the t-plane, is

$$4\pi^2 R^2 N\left(1 - \frac{N}{R^2}\right). \tag{7.310}$$

The generating 2-cycle of $\mathcal{M}_N$ therefore has area

$$[\omega] = 4\pi^2 R^2\left(1 - \frac{N}{R^2}\right) = \pi(A - 4\pi N), \tag{7.311}$$

so the total volume of moduli space is

$$\mathrm{Vol}(\mathcal{M}_N) = \frac{[\omega]^N}{N!} = \frac{\pi^N(A - 4\pi N)^N}{N!}. \tag{7.312}$$

Note that this vanishes at the Bradlow limit.

For one vortex, the volume is $\pi(A - 4\pi)$. This is less than πA, because of the factor $(1 - \frac{1}{R^2})$ from (7.309) which can be interpreted as a curvature effect reducing the inertia of a vortex. For N vortices, the volume is not only less than $\pi^N A^N/N!$, but it is less than $\pi^N(A - 4\pi)^N/N!$. The interpretation is that each vortex occupies a finite area, reducing the space available to other vortices.

To generalize this result to vortices on a Riemann surface X of genus g, with an arbitrary metric, one needs to understand the cohomology ring of $\mathcal{M}_N = X^N/S_N$. Macdonald has given a presentation of the 2-cocycles which are generators of the integral cohomology ring, and their product relations [270]. In [289] it has been shown that the cohomology class of the Kähler 2-form ω is the real linear combination

$$\pi\left(4\pi\sigma + (A - 4\pi N)\eta\right) \tag{7.313}$$

where σ and η are integer cohomology classes directly related to the topological fibration of X^N/S_N with the Jacobian $\mathbb{T}^g$ of X as base and the complex projective space $\mathbb{CP}^{N-g}$ as fibre. Hence $[\omega]$ is known for any 2-cycle, and consequently the volume form $\omega^N/N!$ can be integrated using Macdonald's algebraic formulae to give

$$\text{Vol}(\mathcal{M}_N) = \pi^N \sum_{j=0}^{g} \frac{(4\pi)^j (A - 4\pi N)^{N-j} g!}{j!(N-j)!(g-j)!} \tag{7.314}$$

for $N \geq g$. For $N < g$ one should retain the terms in the sum up to $j = N$. Thus the volume of moduli space depends only on the vortex number, the area of the surface X and the genus of X, but not on the detailed structure of the metric on X. In particular, (7.312) is the volume of $\mathcal{M}_N$ for N vortices on any surface of area A that is topologically a sphere.

Note that for $N > g$, the volume always tends to zero in the Bradlow limit $A \to 4\pi N$. This is consistent with what happens to the fields. Close to the Bradlow limit, B is approximately constant, and ϕ is small. Thus as the vortices move around, the magnetic field hardly changes, and the scalar field changes only a little. There is little kinetic energy associated with this. Nasir has checked, using the explicit metric close to the Bradlow limit, that the leading term

$$\text{Vol}(\mathcal{M}_N) = \frac{2^{2g} \pi^{N+g} (A - 4\pi N)^{N-g}}{(N-g)!} \tag{7.315}$$

gives the correct moduli space volume to leading order in $A - 4\pi N$ [316]. However, even at the Bradlow limit, the holonomy of the connection around non-contractible loops can change significantly, and this is why the volume of moduli space remains finite if $N \leq g$.

7.14.2 Toroidal geometry – the Abrikosov lattice

A physically interesting surface on which to study vortices is the torus $\mathbb{T}$ with a flat metric. Any solution we find can be interpreted as a spatially periodic solution in the plane. We define the torus as a parallelogram with opposite sides identified, of sides u, v and internal angle β as sketched in Fig. 7.19.

The fields need to be periodic only up to a gauge transformation. One may choose a gauge so that the periodicity conditions are

$$\phi(x^1 + u, x^2) = \phi(x^1, x^2) \tag{7.316}$$

$$\phi(x^1 + v\cos\beta, x^2 + v\sin\beta) = \phi(x^1, x^2)e^{-2\pi i N x^1/u} \tag{7.317}$$

$$a_i(x^1 + u, x^2) = a_i(x^1, x^2) \tag{7.318}$$

$$a_i(x^1 + v\cos\beta, x^2 + v\sin\beta) = a_i(x^1, x^2) - \frac{2\pi N}{u}\delta_{i1}, \tag{7.319}$$

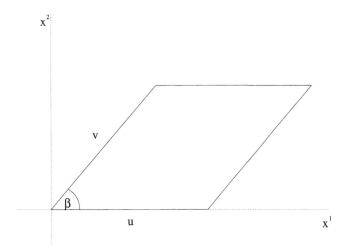

Fig. 7.19. The fundamental parallelogram of the torus.

and such fields have first Chern number $c_1 = N$, because, using Stokes' theorem, one can check that

$$\int_{\mathbb{T}} f_{12} \, d^2x = 2\pi N \,. \tag{7.320}$$

Provided the Bradlow condition $uv \sin\beta > 4\pi N$ is satisfied, there is a moduli space of solutions to the Bogomolny equations, a manifold of the form $\mathcal{M}_N = \mathbb{T}^N/S_N$, with local coordinates the zeros of ϕ, as usual.

When $N = 1$, the moduli space metric is

$$ds^2 = \pi \, dZ \, d\bar{Z} \,, \tag{7.321}$$

proportional to the metric on the underlying torus. This is a consequence of reflection symmetries, which require b, the coefficient of the linear term in the expansion of h, to vanish.

Generally, for N vortices on a torus, we find using (7.314) that the volume of the moduli space is [373, 289]

$$\mathrm{Vol}(\mathcal{M}_N) = \frac{\pi^N (A - 4\pi N)^{N-1} A}{N!} \,. \tag{7.322}$$

In particular, $\mathrm{Vol}(\mathcal{M}_1) = \pi A$, in agreement with (7.321).

Special shapes of tori have more than reflection symmetry. For example if $u = v$ and $\beta = \pi/2$, the torus has square geometry, and if $u = v$ and $\beta = \pi/3$, it has hexagonal geometry. In both these cases, one may consider solutions of the full second order static GL equations, for $\lambda > 1$, with one vortex on the torus. These solutions give the Abrikosov lattices of vortices in Type II superconductors [6], with the density of vortices, and hence of magnetic flux, fixed by the length unit u. (The analogous solutions for $\lambda < 1$ give unstable lattices, since the vortices in the plane tend to coalesce.) It is of interest to know for which geometry the energy is minimized, given the density of vortices. It has been shown that a hexagonal lattice of vortices, with lattice spacing u, has lower energy than a square lattice of vortices with lattice spacing $(\frac{3}{4})^{1/4}u$ (and hence the same density) [238]. So if a large number of vortices penetrate a region of given area A, they tend to form a hexagonal lattice. This is verified experimentally by placing a sample of Type II superconductor in an external magnetic field.

7.14.3 Vortices on the hyperbolic plane

An interesting, integrable case of vortices on a non-compact surface occurs when X is the hyperbolic plane of curvature $-\frac{1}{2}$. This was studied by Strachan, who found a general formula for the metric on the N-vortex moduli space, and showed that the metric was Kähler (an early result of this kind) [384].

In the Poincaré disc model, the metric of the hyperbolic plane is

$$ds^2 = \frac{8}{(1 - |z|^2)^2} \, dz \, d\bar{z} \tag{7.323}$$

where $|z| < 1$. If we set $h = 2g + 2\log\frac{1}{2}(1 - |z|^2)$, the equation (7.291) for h becomes

$$\nabla^2 g - e^{2g} = 2\pi \sum_{r=1}^{N} \delta^2(z - Z_r). \tag{7.324}$$

This is Liouville's equation with sources, which can be solved exactly. The solution is

$$g = -\log\frac{1}{2}(1 - |f|^2) + \frac{1}{2}\log\left|\frac{df}{dz}\right|^2, \tag{7.325}$$

where $f(z)$ is an arbitrary, complex analytic function. From this we can reconstruct the scalar field, finding that

$$|\phi| = \frac{1 - |z|^2}{1 - |f|^2}\left|\frac{df}{dz}\right|. \tag{7.326}$$

A simple choice of phase is to set

$$\phi = \frac{1 - |z|^2}{1 - |f|^2}\frac{df}{dz}. \tag{7.327}$$

Then the first Bogomolny equation, $D_{\bar{z}}\phi = 0$, is satisfied if

$$a_{\bar{z}} = -i\partial_{\bar{z}} \log \left(\frac{1 - |z|^2}{1 - |f|^2} \right).$$ (7.328)

The vortex locations $\{Z_r\}$ are the points where $\frac{df}{dz}$, and hence ϕ, vanishes.

We still need to ensure that $|\phi| = 1$ on the boundary of the disc, the circle $|z| = 1$, and that ϕ has no singularities inside the disc. This requires that $|f| = 1$ on the boundary, and $|f| < 1$ inside. These constraints are satisfied by choosing f to be a Blaschke product of the form

$$f(z) = \prod_{i=1}^{N+1} \left(\frac{z - c_i}{1 - \bar{c}_i z} \right)$$ (7.329)

where $|c_i| < 1 \; \forall i$. Each factor in this product has magnitude less than 1 inside the unit disc, magnitude 1 on the boundary, and its phase is strictly increasing along the boundary. It follows that f has the same three properties. Therefore $|\phi|$ has no singularity in the unit disc, and it follows fairly easily from (7.327) that because $\frac{df}{dz}$ is non-zero on the boundary, $|\phi| = 1$ there, and the radial derivative of $|\phi|$ vanishes too.

It should be noted that this solution does not depend on $N+1$ complex parameters, but only on N, because there is a 1-parameter family of Möbius transformations of f that only produce a gauge transformation of ϕ, and leave the zeros of ϕ fixed.

A simple example of a solution is that representing N coincident vortices at $z = 0$. Set $c_i = 0 \; \forall i$, so $f = z^{N+1}$. Then

$$\phi = \frac{(N + 1)z^N}{|z|^{2N} + |z|^{2N-2} + \cdots + 1}$$ (7.330)

which clearly has an N-fold zero at the origin, satisfies the boundary condition, and has winding number N along the boundary. Any deformation of this solution will also have winding number N, and hence N vortices, provided no parameter c_i crosses the unit circle.

Since the solutions are fairly explicit, one may hope to calculate the metric on moduli space explicitly. This is algebraically complicated for vortex numbers greater than two, because one needs the expansion of h around the vortex zeros, and these are only implicitly given as the zeros of $\frac{df}{dz}$. However, calculations have been done for centred 2-vortices, and geodesics representing vortex scattering have been found, analogous to those shown in Fig. 7.18.

We shall indicate here how the metric on the moduli space $\mathcal{M}_1$ is calculated. Take the Blaschke function

$$f(z) = \left(\frac{z - Z}{1 - \bar{Z}z} \right)^2$$ (7.331)

with Z inside the unit disc. This has a double zero at Z, so $\frac{df}{dz}$ has a simple zero, and therefore there is one vortex at Z. h has an expansion of the type (7.201) about Z, with $b = 2Z/(1 - |Z|^2)$. Using (7.295) we find that the metric is

$$ds^2 = \frac{12\pi}{(1 - |Z|^2)^2} \, dZ \, d\bar{Z} \,, \tag{7.332}$$

just a multiple of the metric on the underlying surface. The effective inertial mass of the vortex is $\frac{3}{2}\pi$, whereas its potential energy (analogous to its rest mass) is π as usual.

It is possible to consider one vortex on the hyperbolic plane with metric

$$ds^2 = \frac{4R^2}{(1 - |z|^2)^2} \, dz \, d\bar{z} \tag{7.333}$$

and hence curvature $-\frac{1}{R^2}$. The metric on the moduli space $\mathcal{M}_1$ can be computed by a symmetry argument, even though ϕ is not explicitly known, and is found to be [363]

$$ds^2 = \left(1 + \frac{1}{R^2}\right) \frac{4\pi R^2}{(1 - |Z|^2)^2} \, dZ \, d\bar{Z} \,. \tag{7.334}$$

This is consistent with the result (7.332) for $R = \sqrt{2}$, and also with the analogous result for a vortex on a sphere.

There is a close connection between vortices on the hyperbolic plane of curvature $-\frac{1}{2}$ and Yang-Mills instantons with $SO(3)$ cylindrical symmetry. This will be clarified in Section 10.1.

7.15 Statistical mechanics of vortices

Consider N vortices at critical coupling on a compact surface X of area A, with the second order dynamics, and suppose the system is raised to a temperature T high enough to allow the vortices to move around, but not high enough to excite the field modes orthogonal to the moduli space $\mathcal{M}_N$ [285]. The effective dynamical system is free, geodesic motion on the moduli space. Since there is a mass gap in the theory, with both the gauge and scalar fields having mass 1, the temperature range we are talking about is $0 < T \ll 1$. Suppose further that we can use classical statistical mechanics. This is valid if the temperature is much greater than a typical quantum energy. For a vortex of mass π on a surface of typical length dimension $\sqrt{A}$, energy quanta are $\sim \frac{\hbar^2}{A}$, so we require $T \gg \frac{\hbar^2}{A}$, where $\hbar$ is Planck's constant (in appropriate units).

In classical statistical mechanics one uses the Gibbs distribution on phase space. For free motion on $\mathcal{M}_N$, the phase space is the cotangent bundle of $\mathcal{M}_N$, with $\{y^i\}$ real coordinates on $\mathcal{M}_N$ and $\{p_i\}$ the conjugate momenta. The partition function is

$$\mathcal{Z} = \frac{1}{(2\pi\hbar)^{2N}} \int e^{-E(\mathbf{p},\mathbf{y})/T} \, d^{2N}y \, d^{2N}p \tag{7.335}$$

where $E(\mathbf{p},\mathbf{y}) = \frac{1}{2}g^{ij}(\mathbf{y})p_i p_j$ and g^{ij} is the (inverse) metric on $\mathcal{M}_N$. The Gaussian integrals over the momenta can be done explicitly, leaving

$$\mathcal{Z} = \left(\frac{T}{2\pi\hbar^2}\right)^N \int_{\mathcal{M}_N} \sqrt{\det(g_{ij}(\mathbf{y}))} \, d^{2N}y \,. \tag{7.336}$$

The last integral is simply the volume of $\mathcal{M}_N$. Let us suppose that the N vortices are moving on a 2-sphere of area A. Then from the result (7.312), we conclude that

$$\mathcal{Z} = \frac{1}{N!}(A - 4\pi N)^N \left(\frac{T}{2\hbar^2}\right)^N . \tag{7.337}$$

Note the presence of the $N!$ factor here, which arises naturally because the vortices are indistinguishable even at the classical level. In the historical treatment of classical statistical mechanics of N point particles, this factor had to be inserted by hand, and its justification is usually based on quantum mechanics.

Let us now suppose that N and A are large, so the 2-sphere is almost flat on the scale of the vortex core size. Take the limit $N \to \infty$, $A \to \infty$ with N/A finite. The free energy $F = -T \log \mathcal{Z}$ is

$$F = -NT \left(\log(A - 4\pi N) - \log N + \log \frac{eT}{2\hbar^2} \right), \tag{7.338}$$

where we have used Stirling's approximation, $\log N! \simeq N \log N - N$. The pressure of the N-vortex system is $P = -\partial F/\partial A$ so

$$P = \frac{NT}{A - 4\pi N} \,. \tag{7.339}$$

This is Clausius' special case of the van der Waals equation of state

$$\left(P + \frac{\alpha N^2}{A^2}\right)(A - \beta N) = NT \tag{7.340}$$

with $\alpha = 0$, $\beta = 4\pi$. The interpretation is that the vortices are interacting, but in a purely geometrical way. The vortices neither attract nor repel, but each vortex occupies space that is unavailable to the others. Equation (7.339) has the virial expansion

$$PA = NT \left(1 + 4\pi\rho + (4\pi)^2\rho^2 + (4\pi)^3\rho^3 + \cdots\right) \tag{7.341}$$

where $\rho = N/A$ is the number density of vortices. To second order in the virial expansion (the term $4\pi\rho$), the equation of state for a gas of hard discs each of area 2π is the same as (7.341). However, (7.341) is an exact result, whereas the higher order terms in the virial expansion for the hard disc gas are different and increasingly hard to calculate [344]. Curiously, the equation of state (7.339) is closely analogous to that for a gas of hard rods in one dimension [178].

One could assume that the N vortices are moving on a different surface, for example, a 2-torus of area A. Then one should use the formula (7.322) for the volume of moduli space. However, in the large N, large A limit it makes no difference to the equation of state.

There is one caveat about these calculations. When the density of vortices is large, and approaches the Bradlow limit $N/A \rightarrow 1/4\pi$, then the scalar field is everywhere close to zero, and the magnetic field close to constant. One now needs to reconsider the field fluctuations orthogonal to moduli space. There is no longer a mass gap of order 1. As a consequence, the singularity in the pressure at $\rho = 1/4\pi$ is probably not real, but smoothed out.

There is another physical situation where the moduli space volume plays an important role. Recall that in the reduced dynamics of the Schrödinger-Chern-Simons vortex system, the moduli space itself is the phase space. Romão has carried out the geometrical quantization of the reduced dynamics and calculated the dimension of the Hilbert space of states [347]. In the semi-classical limit, where $A \rightarrow \infty$ and N/A is fixed, this dimension is asymptotic to $\text{Vol}(\mathcal{M}_N)/(2\pi\hbar)^N$, as one expects on the basis of the Bohr-Sommerfeld quantization ideas.

8
Monopoles

8.1 Dirac monopoles

The idea of magnetic monopoles goes back to the early history of magnetism, since magnets appear to have two poles of opposite strength. But isolated magnetic poles are never seen. In the nineteenth century, it was recognized that electric currents are a source of magnetism, and later, the magnetism of materials was understood as due either to currents at the atomic scale, or to the magnetic dipole moments associated with fundamental particles, like the electron. Modern elementary particle theory has no need for monopoles, and no monopoles have been experimentally confirmed, despite intensive searches throughout the accessible cosmos [161].

The absence of monopoles is built into Maxwell's theory of electromagnetism. The equation $\nabla \cdot \mathbf{b} = 0$ for the magnetic field $\mathbf{b}$ implies there is no source of magnetic flux, and the flux of $\mathbf{b}$ through any closed surface vanishes. Simply inserting a magnetic charge density ρ_{m} and postulating that $\nabla \cdot \mathbf{b} = \rho_{\mathrm{m}}$ leads to contradictions with other equations of electromagnetism, for example, Faraday's law $\nabla \times \mathbf{e} + \frac{\partial \mathbf{b}}{\partial t} = \mathbf{0}$, where $\mathbf{e}$ is the electric field. Taking the divergence of this equation leads to $\frac{\partial}{\partial t}\rho_{\mathrm{m}} = 0$, so the magnetic charge density would be unchanging for all time, which is implausible in an evolving universe. A more subtle objection to a magnetic charge density is that it would forbid the introduction of a covariant vector potential $\mathbf{a}$, satisfying $-\nabla \times \mathbf{a} = \mathbf{b}$. There is ample evidence that although the existence of a vector potential is not essential in classical electromagnetism, it is vital in the formulation of the quantum mechanics of electrically charged particles.

Despite all these arguments, Dirac reconsidered the matter in a famous paper published in 1931 [109]. He showed that the quantum mechanics of an electrically charged particle can be consistently formulated even in the presence of a point magnetic charge, provided the magnitude of the charge obeys a certain condition.

A point magnetic charge of strength g, at rest at the origin, is by definition an object with a magnetic field

$$\mathbf{b} = \frac{g}{4\pi r^2}\hat{\mathbf{x}}.$$

(8.1)

Away from the origin, all the usual vacuum Maxwell equations are satisfied, but there is a delta-function source for the magnetic field

$$\boldsymbol{\nabla} \cdot \mathbf{b} = g\,\delta^3(\mathbf{x})$$

(8.2)

and the flux of the magnetic field out of any closed surface S, with the origin inside, is

$$\int_S \mathbf{b} \cdot d\mathbf{S} = g.$$

(8.3)

There is, in fact, no problem finding a solution of Maxwell's equations for several point monopoles moving along arbitrary world lines. The fields obey the equations away from the world lines, and the singularity along each world line is that obtained from the field (8.1) by a translation and Lorentz boost. We shall present formulae for the fields due to a moving monopole later, in Section 8.12. But for the moment, let us just consider the static field (8.1).

Dirac's argument is essentially mathematical in nature, and quantum mechanics plays only a peripheral role. Dirac insisted that a vector potential for the field (8.1) should, in some sense, exist. Such a vector potential will have a singularity at the origin, but this doesn't matter. The more significant difficulty is that no smooth vector potential can be defined in $\mathbb{R}^3$, even with the origin excluded. For if $\mathbf{b} = -\boldsymbol{\nabla} \times \mathbf{a}$, then by Stokes' theorem

$$\int_S \mathbf{b} \cdot d\mathbf{S} = -\int_S \boldsymbol{\nabla} \times \mathbf{a} = 0$$

(8.4)

for any closed surface S with no boundary, and this contradicts (8.3).

But Dirac realized that the vector potential $\mathbf{a}$ need not be globally well defined. It is sufficient, in the mathematical language that we discussed in Chapter 3, for $\mathbf{a}$ to be a connection. This should be defined in $\mathbb{R}^3$ with the origin, $\mathbf{0}$, removed. It is possible to cover $\mathbb{R}^3 - \{\mathbf{0}\}$ with just two regions. Let us use spherical polar coordinates (r, θ, φ) and introduce a pair of angles θ_0, θ_1 satisfying $0 < \theta_0 < \theta_1 < \pi$. One region is taken to be $0 \leq \theta < \theta_1$, the other $\theta_0 < \theta \leq \pi$, with r, φ taking their full range of values, $0 < r < \infty$ and $0 \leq \varphi \leq 2\pi$. The important property of each of these regions is that they are contractible, hence topologically trivial. In each region there is no difficulty finding a local vector potential for the magnetic field of a monopole. In the first, an example of a suitable potential is

$$\mathbf{a}^{(1)} = \frac{g}{4\pi r}\frac{(-1 + \cos\theta)}{\sin\theta}\mathbf{e}_\varphi \tag{8.5}$$

where $\mathbf{e}_\varphi = (-\sin\varphi, \cos\varphi, 0)$. In the second, a suitable potential is

$$\mathbf{a}^{(2)} = \frac{g}{4\pi r}\frac{(1 + \cos\theta)}{\sin\theta}\mathbf{e}_\varphi. \tag{8.6}$$

Not surprisingly, the first formula becomes singular if extended to include $\theta = \pi$, and the second is singular at $\theta = 0$. These singularities are called Dirac strings, but they are not physical. The gauge transformation relating $\mathbf{a}^{(1)}$ and $\mathbf{a}^{(2)}$ in the region of overlap $\theta_0 < \theta < \theta_1$ (which is not simply connected, but this does not matter) is

$$\mathbf{a}^{(2)} = \mathbf{a}^{(1)} - \nabla\alpha^{(21)} \tag{8.7}$$

where $\alpha^{(21)} = -\frac{g}{2\pi}\varphi$. Now $\nabla\alpha^{(21)}$ is single-valued, but $\alpha^{(21)}$ is not, since $\alpha^{(21)}(\varphi = 2\pi) = \alpha^{(21)}(\varphi = 0) - g$. Does this matter? We need to consider the charges of the fields coupled to the monopole. Suppose there is a field ϕ of unit charge, coupled as in scalar electrodynamics with a covariant derivative $\partial_i\phi - ia_i\phi$. In the presence of the monopole, ϕ should be well defined in each of the regions, and its values where the regions overlap should be related by the appropriate gauge transformation,

$$\phi^{(2)}(\mathbf{x}) = e^{-i\alpha^{(21)}(\mathbf{x})}\phi^{(1)}(\mathbf{x}). \tag{8.8}$$

This equation is consistent, provided $e^{-i\alpha^{(21)}(\mathbf{x})}$ is single-valued, and this requires g to be 2π times an integer.

Further fields may be introduced, having any integer charge n. The analogue of Eq. (8.8) has a factor $e^{-in\alpha^{(21)}(\mathbf{x})}$, and this is single-valued since $e^{ing} = 1$ if $g/2\pi$ is an integer.

Therefore, with the above assumptions, the magnetic charge of the monopole, g, must be an integer multiple of 2π. This conclusion is usually given in a somewhat more general form. Suppose the field ϕ with smallest charge has charge e, its covariant derivative being $\partial_i\phi - iea_i\phi$, and suppose that all other fields have charges ne, with n integral. Then g must be an integer multiple of $2\pi/e$. Now, when a field of charge e is quantized, the particles associated with the field have electric charge $q = -e\hbar$. This can be seen from the operator occurring in the Schrödinger equation for one particle $-i\hbar\partial_i + qa_i$, which is $-i\hbar$ times the covariant derivative $\partial_i - iea_i$. So we have the constraint on the magnetic charge of a monopole

$$gq = -2\pi\hbar N, \quad N \in \mathbb{Z}, \tag{8.9}$$

which is the quantization condition given by Dirac. It remains satisfied if q is the electric charge not of a minimally charged particle, but of any particle with a charge that is an integer multiple of the minimal value.

Dirac subtly interpreted the result as follows. If there is at least one monopole in the universe, of magnetic charge g, then electrically charged particles must all have charges q which are integer multiples of $2\pi\hbar/g$. This is an elegant possible explanation of electric charge quantization. We will take the more mathematical line, which is to set the coupling e and also $\hbar$ to unity, and to postulate that the gauge group of electromagnetism is $U(1)$, which means that a constant gauge transformation $e^{-i\alpha}$ should have no effect on any field if α is an integer multiple of 2π. That implies that fields and particles have integer charges, and it also implies that monopoles have magnetic charges that are integer multiples of 2π.

The Dirac monopole is not a topological soliton because of its singular behaviour at $r = 0$. Naively, it has an infinite mass because the energy density in the magnetic field is proportional to $1/r^4$, and when integrated over $\mathbb{R}^3$ there is a linear divergence as $r \to 0$. This infinity can be regularized by supposing there is some unknown structure at short distances, which gives the monopole less singular fields there, and a finite mass. The classical dynamics of a monopole of charge g in an electromagnetic field is determined by the Lorentz force

$$\mathbf{F} = g(\mathbf{b} - \mathbf{v} \times \mathbf{e})\,, \tag{8.10}$$

which should be compared to the force on an electric particle of charge q

$$\mathbf{F} = q(\mathbf{e} + \mathbf{v} \times \mathbf{b})\,. \tag{8.11}$$

Equation (8.10) is a postulate, consistent with Lorentz invariance.

Quantizing the dynamics of Dirac monopoles and electrically charged particles is rather difficult [440]. There is not yet a quantum field theory of Dirac monopoles, including processes like monopole-antimonopole pair production. These matters can, however, be sensibly considered in theories with monopoles as topological solitons, but it is still difficult to calculate anything.

The Dirac quantization condition can be directly related to the integrality of the first Chern number c_1, as presented in Section 3.4. The vector potentials of the Dirac monopole need first to be reexpressed as coordinate invariant 1-forms. The 1-form in region 1 is

$$a^{(1)} = a_1\, dx^1 + a_2\, dx^2 + a_3\, dx^3\,. \tag{8.12}$$

Transforming to spherical polars, via $x^1 = r\sin\theta\cos\varphi$, $x^2 = r\sin\theta\sin\varphi$, $x^3 = r\cos\theta$, and thus $dx^1 = \sin\theta\cos\varphi\,dr + r\cos\theta\cos\varphi\,d\theta - r\sin\theta\sin\varphi\,d\varphi$, etc., we find

$$a^{(1)} = \frac{g}{4\pi}(-1 + \cos\theta)\,d\varphi, \tag{8.13}$$

and similarly

$$a^{(2)} = \frac{g}{4\pi}(1 + \cos\theta)\,d\varphi. \tag{8.14}$$

The gauge transformation relating $a^{(2)}$ to $a^{(1)}$ is now

$$a^{(2)} = a^{(1)} - d\alpha^{(21)} \tag{8.15}$$

where $d\alpha^{(21)} = -\frac{g}{2\pi}d\varphi$. The field strength, calculated either from $a^{(1)}$ or $a^{(2)}$, is the 2-form

$$f = da = -\frac{g}{4\pi}\sin\theta\,d\theta \wedge d\varphi \tag{8.16}$$

which is simply $-g$ times the normalized area form on the 2-sphere. We see from this that the field of a Dirac monopole is essentially a two-dimensional notion, defined on a 2-sphere; there is no dependence on r, except for the singularity at $r = 0$.

The flux of f through a 2-sphere of any radius is $-g$. But we showed quite generally, in Section 3.4, that for any closed surface X,

$$c_1 = \frac{1}{2\pi}\int_X f \tag{8.17}$$

is an integer, N. So if c_1, evaluated on a 2-sphere enclosing a monopole, is N, the monopole's magnetic charge is $g = -2\pi N$, in agreement with the Dirac quantization condition.

The connection viewpoint is extremely powerful in situations where space has a non-trivial topological structure. The space $\mathbb{R}^3 - \{0\}$ is topologically non-trivial, as is the 2-sphere of fixed radius onto which it retracts, and the Dirac monopole field is a connection on a $U(1)$ bundle over either of these spaces. However, for most of this chapter, we shall be interested in monopoles without singularities, defined in all of $\mathbb{R}^3$. Bundles over $\mathbb{R}^3$ always have a trivial structure, which means that a connection can always be expressed as a smooth gauge potential throughout $\mathbb{R}^3$. There is no need for more than one region to cover $\mathbb{R}^3$. Nevertheless, these monopoles are similar in important ways to the Dirac monopole, and they have a topological character, as we shall see.

We conclude this section by showing that the Dirac monopole of charge $g = -2\pi N$ is spherically symmetric. Since the 2-form field strength f, at any distance r from the origin, is proportional to the area element of the 2-sphere of radius r centred at the origin, the monopole looks spherically symmetric. However, we should check this more carefully, by considering the connection 1-form too.

A basis for the vector fields generating rotations in $\mathbb{R}^3$ is

$$\xi_1 = -\sin\varphi \frac{\partial}{\partial\theta} - \cot\theta\cos\varphi \frac{\partial}{\partial\varphi} \tag{8.18}$$

$$\xi_2 = \cos\varphi \frac{\partial}{\partial\theta} - \cot\theta\sin\varphi \frac{\partial}{\partial\varphi} \tag{8.19}$$

$$\xi_3 = \frac{\partial}{\partial\varphi}, \tag{8.20}$$

with Lie brackets $[\xi_m, \xi_n] = -\varepsilon_{mnp}\xi_p$. The connection 1-form a, on any region, is spherically symmetric if it satisfies the conditions

$$\mathcal{L}_{\xi_m} a = d\alpha_m \tag{8.21}$$

where $\mathcal{L}_{\xi_m}$ denotes the Lie derivative in the direction of ξ_m, and $d\alpha_m$ represents an infinitesimal gauge transformation. This is as in Eq. (2.31), and is also the infinitesimal version of the rotational symmetry condition that we discussed in Section 4.3, appropriate for a $U(1)$ gauge theory.

There is an elegant general expression for the Lie derivative of a 1-form, namely

$$\mathcal{L}_{\xi} a = d(i(\xi)a) + i(\xi)da \tag{8.22}$$

where $i(\xi)$ denotes the interior product of the vector field ξ with the form that follows. Therefore (8.21) can be rewritten as $i(\xi_m)da = d(\alpha_m - i(\xi_m)a)$, or better, as

$$i(\xi_m)f = d\psi_m \tag{8.23}$$

where f is the 2-form field strength and (as in Eq. (2.38)) $\psi_m = \alpha_m - i(\xi_m)a$. The connection is symmetric if for some choice of ψ_m, its field strength obeys (8.23). ψ_m, unlike α_m, is gauge invariant.

We can now easily calculate that for $f = \frac{N}{2}\sin\theta\, d\theta \wedge d\varphi$,

$$i(\xi_1)f = \frac{N}{2}(\cos\theta\cos\varphi\, d\theta - \sin\theta\sin\varphi\, d\varphi) \tag{8.24}$$

$$i(\xi_2)f = \frac{N}{2}(\cos\theta\sin\varphi\, d\theta + \sin\theta\cos\varphi\, d\varphi) \tag{8.25}$$

$$i(\xi_3)f = -\frac{N}{2}\sin\theta\, d\theta. \tag{8.26}$$

This is of the form (8.23), with

$$\psi_1 = \frac{N}{2}\sin\theta\cos\varphi, \quad \psi_2 = \frac{N}{2}\sin\theta\sin\varphi, \quad \psi_3 = \frac{N}{2}\cos\theta. \tag{8.27}$$

So, for all N, the Dirac monopole is spherically symmetric.

Using (8.27) we can find the angular momentum for a particle moving in the background field of the monopole. Suppose the particle has mass

m and unit electric charge (with $\hbar = 1$), and the vector potential of the monopole is **a**. The Lagrangian for the particle is

$$L = \frac{1}{2} m \dot{x}^i \dot{x}^i - a_i(\mathbf{x}) \dot{x}^i , \tag{8.28}$$

as in (2.13). In Cartesian coordinates, the rotation generators and the quantities ψ_m are

$$\xi_m = \varepsilon_{mnp} x^n \frac{\partial}{\partial x^p} , \qquad \psi_m = \frac{N}{2} \frac{x^m}{r} , \tag{8.29}$$

so the particle's conserved angular momentum, according to the general formula (2.37), is [334]

$$\mathbf{l} = m\mathbf{x} \times \dot{\mathbf{x}} + \frac{N}{2} \hat{\mathbf{x}} . \tag{8.30}$$

The actual value of $\mathbf{l}$ depends on the initial data. From (8.30) we deduce that

$$\mathbf{l} \cdot \hat{\mathbf{x}} = \frac{N}{2} , \tag{8.31}$$

the equation of a cone with vertex at the origin. The orbit of the particle is an in-and-out spiralling motion on this cone (intrinsically, it is along a geodesic at constant speed).

To understand how the above discussion of the spherical symmetry of a Dirac monopole relates to the formalism of Section 4.3 we should evaluate the quantities α_m, because from these we can identify the homomorphism λ. Let us take the non-singular expression for the 1-form gauge potential a in region 1,

$$a = \frac{N}{2} (1 - \cos \theta) \, d\varphi . \tag{8.32}$$

Then, from the formula $\alpha_m = \psi_m + i(\xi_m) a$, we find that

$$\alpha_1 = \frac{N}{2} \frac{(1 - \cos \theta)}{\sin \theta} \cos \varphi , \qquad \alpha_2 = \frac{N}{2} \frac{(1 - \cos \theta)}{\sin \theta} \sin \varphi , \qquad \alpha_3 = \frac{N}{2} . \tag{8.33}$$

Let us select the points lying on the positive x^3-axis as the preferred points $\mathbf{x}_0$, one on each S^2-orbit of the rotation group. Their isotropy group is the $SO(2)$ of rotations about the x^3-axis, generated by ξ_3. The constant value of α_3, associated with this generator, suggests that the homomorphism $\lambda : SO(2) \to U(1)$, where $SO(2)$ is the isotropy group and $U(1)$ the gauge group, maps a rotation by χ to a gauge transformation with phase $\frac{1}{2} N \chi$. However, this argument is suspect, since only $d\alpha_3$ occurs in the symmetry equation, and a constant α_3 is annihilated by the d operator. We need to look more closely at α_1 (or α_2).

So, let us consider Eq. (4.54). In the case of an abelian gauge theory, it may be written as

$$(R_{ji}a_j(R\mathbf{x}) - a_i(\mathbf{x}))dx^i = -i\,dg_R(\mathbf{x})g_R^{-1}(\mathbf{x}) = d(\arg g_R(\mathbf{x})) \qquad (8.34)$$

where, using (4.63), $g_R(\mathbf{x}) = \lambda(R_{R\mathbf{x}}^{-1}RR_{\mathbf{x}})$, and $R_{R\mathbf{x}}^{-1}RR_{\mathbf{x}}$ is Wigner's little group element. By using 1-form notation here, we can pass easily between Cartesian and polar coordinates. Let us choose R to be a rotation about the x^1-axis through an infinitesimal angle ε. The left-hand side of (8.34) is, by definition, $\varepsilon\mathcal{L}_{\xi_1}a$, expressed in Cartesian form. It may be evaluated most easily using the Cartesian components of the gauge potential (8.5), and the rotation matrix

$$R = \begin{pmatrix} 1 & 0 & 0 \\ 0 & 1 & -\varepsilon \\ 0 & \varepsilon & 1 \end{pmatrix}. \qquad (8.35)$$

The result can be converted to polar coordinates, and one finds as before that $\mathcal{L}_{\xi_1}a = d\alpha_1$, with α_1 as in (8.33). We now need to evaluate the Wigner little group element and verify that for some choice of λ, the right-hand side of (8.34) is $\varepsilon d\alpha_1$. In polar coordinates, the general point $\mathbf{x}$, of length 1, is

$$\mathbf{x} = (\sin\theta\cos\varphi,\ \sin\theta\sin\varphi,\ \cos\theta), \qquad (8.36)$$

and the special rotation $R_{\mathbf{x}}$, which takes the point $\mathbf{x}_0 = (0,0,1)$ to $\mathbf{x}$, is

$$R_{\mathbf{x}} = \begin{pmatrix} \cos\theta\cos^2\varphi + \sin^2\varphi & (\cos\theta - 1)\cos\varphi\sin\varphi & \sin\theta\cos\varphi \\ (\cos\theta - 1)\cos\varphi\sin\varphi & \cos\theta\sin^2\varphi + \cos^2\varphi & \sin\theta\sin\varphi \\ -\sin\theta\cos\varphi & -\sin\theta\sin\varphi & \cos\theta \end{pmatrix}. \qquad (8.37)$$

This matrix represents a rotation by θ about the axis $(-\sin\varphi,\ \cos\varphi,\ 0)$. The axis is orthogonal to $(0,0,1)$, which is what we require, as the matrix should be generated by an element in the subspace $m \subset so(3)$ orthogonal to the $so(2)$ subalgebra of the isotropy group.

It is now straightforward, but a bit tedious, to evaluate $R_{R\mathbf{x}}^{-1}RR_{\mathbf{x}}$. This is a rotation about the x^3-axis by an infinitesimal angle that is proportional to ε, depending non-trivially on θ and φ. One may find the rotation angle by evaluating the action on $(-\sin\theta\cos\varphi,\ -\sin\theta\sin\varphi,\ \cos\theta)$. $R_{\mathbf{x}}$ maps this point to $(0,0,1)$, which is mapped by R to $(0,-\varepsilon,1)$. $R_{R\mathbf{x}}$, the analogue of the matrix (8.37) for the rotated vector

$$R\mathbf{x} = (\sin\theta\cos\varphi,\ \sin\theta\sin\varphi - \varepsilon\cos\theta,\ \cos\theta + \varepsilon\sin\theta\sin\varphi), \qquad (8.38)$$

is found by expressing the entries of (8.37) in terms of the three components of $\mathbf{x}$, and replacing these by the components of $R\mathbf{x}$. $R_{R\mathbf{x}}^{-1}$ is its transpose. After acting with this on $(0, -\varepsilon, 1)$, one deduces that $R_{R\mathbf{x}}^{-1} R R_{\mathbf{x}}$ is a rotation about the x^3-axis by

$$\varepsilon \frac{(1 - \cos\theta)}{\sin\theta} \cos\varphi . \tag{8.39}$$

By dropping the factor ε and multiplying by $\frac{1}{2}N$, one again obtains α_1. Therefore, at the Lie algebra level, the homomorphism λ is just multiplication by $\frac{1}{2}N$. At the Lie group level, λ maps a rotation by χ about the x^3-axis to a $U(1)$ gauge transformation by $\exp(\frac{1}{2}Ni\chi)$. This is just what was indicated by the value of α_3. In conclusion, the Dirac monopole is spherically symmetric in the sense of Eq. (4.54) for any integer N, but the homomorphism λ depends on N.

The fact that $\exp(\frac{1}{2}Ni\chi) = -1$ when $\chi = 2\pi$, for N odd, is rather strange but not contradictory; it means that a scalar field of unit charge coupled to the monopole has a spinorial character. A quantized scalar particle of unit electric charge, coupled to the monopole, has half integer angular momentum [131, 159].

8.2 Monopoles as solitons

In 1974, 't Hooft [401] and Polyakov [336] made the important discovery that non-abelian gauge theories can have magnetic monopole solutions with no singularities. We have seen in Chapter 4 that the Derrick theorem implies that the pure Yang-Mills equation has no topological soliton solutions in three space dimensions. However, when Yang-Mills fields are coupled to Higgs scalar fields, then topologically stable monopole solutions with finite energy are possible. The core has a rather complicated nature, but the long-range electromagnetic fields are the same as those of a Dirac monopole. These solitons can be regarded as Dirac monopoles embedded in Yang-Mills-Higgs theory, with the singularity smoothed out. They are stable because the magnetic charge has a topological character, and so can not change under any smooth deformation of the field.

It is possible to define Yang-Mills-Higgs theory with any compact Lie group G as the gauge group, and the Higgs field transforming under any finite-dimensional representation of G. In this chapter we shall suppose the Higgs field transforms via the adjoint representation of G. That is, the Higgs field Φ is valued in the Lie algebra of G, and transforms by conjugation. Our main example is the Yang-Mills-Higgs theory with gauge group $SU(2)$. (This theory can be regarded alternatively as having gauge group $SO(3)$ and a Higgs field transforming via the fundamental three-dimensional representation of $SO(3)$.) If the gauge symmetry is spontaneously broken to $U(1)$ by the Higgs mechanism, then there are monopole solutions. For a short time, before the experimental status of the "neutral currents" mediated by the Z boson was clarified, this Georgi-Glashow $SU(2)$ theory [146] was a competitor to the $U(2)$ Glashow-Weinberg-Salam electroweak theory with a complex doublet Higgs field. In the electroweak theory the Higgs field also breaks the gauge symmetry to $U(1)$, leaving just the electromagnetic field massless. Despite this, the electroweak theory has no monopoles; instead it has unstable, sphaleron solutions that we shall consider in Chapter 11.

We shall later briefly consider gauge groups $SU(m)$ for general m. Of particular interest is the $SU(5)$ theory with adjoint Higgs, where the Higgs field spontaneously breaks the $SU(5)$ gauge symmetry to $SU(3) \times U(2)$. This is a simplified version of the simplest Grand Unified Theory (GUT), unifying the gauge groups of QCD and electroweak theory [147]. This theory and more complicated variants with further Higgs fields, and possibly a larger gauge group, have monopole solutions.

Why should one be interested in monopoles, despite the absence of any experimental evidence for them? There are several reasons. If one believes in GUTs then the monopoles are a crucial signal and constraint. It is predicted that monopoles should have been plentifully produced in the very early history of the universe, and although monopoles and antimonopoles would have appeared in roughly equal numbers, they would not all have annhilated by now. Even if most had annihilated, we would still be able to detect the relic radiation produced. Current cosmic data rule out this scenario of a dense sea of monopoles in the early universe, and severely constrains the parameters of GUT models [407]. GUTs themselves can be an approximation to a grander unified theory including gravity – perhaps string theory. Such theories can also have monopoles, so again there are constraints from the cosmic data. One of the motivations for inflationary models of the early universe is to remove almost all primordial monopoles [168]. So one reason, a rather negative one, for studying monopoles in various theories is to constrain such theories so that monopoles do not make an appearance.

A positive reason is that monopoles are a paradigm for solitons in three dimensions, and they are of great mathematical interest and beauty. It has been discovered that the $SU(2)$ gauge theory with an adjoint Higgs can be analysed in great mathematical detail in the Bogomolny-Prasad-Sommerfield (BPS) limit, where the Higgs field is massless. It is now fairly straightforward to construct a variety of multi-monopole solutions and to predict the outcome of various multi-monopole scattering processes. Moreover, although the theory is not physical, it gives considerable insight into the physics of Higgs fields and solitons more generally. For example, study of monopole-antimonopole dynamics helped to uncover and clarify the significance of the sphaleron solution of the electroweak theory. The understanding of monopoles and multi-monopole dynamics has also helped towards the understanding of Skyrmions and their dynamics. Skyrmions are discussed in Chapter 9, where we shall explain the analogy with monopoles in detail.

Finally, monopoles have been under much scrutiny recently, because gauge theories with adjoint Higgs fields have various supersymmetric extensions, and the BPS limit is natural from the supersymmetric point of view. It is now understood that in certain quantized supersymmetric theories with monopoles, there is an exact duality symmetry [304], and the masses of the monopoles can be precisely predicted [429]. This is despite the apparent limitation of perturbative quantum field theory to deal with monopoles. The dynamics of monopoles, and in particular the lowest energy bound states, are also of great interest in supersymmetric theories [372, 371].

In the remainder of this section we shall describe the $SU(2)$ Yang-Mills-Higgs theory and its basic monopole solution. The fields are the $SU(2)$ gauge potential A_μ and the adjoint Higgs field Φ, both valued in the Lie algebra $su(2)$. The covariant derivative of the Higgs field and the Yang-Mills field tensor are, respectively,

$$D_\mu\Phi = \partial_\mu\Phi + [A_\mu, \Phi] \tag{8.40}$$
$$F_{\mu\nu} = \partial_\mu A_\nu - \partial_\nu A_\mu + [A_\mu, A_\nu]. \tag{8.41}$$

We choose the basis $\{t^a = i\tau^a : a = 1, 2, 3\}$ for $su(2)$, where $\{\tau^a\}$ are the Pauli matrices, with commutation relations $[t^a, t^b] = -2\varepsilon_{abc}t^c$ and normalization $\mathrm{Tr}(t^a t^b) = -2\delta^{ab}$. With respect to this basis the Higgs field and gauge potential can be expressed in terms of their component fields as

$$\Phi = \Phi^a t^a, \quad A_\mu = A_\mu^a t^a. \tag{8.42}$$

The field tensor automatically satisfies the Bianchi (or Jacobi) identity

$$\varepsilon^{\sigma\tau\mu\nu} D_\tau F_{\mu\nu} = 0, \tag{8.43}$$

where $\varepsilon^{\sigma\tau\mu\nu}$ is the alternating tensor in Minkowski space (with $\varepsilon^{0123} = 1$).

The theory has the Lorentz invariant Lagrangian density

$$\mathcal{L} = \frac{1}{8}\mathrm{Tr}(F_{\mu\nu}F^{\mu\nu}) - \frac{1}{4}\mathrm{Tr}(D_\mu\Phi D^\mu\Phi) - \frac{\lambda}{4}(1 - |\Phi|^2)^2 \tag{8.44}$$

where $|\Phi|^2 = -\frac{1}{2}\mathrm{Tr}\,\Phi^2$ is the non-negative squared norm of the Higgs field. It will be convenient to split the Lagrangian

$$L = \int \mathcal{L}\, d^3x \tag{8.45}$$

into its kinetic energy and potential energy parts, $L = T - V$, where

$$T = \int \left(-\frac{1}{4}\mathrm{Tr}(E_i E_i) - \frac{1}{4}\mathrm{Tr}(D_0\Phi D_0\Phi) \right) d^3x \qquad (8.46)$$

and

$$V = \int \left(-\frac{1}{8}\mathrm{Tr}(F_{ij}F_{ij}) - \frac{1}{4}\mathrm{Tr}(D_i\Phi D_i\Phi) + \frac{\lambda}{4}(1 - |\Phi|^2)^2 \right) d^3x, \qquad (8.47)$$

and the integrations are over $\mathbb{R}^3$. Here $E_i = F_{0i}$ is the $SU(2)$ electric field, and we shall frequently use the notation $B_i = -\frac{1}{2}\varepsilon_{ijk}F_{jk}$ for the $SU(2)$ magnetic field. We see from the expression for V that the classical vacuum, which minimizes V, is a field configuration with $|\Phi|^2 = 1$ and Φ covariantly constant, so $D_i\Phi = 0$; also $F_{ij} = 0$, which means that the gauge potential is pure gauge, $A_i = -\partial_i g\, g^{-1}$ for some $SU(2)$-valued function $g(\mathbf{x})$. By a gauge transformation, we can make A_i vanish, and then Φ is a constant. By a further global gauge transformation $\Phi \mapsto g_0\Phi g_0^{-1}$, with g_0 constant, we can bring Φ to the standard form $\Phi = t^3$.

Because Φ has a non-zero vacuum expectation value, the $SU(2)$ gauge symmetry is spontaneously broken to $U(1)$. The unbroken group is associated with gauge transformations of the form $g(\mathbf{x})$ which satisfy the equation $g(\mathbf{x})t^3 g(\mathbf{x})^{-1} = t^3$. Such $g(\mathbf{x})$ are in the $U(1)$ subgroup generated by t^3. This can be restated in a gauge invariant way: if $\{\Phi(\mathbf{x}), A_i(\mathbf{x})\}$ is the vacuum field in an arbitrary gauge, the unbroken part of the gauge symmetry consists of gauge transformations $g(\mathbf{x})$ which are at each point in the subgroup of $SU(2)$ generated by $\Phi(\mathbf{x})$. Such gauge transformations preserve Φ, but change A_i while preserving both $D_i\Phi = 0$ and $F_{ij} = 0$. As a consequence of the spontaneous symmetry breaking, the theory (when perturbatively quantized) has a massless photon associated with the unbroken $U(1)$, and two massive gauge particles W^+ and W^-. The masses can be calculated by diagonalizing $-\frac{1}{4}\mathrm{Tr}([A_\mu, t^3][A_\mu, t^3])$, the adjoint Higgs analogue of the expression (2.165).

We can not insist, as a general boundary condition, that $\Phi \to t^3$ in all directions as $r \to \infty$, as that would exclude any non-trivial topological structure in the Higgs field at infinity. However, it will turn out to be useful when we come to discuss monopoles to impose as a boundary condition $\Phi(0, 0, x^3) \to t^3$ as $x^3 \to \infty$. We shall also define the class of based gauge transformations to be those satisfying $g(0, 0, x^3) \to 1$ as $x^3 \to \infty$. That leaves a residual global action of the unbroken $U(1)$ gauge group on fields, by the transformations $g_0 = \exp(\alpha t^3)$ with α real, which preserve the boundary condition. This $U(1)$ acts non-trivially on generic fields but leaves the vacuum invariant. Fields that are identified modulo based gauge transformations are said to be framed, and fields that differ by an element of the residual global $U(1)$ (possibly combined with a based

gauge transformation) are said to differ in their framing. If one quotients
out by all gauge transformations one gets unframed fields.

The field equations obtained from the Lagrangian density (8.44) are

$$
\begin{aligned}
D_\mu D^\mu \Phi &= \lambda(1 - |\Phi|^2)\Phi & (8.48) \\
D_\mu F^{\mu\nu} &= [D^\nu \Phi, \Phi]. & (8.49)
\end{aligned}
$$

The general solution of these nonlinear PDEs is not known. It is worthwhile to write down the linearization of these equations, around the vac-
uum. So let $\Phi = (1+\phi)t^3$ and $A_\mu = W_\mu^1 t^1 + W_\mu^2 t^2 + a_\mu t^3$, where ϕ, W_μ^1, W_μ^2
and a_μ are all small. We have fixed the gauge to eliminate the small
coefficient functions of t^1 and t^2 in Φ. The equations reduce to

$$
\begin{aligned}
\partial_\mu \partial^\mu \phi &= -2\lambda\phi & (8.50) \\
\partial_\mu(\partial^\mu W^{1\nu} - \partial^\nu W^{1\mu}) &= -4W^{1\nu} & (8.51) \\
\partial_\mu(\partial^\mu W^{2\nu} - \partial^\nu W^{2\mu}) &= -4W^{2\nu} & (8.52) \\
\partial_\mu(\partial^\mu a^\nu - \partial^\nu a^\mu) &= 0. & (8.53)
\end{aligned}
$$

From these wave equations we read off that the Higgs particle has mass
$\sqrt{2\lambda}$, the W particles have mass 2, and the photon is massless.

't Hooft and Polyakov independently found the static solution of the
field equations of this theory, representing a magnetic monopole [401, 336].
We shall present this solution first and then discuss the topological reason
for its existence. Static fields obey

$$
\begin{aligned}
D_i D_i \Phi &= -\lambda(1 - |\Phi|^2)\Phi & (8.54) \\
D_i F_{ij} &= -[D_j \Phi, \Phi], & (8.55)
\end{aligned}
$$

the equations for a stationary point of the potential energy V. To solve
these, one may try fields of the spherically symmetric and reflection symmetric form

$$
\begin{aligned}
\Phi &= h(r)\frac{x^a}{r}t^a & (8.56) \\
A_i &= -\frac{1}{2}(1 - k(r))\,\varepsilon_{ija}\frac{x^j}{r^2}t^a & (8.57)
\end{aligned}
$$

where $h(r)$ and $k(r)$ are functions just of the distance from the origin,
r. These fields are spherically symmetric in the sense that a rotation has
the same effect as a spatially independent gauge transformation. The
gauge transformation paired with the rotation R is $D(R) = R$. (This
makes immediate sense if the gauge group is $SO(3)$. If the gauge group is
$SU(2)$ then there are two elements of $SU(2)$ corresponding to $D(R)$, but

they act in the same way on the fields.) For fields of this form, a simple calculation shows that the equations (8.54) and (8.55) simplify to

$$\frac{d^2h}{dr^2} + \frac{2}{r}\frac{dh}{dr} = \frac{2}{r^2}k^2h - \lambda(1 - h^2)h \tag{8.58}$$

$$\frac{d^2k}{dr^2} = \frac{1}{r^2}(k^2 - 1)k + 4h^2k. \tag{8.59}$$

These ODEs can not be solved analytically for general values of λ, but a numerical solution is straightforward. The boundary conditions to be imposed at the origin are that $h(0) = 0$ and $k(0) = 1$, so as to avoid a singularity. Also h should tend to 1 and k tend to 0 as $r \to \infty$, so that asymptotically $|\Phi|^2 = 1$ and both $D_i\Phi$ and F_{ij} vanish, to ensure that the solution has finite energy. Note that $\Phi(0, 0, x^3) \to t^3$ as $x^3 \to \infty$.

The existence, though not uniqueness, of the solution is rigorously established [301] for general $\lambda \geq 0$. But numerically it seems clear that there is just one smooth solution for each non-negative value of λ. This was discovered by 't Hooft and Polyakov, and a more systematic study over a large range of values of λ was carried out by Bogomolny and Marinov [57]. The forms of $h(r)$ and $k(r)$, for a few values of λ, are shown in Fig. 8.1. Figure 8.2 gives the energy, or mass M of the monopole, as a function of λ.

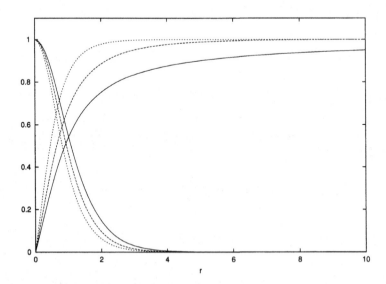

Fig. 8.1. The monopole profile functions $h(r)$ and $k(r)$ for $\lambda = 0$ (solid curves), $\lambda = 0.1$ (dashed curves), and $\lambda = 1.0$ (dotted curves).

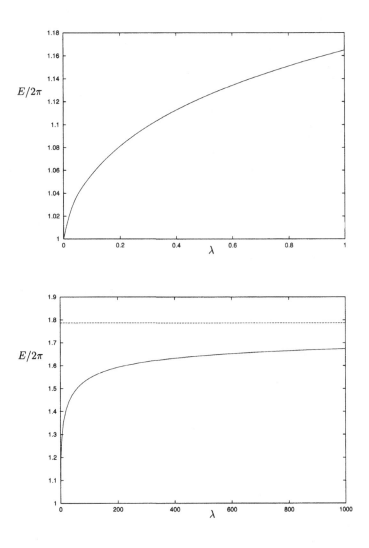

Fig. 8.2.   The monopole energy (in units of 2π) as a function of λ for $0 \leq \lambda \leq 1$, and for $0 \leq \lambda \leq 1000$, with the asymptotic value $E(\lambda = \infty) = 2\pi \times 1.787$ marked as a dashed line.

For small λ, the mass has an expansion of the form [234, 143]

$$M = 2\pi \left(1 + \frac{1}{2}\delta + \frac{1}{2}\delta^2 \log \delta + 0.7071\, \delta^2 + \cdots \right) \tag{8.60}$$

whereas the asymptotic mass for large λ is [234]

$$M = 2\pi \left(1.787 - \frac{2.228}{\delta} + \cdots \right). \tag{8.61}$$

Here $\delta = \sqrt{\lambda/2}$, which is the ratio of the Higgs mass to the W mass.

To evaluate the magnetic charge of these solutions we need to have a definition of the magnetic field. In a pure $SU(2)$ Yang-Mills theory there is no unambiguous definition, but in a spontaneously broken theory, with unbroken group $U(1)$, things are different. Provided the fields are close to the vacuum, a magnetic field can be defined. The analysis that follows applies to any finite energy, possibly time dependent solution of the Yang-Mills-Higgs equations, and in particular, to solutions of the Bogomolny equation that we will consider below [275].

Suppose in some region of space-time the Higgs field can be expressed as $\Phi = h\hat{\Phi}$, where h is positive and $|\hat{\Phi}|^2 = 1$. $\hat{\Phi}$ is well defined provided $|\Phi|$ is nowhere zero in the region. Suppose also that $D_\mu \hat{\Phi} = 0$. We shall now show that these assumptions imply that the Yang-Mills-Higgs theory abelianizes, and the field equations become a version of Maxwell's equations. (One might have imposed the condition $D_\mu \Phi = 0$, but this is unnecessarily restrictive.) The condition $D_\mu \hat{\Phi} = 0$ can be solved for the gauge potential. The general solution is

$$A_\mu = \frac{1}{4}[\partial_\mu \hat{\Phi}, \hat{\Phi}] + a_\mu \hat{\Phi} \tag{8.62}$$

where a_μ is an arbitrary smooth 4-vector function. The field tensor is therefore

$$F_{\mu\nu} = \left(\frac{1}{8}\mathrm{Tr}([\partial_\mu \hat{\Phi}, \partial_\nu \hat{\Phi}]\hat{\Phi}) + \partial_\mu a_\nu - \partial_\nu a_\mu \right) \hat{\Phi} \tag{8.63}$$

and so is in the direction $\hat{\Phi}$ in $su(2)$. The covariant derivative of Φ reduces to

$$D_\mu \Phi = (\partial_\mu h)\hat{\Phi}. \tag{8.64}$$

Let us define

$$f_{\mu\nu} = -\frac{1}{2}\mathrm{Tr}(F_{\mu\nu}\hat{\Phi}) = \frac{1}{8}\mathrm{Tr}([\partial_\mu \hat{\Phi}, \partial_\nu \hat{\Phi}]\hat{\Phi}) + \partial_\mu a_\nu - \partial_\nu a_\mu \tag{8.65}$$

to be the Maxwell field tensor. With our assumptions, we now see that the Yang-Mills-Higgs equations (8.48) and (8.49) reduce to

$$\partial_\mu \partial^\mu h = \lambda(1 - h^2)h \tag{8.66}$$

$$\partial_\mu f^{\mu\nu} = 0. \tag{8.67}$$

The first equation is a nonlinear scalar wave equation which simplifies further if h is close to 1. The second is the inhomogeneous Maxwell equation, but with vanishing current source. Also, $f_{\mu\nu}$ satisfies the homogeneous Maxwell equation (the abelian Bianchi identity)

$$\varepsilon^{\sigma\tau\mu\nu}\partial_\tau f_{\mu\nu} = 0. \tag{8.68}$$

This is because, from (8.65),

$$\varepsilon^{\sigma\tau\mu\nu}\partial_\tau f_{\mu\nu} = \frac{1}{8}\varepsilon^{\sigma\tau\mu\nu}\mathrm{Tr}([\partial_\mu\hat{\Phi}, \partial_\nu\hat{\Phi}]\partial_\tau\hat{\Phi}). \tag{8.69}$$

Since $\hat{\Phi}$ is a unit element of $su(2)$, $\partial_\mu\hat{\Phi}$, $\partial_\nu\hat{\Phi}$ and $\partial_\tau\hat{\Phi}$ (with μ, ν, τ distinct) are linearly dependent, being all orthogonal to $\hat{\Phi}$, and therefore (8.68) follows. For static fields, satisfying our assumptions, we can interpret

$$b_i = -\frac{1}{2}\varepsilon_{ijk}f_{jk} = -\frac{1}{2}\varepsilon_{ijk}\left(\frac{1}{8}\mathrm{Tr}([\partial_j\hat{\Phi}, \partial_k\hat{\Phi}]\hat{\Phi}) + \partial_j a_k - \partial_k a_j\right) \tag{8.70}$$

as the magnetic field, and it satisfies $\nabla \times \mathbf{b} = 0$ and $\nabla \cdot \mathbf{b} = 0$.

For the 't Hooft-Polyakov monopole solution, it can be checked that as $r \to \infty$, $D_i\hat{\Phi}$ tends to zero exponentially fast, and also $h \to 1$. Therefore, outside the core region of the monopole, which has a radius of order 1, the fields abelianize and there is a radial magnetic field. One may calculate the magnetic field to be

$$b_i = -\frac{x^i}{2r^3}. \tag{8.71}$$

This is the magnetic field of a magnetic monopole of charge -2π.

One might ask where the source of the magnetic field is. If one uses the formula (8.70) throughout space, then the answer is that there is a point source at the origin, as for a Dirac monopole, but other formulae for the magnetic field have been suggested, which imply that there is a smooth magnetic charge density in the core of the monopole. However, from a mathematical point of view, there is no unambiguous answer, because the magnetic field and hence its divergence are simply not well defined in a region where the full $SU(2)$ nature of the Yang-Mills-Higgs fields manifests itself – as in the core region of the monopole. Physically also, there is no unambiguous way to measure the charge density, as Coleman has argued [87]. Only the total charge, and its assignment in a general way to the core of the monopole, makes sense.

We shall now investigate the topological character of the monopole and its magnetic charge. Consider any smooth finite energy field configuration (at a given time) which approaches the vacuum at spatial infinity, that is, $|\Phi|^2$ approaches 1, and $D_i\hat{\Phi}$ vanishes. The magnetic field is given by the formula (8.70) and we may integrate over S^2_∞, the 2-sphere at infinity, to find the total flux, which equals the magnetic charge g. In differential form notation,

$$g = -\int_{S^2_\infty} f \,, \tag{8.72}$$

where

$$f = \frac{1}{8}\mathrm{Tr}([d\hat{\Phi}, d\hat{\Phi}]\hat{\Phi}) + da \,. \tag{8.73}$$

By Stokes' theorem, the contribution of a vanishes, so

$$g = -\frac{1}{8}\int_{S^2_\infty} \mathrm{Tr}([d\hat{\Phi}, d\hat{\Phi}]\hat{\Phi}) \,. \tag{8.74}$$

Now $\hat{\Phi}$ restricts to a map $\hat{\Phi} : S^2_\infty \mapsto S^2$, where the target is the unit sphere in $su(2)$. This map has some degree N, and it is easy to verify that the right-hand side of (8.74) is -2π times this. Therefore $g = -2\pi N$. N is called the monopole number.

So, finite energy implies that the asymptotic Higgs field $\hat{\Phi}$ has associated with it a topological charge, its degree N, and the magnetic charge of the field configuration is $-2\pi N$. The correlation between these things is because $D_i\hat{\Phi} = 0$ asymptotically, which relates the gauge and Higgs fields. N is also the number of zeros of the Higgs field in $\mathbb{R}^3$ (assuming the zeros are isolated), counted with multiplicity. In the special case of the monopole solution with spherical symmetry, with fields of the form (8.56) and (8.57), it is easy to see that $\hat{\Phi} : S^2_\infty \mapsto S^2$ is the identity map, of degree 1, and hence the magnetic charge is -2π. The Higgs field has its single zero at the origin.

If the fields vary smoothly with time, which they do if they satisfy the field equations, then the integer N can not change. The magnetic charge is therefore topologically conserved, and that is why the monopole is a topological soliton.

The charge quantization condition $g = -2\pi N$ looks similar to the Dirac quantization condition. Let us consider more carefully why this is so [157]. Suppose we try to go to the unitary gauge. This is the gauge where $\hat{\Phi} = t^3$, a constant. But clearly, if $\hat{\Phi} : S^2_\infty \mapsto S^2$ has degree N, then there is no smooth transformation which changes $\hat{\Phi}$ to a constant map. We can, however, transform to the unitary gauge separately over two regions of S^2_∞, which together cover S^2_∞ and overlap on the equator. Let $g^{(1)}$ be the gauge transformation in region 1 and $g^{(2)}$ the gauge transformation in region 2. On the overlap of the regions $\hat{\Phi} = t^3$ after either of the gauge transformations. So $g^{(1)}g^{(2)-1}$ preserves $\hat{\Phi} = t^3$ and therefore lies in the $U(1)$ subgroup of $SU(2)$ generated by t^3. Let us therefore write, on the equator of S^2_∞,

$$g^{(1)}g^{(2)-1} = \exp(\alpha(\varphi)t^3) \tag{8.75}$$

where φ is the azimuthal coordinate. Since $g^{(1)}$ and $g^{(2)}$ are both well defined in their own regions, $\alpha(2\pi) = \alpha(0) + 2\pi\tilde{N}$ for some integer $\tilde{N}$. Here we have used the result that the smallest positive α for which $\exp(\alpha t^3) = 1$ is $\alpha = 2\pi$.

Next, let us compare the abelian gauge potentials on the overlap. Before the gauge transformations, a was well defined over the whole S_∞^2. After the gauge transformations, one obtains $a^{(1)}$ and $a^{(2)}$, which on the overlap are related by the $U(1)$ transformation $g^{(1)}g^{(2)-1}$. Therefore

$$a^{(1)} - a^{(2)} = -d\alpha(\varphi) \qquad (8.76)$$

on the equator. Also, since $\hat{\Phi}$ is now constant, the formula (8.70) implies that the 2-form field strength is $f = da^{(1)} = da^{(2)}$. We therefore see that the total magnetic flux is determined as for a Dirac monopole. It depends not on the details of the field, but simply on the amount by which α increases around the equator. The total magnetic flux through S_∞^2, and hence the magnetic charge, is $-2\pi\tilde{N}$. It is easy to show directly that $\tilde{N}$, as determined from the form of $g^{(1)}g^{(2)-1}$, is the original degree of $\hat{\Phi}$, even though there is considerable ambiguity in $g^{(1)}$ and $g^{(2)}$ themselves. So $\tilde{N} = N$, and $g = -2\pi N$ as before.

Because monopoles in this theory have magnetic charges that are integer multiples of 2π, the Dirac quantization condition (8.9) implies that the minimal unit of electric charge is 1. If the gauge group is $SU(2)$ (but not if it is $SO(3)$) this charge occurs when the gauge and Higgs field are coupled to a further $SU(2)$ doublet field. The $W^\pm$ particles have charges ± 2, in our units.

Let us conclude this section with a brief summary of some of the further properties of the 't Hooft-Polyakov magnetic monopole.

It has been shown by numerically assisted analysis that, for a large range of values of λ, any small deformation increases its energy [25]. Therefore, the monopole is a local minimum of the Yang-Mills-Higgs energy in the sector with $N = 1$, and hence stable. However, there is no proof that it is a global minimum of the energy except in the limit $\lambda = 0$ (see Section 8.3).

The monopole centre can be shifted to an arbitrary location, and the framing can also be changed, so the monopole has four collective coordinates. In many ways the monopole behaves like a point particle, despite its finite core size. In particular, it has no rotational moment of inertia. This is because the fields are spherically symmetric in the sense described earlier, so a rotation has no physical effect at all. It follows that in the quantized field theory, if there are no additional fields, the monopole has spin 0.

There is a second solution of the equations (8.54) and (8.55), obtained by reversing the sign of Φ. This is the antimonopole, which has the same mass but opposite magnetic charge to the monopole. The antimonopole can also be obtained from the monopole by inversion in the origin, $\mathbf{x} \mapsto -\mathbf{x}$.

Because of the Poincaré invariance of the theory, the monopole can be set into motion along any line in space, and there are the usual relations between energy, momentum and the rest mass. There have been no substantial, purely numerical studies of multi-monopole motion in this particular theory, but initial fields could be set up describing several monopoles in relative motion. The force between well separated monopoles is primarily due to the magnetic Coulomb force. Two monopoles each with magnetic charge $g = -2\pi$, and separated by a distance R, will experience a repulsive force of π/R^2. A monopole attracts an antimonopole with a force of the same magnitude.

Although the Coulomb repulsion of monopoles is expected, it is nevertheless a remarkable result. Recall that in electromagnetism, an electrically charged particle is a source of a Coulomb electric field, but that it requires an additional postulate – the Lorentz force law – to say that another charge experiences the Coulomb force. The $SU(2)$ Yang-Mills-Higgs theory, through its field equations, not only supports magnetic monopole solutions but also predicts the forces they exert on each other. The theory, like others with solitons, is a theory of particle structure and particle interactions.

There is more than one way to calculate the force between monopoles, other than by simulating the time dependent dynamics. One may construct a static approximation to a two-monopole field. This is not too difficult if the cores do not overlap. The field outside the cores is taken to be a superposition of the usual magnetic fields (easiest to do in patches, in the unitary gauge). It is possible to estimate the minimal energy assuming that the zeros of the Higgs field are constrained to a fixed separation. The gradient of this energy is the force between the monopoles [273].

Alternatively, using the energy-momentum tensor of the Yang-Mills-Higgs theory, one can calculate the net momentum flux into a sphere enclosing one of the monopoles, and this can be identified with the force acting on that monopole [158]. It is found that the force is the magnetic Coulomb force. It is conceivable that if the fields were now allowed to evolve according to the field equations, the momentum would flow into a variety of field modes inside the sphere. In practice, the monopole starts to accelerate rigidly. The consistency of this has been established by showing that an accelerated one-monopole solution is possible so long as, asymptotically, there is both the Coulomb field of the monopole, and the additional constant or approximately constant magnetic field produced by other, more distant monopoles [275]. This is discussed in more detail in Section 8.10.

8.3 Bogomolny-Prasad-Sommerfield monopoles

Shortly after 't Hooft and Polyakov's discovery of a magnetic monopole in Yang-Mills-Higgs theory, Prasad and Sommerfield found an analytic form for the solution in the special case $\lambda = 0$ [340]. Apparently, Prasad and Sommerfield were attempting to fit the numerical solution with simple analytic functions when they discovered that what they thought was an approximate solution was in fact exact. The equations which they were considering were

$$\frac{d^2 h}{dr^2} + \frac{2}{r}\frac{dh}{dr} = \frac{2}{r^2}k^2 h \tag{8.77}$$

$$\frac{d^2 k}{dr^2} = \frac{1}{r^2}(k^2 - 1)k + 4h^2 k\,, \tag{8.78}$$

the $\lambda = 0$ case of (8.58) and (8.59). These equations have the solution

$$h(r) = \coth 2r - \frac{1}{2r}$$

$$k(r) = \frac{2r}{\sinh 2r}\,. \tag{8.79}$$

The energy (rest mass) of the monopole in this special case is 2π.

The limit $\lambda \to 0$ of the Lagrangian is somewhat strange. For any positive value of λ, finite energy can only occur if $|\Phi| = 1$ asymptotically. When $\lambda = 0$, the condition of finite energy no longer constrains the asymptotic value of $|\Phi|$; nevertheless one may impose as a boundary condition $|\Phi| = 1$. More generally, one can impose the boundary condition $|\Phi| = v$, where v is an arbitrary positive constant. The vacuum field is $\Phi = vt^3$. It is not compatible with finite energy for $|\Phi|$ to approach different values in different directions as one moves off to infinity, because of the contribution of the derivatives of Φ. The value of v also can not change with time. However, by a simple rescaling, v can be reset to unity. One needs to replace Φ by $\frac{1}{v}\Phi$, and rescale lengths and energies by v. We shall fix $v = 1$ from now on.

The perturbative quantization of the theory around the vacuum $\Phi = t^3$ gives a spectrum of particles. For $\lambda = 0$, and calculating at tree level, one finds W^+ and W^- gauge particles with mass 2 and electric charges ± 2, a massless photon, and a neutral Higgs particle that is also massless. The monopole solution of Prasad and Sommerfield reflects the masslessness of the Higgs particle in that the Higgs field approaches its vacuum value rather slowly. The expansion of $h(r)$ for large r is

$$h = 1 - \frac{1}{2r} + O(e^{-4r}). \tag{8.80}$$

Bogomolny gave a much deeper understanding of the $\lambda = 0$ limit of the Yang-Mills-Higgs theory [56]. Bogomolny noted that the energy of a static field

$$E = -\frac{1}{4} \int \left(\text{Tr}(B_i B_i) + \text{Tr}(D_i \Phi D_i \Phi) \right) d^3x, \tag{8.81}$$

where $B_i = -\frac{1}{2}\varepsilon_{ijk}F_{jk}$, can be rewritten as

$$E = -\frac{1}{4} \int \text{Tr}(B_i + D_i \Phi)(B_i + D_i \Phi) \, d^3x + \frac{1}{2} \int \partial_i(\text{Tr}(B_i \Phi)) \, d^3x. \tag{8.82}$$

To obtain this expression we have made use of the following identity $\partial_i(\text{Tr}(B_i \Phi)) = \text{Tr}((D_i B_i)\Phi) + \text{Tr}(B_i D_i \Phi) = \text{Tr}(B_i D_i \Phi)$, which depends on the Bianchi identity $D_i B_i = 0$. The second integral in (8.82) can be expressed as a surface integral over the 2-sphere at infinity,

$$E = -\frac{1}{4} \int_{\mathbb{R}^3} \text{Tr}(B_i + D_i \Phi)(B_i + D_i \Phi) \, d^3x - \int_{S^2_\infty} b_i \, dS^i, \tag{8.83}$$

where b_i is the abelian magnetic field introduced earlier, which is equal to $-\frac{1}{2}\text{Tr}(B_i \Phi)$ on the 2-sphere at infinity. Recalling the quantization of magnetic flux, one obtains

$$E = -\frac{1}{4} \int_{\mathbb{R}^3} \text{Tr}(B_i + D_i \Phi)(B_i + D_i \Phi) \, d^3x + 2\pi N \tag{8.84}$$

where N is the monopole number. For $N > 0$ there is therefore the non-trivial energy bound

$$E \geq 2\pi N, \tag{8.85}$$

with equality if

$$B_i = -D_i \Phi. \tag{8.86}$$

Equation (8.85) is the Bogomolny energy bound, and (8.86) is the Bogomolny equation for this theory.

Bogomolny showed that the Prasad-Sommerfield monopole solution satisfies Eq. (8.86), and has monopole number 1, hence explaining why its energy is 2π. If one inserts into the Bogomolny equation the spherically symmetric ansatz for the fields (8.56) and (8.57), one obtains the coupled first order equations

$$\frac{dh}{dr} = \frac{1}{2r^2}(1 - k^2) \tag{8.87}$$

$$\frac{dk}{dr} = -2hk. \tag{8.88}$$

These can be simplified by the substitution $h = \tilde{h} - \frac{1}{2r}$, $k = 2r\tilde{k}$, and then integrated to give, as before, the solution (8.79).

The second order field equations for static fields (8.54) and (8.55) are the condition for a stationary point of the energy, whereas the first order Bogomolny equation is the condition for the global minimum of the energy in the sector with monopole number N. A minimum should be a stationary point (in a smooth function space), and we now verify this. If the Bogomolny equation $B_i = -D_i\Phi$ is satisfied, then firstly

$$D_i D_i \Phi = -D_i B_i = 0 \,, \tag{8.89}$$

using the Bianchi identity, and secondly

$$
\begin{aligned}
D_i F_{ij} = -\varepsilon_{ijk} D_i B_k = \varepsilon_{ijk} D_i D_k \Phi &= \frac{1}{2}\varepsilon_{ijk}[D_i, D_k]\Phi \\
&= \frac{1}{2}\varepsilon_{ijk}[F_{ik}, \Phi] \\
&= [B_j, \Phi] \\
&= -[D_j\Phi, \Phi] \,, \tag{8.90}
\end{aligned}
$$

where we have used the basic result that the commutator of two covariant derivatives gives the field tensor. So both field equations are satisfied.

For negative N there is an analogous bound, whose derivation requires changes of sign in (8.84). The result is that

$$E \geq 2\pi|N| \tag{8.91}$$

with equality if $B_i = D_i\Phi$.

The Prasad-Sommerfield solution has charge $N = 1$. Are there solutions of the Bogomolny equation with $N > 1$? Spherically symmetric solutions do not exist for $N > 1$. A physical argument, suggesting the existence of further solutions, is as follows. One may calculate the forces between two monopoles, or between a monopole $(N = 1)$ and an anti-monopole $(N = -1)$, in the case where $\lambda = 0$. This is done by constructing an approximate field which superposes two well separated monopoles, and then calculating the value of the acceleration of each monopole that is compatible with the field equations. It is found that for two monopoles the acceleration is zero, whereas for a monopole and antimonopole it is double the value that occurs if $\lambda \neq 0$ [275]. The reason for the change is the masslessness of the scalar Higgs field, leading to the long-range, Coulomb tail (8.80). A monopole can therefore be thought of as having a scalar Coulomb charge 2π in addition to its magnetic charge. Scalar charges, like gravitational masses, are always positive, and scalar interactions are always attractive. At separation R, the scalar attraction has strength π/R^2, both for monopoles and antimonopoles. This cancels the Coulomb magnetic repulsion of two monopoles, and doubles the magnetic attraction of a monopole and antimonopole.

The absence of a force between two monopoles, when $\lambda = 0$, suggests that two-monopole static solutions of the Bogomolny equation might exist, with the separation being an adjustable parameter. Such solutions would all have the same energy 4π, independently of the separation, and this is of course compatible with the absence of forces.

Two-monopole solutions do indeed exist, as do N-monopole solutions for any integer $N > 0$. They were originally very difficult to find, and this difficulty attracted a deep investigation of the Bogomolny equation by mathematicians.

One observation inspired various efforts to understand BPS monopoles. This is that the Bogomolny equation is the time independent version of the self-dual Yang-Mills equation [77, 268]. A pure Yang-Mills gauge field in four-dimensional Euclidean space, with coordinates $\{x^\mu : \mu = 1, 2, 3, 4\}$, is self-dual if

$$F_{\mu\nu} = \frac{1}{2}\varepsilon_{\mu\nu\sigma\tau}F_{\sigma\tau}, \tag{8.92}$$

where $\varepsilon_{\mu\nu\sigma\tau}$ is the totally antisymmetric tensor in four dimensions (with $\varepsilon_{1234} = -1$). Equation (8.92) can be written alternatively as

$$F_{4i} = \frac{1}{2}\varepsilon_{ijk}F_{jk}. \tag{8.93}$$

Finite-action solutions of this equation are known as multi-instantons, and they minimize the action for a given value of the topological charge of the Yang-Mills field, the second Chern number; see Chapter 10 for a detailed discussion. Suppose now that the Yang-Mills gauge potential A_μ is independent of the Euclidean time x^4. Then Eq. (8.93) simplifies to

$$-\partial_i A_4 - [A_i, A_4] = \frac{1}{2}\varepsilon_{ijk}F_{jk}. \tag{8.94}$$

If we now identify $-A_4$ with the Higgs field Φ, then (8.94) becomes the Bogomolny equation in three dimensions, $D_i\Phi = \frac{1}{2}\varepsilon_{ijk}F_{jk}$.

This identification makes a lot of sense. From a three-dimensional point of view, A_4 is a scalar field. Gauge transformations $g(x)$ in four dimensions simplify if they are assumed to be independent of x^4, and A_4 then transforms in the same way as Φ, namely $\Phi \mapsto g\Phi g^{-1}$.

A static monopole solution, then, is a time independent self-dual gauge field. (This analogy does not extend to dynamical Yang-Mills-Higgs fields in Minkowski space-time.) Its four-dimensional action is infinite, because of the time translation invariance, and the boundary conditions are different from those of a finite-action instanton. Nevertheless, of the various techniques developed to find instantons, some can be adapted to the time independent case, and used to construct monopoles. In fact, only the simplest Prasad-Sommerfield solution, with $N = 1$, was rederived in this way initially [276], but more sophisticated techniques led to progress with the construction of both the instantons and monopoles.

Two important results concerning BPS monopoles were established, before any explicit solutions for $N > 1$ were found. First, the dimension of the space of solutions of charge N was calculated. It is necessary to assume that at least one solution of charge N exists, and that the space of solutions is a manifold. The tangent space to this manifold is the space of solutions of the linearized Bogomolny equation. Let $\{\Phi, A_i\}$ be the presumed N-monopole solution, and consider fields $\{\Phi + \phi, A_i + a_i\}$, where ϕ and a_i are small. Substituting in the Bogomolny equation and linearizing gives

$$\frac{1}{2}\varepsilon_{ijk}(D_j^A a_k - D_k^A a_j) = D_i^A \phi + [a_i, \Phi]. \qquad (8.95)$$

Note that these equations have infinitely many solutions of the form $\phi = [\alpha, \Phi]$, $a_i = -D_i^A \alpha$, with α valued in $su(2)$, which simply represent infinitesimal gauge transformations of the original solution, and do not physically change it. To avoid these, one requires that the deformations $\{\phi, a_i\}$ are orthogonal to infinitesimal gauge transformations, in the sense that

$$\int \left(-\mathrm{Tr}(a_i D_i^A \alpha) + \mathrm{Tr}(\phi[\alpha, \Phi])\right) d^3x = 0 \qquad (8.96)$$

for any (compactly supported) α. Integrating by parts, and rearranging, one sees that this orthogonality requirement is that

$$D_i^A a_i + [\Phi, \phi] = 0, \qquad (8.97)$$

which is known as the background gauge condition. The true deformation space of the monopole is the space of solutions of the combined equations (8.95) and (8.97) subject to suitable boundary conditions (ϕ and a_i decaying towards infinity).

E. Weinberg made the first calculation of the dimension of this space [422], using methods used to establish the Atiyah-Singer index theorem. There are some subtleties because the problem is set up on $\mathbb{R}^3$, which is non-compact, and the boundary conditions are not strong enough to compactify $\mathbb{R}^3$. Nevertheless the dimension is determined purely by the topological charge N of the background field $\{\Phi, A_i\}$, and it is $4N - 1$. For framed monopoles, the dimension is $4N$.

The result $4N$ is rather surprising. The force argument suggests that solutions of the Bogomolny equation with N monopoles at N arbitrary positions in $\mathbb{R}^3$ might exist, but that would give only a $3N$-dimensional space. In fact, the monopoles each have an additional phase parameter, making $4N$ all together.

The second result was a tour de force of analysis by Taubes, in which the existence of N-monopole solutions of the Bogomolny equation, for all $N > 0$, was established [223]. Taubes constructed a field configuration which is a superposition of N well separated unit charge monopoles, with the magnetic fields and the difference of the Higgs fields from the vacuum linearly superposed outside the core region of each monopole. This gives an approximate solution of the Bogomolny equation. He then showed that close to this approximate solution there is an exact solution. A technical difficulty, that Taubes overcame, is to ensure that as the exact solution is approached, the monopoles do not simply drift off to infinity. Morally speaking, Taubes' method establishes a continuous family of solutions, parametrized by the locations of the N well separated monopoles, but more precisely, it only establishes the existence of a discrete but fairly dense subset of such solutions. The construction throws some light on the additional phase parameter associated with each monopole. It has to do with an approximate $U(1)$ gauge invariance in the way a single monopole is glued into the abelian background field produced by the other monopoles. However, the global structure of the $4N$-dimensional space of solutions is not greatly clarified.

Taubes' result, combined with the index calculations, establishes the existence of a $4N$-dimensional manifold $\mathcal{M}_N$ of gauge inequivalent solutions to the Bogomolny equation, for monopole number N. $\mathcal{M}_N$ is known as the moduli space of N-monopole solutions. Coordinates on it are referred to equivalently as moduli, parameters, or collective coordinates for the monopoles. $\mathcal{M}_N$ turns out to be connected, although it takes further analysis to establish this.

8.4 Dyons

In addition to magnetic monopole solutions, the Yang-Mills-Higgs theory with an adjoint Higgs field has dyon solutions. By definition, a dyon is a particle or soliton with both magnetic and electric charge. The name was coined by Schwinger [369]. Dyons are not strictly static, although they are stationary in certain gauges, and they have non-zero kinetic energy.

Julia and Zee [229] showed that there is a generalization of the 't Hooft-Polyakov monopole which is electrically charged. The time component of the gauge potential, A_0, is non-vanishing and of the form

$$A_0 = j(r)\frac{x^a}{r}t^a. \tag{8.98}$$

The spatial components of the gauge potential, and the Higgs field, have the same form as in (8.56) and (8.57), but h and k are modified. The Yang-Mills electric field

$$E_i = \partial_0 A_i - \partial_i A_0 + [A_0, A_i] \tag{8.99}$$

simplifies to $E_i = -D_i A_0$ because the fields are time independent.

Asymptotically, $D_\mu \hat{\Phi} = 0$ still decays exponentially fast, so the $U(1)$ electric field e_i can be defined as the projection on to the Higgs field of the non-abelian electric field $e_i = -\frac{1}{2}\text{Tr}(E_i \hat{\Phi})$.

Julia and Zee found a 1-parameter family of solutions, all of which have the same magnetic charge $g = -2\pi$, but a variable electric charge, whose strength q is seen from the form of the asymptotic electric field

$$\mathbf{e} = \frac{q}{4\pi r^2} \hat{\mathbf{x}}. \tag{8.100}$$

These solutions were found numerically for a number of values of q, the energy increasing with $|q|$. Dyons are therefore more massive than monopoles but there is no simple formula for the dependence of mass on electric charge.

Prasad and Sommerfield, we recall, found the exact monopole solution in the limit where the Higgs coupling constant λ vanishes. They also found analytic formulae for the fields of a dyon, and a simple expression for the mass, in this limit. One can rederive their results using a type of Bogomolny argument, as follows [89].

Suppose that the fields are time independent, and that $D_0 \Phi = 0$. There is no particular reason for this last assumption, but we shall find that it is satisfied by the dyon solution. The energy of the fields is

$$E = -\frac{1}{4} \int \left(\text{Tr}(E_i E_i) + \text{Tr}(B_i B_i) + \text{Tr}(D_i \Phi D_i \Phi) \right) d^3x. \tag{8.101}$$

This may be rewritten as

$$\begin{aligned} E = {} & -\frac{1}{4} \int \text{Tr}(E_i + \sin\mu\, D_i \Phi)(E_i + \sin\mu\, D_i \Phi)\, d^3x \\ & -\frac{1}{4} \int \text{Tr}(B_i + \cos\mu\, D_i \Phi)(B_i + \cos\mu\, D_i \Phi)\, d^3x \\ & +\frac{1}{2} \sin\mu \int \text{Tr}(E_i D_i \Phi)\, d^3x + \frac{1}{2}\cos\mu \int \text{Tr}(B_i D_i \Phi)\, d^3x \end{aligned} \tag{8.102}$$

where μ is an arbitrary constant angle. Physical fields must satisfy Gauss' law, which takes the form $D_i E_i = 0$ if $D_0 \Phi = 0$. The last two terms can therefore be expressed as surface integrals over the 2-sphere at infinity, so

$$E = -\frac{1}{4} \int \text{Tr}(E_i + \sin\mu\, D_i \Phi)(E_i + \sin\mu\, D_i \Phi)\, d^3x$$

$$-\frac{1}{4}\int \mathrm{Tr}(B_i + \cos\mu\, D_i\Phi)(B_i + \cos\mu\, D_i\Phi)\, d^3x$$

$$-\sin\mu\int_{S^2_\infty} e_i\, dS^i - \cos\mu\int_{S^2_\infty} b_i\, dS^i \qquad (8.103)$$

where e_i and b_i are the asymptotic electric and magnetic fields. If g and q are both negative, we therefore have the Bogomolny bound on the energy

$$E \geq |q|\sin\mu + |g|\cos\mu. \qquad (8.104)$$

Equality occurs if

$$E_i \;=\; -\sin\mu\, D_i\Phi \qquad (8.105)$$
$$B_i \;=\; -\cos\mu\, D_i\Phi. \qquad (8.106)$$

Given the Prasad-Sommerfield monopole solution, it is straightforward to solve this pair of equations, for any μ. Set $A_0 = \sin\mu\,\Phi$, and set A_i and $\widetilde\Phi = \cos\mu\,\Phi$ to be spatially rescaled monopole fields. To have the correct asymptotic behaviour for Φ it is necessary that $|\widetilde\Phi| \to \cos\mu$ as $r \to \infty$. The dyon solution is therefore (8.56), (8.57) and (8.98) with

$$h(r) \;=\; \coth(2r\cos\mu) - \frac{1}{2r\cos\mu} \qquad (8.107)$$

$$k(r) \;=\; \frac{2r\cos\mu}{\sinh(2r\cos\mu)} \qquad (8.108)$$

$$j(r) \;=\; \coth(2r\cos\mu)\sin\mu - \frac{\tan\mu}{2r}. \qquad (8.109)$$

It is easy to check that $D_0\Phi = 0$ for these fields, and that Gauss' law $D_i E_i = 0$ is satisfied.

The asymptotic Higgs field is the same as for the monopole, and has the same topology. It satisfies the boundary condition $\Phi(0,0,x^3) \to t^3$ as $x^3 \to \infty$. Since $D_i\Phi \to 0$ asymptotically, the magnetic charge has the unchanged value $g = -2\pi$. From (8.105) and (8.106) we deduce that the electric charge is $q = g\tan\mu$, so $\sin\mu = |q|/(g^2 + q^2)^{1/2}$ and $\cos\mu = |g|/(g^2 + q^2)^{1/2}$. It follows from (8.104) that the dyon has energy, or mass,

$$M = (g^2 + q^2)^{1/2}. \qquad (8.110)$$

Although the dyon fields are time independent, there is net kinetic energy, because A_0 and hence E_i are non-vanishing. We can see this more explicitly by going to the gauge $A_0 = 0$. This is achieved by the time dependent gauge transformation

$$g(t,\mathbf{x}) = \exp(tA_0(\mathbf{x})). \qquad (8.111)$$

Φ is unchanged, but A_i acquires a rather complicated time dependence, while remaining spherically symmetric. Notice that the gauge transformation (8.111) does not satisfy the condition $g(0, 0, x^3) \to 1$ as $x^3 \to \infty$ at all times, but only if t is an integer multiple of $2\pi/\sin\mu$. The framing of the fields is therefore steadily rotating, and this is the origin of the kinetic energy. Fields separated in time by the period $2\pi/\sin\mu$ differ by a based gauge transformation, and are hence physically the same. The dyon therefore appears in this gauge to be periodic with period $2\pi/\sin\mu$.

(Note that the gauge transformation at time $t = 2\pi/\sin\mu$ is a map from $\mathbb{R}^3 \mapsto SU(2)$, satisfying the boundary condition $g \to 1$ as $|\mathbf{x}| \to \infty$, but which can not be smoothly deformed to the identity $g = 1$ everywhere while preserving this boundary condition. It is therefore topologically non-trivial.)

We shall see below how the idea of a dyon as a periodic solution is built into the moduli space picture of monopole dynamics. The periodicity also implies that the electric charge of a dyon is quantized, in a manner compatible with the Dirac quantization condition (8.9), when one considers the quantum dynamics of monopoles.

8.5 The Nahm transform

Direct construction of solutions of the Bogomolny equation with monopole numbers greater than 1 is very difficult. To circumvent this problem a number of brilliant ideas have been put forward for transforming a monopole into an alternative mathematical structure, which can be more concretely constructed. In this section, and the following one, we describe in detail the transformation of BPS monopoles discovered by Nahm [313]. Other approaches will be discussed in subsequent sections.

Many solutions of the Bogomolny equation have been discovered using the Nahm transform, although this does not mean that analytic expressions for the gauge and Higgs field are known.

The Nahm transform is a two-way transformation, like the Fourier transform. It takes monopole solutions of the Bogomolny equation

$$B_i = -D_i\Phi \qquad (8.112)$$

defined in $\mathbb{R}^3$, and satisfying certain boundary conditions, to solutions of the Nahm equation

$$\frac{dT_i(s)}{ds} = \frac{1}{2}\varepsilon_{ijk}\left[T_j(s), T_k(s)\right]. \qquad (8.113)$$

Here $T_1(s)$, $T_2(s)$ and $T_3(s)$ are matrices defined on the interval $-1 \le s \le 1$, and subject to certain symmetry and antihermiticity conditions. The matrices are smooth on the interval but diverge at the endpoints in such a way that they have simple poles there. If the monopole has charge N then the Nahm matrices are $N \times N$ matrices.

The inverse transformation takes a solution of the Nahm equation to a solution of the Bogomolny equation. Moreover, acting on a monopole, the Nahm transform followed by its inverse gives back the same monopole.

Neither the Nahm transform nor its inverse are easily performed explicitly. Nevertheless, the transform is valuable for a number of reasons. The first is that, given a solution of the Nahm equation satisfying the various subsidiary conditions, it can be shown that the gauge potential and Higgs field it transforms to give a smooth solution of the Bogomolny equation.

Secondly, it is possible to solve the Nahm equation analytically in a number of non-trivial cases. It is then also possible to carry out the (inverse) Nahm transform numerically. In this way a number of interesting monopoles of various charges have been constructed numerically. It is particularly convenient to find the magnitude of the Higgs field, and from this the energy density of the monopole can be calculated and displayed by making use of Ward's formula [413] for the energy density

$$\mathcal{E} = \frac{1}{2}\nabla^2|\Phi|^2. \tag{8.114}$$

This formula follows from the fact that for solutions of the Bogomolny equation, the two terms in the energy density (the integrand of (8.81)) are identically equal, so $\mathcal{E} = -\frac{1}{2}\text{Tr}(D_i\Phi D_i\Phi)$. Using the covariant Leibniz rule twice, and the field equation $D_i D_i \Phi = 0$, one can reexpress the energy density as $\mathcal{E} = \partial_i(-\frac{1}{2}\text{Tr}(\Phi D_i\Phi)) = \nabla^2(-\frac{1}{4}\text{Tr}(\Phi\Phi))$, which is equivalent to (8.114).

Thirdly, symmetries of monopoles imply certain algebraic constraints on the Nahm matrices. Analytic solutions of the Nahm equation are easier to obtain in these cases. Some of the symmetries are rather surprising. The most symmetric monopole solutions of a given charge N are usually solutions in which N single monopoles have coalesced and lost their individual identities. These are solutions for which Taubes' approach gives no information.

Fourthly, the Nahm transform and its inverse are isometries. We shall see that there is a natural metric on the moduli space of monopoles, and there is a similar metric on the moduli space of solutions of the Nahm equation. The metrics have been shown to be the same, by Nakajima [314]. The metric can be calculated explicitly from the Nahm data in a number of cases where direct calculation of the metric on the monopole moduli space has not been possible.

We shall now discuss the transformation from a BPS monopole to Nahm matrices. The following aims to explain the main point, but for a mathematically more complete presentation, see refs. [184, 97]. Our presentation follows that of Corrigan and Goddard [97].

Consider a Dirac spinor field $\Psi(\mathbf{x})$, transforming via the fundamental representation of the gauge group $SU(2)$, and coupled to the gauge field $\mathbf{A}$ and Higgs field Φ of a monopole of charge N. Ψ can be written as a pair of Weyl two-component spinors

$$\begin{pmatrix} \Psi^- \\ \Psi^+ \end{pmatrix}. \tag{8.115}$$

Let S^- and S^+ denote the spaces of such Weyl spinor fields.

Motivated by the Dirac operator in the background of a four-dimensional instanton, Nahm introduced the Dirac equation in three dimensions

$$\begin{pmatrix} 0 & i(\boldsymbol{\tau} \cdot \mathbf{D} - i\Phi - s) \\ i(\boldsymbol{\tau} \cdot \mathbf{D} + i\Phi + s) & 0 \end{pmatrix} \begin{pmatrix} \Psi^- \\ \Psi^+ \end{pmatrix} = 0 \tag{8.116}$$

where $\boldsymbol{\tau}$ are Pauli matrices, $\mathbf{D}$ is the gauge covariant derivative and s is a constant real parameter. Of course, (8.116) reduces to the pair of equations

$$D\Psi^- = 0 \tag{8.117}$$
$$D^\dagger \Psi^+ = 0 \tag{8.118}$$

where $D = i(\boldsymbol{\tau} \cdot \mathbf{D} + i\Phi + s)$ and $D^\dagger = i(\boldsymbol{\tau} \cdot \mathbf{D} - i\Phi - s)$ is its adjoint.

Let $\ker D$ and $\ker D^\dagger$ denote the vector spaces of normalizable solutions of (8.117) and (8.118) respectively. The possibility that these vector spaces are non-trivial depends on the asymptotic eigenvalues of $i\Phi + s$, which are $1 + s$ and $-1 + s$. These eigenvalues must have opposite sign, so that solutions can decay in all directions. Therefore s is restricted to the interval $-1 < s < 1$. It can be proved using the Atiyah-Singer index theorem that for s in this interval

$$\dim \ker D - \dim \ker D^\dagger = N. \tag{8.119}$$

But, as we now show, $\ker D^\dagger$ is trivial, so $\ker D$ has dimension N. We shall refer to the normalizable solutions of $D\Psi^- = 0$ as zero modes of D.

To see that $\ker D^\dagger$ is trivial consider the operator $DD^\dagger : S^+ \mapsto S^+$. A simple calculation shows that

$$\begin{aligned} DD^\dagger &= -(\boldsymbol{\tau} \cdot \mathbf{D} + i\Phi + s)(\boldsymbol{\tau} \cdot \mathbf{D} - i\Phi - s) \\ &= -\mathbf{D} \cdot \mathbf{D} + \Phi\Phi^\dagger + s^2 + i\boldsymbol{\tau} \cdot (\mathbf{B} + \mathbf{D}\Phi) \end{aligned} \tag{8.120}$$

using $\tau_i \tau_j = \delta_{ij} + i\varepsilon_{ijk}\tau_k$, and the commutation relation $[D_i, D_j] = F_{ij}$. For fields satisfying the Bogomolny equation $\mathbf{B} + \mathbf{D}\Phi = 0$, therefore,

$$DD^\dagger = -\mathbf{D} \cdot \mathbf{D} + \Phi\Phi^\dagger + s^2 \tag{8.121}$$

and this is a positive operator, since $\langle \Psi^+, -\mathbf{D}{\cdot}\mathbf{D}\Psi^+ \rangle = \langle \mathbf{D}\Psi^+, \mathbf{D}\Psi^+ \rangle > 0$. A positive operator has no normalizable zero modes, and if $DD^\dagger$ has no zero modes, then $D^\dagger$ has no zero modes.

We therefore have the following picture. We may split S^- into the direct sum $S^- = S_0^- \oplus \tilde{S}^-$ where S_0^- is the N-dimensional kernel of D, and $\tilde{S}^-$ is its orthogonal complement. The operators D and $D^\dagger$ act schematically as in Fig. 8.3.

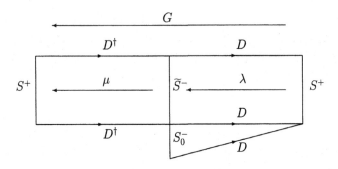

Fig. 8.3.   A sketch illustrating the action of the operators D and $D^\dagger$, and their Green's functions.

Since $DD^\dagger : S^+ \mapsto S^+$ is invertible, $D : \tilde{S}^- \mapsto S^+$ and $D^\dagger : S^+ \mapsto \tilde{S}^-$ are both invertible. Let us introduce Green's functions for these operators. The most important is the Green's function $G(\mathbf{x}, \mathbf{y})$ of the operator $DD^\dagger$ satisfying

$$(-\mathbf{D} \cdot \mathbf{D} + \Phi\Phi^\dagger + s^2)G(\mathbf{x}, \mathbf{y}) = \delta(\mathbf{x} - \mathbf{y}). \qquad (8.122)$$

Let us also introduce a Green's function $\lambda(\mathbf{x}, \mathbf{y})$ for the restricted operator $D : \tilde{S}^- \mapsto S^+$. A solution of the equation

$$D\Psi^- = \chi^+ \qquad (8.123)$$

where $\Psi^- \in \tilde{S}^-$ is then $\Psi^- = \int \lambda(\mathbf{x}, \mathbf{y})\chi^+(\mathbf{y})\, d^3y$. The general solution of (8.123) has an additional piece in $\ker D$. Similarly, let the Green's function of $D^\dagger$ be $\mu(\mathbf{x}, \mathbf{y})$.

Both λ and μ can be expressed in terms of G. Formally, $\lambda = D^\dagger G$ and $\mu = GD$. Clearly $D^\dagger G$ is a (matrix-valued) function. The expression GD can also be made into a function by integrating by parts wherever μ occurs in an integral. Then $\mu = G\overleftarrow{D}$, which means

$$\mu(\mathbf{x}, \mathbf{y}) = i\tau \cdot (-\nabla_{\mathbf{y}}G + G\mathbf{A}) - G\Phi + iGs. \qquad (8.124)$$

By construction, $D^\dagger\mu$ is the identity operator on $\tilde{S}^-$. Let $\{\Psi_a^0 : 1 \le a \le N\}$ be an orthonormal basis of zero modes of D; that is,

$$\int \Psi_a^{0\dagger}(\mathbf{x})\Psi_b^0(\mathbf{x})\, d^3x = \delta_{ab}\,. \tag{8.125}$$

Then

$$\delta(\mathbf{x}-\mathbf{y}) - \Psi_a^0(\mathbf{x})\Psi_a^{0\dagger}(\mathbf{y}) \tag{8.126}$$

is the projection operator from S^- to $\tilde{S}^-$, and it is the identity operator on $\tilde{S}^-$. Therefore

$$D_{\mathbf{x}}^\dagger\mu(\mathbf{x},\mathbf{y}) = \delta(\mathbf{x}-\mathbf{y}) - \Psi_a^0(\mathbf{x})\Psi_a^{0\dagger}(\mathbf{y})\,, \tag{8.127}$$

and this is the most important result concerning these Green's functions.

The Nahm matrices are constructed from the zero modes $\Psi_a^0(\mathbf{x})$. These zero modes depend on the parameter s, and we may specify more carefully how we choose the basis Ψ_a^0 by requiring that

$$\int \Psi_a^{0\dagger}(\mathbf{x})\frac{\partial}{\partial s}\Psi_b^0(\mathbf{x})\, d^3x = 0\,. \tag{8.128}$$

(More geometrically, the left-hand side is an $SO(N)$ connection on the s-axis, but a connection in one dimension can be gauged to zero.) The formula for the Nahm matrices T_i is then

$$(T_i)_{ab} = -i\int x^i \Psi_a^{0\dagger}(\mathbf{x})\Psi_b^0(\mathbf{x})\, d^3x\,, \tag{8.129}$$

i.e., three $N \times N$ matrices, depending on s. To show that these obey the Nahm equation, we use the Green's function identities established above. First, we have

$$(T_iT_j)_{ab} = -\int x^i y^j \Psi_a^{0\dagger}(\mathbf{x})\Psi_c^0(\mathbf{x})\Psi_c^{0\dagger}(\mathbf{y})\Psi_b^0(\mathbf{y})\, d^3x\, d^3y$$

$$= -\int x^i y^j \Psi_a^{0\dagger}(\mathbf{x})(\delta(\mathbf{x}-\mathbf{y}) - D_{\mathbf{x}}^\dagger\mu(\mathbf{x},\mathbf{y}))\Psi_b^0(\mathbf{y})\, d^3x\, d^3y$$

$$= -\int x^i x^j \Psi_a^{0\dagger}(\mathbf{x})\Psi_b^0(\mathbf{x})\, d^3x - \int (D_{\mathbf{x}}(x^i\Psi_a^0(\mathbf{x})))^\dagger\mu(\mathbf{x},\mathbf{y})y^j\Psi_b^0(\mathbf{y})\, d^3x\, d^3y$$

$$= -\int x^i x^j \Psi_a^{0\dagger}(\mathbf{x})\Psi_b^0(\mathbf{x})\, d^3x + i\int \Psi_a^{0\dagger}(\mathbf{x})\tau_i\mu(\mathbf{x},\mathbf{y})y^j\Psi_b^0(\mathbf{y})\, d^3x\, d^3y\,,$$

$$\tag{8.130}$$

using (8.127), then integrating by parts, and finally using the fact that Ψ_a^0 is in the kernel of D. Now, replacing $\mu(\mathbf{x},\mathbf{y})$ by the operator $G(\mathbf{x},\mathbf{y})D_{\mathbf{y}}$, and again noting that $D_{\mathbf{y}}$ annihilates $\Psi_b^0(\mathbf{y})$, we obtain

$$(T_iT_j)_{ab} = -\int x^i x^j \Psi_a^{0\dagger}(\mathbf{x})\Psi_b^0(\mathbf{x})\, d^3x - \int \Psi_a^{0\dagger}(\mathbf{x})\tau_i\tau_j G(\mathbf{x},\mathbf{y})\Psi_b^0(\mathbf{y})\, d^3x\, d^3y. \tag{8.131}$$

The commutator of T_i and T_j is therefore

$$[T_i, T_j]_{ab} = -2i\varepsilon_{ijk} \int \Psi_a^{0\dagger}(\mathbf{x}) \tau_k G(\mathbf{x}, \mathbf{y}) \Psi_b^0(\mathbf{y}) \, d^3x \, d^3y \,. \tag{8.132}$$

We need to compare this with the s-derivative of the Nahm matrix T_k

$$\left(\frac{dT_k}{ds}\right)_{ab} = -i \int x^k \frac{\partial \Psi_a^{0\dagger}(\mathbf{x})}{\partial s} \Psi_b^0(\mathbf{x}) \, d^3x - i \int x^k \Psi_a^{0\dagger}(\mathbf{x}) \frac{\partial \Psi_b^0(\mathbf{x})}{\partial s} \, d^3x \,. \tag{8.133}$$

Taking the s-derivative of $D_{\mathbf{x}} \Psi_a^0(\mathbf{x}) = 0$, we have

$$D \frac{\partial \Psi_a^0}{\partial s} + i\Psi_a^0 = 0 \,. \tag{8.134}$$

In this equation, $i\Psi_a^0$ should be interpreted as in S^+ and $\partial \Psi_a^0 / \partial s$ in S^-. The choice of basis implied by (8.128) means that $\partial \Psi_a^0 / \partial s$ is in $\widetilde{S}^-$, so we may invert (8.134) using the Green's function $\lambda(\mathbf{x}, \mathbf{y}) = D_{\mathbf{x}}^\dagger G(\mathbf{x}, \mathbf{y})$. Therefore,

$$\frac{\partial \Psi_a^0(\mathbf{x})}{\partial s} = -i \int D_{\mathbf{x}}^\dagger G(\mathbf{x}, \mathbf{y}) \Psi_a^0(\mathbf{y}) \, d^3y \,. \tag{8.135}$$

Substituting this and its adjoint into (8.133), and integrating by parts, and again remembering that $D_{\mathbf{x}}$ annihilates $\Psi_a^0(\mathbf{x})$, we obtain finally

$$\left(\frac{dT_k}{ds}\right)_{ab} = -2i \int \Psi_a^{0\dagger}(\mathbf{x}) \tau_k G(\mathbf{x}, \mathbf{y}) \Psi_b^0(\mathbf{y}) \, d^3x \, d^3y \,. \tag{8.136}$$

Equations (8.132) and (8.136) show that the Nahm equation (8.113) is satisfied.

Let us conclude by stating more precisely the properties of the Nahm matrices which emerge from this transformation [313]. Given a charge N BPS monopole, the matrices T_i obey the following:

(i) The Nahm equation

$$\frac{dT_i}{ds} = \frac{1}{2} \varepsilon_{ijk} [T_j, T_k] \tag{8.137}$$

(ii) $T_i(s)$ is regular on $-1 < s < 1$, but has simple pole behaviour at $s = 1$ and $s = -1$. Near $s = 1$,

$$T_i(s) \sim \frac{R_i}{s-1} + O(1) \tag{8.138}$$

where the matrix residues $\{R_i : i = 1, 2, 3\}$ define the irreducible representation of $su(2)$ of dimension N, with commutation relations $R_1 = -[R_2, R_3]$, etc.

(iii)

$$T_i(s)^\dagger = -T_i(s) \qquad (8.139)$$

(iv)

$$T_i(-s) = T_i^t(s) \,, \qquad (8.140)$$

where the superscript t denotes the transpose.

Comments: The normalizable zero modes of the operator D are less and less localized around the monopole as s approaches ± 1; hence the moments which define the Nahm matrices (8.129) diverge as $s \to \pm 1$. Property (ii) states that this divergence gives a simple pole. Substituting (8.138) into the Nahm equation, the s^{-2} term is $R_i = -\frac{1}{2}\varepsilon_{ijk}[R_j, R_k]$, so $\{R_i\}$ defines an N-dimensional representation of $su(2)$. The irreducibility of the representation is a consequence of having the correct number of zero modes. It is often convenient to express Nahm data in a basis in which property (iv) is not manifest. The existence of a basis in which relation (iv) is explicitly satisfied can then be guaranteed by the properties of other basis independent objects, such as the spectral curve which we introduce later.

Let us now, more briefly, describe the inverse Nahm transform. Structurally it is rather similar to the Nahm transform. Start with a triple of $N \times N$ Nahm matrices $\{T_j(s) : -1 < s < 1\}$ obeying the conditions (8.137)–(8.140). Then consider the one-dimensional Weyl equations on $-1 < s < 1$

$$\left(1_{2N}\frac{d}{ds} + iT_j(s) \otimes \tau_j - 1_N \otimes x^j\tau_j\right) v^-(s) \;=\; 0 \qquad (8.141)$$

$$\left(-1_{2N}\frac{d}{ds} + iT_j(s) \otimes \tau_j - 1_N \otimes x^j\tau_j\right) v^+(s) \;=\; 0 \qquad (8.142)$$

where $\mathbf{x}$ is now a parameter. The Nahm equation implies that

$$\left(1_{2N}\frac{d}{ds} + iT_j(s) \otimes \tau_j - 1_N \otimes x^j\tau_j\right)\left(-1_{2N}\frac{d}{ds} + iT_j(s) \otimes \tau_j - 1_N \otimes x^j\tau_j\right)$$

$$= -1_{2N}\frac{d^2}{ds^2} + (iT_j - 1_N x^j)(iT_j - 1_N x^j) \otimes 1_2 \qquad (8.143)$$

which is a positive operator. It follows that (8.142) has no solutions. An analysis of the boundary conditions on the Nahm data, and in particular the fact that the residues form the irreducible representation, reveals that (8.141) has precisely two solutions which do not diverge at either $s = 1$ or $s = -1$. Let us choose an orthonormal basis for these solutions $\{v_a(s) : a = 1, 2\}$, satisfying

$$\int_{-1}^{1} v_a^\dagger(s)v_b(s) \; ds = \delta_{ab} \,, \qquad (8.144)$$

and varying smoothly with the parameter $\mathbf{x}$. From these we define a Higgs field and gauge potential in $\mathbb{R}^3$, with matrix components

$$\Phi(\mathbf{x})_{ab} = i \int_{-1}^{1} s v_a^\dagger(s) v_b(s) \, ds \tag{8.145}$$

$$A_i(\mathbf{x})_{ab} = \int_{-1}^{1} v_a^\dagger(s) \frac{\partial}{\partial x^i} v_b(s) \, ds . \tag{8.146}$$

It can be shown not only that $\{\Phi, A_i\}$ obey the Bogomolny equation and boundary conditions for a charge N monopole, but also that the fields are smooth functions of $\mathbf{x}$. Further, the Nahm transform followed by its inverse gives back the monopole one starts with.

The gauge arbitrariness of a monopole arises because of the ambiguity in the choice of basis $\{v_a\}$ as $\mathbf{x}$ varies. In three dimensions, there is no analogue of the condition (8.128), so we can not simply make A_i vanish. However, we may choose a basis so that one component, say A_1, vanishes. This is an axial gauge choice.

8.6 Construction of monopoles from Nahm data

The Nahm transform clearly provides a powerful tool for constructing N-monopole solutions. Even so, only a limited number of solutions with a high degree of geometrical symmetry have been found using it. We shall first describe the Nahm data for $N = 1$ and $N = 2$ solutions; here, all solutions are obtained. We shall then consider a class of $N = 4$ solutions with tetrahedral symmetry, and briefly mention some further examples with $N = 3$ and $N = 7$. Although the Nahm data are known analytically in all these cases, it is necessary in most of them to use numerical calculations to apply the inverse Nahm transform and construct the Higgs (and gauge) fields.

For $N = 1$, the Nahm matrices are 1×1 matrices, with trivial commutators. The Nahm equation is solved by $T_i(s) = ic_i$, where c_i is a constant vector. This constant solution satisfies all the requirements (8.137)–(8.140) provided c_i is real. The only one-dimensional representation of $su(2)$ is the trivial one, so there are no poles at $s = 1$ or $s = -1$. It is fairly clear from the form of the operator $1_{2N} \frac{d}{ds} + iT_j(s) \otimes \tau_j - 1_N \otimes x^j \tau_j$ that a shift of c_i corresponds to a translation in $\mathbb{R}^3$. The centred $N = 1$ monopole corresponds to $c_i = 0$.

It is remarkable that the trivial Nahm data, $T_i(s) = 0$, give a monopole. We now demonstrate this. The equation (8.141) reduces to

$$\left(\frac{d}{ds} - \boldsymbol{\tau} \cdot \mathbf{x} \right) v(s) = 0 \tag{8.147}$$

which we must solve for all $\mathbf{x}$. It is straightforward to integrate (8.147), since $\boldsymbol{\tau} \cdot \mathbf{x}$ is independent of s. We find

$$
\begin{aligned}
v(s) &= \exp(s\boldsymbol{\tau} \cdot \mathbf{x})v(0) \\
&= (\cosh sr + \sinh sr \, \hat{\mathbf{x}} \cdot \boldsymbol{\tau})v(0)
\end{aligned}
\tag{8.148}
$$

where $r = |\mathbf{x}|$ and $\hat{\mathbf{x}} = \mathbf{x}/r$. Two independent solutions are

$$
v_1(s) = \sqrt{\frac{r}{\sinh 2r}}(\cosh sr + \sinh sr \, \hat{\mathbf{x}} \cdot \boldsymbol{\tau}) \begin{pmatrix} 1 \\ 0 \end{pmatrix}
\tag{8.149}
$$

$$
v_2(s) = \sqrt{\frac{r}{\sinh 2r}}(\cosh sr + \sinh sr \, \hat{\mathbf{x}} \cdot \boldsymbol{\tau}) \begin{pmatrix} 0 \\ 1 \end{pmatrix}.
\tag{8.150}
$$

These obey the orthogonality condition

$$
\int_{-1}^{1} v_1^{\dagger}(s)v_2(s) \, ds = 0
\tag{8.151}
$$

since $\cosh sr \sinh sr$ integrates to zero on $[-1,1]$. The factor $\sqrt{r/\sinh 2r}$ ensures the normalization conditions

$$
\int_{-1}^{1} v_1^{\dagger}(s)v_1(s) \, ds = \int_{-1}^{1} v_2^{\dagger}(s)v_2(s) \, ds = 1.
\tag{8.152}
$$

We can now reconstruct the Higgs field Φ. Its matrix components are

$$
\Phi_{ab} = i \int_{-1}^{1} s v_a^{\dagger}(s)v_b(s) \, ds.
\tag{8.153}
$$

Only the terms linear in $\hat{\mathbf{x}} \cdot \boldsymbol{\tau}$ contribute to this integral. One finds

$$
\begin{aligned}
\Phi_{ab} &= i\frac{r}{\sinh 2r}(\hat{\mathbf{x}} \cdot \boldsymbol{\tau})_{ab} \int_{-1}^{1} 2s \cosh sr \sinh sr \, ds \\
&= i\frac{r}{\sinh 2r}(\hat{\mathbf{x}} \cdot \boldsymbol{\tau})_{ab} \left(\frac{\cosh 2r}{r} - \frac{\sinh 2r}{2r^2} \right) \\
&= i\left(\coth 2r - \frac{1}{2r} \right)(\hat{\mathbf{x}} \cdot \boldsymbol{\tau})_{ab},
\end{aligned}
\tag{8.154}
$$

which is the Higgs field of the Prasad-Sommerfield solution. A similar calculation, using (8.146), recovers the gauge potential.

The Nahm data for $N = 2$ monopoles can also be given in closed form [69]. It is known that any $N = 2$ monopole has a $\mathbb{Z}_2 \times \mathbb{Z}_2$ symmetry, this being the group of 180° rotations about three perpendicular axes, together with the identity. If the monopole is centred at the origin and oriented suitably, then it is invariant under 180° rotations about the three Cartesian axes, and the Nahm data simplify.

The Nahm matrices are antihermitian 2×2 matrices, and therefore linear combinations of the matrices $\{i\tau_j\}$. Because of the $\mathbb{Z}_2 \times \mathbb{Z}_2$ symmetry, they can be expressed as

$$T_1(s) = \frac{i}{2}f_1(s)\tau_1, \quad T_2(s) = \frac{i}{2}f_2(s)\tau_2, \quad T_3(s) = -\frac{i}{2}f_3(s)\tau_3. \quad (8.155)$$

The conditions (8.137)–(8.140) reduce to

$$\frac{df_1}{ds} = f_2 f_3, \quad \frac{df_2}{ds} = f_3 f_1, \quad \frac{df_3}{ds} = f_1 f_2 \quad (8.156)$$

with f_1, f_2 and f_3 having simple poles of residue ± 1 at both $s = 1$ and $s = -1$; also $f_i(s) = f_i(-s)$ and $f_i(s)$ is real for $-1 < s < 1$.

Equations (8.156) are the well known Euler equations for a rigid body. There are three constants of integration, independent of s,

$$f_2^2 - f_1^2 = c_{21}^2, \quad f_1^2 - f_3^2 = c_{13}^2, \quad f_2^2 - f_3^2 = c_{23}^2, \quad (8.157)$$

where we have chosen to order the functions so that $f_2^2 \geq f_1^2 \geq f_3^2$. The constants are related by $c_{21}^2 + c_{13}^2 = c_{23}^2$

There is a scaling symmetry of Eqs. (8.156) so that

$$f_j(s) = L F_j(u), \quad u = L(s + s_0) \quad (8.158)$$

is again a solution of Eqs. (8.156), where L and s_0 are arbitrary constants, if the F_j satisfy the Euler equations $\frac{dF_1}{du} = F_2 F_3$, and cyclically.

We fix the scaling symmetry by setting the constants to be

$$c_{21}^2 = L^2 k^2, \quad c_{13}^2 = L^2(1 - k^2), \quad c_{23}^2 = L^2, \quad (8.159)$$

where $k \in [0, 1]$ to be consistent with our choice of ordering. Using these, we can express F_1 and F_3 in terms of F_2 as

$$F_1^2 = F_2^2 - k^2, \quad F_3^2 = F_2^2 - 1, \quad (8.160)$$

leaving the equation for F_2

$$\left(\frac{dF_2}{du}\right)^2 = (F_2^2 - k^2)(F_2^2 - 1). \quad (8.161)$$

Writing $F_2 = -1/y$, this becomes

$$\left(\frac{dy}{du}\right)^2 = (1 - y^2)(1 - k^2 y^2), \quad (8.162)$$

which is the standard form of the equation satisfied by the Jacobi elliptic function $y(u) = \text{sn}_k(u)$, where k is the modulus.

This gives the following elliptic solution of the Nahm equation,

$$f_1 = \frac{-L\,\mathrm{dn}_k(u)}{\mathrm{sn}_k(u)}, \quad f_2 = \frac{-L}{\mathrm{sn}_k(u)}, \quad f_3 = \frac{-L\,\mathrm{cn}_k(u)}{\mathrm{sn}_k(u)}. \tag{8.163}$$

The function $\mathrm{sn}_k(u)$ has zeros at $u = 0$ and $u = 2K_k$, where

$$K_k = \int_0^{\frac{1}{2}\pi} \frac{d\theta}{\sqrt{1 - k^2 \sin^2\theta}} \tag{8.164}$$

is the complete elliptic integral of the first kind. Therefore, choosing $L = K_k$ and $s_0 = 1$, the functions f_i have the required poles at $s = \pm 1$. All the Nahm conditions are now satisfied, so we have a 1-parameter family of Nahm data parametrized by $k \in [0, 1)$.

From the Nahm data, the monopole fields, and in particular the Higgs field and hence the energy density, can be reconstructed via the Nahm construction. In practice it is convenient to do this numerically. Figure 8.4 exhibits the energy density for various values of k.

Fig. 8.4. Energy density isosurfaces for the $N = 2$ monopole solution with elliptic modulus given by (a) $k = 0.99$, (b) $k = 0.7$, (c) $k = 0$.

The parameter k is a measure of the splitting of the $N = 2$ monopole into two unit charge monopoles; as $k \to 1$ the separation tends to infinity. When $k = 0$,

$$f_1(s) = f_2(s) = \frac{-\pi}{2\sin(\pi(s+1)/2)}, \quad f_3(s) = \frac{-\pi}{2\tan(\pi(s+1)/2)}, \tag{8.165}$$

and since $f_1 = f_2$ the fields are axially symmetric about the x^3-axis, giving a toroidal $N = 2$ monopole.

For all values of k, analytic formulae for the Higgs field can be found on the Cartesian axes. These involve theta functions and are given in [69]. The zeros of the Higgs field lie on the x^2-axis. Their approximate locations, for small k, are at $x^2 = \pm(24 - 2\pi^2)^{-1/2}k$, whereas as $k \to 1$ they are at $x^2 = \pm\frac{1}{2}K_k$. When $k = 0$, the Higgs field has a double zero at the origin; and on the x^3-axis,

$$|\Phi| = \left| \tanh(2x^3) - \frac{16x^3}{16(x^3)^2 + \pi^2} \right|, \tag{8.166}$$

whereas in the (x^1, x^2) plane,

$$|\Phi| = \left| \frac{2\pi^2 \cosh a \, (\sinh a - a \cosh a)}{a(4a^2 - \pi^2 \sinh^2 a)} - 1 \right| \qquad (8.167)$$

where $a = \frac{1}{2}\sqrt{16((x^1)^2 + (x^2)^2) - \pi^2}$. These formulae are due to Ward [413].

Few explicit solutions of the Nahm equation are known for monopole number $N > 2$. One family of solutions is modelled on the $N = 2$ solutions, and exists for all N. Here

$$T_1(s) = -\frac{1}{2} f_1(s) \rho_1, \quad T_2(s) = -\frac{1}{2} f_2(s) \rho_2, \quad T_3(s) = \frac{1}{2} f_3(s) \rho_3 \quad (8.168)$$

where $\{\rho_i : i = 1, 2, 3\}$ is a basis of matrices for the N-dimensional irreducible representation of $su(2)$, with $[\rho_i, \rho_j] = 2\varepsilon_{ijk}\rho_k$, and $\{f_i(s) : i = 1, 2, 3\}$ are the same as in the $N = 2$ case. In general, this solution describes a string of N unit charge monopoles equally spaced along the x^2-axis [121]. In the special case $k = 0$, $f_1 = f_2$ and the solution gives a toroidal monopole of charge N, axially symmetric about the x^3-axis.

Further solutions of the Nahm equation are known for $N = 3, 4, 5, 7$; they are obtained by imposing Platonic symmetries. Below we shall describe two examples with $N = 4$. The first gives a monopole with octahedral symmetry. The second gives a 1-parameter family of monopoles with tetrahedral symmetry, and for a special parameter value the octahedral monopole is recovered.

Let us first explain how the symmetry of an N-monopole is reflected in its Nahm data. Naively, one sees from (8.142) that a rotation $R \in SO(3)$, represented by the matrix R_{ij}, acts on the Nahm data by $T_i \mapsto R_{ij} T_i$. But it is possible to conjugate the Nahm matrices by a fixed element of $SU(N)$, preserving the Nahm equation. Such a conjugation arises from a basis change of the zero modes of the operator D. So Nahm data are said to be symmetric under a subgroup $K \subset SO(3)$ if for each $R \in K$, there is a matrix $M(R) \in SU(N)$ such that

$$R_{ij} T_i = M(R) T_i M(R)^{-1}. \qquad (8.169)$$

This equation is trivially satisfied for all R by the $N = 1$ Nahm data $T_i = 0$, so these Nahm data and the resulting monopole are spherically symmetric. The $N = 2$ Nahm data (8.155) satisfy (8.169) if $M(R)$ is taken to be $i\tau_j$ when R is a rotation by $180°$ about the x^j-axis.

In general, the matrices $M(R)$ are highly restricted by the pole behaviour of T_i near $s = 1$ and $s = -1$. The residues are rotated by R and since the residues define the N-dimensional irreducible representation of

$SU(2)$, $M(R)$ must be the matrix representing R in this irreducible representation. We may therefore think of the Nahm matrices as lying in the tensor product space

$$\mathbb{R}^3 \otimes su(N) \tag{8.170}$$

and transforming under $SO(3)$ via the representation

$$\underline{3} \otimes (\underline{N} \otimes \underline{N})_o \tag{8.171}$$

where $\underline{3}$ is the defining representation of $SO(3)$ and $\underline{N}$ is the complex N-dimensional representation, with o denoting the traceless part.

This tensor product representation of $SO(3)$ can be restricted to any subgroup K. K-invariant Nahm data are constructed from the subspace of $\mathbb{R}^3 \otimes su(N)$ which transforms trivially under $K \subset SO(3)$. By the standard Clebsch-Gordon rules for $SO(3)$ representations, (8.171) can be decomposed into $SO(3)$ irreducibles as

$$\underline{3} \otimes (\underline{2N-1} \oplus \underline{2N-3} \oplus \cdots \oplus \underline{5} \oplus \underline{3}) \tag{8.172}$$
$$= \underline{2N+1} \oplus \underline{2N-1} \oplus \cdots \oplus \underline{5}$$
$$\oplus \underline{2N-1} \oplus \underline{2N-3} \oplus \cdots \oplus \underline{3}$$
$$\oplus \underline{2N-3} \oplus \underline{2N-5} \oplus \cdots \oplus \underline{1}.$$

This shows that there is always one $SO(3)$ invariant, which is automatically a K-invariant, but there can be further K-invariants inside the other irreducible representations of $SO(3)$ that are present here.

The very simplest idea is to try to construct $SO(3)$-invariant Nahm data from the $\underline{1}$ in (8.172). This means that the matrices T_i are proportional to the matrices ρ_i we introduced earlier, with the coefficient functions being related. Explicitly,

$$T_1(s) = -\frac{1}{2}f(s)\rho_1, \quad T_2(s) = -\frac{1}{2}f(s)\rho_2, \quad T_3(s) = \frac{1}{2}f(s)\rho_3. \tag{8.173}$$

But the Nahm equation then reduces to $\frac{df}{ds} = f^2$, whose solution is $f(s) = -1/(s+s_0)$. This can not have poles at both $s = 1$ and $s = -1$, for any choice of s_0. So there are no $SO(3)$-symmetric solutions for $N > 1$.

The next simplest situation occurs when N is just large enough for there to be one trivial K-singlet in addition to the $\underline{1}$ when the representations in (8.172) are decomposed into irreducible representations of K; in other words, the representation $\underline{2N+1}$ of $SO(3)$ has a trivial K-singlet, but the lower dimensional representations do not.

For example, if K is the octahedral group O, the critical value is $N = 4$, because the nine-dimensional representation of $SO(3)$ has a trivial O-singlet in it, but the smaller non-trivial representations of $SO(3)$ do not.

It is therefore possible to find two sets of 4×4 matrices $\{\rho_i, \chi_i : i = 1, 2, 3\}$ such that the Nahm data

$$T_i(s) = f(s)\rho_i + g(s)\chi_i \tag{8.174}$$

are octahedrally symmetric. The matrices χ_i can be found explicitly – the method is explained in refs. [187, 194]. Because of the symmetry, the commutators of the matrices ρ_i and χ_i close on themselves. One finds

$$[\rho_1, \rho_2] = 2\rho_3 \,, \quad [\chi_1, \chi_2] = -48\rho_3 - 8\chi_3 \,, \tag{8.175}$$

$$[\chi_1, \rho_2] + [\rho_1, \chi_2] = -6\chi_3 \,, \tag{8.176}$$

and the cyclic permutations of these relations. Therefore, for Nahm data of the form (8.174), the Nahm equation reduces to

$$\frac{df}{ds} = 2f^2 - 48g^2 \tag{8.177}$$

$$\frac{dg}{ds} = -6fg - 8g^2 \,. \tag{8.178}$$

Remarkably, these equations can be solved in terms of a Weierstrass elliptic function. First note that the following combination,

$$\mu^4 = g(f + 3g)(f - 2g)^2 \,, \tag{8.179}$$

is a conserved quantity for the system (8.177)–(8.178). This is easily verified by differentiation and direct substitution of the expressions for the derivatives of f and g. Exchanging f for a new function w, via the substitution $f = (5w^2 - 3)g$, results in the expression

$$\mu^4 = 5^3 g^4 w^2 (w^2 - 1)^2 \,. \tag{8.180}$$

This can be used to eliminate g in terms of w and produces the equation

$$\left(\frac{dw}{ds}\right)^2 = 16\sqrt{5}\mu^2 w(w^2 - 1) \,. \tag{8.181}$$

In terms of the scaled independent variable

$$u = 2\sqrt{5}\mu^2(s + s_0) \,, \tag{8.182}$$

the solution of (8.181) is given by $w(s) = \wp(u)$, where $\wp(u)$ is the Weierstrass function satisfying

$$\wp'(u)^2 = 4\wp(u)^3 - 4\wp(u) \,. \tag{8.183}$$

The functions in (8.174) are therefore

$$f(u) = \frac{2\mu(5\wp^2(u) - 3)}{5^{3/4}\wp'(u)}, \quad g(u) = \frac{2\mu}{5^{3/4}\wp'(u)}. \tag{8.184}$$

The correct linear relation (8.182) between u and s is determined by requiring that the Nahm matrices have poles at $s = -1$ and $s = 1$, and no poles in between. Note that the period lattice of the Weierstrass function satisfying (8.183) is a square lattice, since there is the symmetry $u \mapsto iu$, $\wp \mapsto -\wp$. Let the real and imaginary periods be 2ω and $2i\omega$. It can be checked that

$$\omega = \int_0^1 \frac{dt}{\sqrt{1 - t^4}} = \frac{1}{4\sqrt{2\pi}} \left(\Gamma\left(\frac{1}{4}\right)\right)^2 \tag{8.185}$$

where Γ is the standard gamma function. Since $\wp$ has a double pole at $u = 0$, and $\wp'$ a triple pole, f has a simple pole at $u = 0$ whereas g is regular. The Nahm matrices T_i therefore have a simple pole at $u = 0$, and they have the required residues, coming from ρ_i. Wherever $\wp' = 0$, f and g both have simple poles. These are at the half-period points, ω, $i\omega$ and $(1 + i)\omega$, where respectively $\wp = 1, -1$ and 0. The residues of the Nahm matrices can be evaluated, and it is seen that only at $(1+i)\omega$, where $\wp = 0$, do the residues define an irreducible four-dimensional representation of $SU(2)$.

The Nahm matrices are regular for $-1 < s < 1$, and they have the required poles at $s = \pm 1$, provided we arrange that $s = -1$ at $u = 0$ and $s = 1$ at $u = (1 + i)\omega$. Thus

$$u = \frac{(1 + i)\omega}{2}(s + 1) \tag{8.186}$$

with the associated constant being

$$\mu = \frac{(1 + i)\omega}{5^{1/4}4}. \tag{8.187}$$

Combining all the above results gives Nahm data of an octahedrally symmetric $N = 4$ monopole. These Nahm data can be used to find the Higgs field and energy density. A picture of a constant energy density surface is shown in Fig. 8.5.

A more general solution has been found, by relaxing the octahedral symmetry O to tetrahedral symmetry T. There are four sets of T-singlets in (8.172) for $N = 4$, though one of these sets plays no role, so the ansatz for the Nahm matrices is of the form

$$T_i(s) = f(s)\rho_i + g(s)\psi_i + h(s)\zeta_i. \tag{8.188}$$

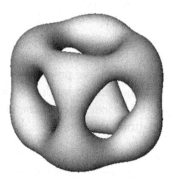

Fig. 8.5.   Energy density isosurface for the $N = 4$ monopole with octahedral symmetry.

The Nahm equation becomes a triple of nonlinear ODEs which may still be solved in terms of a Weierstrass function $\wp(u)$, but now obeying the more general equation

$$\wp'(u)^2 = 4\wp(u)^3 - 4\wp(u) + 12a^2 . \tag{8.189}$$

For each real constant a in the range $(-3^{-5/4}\sqrt{2}, 3^{-5/4}\sqrt{2})$, the solution satisfies all requirements. Near the ends of this interval the $N = 4$ monopole constructed from these Nahm data looks like four well separated unit charge monopoles on the vertices of a tetrahedron, but at $a = 0$ it is the octahedrally symmetric monopole discussed earlier. See Fig. 8.6 for pictures of these monopoles.

8.7 Spectral curves

We recall the properties of a matrix differential equation in Lax form. Let $A(s)$, $B(s)$ be two complex-valued $N \times N$ matrices, depending on s, and suppose

$$\frac{dA}{ds} = [A, B] . \tag{8.190}$$

(This is a well posed linear equation for A if $B(s)$ is given, or a well posed nonlinear equation if B is some specified function of A.) The quantities $c_n = \mathrm{Tr}(A^n)$ are constants independent of s because

$$\frac{dc_n}{ds} = n \, \mathrm{Tr}\left(A^{n-1}\frac{dA}{ds}\right)$$

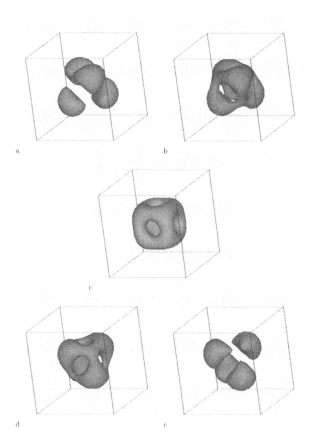

Fig. 8.6. Energy density isosurfaces for a family of $N = 4$ monopoles with tetrahedral symmetry.

$$\begin{aligned} &= \; n\,\mathrm{Tr}(A^{n-1}(AB - BA)) \\ &= \; 0 \end{aligned} \qquad (8.191)$$

using the cyclic property of the trace. It follows that all symmetric polynomials in the (possibly repeated) eigenvalues of A are independent of s, whether or not A can be diagonalized, and hence that the defining polynomial (in η) for the eigenvalues

$$\det(\eta\,1_N + A(s)) \qquad (8.192)$$

is independent of s. Geometrically, the point set in the η-plane defined by

$$\det(\eta\,1_N + A(s)) = 0 \qquad (8.193)$$

is independent of s, and each coefficient of this degree N polynomial in η may be regarded as a constant of integration for the solution $A(s)$ of (8.190).

The Nahm equation can be written in Lax form. Note first that it follows from the Nahm equation that

$$\frac{d}{ds}(T_1 + iT_2) = [T_1 + iT_2, -iT_3] \tag{8.194}$$

so $\det(\eta\,1_N + T_1(s) + iT_2(s))$ is independent of s. But Eq. (8.194) uses only two components of the Nahm equation, and it does not capture the intrinsic $SO(3)$ invariance. Fortunately, we may introduce a complex parameter ξ and consider the linear combination of Nahm matrices

$$T = T_1 + iT_2 - 2iT_3\xi + (T_1 - iT_2)\xi^2\,. \tag{8.195}$$

Introducing also

$$T^+ = -iT_3 + (T_1 - iT_2)\xi \tag{8.196}$$

we can verify that the entire Nahm equation is equivalent to

$$\frac{dT}{ds} = [T, T^+] \tag{8.197}$$

if this last equation holds for all ξ.

Therefore, for a solution of the Nahm equation, the geometrical object defined by the equation

$$P(\xi, \eta) \equiv \det(\eta\,1_N + (T_1 + iT_2) - 2iT_3\xi + (T_1 - iT_2)\xi^2) = 0 \tag{8.198}$$

provides constants of integration, independent of s. For each ξ, the set of points η satisfying (8.198) is the set of eigenvalues of $-T$. When (8.198) is expanded out, for any given solution of the Nahm equation, it is an Nth-order polynomial in η where the coefficients are polynomials in ξ. More precisely, (8.198) has the structure

$$\eta^N + a_1(\xi)\eta^{N-1} + \cdots + a_N(\xi) = 0 \tag{8.199}$$

where $a_r(\xi)$ is of degree at most $2r$. The equation (8.198), or equivalently (8.199), defines an algebraic curve in the complex (ξ, η) space. This is known as the spectral curve of the solution of the Nahm equation. If the Nahm data are those of an N-monopole, then (8.199) is called the spectral curve of the monopole. The curve has one complex dimension, but as a real object it is a surface – a Riemann surface.

It is straightforward to deduce the consequence of the reality condition on the Nahm data, which is that, for all r

$$a_r(\xi) = (-1)^r \xi^{2r} \overline{a_r(-1/\bar{\xi})}\,. \tag{8.200}$$

Condition (8.200) imposes simple reality constraints on the coefficients of the polynomials $a_r(\xi)$. For example, $a_1(\xi)$ must have the form

$$a_1(\xi) = -(c_2 - ic_1) + 2c_3\xi + (c_2 + ic_1)\xi^2 \tag{8.201}$$

where c_1, c_2 and c_3 are real.

Let us now give some examples of spectral curves of monopoles. We start by recalling that the Nahm data for an $N = 1$ monopole centred at (c_1, c_2, c_3) are $T_i = ic_i$, and hence the spectral curve of the monopole is

$$\eta - (c_2 - ic_1) + 2c_3\xi + (c_2 + ic_1)\xi^2 = 0 \tag{8.202}$$

which has the reality property we expect.

For the centred and oriented $N = 2$ monopoles discussed in the previous section the spectral curve is

$$\det\left(\eta 1_N + (\tfrac{i}{2}f_1\tau_1 - \tfrac{1}{2}f_2\tau_2) - f_3\tau_3\xi + (\tfrac{i}{2}f_1\tau_1 + \tfrac{1}{2}f_2\tau_2)\xi^2\right) = 0. \tag{8.203}$$

On evaluating the determinant this becomes

$$\eta^2 + \frac{1}{4}(f_1^2 - f_2^2)(1 + \xi^4) + \frac{1}{2}(f_1^2 + f_2^2 - 2f_3^2)\xi^2 = 0 \tag{8.204}$$

which involves just the constants of integration c_{21}^2, c_{23}^2 and c_{13}^2, and reduces to

$$\eta^2 - \frac{K_k^2}{4}\left(k^2(1 + \xi^4) - 2(2 - k^2)\xi^2\right) = 0 \tag{8.205}$$

for the solution (8.163). For the axially symmetric 2-monopole, with $k = 0$, the spectral curve is

$$\eta^2 + \frac{1}{4}\pi^2\xi^2 = 0. \tag{8.206}$$

As mentioned earlier, there are axially symmetric N-monopole solutions for all N. Their spectral curves were first obtained by Hitchin [183]. For N odd the curve is

$$\eta(\eta^2 + \pi^2\xi^2)(\eta^2 + 4\pi^2\xi^2)\cdots\left(\eta^2 + \left(\frac{N-1}{2}\right)^2\pi^2\xi^2\right) = 0. \tag{8.207}$$

For N even,

$$\left(\eta^2 + \frac{1}{4}\pi^2\xi^2\right)\left(\eta^2 + \frac{9}{4}\pi^2\xi^2\right)\cdots\left(\eta^2 + \left(\frac{N-1}{2}\right)^2\pi^2\xi^2\right) = 0. \tag{8.208}$$

The $N = 4$ octahedral monopole has a spectral curve of the form

$$\eta^4 + c(\xi^8 + 14\xi^4 + 1) = 0 \tag{8.209}$$

where $c = -960\mu^4$. μ^4 is the conserved quantity of the reduced system given by Eq. (8.179), which must take the value given by (8.187) and (8.185), so

$$c = \frac{3}{1024\pi^2}\left(\Gamma\left(\frac{1}{4}\right)\right)^8. \tag{8.210}$$

This coefficient determines the scale of the spectral curve and hence of the monopole. $N = 4$ monopoles with tetrahedral symmetry have spectral curves of the slightly more general form

$$\eta^4 + ic_1\xi(\xi^4 - 1)\eta + c_2(\xi^8 + 14\xi^4 + 1) = 0 \tag{8.211}$$

where c_1 and c_2 are real constants of integration of the reduced equations. The explicit values are given by

$$c_1 = 36a\kappa^3, \quad c_2 = 3\kappa^4 \tag{8.212}$$

where a is the free parameter in the range $(-3^{-5/4}\sqrt{2}, 3^{-5/4}\sqrt{2})$ and κ, which is a function of a, is the real half-period of the elliptic function satisfying (8.189). For the special value $a = 0$, the spectral curve of the octahedral monopole with $c_1 = 0$ and $c_2 = c$ is recovered.

Two more examples of highly symmetric monopoles are the $N = 3$ monopole with tetrahedral symmetry and the $N = 7$ monopole with icosahedral symmetry. Their spectral curves are, respectively,

$$\eta^3 + \frac{\Gamma\left(\frac{1}{6}\right)^3\Gamma\left(\frac{1}{3}\right)^3}{48\sqrt{3}\pi^{3/2}}i\xi(\xi^4 - 1) = 0 \tag{8.213}$$

and

$$\eta^7 + \frac{\Gamma\left(\frac{1}{6}\right)^6\Gamma\left(\frac{1}{3}\right)^6}{64\pi^3}\xi(\xi^{10} + 11\xi^5 - 1)\eta = 0. \tag{8.214}$$

These curves were calculated [187, 195] by explicitly solving the Nahm equation in a similar manner as described above for the octahedral $N = 4$ monopole. Since the Nahm data are known, we can numerically compute the energy density of these monopoles, producing the surfaces displayed in Fig. 8.7. A numerical construction of the Higgs field of the tetrahedral $N = 3$ monopole reveals [196] that the number of zeros of the Higgs field is greater than three. In Fig. 8.8 we plot the three components of the Higgs field along a line which passes through the origin and a vertex of the tetrahedron. We see that there are two zeros of the Higgs field along this line; the first is at the origin and the second is associated with a vertex of the tetrahedron. By tetrahedral symmetry there are five zeros in total, four on the vertices of a tetrahedron and one at the origin. The zero at the origin has negative multiplicity (termed an antizero), that is, the

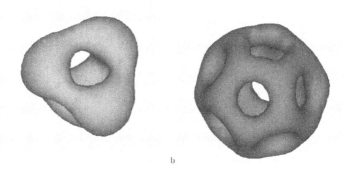

Fig. 8.7. Energy density isosurfaces for (a) the $N = 3$ monopole with tetrahedral symmetry, (b) the $N = 7$ monopole with icosahedral symmetry.

local winding around this point is that of an antimonopole. So the total number of zeros when counted with multiplicity is three, as it must be for a monopole with $N = 3$. There is a particular evolution of three monopoles which instantaneously forms the tetrahedral 3-monopole. The zeros of the Higgs field can be tracked during this evolution, producing a consistent, though elaborate, picture of their dynamics, including zero-antizero pair production and annihilation [196]. Other Platonic monopoles also appear to contain antizeros [391] but this phenomenon is still not well understood.

We close this section by explaining the geometrical significance of spectral curves. The derivation of a spectral curve from Nahm data does not make this very clear, but the same curve arises in other approaches to the Bogomolny equation for monopoles, for example, in the twistor space methods of Ward *et al.*, and in Hitchin's approach [183, 184] to monopoles based on scattering data. We discuss the geometry here, and the scattering data in the next section.

Hitchin's approach requires that (ξ, η) are coordinates on the tangent bundle of the Riemann sphere, denoted $\mathrm{T}\mathbb{CP}^1$. That is, ξ is the standard inhomogeneous, complex coordinate on the Riemann sphere $\mathbb{CP}^1$, and η is the complex coordinate in the tangent plane to the sphere at ξ. (η is normalized so that the tangent vector from ξ to an infinitesimally close point $\xi + \Delta\xi$ on the sphere is $\eta = \Delta\xi$.) The tangent bundle to the Riemann sphere can be interpreted as the space of oriented straight lines in $\mathbb{R}^3$ – Hitchin's mini-twistor space. This is because an oriented line is specified by giving, first, its direction in $\mathbb{R}^3$, which defines a unit vector or point on the sphere; and, second, the point of intersection with a (complex) plane

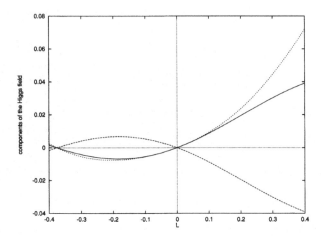

Fig. 8.8. The three components of the Higgs field along a line which passes through the origin and a vertex of the tetrahedron associated with the $N = 3$ monopole with tetrahedral symmetry.

orthogonal to the direction vector. In more conventional terms, the point (x^1, x^2, x^3) lies on the line (ξ, η) if

$$\eta - (x^2 - ix^1) + 2x^3\xi + (x^2 + ix^1)\xi^2 = 0. \qquad (8.215)$$

Note that ξ is a dimensionless, angular variable, whereas η is a linear variable that scales with $|\mathbf{x}|$.

This interpretation of ξ and η means that we should include the value $\xi = \infty$, since that is just the direction opposite to $\xi = 0$. To investigate the neighbourhood of $\xi = \infty$ we make the change of variables

$$\xi = \frac{1}{\widetilde{\xi}}, \quad \eta = -\frac{\widetilde{\eta}}{\widetilde{\xi}^2}, \qquad (8.216)$$

where the second formula is motivated by the derivative of the first. In this way, any spectral curve becomes a curve in $\mathrm{T}\mathbb{CP}^1$. It is in fact compact in $\mathrm{T}\mathbb{CP}^1$. Clearly, while ξ is finite, the N roots η of $P(\xi, \eta) = 0$ are finite. As $\xi \to \infty$, the roots η diverge, but in the coordinates $(\widetilde{\xi}, \widetilde{\eta})$ the N roots $\widetilde{\eta}$ are finite in a neighbourhood of $\widetilde{\xi} = 0$.

Not all compact curves of the form (8.199), and satisfying the reality conditions (8.200), occur as the spectral curves of monopoles. This is because their Nahm data are constrained to satisfy the boundary conditions given in (8.138). Hitchin has given a criterion for a curve to be a spectral curve, in terms of the triviality of the restriction to the curve of a standard bundle over the twistor space $T\mathbb{CP}^1$. This criterion can be made more explicit in terms of integrals of holomorphic differentials around closed 1-cycles on the spectral curve. Such integral constraints appeared in the investigations of Corrigan and Goddard [96]. They were given in a general form by Ercolani and Sinha [121], extending an analysis of the spectral curves of $N = 2$ monopoles by Hurtubise [199].

Recently, the constraints found by Ercolani and Sinha have been somewhat simplified, and the connection to the work of Corrigan and Goddard clarified [192]. The general holomorphic differential on a curve $P(\xi, \eta) = 0$ of the form (8.199) in $T\mathbb{CP}^1$ can be written as

$$\Omega = \frac{(\beta_0 \eta^{N-2} + \beta_1(\xi)\eta^{N-3} + \cdots + \beta_{N-2}(\xi))\, d\xi}{\partial P/\partial \eta}, \tag{8.217}$$

where each β_j is a polynomial of degree at most $2j$ with arbitrary coefficients. As there are $(N-1)^2$ arbitrary coefficients in total it follows that the curve, and hence any spectral curve of a monopole of charge N, has genus $g = (N-1)^2$, a result of Hitchin [183]. The condition for a curve to be a spectral curve, given in [192], is as follows. There must exist a primitive 1-cycle C on the curve (i.e. not a multiple of another 1-cycle), such that for all Ω of the form (8.217)

$$\oint_C \Omega = -2\beta_0. \tag{8.218}$$

Not only the spectral curve but also the homology class represented by this cycle C give a geometrical characterization of the monopole. However, the physical significance of C is not really understood.

The geometrical interpretation of the spectral curve of a monopole in terms of lines in $\mathbb{R}^3$ allows one to anticipate the form of the spectral curve of a symmetric monopole, and conversely to rule out the existence of monopoles with given symmetries for small values of N. A rotation R about the origin in $\mathbb{R}^3$ is represented by a Möbius transformation on ξ, and the derivative of this on η. The $SU(2)$ Möbius transformation

$$\xi \mapsto \frac{(d+ic)\xi + (b-ia)}{-(b+ia)\xi + d - ic}, \quad \eta \mapsto \frac{\eta}{(-(b+ia)\xi + d - ic)^2} \tag{8.219}$$

corresponds to a rotation by θ about the unit direction (n_1, n_2, n_3), where

$$n_1 \sin\frac{\theta}{2} = a, \quad n_2 \sin\frac{\theta}{2} = b, \quad n_3 \sin\frac{\theta}{2} = c, \quad \cos\frac{\theta}{2} = d. \tag{8.220}$$

If this transformation leaves the spectral curve equation $P(\xi, \eta) = 0$ invariant (i.e. if it multiplies the left-hand side only by a constant) then the curve is invariant under this rotation. A monopole invariant under a subgroup $K \subset SO(3)$ has a spectral curve which is invariant in this sense under K.

It is not difficult to directly compute the consequences of K-invariance if K is an uncomplicated subgroup of $SO(3)$. For example, the spectral curve of an N-monopole with cyclic symmetry C_N about the x^3-axis must be invariant under

$$\xi \mapsto e^{2\pi i/N}\xi, \quad \eta \mapsto e^{2\pi i/N}\eta \tag{8.221}$$

and this implies that all terms in $P(\xi, \eta)$ have the same total degree, mod N. Since the leading term is η^N, all terms must be of degree 0 mod N.

For the Platonic groups, the symmetry groups T, O and Y of the tetrahedron, octahedron/cube and icosahedron/dodecahedron, respectively, the construction of invariant curves in $T\mathbb{CP}^1$ is facilitated by using the Klein polynomials. Recall that these are discussed and listed in Chapter 6, and consist of invariant polynomials in ξ associated with the vertices, edge mid-points and face centres of the Platonic solids.

Using the tetrahedral Klein polynomials, one sees that a tetrahedrally symmetric monopole with $N = 3$ must have a spectral curve of the form (8.213), involving $\mathcal{T}_e$. The coefficient of $\mathcal{T}_e$ was recalculated in [192], using the constraint (8.218) and it was found that the cycle C is the unique primitive cycle that is invariant under the action of the tetrahedral group.

Similarly, it can be shown that no octahedrally symmetric monopole can exist for $N = 2$ or $N = 3$, and that if an octahedrally symmetric $N = 4$ monopole exists, then when it is centred and suitably oriented, it must have a spectral curve of the form

$$\eta^4 + c(\xi^8 + 14\xi^4 + 1) = 0 \tag{8.222}$$

with c real, where we recognize the Klein polynomial $\mathcal{O}_f$ associated with the faces of an octahedron (see Eq. (6.45)). We have seen above that by solving the Nahm equation, it is possible to show that a unique octahedrally symmetric 4-monopole does exist, with the special value of c given in (8.210).

Similar arguments for other monopoles with Platonic symmetries and N small are given in [187], and developed in [195]. In these examples, the Nahm equation is again solved in terms of elliptic functions. This has a simple explanation in terms of the geometry of the associated spectral curves. As mentioned above, the spectral curve of an N-monopole has genus $(N - 1)^2$, but if the monopole is symmetric under a group K then the relevant quantity is $\tilde{g}$, the genus of the quotient of the spectral curve by the symmetry group. This genus can be less than $(N - 1)^2$, and in particular if $\tilde{g} = 1$, as it is for the above Platonic examples, then the Nahm equation can be solved in terms of elliptic functions.

The geometrical interpretation of spectral curves also helps with the understanding of a unit charge monopole and configurations of N well separated monopoles. Recall that the point (x^1, x^2, x^3) lies on the line with $\mathrm{T}\mathbb{CP}^1$ coordinates (ξ, η) if Eq. (8.215) is satisfied. Conversely, given a fixed point (x^1, x^2, x^3), the solutions (ξ, η) of (8.215) determine the set of all oriented lines through it. Now we saw from the Nahm equation that the spectral curve of an $N = 1$ monopole with centre (c_1, c_2, c_3) is

$$\eta - (c_2 - ic_1) + 2c_3\xi + (c_2 + ic_1)\xi^2 = 0, \tag{8.223}$$

so this spectral curve consists precisely of all the oriented lines through the monopole centre.

We shall refer to the spectral curve (8.223) as the star at (c_1, c_2, c_3). Not surprisingly, an N-monopole which consists of N well separated unit charge monopoles (whose existence was established by Taubes) has a spectral curve which is approximately a product of stars. As an example, consider the tetrahedrally symmetric $N = 4$ monopoles. Among these are configurations with unit charge monopoles at the vertices of a large tetrahedron, say at (b, b, b), $(b, -b, -b)$, $(-b, b, -b)$ and $(-b, -b, b)$. The product of stars at these points is

$$\eta^4 + 16ib^3\xi(\xi^4 - 1)\eta + 4b^4(\xi^8 + 14\xi^4 + 1) = 0. \tag{8.224}$$

This is of the same form as the genuine spectral curve of the tetrahedrally symmetric 4-monopole (8.211), except that the constants (c_1, c_2), which are given by (8.212), are not exactly expressible as $(16b^3, 4b^4)$. However, this parametrization becomes increasingly accurate in the limit $b \to \pm\infty$.

8.8 Rational maps and monopoles

We have shown how Nahm data and spectral curves can be used to encode information about a monopole. Yet another transformation converts a monopole into a rational map from the Riemann sphere to itself. Let z be the standard complex coordinate on S^2. Recall from Chapter 6 that a rational map $R : S^2 \mapsto S^2$ is given by a function $z \mapsto R(z)$ where $R(z)$ is a ratio of polynomials

$$R(z) = \frac{p(z)}{q(z)} \tag{8.225}$$

and p and q have no common root (i.e. no common linear factor). A rational map, being a function of z alone, is holomorphic. Poles of R are simply the points mapped to ∞ on the target S^2, and a change of coordinates shows that there is no singularity at ∞ on the domain S^2. R is a map of topological degree N if either p or q has degree N, and the other polynomial has degree N or less. This is because these conditions imply that the equation

$$\frac{p(z)}{q(z)} = c \qquad (8.226)$$

has N roots, counted with multiplicity, for generic values of c.

We shall describe below, in detail, how a monopole is transformed to a rational map. Here let us point out one remarkable feature of this transformation, namely, it is one-to-one and onto. The moduli space of framed charge N monopoles is diffeomorphic to a space of based rational maps. Given a rational map, there is a unique monopole which corresponds to it. Moreover, some of the monopole's properties – for example, some of its symmetries – can be determined directly from the map. In principle, the monopole fields can be reconstructed from the map. This reconstruction has been implemented numerically, and we describe the procedure at the end of this section.

Actually, two different ways are known of transforming a monopole to a rational map. The first is due to Donaldson [110], the second to Jarvis (following a suggestion of Atiyah) [225]. This reflects the one weakness of the transformation, namely, that it does not respect all the Euclidean symmetries of $\mathbb{R}^3$. To define Donaldson's map for a monopole it is necessary to choose a direction in $\mathbb{R}^3$ – following convention we shall choose the positive x^3 direction. Another rational map would be obtained from the same monopole if another direction were chosen, but it is not known how to compute one map from the other. More formally, there exists an $SO(3)$ action on the space of Donaldson rational maps, but this action is not known explicitly. Only the action of the $SO(2)$ subgroup which preserves the x^3-axis is known. In a similar way, the Jarvis map depends on the choice of an origin in $\mathbb{R}^3$, and it is not known how the map changes if the origin is shifted. The relationship between Donaldson maps and Jarvis maps is also unknown, except in certain limiting cases.

Starting with the Higgs and gauge fields of a monopole $\{\Phi, A_i\}$, the Donaldson map is obtained as scattering data for Hitchin's equation [183]

$$(D_3 - i\Phi)v = 0. \qquad (8.227)$$

(In fact, Donaldson originally established a natural bijection between rational maps and Nahm data of monopoles. The direct relationship to monopole fields via Hitchin's equation was pointed out by Hurtubise [200].) $v(x^3)$ is a complex two-component ($SU(2)$-spinor) function defined along a line in the x^3 direction. D_3 is the covariant derivative and Φ the Higgs field along this line. The line is labelled by the Cartesian coordinates (x^1, x^2) that are constant along it. We combine these into the complex parameter $z = x^1 + ix^2$.

If Φ is the Higgs field of an N-monopole, then for large $|x^3|$ it is gauge equivalent to

$$\Phi = \left(1 - \frac{N}{2|x^3|}\right) i\tau_3 + O\left(\frac{1}{|x^3|^2}\right). \tag{8.228}$$

Asymptotically, $i\Phi$ has eigenvalues ± 1, so Hitchin's equation has solutions that are exponentially growing or decaying as $|x^3| \to \infty$. Since (8.227) is a first order equation for a two-component function, the solution space has two complex dimensions, and the generic solution grows exponentially as both $x^3 \to \infty$ and $x^3 \to -\infty$.

It is always possible to find one solution that decays exponentially as $x^3 \to \infty$. This solution is unique up to a multiplicative constant. We may normalize it so that in the gauge (8.228) it has the asymptotic form

$$v(x^3) \sim \binom{1}{0} (x^3)^{N/2} e^{-x^3}. \tag{8.229}$$

The power correction to the exponential is due to the long range $N/(2|x^3|)$ term in (8.228). This same solution, in the same gauge, has the asymptotic form as $x^3 \to -\infty$

$$v(x^3) \sim a \binom{1}{0} |x^3|^{N/2} e^{-x^3} + b \binom{0}{1} |x^3|^{N/2} e^{x^3} \tag{8.230}$$

where a and b are constants. The overall normalization of this solution is not interesting; the important quantity is the ratio b/a. Repeating this analysis for all lines in the x^3 direction we obtain functions $a(z, \bar{z})$ and $b(z, \bar{z})$, and the function of most interest is

$$R(z, \bar{z}) = \frac{b(z, \bar{z})}{a(z, \bar{z})} \tag{8.231}$$

which is independent of normalization, and can, if desired, be defined in a gauge invariant way.

So far we have made no use of the Bogomolny equation, $\frac{1}{2}\varepsilon_{ijk}F_{jk} = D_i\Phi$. But now observe that the operator $D_{\bar{z}} = \frac{1}{2}(D_1 + iD_2)$ commutes with Hitchin's operator $D_3 - i\Phi$ if the Bogomolny equation is satisfied, since

$$[D_{\bar{z}}, D_3 - i\Phi] = \frac{1}{2}(F_{13} + iF_{23} - iD_1\Phi + D_2\Phi) = 0. \tag{8.232}$$

Asymptotically, as $x^3 \to -\infty$, $D_{\bar{z}}$ approaches the operator $\partial_{\bar{z}}$ in the gauge which we have been using. It follows that

$$\partial_{\bar{z}} R = 0 \tag{8.233}$$

so that R is a holomorphic function of z, except possibly for poles where a vanishes. For large $|z|$, Hitchin's equation simplifies as the gauge and Higgs fields approach their vacuum values. b tends to zero as $|z| \to \infty$, so $R \to 0$. A function R with these properties is necessarily a rational function of z,

$$R(z) = \frac{p(z)}{q(z)}, \tag{8.234}$$

although it is not necessarily true that $p = b$ and $q = a$. The boundary condition implies that we can include $z = \infty$ in the domain of the map, and that R is a based map, satisfying $R(\infty) = 0$.

By considering special cases (e.g. the axially symmetric N-monopole), and by a continuity argument, it can be verified that for an N-monopole, R is of degree N. Therefore, q is a polynomial of degree N and p is a polynomial of degree $N-1$ or less. R can be expressed in the normalized form

$$R(z) = \frac{p_1 z^{N-1} + \cdots + p_N}{z^N + q_1 z^{N-1} + \cdots + q_N}. \tag{8.235}$$

The maps (8.235) manifestly have $2N$ complex, or $4N$ real, parameters.

It is helpful to be able to relate the Donaldson map of a monopole to other information about the monopole. First, consider the N roots of the polynomial $q(z)$ (the poles of R). These parametrize the lines in the x^3 direction along which Hitchin's equation has a normalizable solution (one which decays in both directions asymptotically). Hitchin originally defined the spectral curve of a monopole as the set of all lines in $\mathbb{R}^3$ where the equation

$$(D_t - i\Phi)v = 0 \tag{8.236}$$

has a normalizable solution (t is arc length along the line). Such lines are called spectral lines of the monopole. This set of lines is a complex curve in $T\mathbb{CP}^1$, in fact, the spectral curve. Later, it was established that the spectral curve defined via the Nahm data (8.198) is the same. The Donaldson rational map determines the spectral lines in the x^3 direction, that is, in the direction $\xi = 0$. The polynomial $P(0, \eta)$, obtained by setting $\xi = 0$ in the spectral curve equation, is therefore the same as $q(\eta)$ (up to an arbitrary constant multiple). Similarly, the spectral curve determines the denominator of the Donaldson rational map for lines in any direction, but there is no simple algorithm for determining the numerator.

We know, heuristically, that spectral lines pass through the central region of monopoles, near where the energy is concentrated. If the N roots of q are distinct and well separated (separation $\gg 1$), then the N-monopole consists of N unit charge monopoles whose centres are (approximately) on the spectral lines in the x^3 direction labelled by these roots. It is also possible to establish where the monopole centres are along the lines, in terms of the rational map. Let the roots of $q(z)$ be $\{z_i : 1 \leq i \leq N\}$. Then the ith monopole is located at [187]

$$(x_i^1 + ix_i^2, x_i^3) = \left(z_i, \frac{1}{2} \log |p(z_i)|\right) \tag{8.237}$$

and it is consistent to say that the phase of the ith monopole is given by $\varphi_i = \frac{1}{2} \arg p(z_i)$. Note that $p(z_i) \neq 0$, since p and q have no common factors.

For the complementary case of monopoles strung out in well separated clusters along (or nearly along) the x^3-axis, the large z expansion of the rational map $R(z)$ is [17, 195]

$$R(z) \sim \frac{e^{2x+i\beta}}{z^L} + \frac{e^{2y+i\gamma}}{z^{2L+M}} + \cdots \tag{8.238}$$

where L is the charge of the topmost cluster with x its elevation above the (x^1, x^2) plane and M is the charge of the next highest cluster with elevation y.

Although the formula (8.237) is only valid for well separated monopoles, it suggests a definition for the centre of mass of an N-monopole, $\mathbf{X}$, and for its overall phase, χ, namely

$$(X^1 + iX^2, X^3) = \frac{1}{N} \left(\sum_{i=1}^{N} z_i, \frac{1}{2} \sum_{i=1}^{N} \log |p(z_i)|\right) \tag{8.239}$$

$$\chi = \frac{1}{2N} \sum_{i=1}^{N} \arg p(z_i). \tag{8.240}$$

Equivalently $X^1 + iX^2 = -q_1/N$, where q_1 is the coefficient of z^{N-1} in $q(z)$, and

$$X^3 + i\chi = \frac{1}{2N} \log \prod_{i=1}^{N} p(z_i). \tag{8.241}$$

It can be shown that these definitions make sense even if the roots of q are not well separated. The quantity $\prod_{i=1}^{N} p(z_i)$ is another expression for the resultant of the polynomials p and q that was defined by Eq. (6.15), so it has a good limit as roots of q coincide.

A centred N-monopole is one for which $\mathbf{X} = \mathbf{0}$. A strongly centred monopole may be defined as one for which $\chi = 0 \mod \pi/N$ as well. A strongly centred monopole has $q_1 = 0$ and unit resultant. Using twistorial arguments, it can be established that the strongly centred monopoles are a globally well defined subset of monopoles, not depending on the choice of a direction in $\mathbb{R}^3$.

Note that multiplying the rational map of an N-monopole by a constant phase factor $e^{i\alpha}$ ($\neq 1$) gives the rational map of a different monopole, although they differ only in their framing. The resultant of the map changes by a factor $e^{iN\alpha}$, since each number $p(z_i)$ is multiplied by $e^{i\alpha}$. Therefore, multiplying the rational map of a strongly centred N-monopole by $e^{2\pi i/N}$ gives another such monopole, differing only in its framing. This $\mathbb{Z}_N$ action on strongly centred monopoles is topologically significant, as we shall see.

In studying the symmetries of a monopole using its Donaldson map it is necessary to restrict to only those symmetries which preserve the chosen scattering direction – in this case the x^3 direction. The most obvious example is a rotation around the x^3-axis, whose action on the rational map parameter is $z \mapsto e^{i\theta}z$, where θ is the angle of rotation. The rational map (and hence the monopole) is symmetric under such a rotation if only the framing changes. In other words

$$R(e^{i\theta}z) = e^{i\alpha}R(z) \tag{8.242}$$

for some real constant α, depending on θ.

Another symmetry which is compatible with the fixed scattering direction is the reflection $\sigma : (x^1, x^2, x^3) \mapsto (x^1, x^2, -x^3)$. It was shown in [187] that the action of this reflection on the degree N Donaldson map $R = p/q$ is

$$\sigma : \frac{p}{q} \mapsto \frac{\tilde{p}}{q} \tag{8.243}$$

where $\tilde{p}$ is the unique degree $N-1$ polynomial in z such that

$$p\tilde{p} = 1 \quad \mathrm{mod}\ q. \tag{8.244}$$

We shall make use of both the above rotation and reflection symmetries shortly, when we identify various totally geodesic submanifolds as the fixed point sets of certain group actions.

The following are examples of Donaldson rational maps. A 1-monopole has a map

$$R(z) = \frac{p_1}{z + q_1}. \tag{8.245}$$

This is strongly centred if $R(z) = 1/z$. A 2-monopole has a map of the general form

$$R(z) = \frac{p_1 z + p_2}{z^2 + q_1 z + q_2} \tag{8.246}$$

and is strongly centred if $q_1 = 0$ and $(p_1(i\sqrt{q_2})+p_2)(p_1(-i\sqrt{q_2})+p_2) = 1$, which reduces to the condition involving the resultant

$$p_1^2 q_2 + p_2^2 = 1. \tag{8.247}$$

The strongly centred axisymmetric 2-monopole, with its symmetry axis the x^3-axis, has rational map

$$R(z) = \frac{1}{z^2}. \tag{8.248}$$

This map satisfies (8.242) for any angle θ, with $\alpha = -2\theta$. It is also symmetric under the action of the reflection symmetry σ, since $\tilde{p} = p = 1$. Similarly, the map $R(z) = 1/z^N$ is that of a strongly centred axisymmetric N-monopole, with $\alpha = -N\theta$ being the appropriate phase in (8.242). These Donaldson rational maps make it easy to see that axisymmetric N-monopoles exist for all N, and that they are essentially unique. Moreover, for each N, the rational map is about the most elementary function possible. It is extraordinary that the fields of the axisymmetric monopoles, which are difficult to compute, transform into these completely elementary rational maps.

The existence of classes of cyclically symmetric monopoles can also be easily shown using Donaldson maps. Consider the degree N rational maps which are invariant under the cyclic group of rotations about the x^3-axis, C_N. Such maps are of the form

$$R(z) = \frac{az^l}{z^N - b} \tag{8.249}$$

where l is any integer in the range $0 \le l \le N - 1$, and where a and b are complex constants. Strong centring determines a in terms of b (up to some discrete phase choice). The phase of b can be changed by a rotation about the x^3-axis (not in C_N). That leaves $|b|$ as the only interesting parameter.

The monopoles corresponding to maps of the form (8.249) are not quite trivial to describe. For large $|b|$, there are N well separated, unit charge monopoles lying at the vertices of a regular N-gon (in the plane $x^3 = 0$, if the monopole is strongly centred). The relative phases of the monopoles depend on the value of l. If $l = 0$, the strongly centred monopoles have rational maps

$$R(z) = \frac{1}{z^N - b} \tag{8.250}$$

and as $b \to 0$, this tends to the map of the axisymmetric N-monopole. The N-gon of unit charge monopoles contracts onto a toroidal configuration.

If $l \ne 0$, then $b = 0$ is not allowed, as the numerator and denominator of R would have a common factor of z^l. As $b \to 0$, the monopoles (or some subset) must go off to infinity. In fact, as $b \to 0$, the map (8.249) tends, naively, to the form a/z^{N-l}, although not uniformly in z. This is the map of an axisymmetric $(N - l)$-monopole. The coefficient a, determined by b

if the monopole is strongly centred, has a value implying that the $(N-l)$-monopole is moving to infinity along the positive x^3-axis. It can be shown that there is a further axisymmetric l-monopole moving to infinity in the opposite direction. So, as b decreases from infinity to zero, N unit charge monopoles on the vertices of an N-gon approach each other in the (x^1, x^2) plane, they then somehow coalesce, and split up into two approximately axisymmetric clusters moving in opposite directions along the x^3-axis, one of charge $N - l$, the other of charge l.

Details of the fields of these cyclically symmetric monopoles are not known for general N and l, because the Nahm equation, even with cyclic symmetry imposed, is not easily solved. For $N = 3$, with $l = 0, 1, 2$, approximate Nahm data have been obtained and used to numerically calculate the Higgs field and plot the energy density. The cases $l = 1$ and $l = 2$ are related by the reflection $x^3 \mapsto -x^3$, so we consider just the former. Figure 8.9 shows a sequence of energy density isosurfaces as b varies.

The Jarvis rational maps are constructed in a rather similar way to the Donaldson maps. First, choose a point in $\mathbb{R}^3$ – we take this to be the origin, $\mathbf{0}$. It is convenient here to redefine the class of based gauge transformations to be those for which $g(\mathbf{0}) = 1$. That leaves a residual global $SU(2)$ group acting on fields. Now take an N-monopole solution of the Bogomolny equation, and consider Hitchin's equation along each radial line from this point

$$(D_r - i\Phi)v = 0. \tag{8.251}$$

D_r is the radial covariant derivative, and $v(r)$, $0 \le r < \infty$, is a two-component complex function. Select the solution $v(r) = \begin{pmatrix} v_1(r) \\ v_2(r) \end{pmatrix}$ that is exponentially decaying as $r \to \infty$. This is unique up to a multiplicative constant. From the value of this solution at the origin, $\begin{pmatrix} v_1(0) \\ v_2(0) \end{pmatrix}$, we define the ratio

$$R = \frac{v_1(0)}{v_2(0)}. \tag{8.252}$$

Now the lines from the origin are labelled by their direction, a point on the Riemann sphere, labelled as usual by the complex coordinate $z = \tan \frac{\theta}{2} e^{i\varphi}$. By defining R as in (8.252) for all lines we obtain a function $R(z, \bar{z})$.

The definition of R requires no gauge fixing. A gauge transformation $g(\mathbf{x})$ which is smooth in $\mathbb{R}^3$ replaces $\begin{pmatrix} v_1(0) \\ v_2(0) \end{pmatrix}$ by $g_0 \begin{pmatrix} v_1(0) \\ v_2(0) \end{pmatrix}$ where g_0 is the $SU(2)$ gauge transformation matrix at the origin. The same g_0

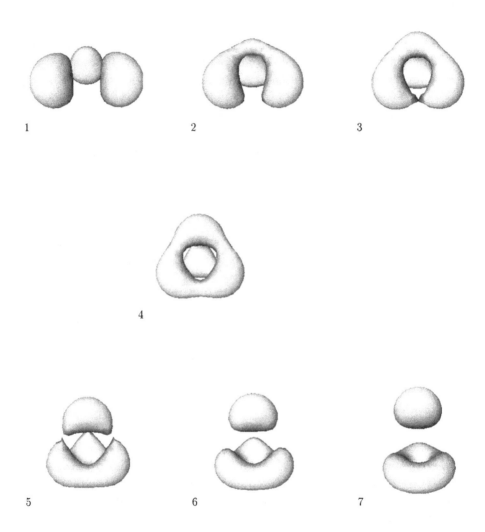

Fig. 8.9. Energy density isosurfaces for a family of $N = 3$ monopoles with cyclic C_3 symmetry.

occurs for all directions z. Thus based gauge transformations have no effect on R, but the residual global $SU(2)$ acts non-trivially by an $SU(2)$ Möbius transformation

$$R \mapsto \frac{\alpha R + \beta}{-\bar{\beta}R + \bar{\alpha}}, \qquad |\alpha|^2 + |\beta|^2 = 1. \tag{8.253}$$

The Bogomolny equation implies that the angular covariant derivative $D_{\bar{z}}$ commutes with Hitchin's radial operator $D_r - i\Phi$. In a gauge where there are no field singularities at the origin, the operator $D_{\bar{z}}$ reduces to $\partial_{\bar{z}}$ as $r \to 0$. It follows that R is independent of $\bar{z}$, and is therefore a holomorphic map $R : S^2 \mapsto S^2$. Note that the direction labelled by $z = \infty$ is not special, and neither is the value $R = \infty$, as this can be changed by a Möbius transformation. The Jarvis map of an N-monopole is therefore an *unbased* rational map of degree N, of the form

$$R(z) = \frac{p(z)}{q(z)}. \tag{8.254}$$

p and q are polynomials of degree no greater than N, with no common factor, and one of them, at least, has degree N.

The space of such maps has real dimension $4N + 2$, and this is the same as the dimension of the moduli spaces of fully framed N-monopoles. Here, fully framed means that we quotient the solution space of the Bogomolny equation by gauge transformations that are the identity at the origin. The three-dimensional group of global gauge transformations $SU(2)/\pm 1$ still acts non-trivially. Jarvis has proved that the space of fully framed monopoles and unbased rational maps are diffeomorphic.

The naturalness of the construction of the Jarvis map implies that any rotational symmetry of a monopole, about the origin, is captured by the map. Conversely, the existence of maps with certain rotational symmetries implies the existence of monopoles with the same symmetries. Therefore, using the Jarvis maps, it is easier than before to establish the existence of monopoles with three-dimensional rotational symmetries.

We have already discussed the action of rotations on the space of rational maps in Chapter 6, but let us recall the main points. An $SO(3)$ rotation is realized by an $SU(2)$ Möbius transformation of z. A rational map $R : S^2 \mapsto S^2$ is invariant under a subgroup $K \subset SO(3)$ if there is a set of Möbius transformation pairs $\{k, M_k\}$ with $k \in K$ acting on the domain S^2 and M_k acting on the target S^2, such that

$$R(k(z)) = M_k R(z). \tag{8.255}$$

The transformation M_k should represent k in the sense that $M_{k_1} M_{k_2} = M_{k_1 k_2}$.

The Jarvis rational map of a centred $N = 1$ monopole, with a standard framing, is $R(z) = z$. This is $SO(3)$-invariant with $M_k = k$, since $R(k(z)) = k(z)$ for any $k \in SO(3)$. Hence the monopole is spherically symmetric.

From Chapter 6 we recall that there is a (unique up to orientation) degree 3 rational map with tetrahedral symmetry, given by

$$R(z) = \frac{\sqrt{3}iz^2 - 1}{z(z^2 - \sqrt{3}i)}.$$

(8.256)

Identifying this as a Jarvis map re-establishes the existence of an $N = 3$ monopole, symmetric under the tetrahedral group T.

Also recall from Chapter 6 the 1-parameter family of degree 4 maps

$$R(z) = c\frac{z^4 + 2\sqrt{3}iz^2 + 1}{z^4 - 2\sqrt{3}iz^2 + 1}$$

(8.257)

where $c \neq 0, \infty$ is a complex constant. These maps are tetrahedrally symmetric, the numerator and denominator being tetrahedral Klein polynomials, whose roots (regarded as points on the Riemann sphere) lie on the vertices of two dual tetrahedra. When $c = 1$, $R(z)$ has the additional 90° rotational symmetry

$$R(iz) = \frac{1}{R(z)},$$

(8.258)

so R has octahedral symmetry in this case. The existence of these Jarvis maps therefore shows that there is a 1-parameter family of tetrahedrally symmetric 4-monopoles (the phase of c only affects the framing), one of which has octahedral symmetry. These monopoles are, of course, just the ones we discussed in Section 8.6.

As another example, we mention that there is essentially a unique icosahedrally symmetric degree 7 rational map. In a certain orientation it has the form

$$R(z) = \frac{z^5 - 3}{z^2(3z^5 + 1)}.$$

(8.259)

The 7-monopole corresponding to this map is the one found by Houghton and Sutcliffe. Further examples of symmetric rational maps are discussed in ref. [193]. From some of these can be inferred the existence of symmetric monopoles which have not been obtained yet in any other way, for example, $N = 11$ and $N = 17$ monopoles with icosahedral symmetry.

To conclude this section we describe the construction of the monopole fields from the Jarvis rational map. The starting point is to write the Bogomolny equation in terms of the spherical coordinates $r, z, \bar{z}$ and observe that a (complex) gauge can always be chosen so that

$$\Phi = -iA_r = -\frac{i}{2}H^{-1}\partial_r H, \quad A_z = H^{-1}\partial_z H, \quad A_{\bar{z}} = 0$$

(8.260)

where $H(r, z, \bar{z}) \in SL(2, \mathbb{C})$ is a hermitian matrix with unit determinant. The Bogomolny equation is then equivalent to the single equation for H

$$\partial_r \left(H^{-1} \partial_r H \right) + \frac{(1 + |z|^2)^2}{r^2} \partial_{\bar{z}} \left(H^{-1} \partial_z H \right) = 0 . \tag{8.261}$$

As we now explain, solutions of this equation are determined by the rational map, which specifies the boundary condition on H as $r \to \infty$. Recall that on the 2-sphere at infinity the Higgs field boundary condition is $\Phi = \Phi^\infty$, where Φ^∞ is in the gauge orbit of $i\tau_3 = \text{diag}(i, -i)$.

A 2×2 hermitian matrix H with unit determinant can always be written in the form

$$H = \exp\left\{ \frac{w}{2} (2P - 1_2) \right\} \tag{8.262}$$

where w is real and P is a 2×2 hermitian projector, that is, $P^\dagger = P = P^2$. A motivation for introducing projectors is that they provide a useful formulation of similar equations that arise in the context of Skyrmions. Examining the boundary condition on the Higgs field and comparing this behaviour with Eq. (8.260) for the Higgs field in terms of H, we find that the leading order behaviour for large r is that the profile function w is independent of the angular coordinates $z, \bar{z}$ and the projector P is a function only of the angular coordinates. We now examine the behaviour of the functions $w(r)$ and $P(z, \bar{z})$, valid for large r, in more detail.

Computing the Higgs field we obtain

$$\Phi = -\frac{i}{2} H^{-1} \partial_r H = -\frac{i}{4} \frac{dw}{dr} (2P - 1_2) \tag{8.263}$$

with squared magnitude

$$|\Phi|^2 = \frac{1}{16} \left(\frac{dw}{dr} \right)^2 = 1 + O\left(\frac{1}{r} \right) . \tag{8.264}$$

Integrating this equation for w we obtain

$$w(r) = -4r + O(\log r) . \tag{8.265}$$

On substituting the form (8.262) into Eq. (8.261) and using the asymptotic expression (8.265) we find that there is a growing term in Eq. (8.261) in a large r expansion which is proportional to $P \partial_{\bar{z}} \partial_z P + \partial_{\bar{z}} P \partial_z P$, or equivalently $\partial_{\bar{z}}(P \partial_z P)$. Since this term must vanish, the projector P must satisfy

$$\partial_{\bar{z}}(P \partial_z P) = 0 . \tag{8.266}$$

It can be proved that all solutions of (8.266) are of the form

$$P = \frac{\mathbf{f} \, \mathbf{f}^\dagger}{|\mathbf{f}|^2} \tag{8.267}$$

where $\mathbf{f}$ is a two-component column vector whose entries are holomorphic functions of z. Note that multiplication of $\mathbf{f}$ by an overall factor does not change P, so that $\mathbf{f}$ is an element of $\mathbb{CP}^1$, and we may write $\mathbf{f}(z) = (1, R(z))^{\mathrm{t}}$, where $R(z)$ is a rational map, in fact, the Jarvis map.

Substituting the asymptotic behaviour (8.265) into Eq. (8.263) we obtain the expression for the Higgs field on the 2-sphere at infinity

$$\Phi^\infty = i(2P - 1_2) = \frac{i}{1 + |R|^2} \begin{pmatrix} 1 - |R|^2 & 2\bar{R} \\ 2R & |R|^2 - 1 \end{pmatrix} . \qquad (8.268)$$

The monopole charge, N, is the winding number of this map, which is equal to the degree of the rational map $R(z)$. Thus we conclude that the boundary condition on H is determined in this simple and explicit way [207] in terms of the degree N rational map $R(z)$.

Note that (8.268) gives us an explicit expression for the Higgs field at infinity in terms of the rational map. Naively, one may think that this does not contain very much information, since for example it is always possible to choose a (singular) gauge in which the Higgs field at infinity is diagonal and constant. However, the important point is that our expression is given in an explicit *known* gauge, and therefore we have removed the gauge freedom and are left with the physical information in the Higgs field – and the fact that it is rational.

We still need to prove the equivalence between the rational map $\mathbf{f} = (1, R(z))^{\mathrm{t}}$ and the map $R(z)$ introduced earlier as the scattering data associated with the solution of Hitchin's equation along radial lines. We do this now.

In a unitary gauge there is a basis of solutions to Hitchin's equation (8.227) which have the leading order, large r behaviour

$$v(r) \sim e^{-\lambda_j r} u_j \qquad (8.269)$$

where λ_j is an eigenvalue of $-i\Phi^\infty$ and u_j is the corresponding eigenvector. Of course, $\lambda_1 = -\lambda_2 = 1$, and the scattering map is determined by the decaying solution, or more fundamentally by the solution associated with the $\lambda_1 = 1$ eigenspace. Recall that the scattering map is obtained by evaluating this solution at the origin $r = 0$. Now, in the gauge (8.260), Hitchin's equation is trivialized to $\partial_r v = 0$, so the solutions are r-independent and hence the scattering map is the eigenvector of $-i\Phi^\infty$ with eigenvalue 1. Thus all that remains to be shown is that $\mathbf{f}$ is the eigenvector of $-i\Phi^\infty$ with eigenvalue 1. Using the explicit expression (8.268) and the definition of the projector (8.267), this is elementary, as

$$-i\Phi^\infty \mathbf{f} = (2P - 1_2)\mathbf{f} = \left(\frac{2\mathbf{f}\,\mathbf{f}^\dagger}{|\mathbf{f}|^2} - 1_2 \right) \mathbf{f} = \mathbf{f} . \qquad (8.270)$$

The construction of a monopole from its rational map is now clear. Choose a rational map $R(z)$ and then compute the solution of Eq. (8.261) satisfying the boundary condition that for large r

$$H \sim \exp\left\{ \frac{2r}{1+|R|^2} \left(\begin{matrix} |R|^2 - 1 & -2\bar{R} \\ -2R & 1 - |R|^2 \end{matrix} \right) \right\}. \tag{8.271}$$

Obviously this construction is not easy to implement explicitly in practice, since it still requires the solution of a nonlinear partial differential equation. In this sense it is not as powerful as, say, the Nahm construction. The advantage is that for the rational map construction the data are free, in that any rational map is allowed, whereas in the Nahm construction the Nahm data must satisfy complicated constraints, making it difficult to find explicit Nahm data. There is always an inherent difficulty associated with solving the Bogomolny equation and the difference between these alternative constructions is whether the main difficulty resides in performing the construction or specifying the data upon which the construction is performed.

There are simplifying special cases for which we are able to perform the construction explicitly, the easiest example being the rational map $R = z$, which corresponds to the spherically symmetric $N = 1$ monopole. In this case the asymptotic dependence, $w(r)$ and $P(z, \bar{z})$, is valid for all r and substituting (8.262) into (8.261) gives the following ordinary differential equation for the profile function

$$\frac{d^2 w}{dr^2} + \frac{2}{r^2}(1 - e^w) = 0. \tag{8.272}$$

The large r behaviour $w(r) \sim -4r$, together with the condition $w(0) = 0$, which is required for H to be well defined at the origin, determines the unique solution of (8.272) to be

$$w(r) = 2 \log \left(\frac{2r}{\sinh 2r} \right). \tag{8.273}$$

This gives the $N = 1$ monopole fields. Note that there is no freedom in the profile function once the rational map has been specified.

Given any rational map $R(z)$, the solution of Eq. (8.261) satisfying the boundary condition (8.271) can be obtained numerically by introducing an auxiliary time variable, t, choosing a fairly arbitrary initial H, and then solving the gradient flow equation

$$H^{-1} \partial_t H = \partial_r \left(H^{-1} \partial_r H \right) + \frac{(1 + |z|^2)^2}{r^2} \partial_{\bar{z}} \left(H^{-1} \partial_z H \right). \tag{8.274}$$

As $t \to \infty$, the solution converges to a static, that is t-independent, solution of the original elliptic equation (8.261). This approach has been implemented in [208], where further details can be found, together with the results of the algorithm when applied to some of the symmetric rational maps discussed earlier.

8.9 Alternative monopole methods

In previous sections we discussed the Nahm transform in great detail, and also the relationship between monopoles and rational maps. However, historically, the first $N = 2$ monopole solutions were constructed using other sophisticated methods, which we now briefly describe.

One of these alternative approaches is due to Ward. It is known [412] that a holomorphic vector bundle can be associated with a self-dual Yang-Mills field in $\mathbb{R}^4$. The base space of the bundle is the twistor space of all null, self-dual planes in a complexified version of $\mathbb{R}^4$. The bundle is trivialized over two patches of the base space, and is completely determined by a transition function on the overlap between these patches. Ward exploited the observation that BPS monopoles are self-dual Yang-Mills fields which are invariant under translation in the x^4 direction to show that a monopole can also be associated with a holomorphic bundle. Again, the monopole is essentially determined by a transition function on the overlap of two patches. Ward [413] found a transition function for an $N = 2$ monopole with axial symmetry, and was also able to implement the steps required to reconstruct the solution of the Bogomolny equation. Prasad and Rossi [339, 338] extended the construction to obtain the axisymmetric monopole for all N.

Ward's description of this procedure was in terms of bundles over $\mathbb{CP}^3$ which have a special form to obtain the required x^4-independence of the gauge fields. However, as later described by Hitchin [183], the dimensional reduction can be made at the twistor level too, to obtain a direct correspondence between monopoles and bundles over the mini-twistor space $T\mathbb{CP}^1$, which we have described earlier in the section on spectral curves. It is helpful in connecting with other approaches if we adopt this reduced description.

As earlier, on $T\mathbb{CP}^1$ let ξ be the standard inhomogeneous coordinate on the base $\mathbb{CP}^1$ and η the complex fibre coordinate, with these twistor coordinates being related to the space coordinates (x^1, x^2, x^3) via the relation (8.215).

Monopoles correspond to certain rank two vector bundles over $T\mathbb{CP}^1$, which may be characterized by a 2×2 patching matrix which relates the local trivializations over the two patches $U_1 = \{\xi : |\xi| \le 1\}$ and $U_2 = \{\xi : |\xi| \ge 1\}$. For charge N monopoles the patching matrix, F, may be taken to have the Atiyah-Ward form [23]

$$F = \begin{pmatrix} \xi^N & \Gamma \\ 0 & \xi^{-N} \end{pmatrix}.$$ (8.275)

To extract the Higgs and gauge fields from the bundle requires the patching matrix to be split as $F = H_2 H_1^{-1}$ on the overlap $U_1 \cap U_2$, where H_1 and H_2 are regular and holomorphic in the patches U_1 and U_2 respectively.

For a patching matrix of the Atiyah-Ward form this splitting can be done by a contour integral. From the Taylor-Laurent coefficients

$$\Delta_p = \frac{1}{2\pi i} \oint_{|\xi|=1} \Gamma \, \xi^{p-1} \, d\xi$$ (8.276)

the Higgs and gauge fields can be computed. For example, there is the elegant formula [337]

$$|\Phi|^2 = 1 - \frac{1}{4} \nabla^2 \log D$$ (8.277)

where D is the determinant of the $N \times N$ banded matrix with entries

$$D_{pq} = \Delta_{p+q-N-1}, \qquad 1 \le p, q \le N.$$ (8.278)

For charge N monopoles the function Γ in the Atiyah-Ward ansatz (8.275) has the form [420, 96]

$$\Gamma = \frac{\xi^N}{P(\xi, \eta)} \left(e^{(-x^2 - ix^1)2\xi - 2x^3} + (-1)^N e^{(-x^2 + ix^1)2\xi^{-1} + 2x^3} \right)$$ (8.279)

where $P(\xi, \eta) = 0$ is the spectral curve.

As an example, we have seen earlier that the $N = 1$ monopole located at the origin has the spectral curve $P(\xi, \eta) \equiv \eta = 0$. In this case the contour integral (8.276) gives

$$\Delta_0 = \frac{\sinh 2r}{r}.$$ (8.280)

Since $N = 1$, the determinant is $D = \Delta_0$ and (8.277) gives

$$|\Phi|^2 = 1 - \frac{1}{4} \nabla^2 \log \left(\frac{\sinh 2r}{r} \right) = \left(\coth 2r - \frac{1}{2r} \right)^2,$$ (8.281)

reproducing again the Prasad-Sommerfield solution.

Ward's original construction of the axially symmetric $N = 2$ monopole [413] corresponds to the choice

$$P(\xi, \eta) \equiv \eta^2 + \frac{1}{4} \pi^2 \xi^2 = 0$$ (8.282)

which we recognize as the spectral curve of the axially symmetric 2-monopole. The concept of a spectral curve had not yet been introduced at the time of Ward's construction, so he had to derive this particular curve as part of his solution.

Ward [414] was also able to derive the general (up to translation and orientation) $N = 2$ spectral curve (8.205) and reconstruct certain properties of the separated 2-monopole. One weakness of Ward's approach is that the fields reconstructed from the holomorphic bundle formally satisfy the Bogomolny equation, but in general it is not possible to prove that the fields are free from singularities – though in special cases this can be done by a continuity argument. The extension to $N > 2$ is also difficult.

At around the same time that Ward produced his two-monopole solutions using twistor methods, a more traditional integrable systems approach was taken by Forgács, Horváth and Palla [135], and the same results obtained. This method makes use of the fact that the Bogomolny equation (in a suitable formulation) can be written as the compatibility condition of an overdetermined linear system. The linear system can be solved in terms of projectors and the corresponding Higgs and gauge fields extracted. Although the general $N = 1$ and $N = 2$ monopoles have been constructed using this approach, it is again difficult to extend this method to $N > 2$.

8.10 Monopole dynamics

In the previous sections we have discussed at length the static multi-monopole solutions of the Bogomolny equation. It is time to say something about multi-monopole dynamics, and monopole-antimonopole dynamics. We shall concentrate on the BPS limit, $\lambda = 0$, where most is known, and which is also the most interesting case, and make a few remarks later about the general case where the Higgs field is massive.

Fundamentally, the field dynamics is governed by the second order, time dependent field equations (8.48) and (8.49). Certain kinds of initial data can be interpreted as a nonlinear superposition of monopoles and antimonopoles, possibly together with some background radiation. Such an interpretation is more art than science, but it becomes fairly clear-cut for well separated monopoles and antimonopoles. Here, each soliton can be compared with the exact solution of a Lorentz boosted, unit charge monopole or antimonopole, and its position and speed can be determined.

Things are clearest if each monopole core can be surrounded by a ball of radius $R \gg 1$ (the monopole core size is of order 1), and $R \ll s$, where s is the minimal separation between a pair of monopoles or antimonopoles. Furthermore, the field inside the ball is that of a Lorentz boosted monopole modified by an amount of order $1/s$. The field outside the balls is a small modification of the vacuum field, which can be expressed approximately as a linear superposition of the long-range fields of the various monopoles, possibly together with some small amplitude radiation.

In this situation, one may expect that the monopole motion is fairly well defined. Each monopole experiences a force due to the combined effect of the others, and it will accelerate. The fields far from the monopoles obey the linearized version of the field equations. Here, the electromagnetic and scalar fields dominate, as these are long range in the BPS limit. The remaining fields (associated with the massive $W^\pm$ particles) are exponentially small and can be neglected. For slowly moving monopoles, the fields will be a superposition of the quasi-static fields due to the monopoles – these being the instantaneous fields of the monopoles at their current positions – with small corrections due to their motion. In addition there will be some superposed radiation which is, even if not present initially, inevitably produced when monopoles accelerate.

The challenge is to compute the accelerations in terms of the monopole separations and their relative velocities, and hence predict the monopole trajectories. Practical calculations are easiest if the monopole velocities remain modest compared to the speed of light. Since the accelerations are always small for well separated monopoles, this regime is maintained for some time. Much more challenging is to predict, or understand, what happens if monopoles or antimonopoles come close together, either because of the initial conditions or because of the forces acting.

The field equations are at the limit of what can be successfully simulated numerically. Scalar field dynamics in $3 + 1$ dimensions can be simulated, but non-abelian gauge field dynamics involves more degrees of freedom, in addition to other complications, and as far as we know there has been no serious simulation of multi-monopole dynamics in Yang-Mills-Higgs theory. There has been some numerical study of monopole interactions – for example, the force between two monopoles at rest has been estimated by finding the static field that minimizes their energy when their positions are fixed by a Lagrange multiplier constraint [273]. However, most progress has come from analytical work.

The static forces in the BPS limit between well separated monopoles (or a monopole and antimonopole) were calculated in ref. [275]. The asymptotic field of a unit charge monopole at rest is known. There is a magnetic Coulomb field, and a long-range Higgs field

$$\mathbf{b} \sim -\frac{1}{2r^2}\hat{\mathbf{x}}, \qquad |\Phi| \sim 1 - \frac{1}{2r} \tag{8.283}$$

where r is the distance from the monopole. A second monopole, well separated from the first, and at distance s, responds only to these asymptotic fields. In fact, it responds only to the magnetic field and to the gradient of the Higgs field, both of which have magnitude $1/2s^2$. Moreover, near the second monopole the spherical character of the first monopole's fields is irrelevant. It is a sufficient approximation to assume that the second monopole is embedded in a constant magnetic field and a linearly varying Higgs field.

Next one seeks a local solution for the second monopole which has a constant acceleration, and whose asymptotic field is a superposition of the usual field of the second monopole and the additional field due to the first. The acceleration is assumed to be of order $1/s^2$, and the calculation is carried out to this order, with radiation being neglected.

Let us fix the origin to be the centre of the second monopole at the initial time $t = 0$, and suppose that the monopole is initially at rest. If it then accelerates rigidly, with acceleration $\mathbf{a}$, the fields will have the form

$$\Phi(t, \mathbf{x}) = \Phi\left(\mathbf{x} - \frac{1}{2}\mathbf{a}t^2\right) \tag{8.284}$$

$$A_i(t, \mathbf{x}) = A_i\left(\mathbf{x} - \frac{1}{2}\mathbf{a}t^2\right) \tag{8.285}$$

$$A_0(t, \mathbf{x}) = t\mathbf{a} \cdot \mathbf{A}\left(\mathbf{x} - \frac{1}{2}\mathbf{a}t^2\right). \tag{8.286}$$

The first two of these equations are unsurprising. The last results from Lorentz boosting the static field to velocity $t\mathbf{a}$ at time t. This ansatz for the fields is consistent, because, inserted in the field equations, it leads to the static equations

$$D_i D_i \Phi + a_i D_i \Phi = 0, \qquad (D_i + a_i)F_{ij} = -[D_j\Phi, \Phi] \tag{8.287}$$

with corrections of $O(|\mathbf{a}|^2)$ which we neglect. Remarkably, both equations (8.287) are satisfied if the fields satisfy the modified Bogomolny equation

$$B_i + D_i\Phi + a_i\Phi = 0. \tag{8.288}$$

Provided that this equation has a solution representing the second monopole in the background field of the first, one can calculate $\mathbf{a}$. One may assume, as in Section 8.2, that asymptotic fields obey $D_i\hat{\Phi} = 0$, so $B_i = b_i\hat{\Phi}$ and $D_i\Phi = (\partial_i|\Phi|)\hat{\Phi}$. Hence, asymptotically, (8.288) reduces to

$$b_i + \partial_i|\Phi| + a_i = 0 \tag{8.289}$$

(where $|\Phi|$ is approximated by 1 in $a_i\Phi$). The spherically symmetric contribution to $b_i + \partial_i|\Phi|$ from the second monopole vanishes. If the first soliton is a monopole too, then $b_i + \partial_i|\Phi|$ vanishes completely and $\mathbf{a} = \mathbf{0}$. If the first soliton is an antimonopole, then its fields b_i and $\partial_i|\Phi|$ are equal (since $\partial_i |\Phi|$ is unchanged but b_i has the opposite sign). Therefore, for a monopole-antimonopole pair the monopole experiences an acceleration

$$\mathbf{a} = -2\mathbf{b} \tag{8.290}$$

where $\mathbf{b}$ is the magnetic field produced by the antimonopole (which is radially outwards) at the location of the monopole. In terms of the separation s,

$$|\mathbf{a}| = \frac{1}{s^2} \tag{8.291}$$

and the acceleration is towards the antimonopole. Similarly the antimonopole accelerates towards the monopole. Since a monopole has mass 2π, the result (8.291) can be interpreted as an attractive force of magnitude $2\pi/s^2$ between the monopole and antimonopole.

Bak, Lee and Lee [28, 29] have refined this calculation by explicitly solving (8.288) to linear order in $|\mathbf{a}|$. This requires solving a linear equation in the background of the Prasad-Sommerfield monopole. The solution has the asymptotic form that we assumed above in the calculation of the acceleration.

It is not a surprise that for two monopoles, in the BPS limit, there is no acceleration. After all, we know that the usual Bogomolny equation has static solutions representing well separated monopoles. The precise form of (8.291) is more surprising. Since the monopole has charge $g = -2\pi$, we expect a repulsive Coulomb force $g^2/4\pi s^2 = \pi/s^2$ between monopoles, and an attraction of the same magnitude between a monopole and antimonopole. In fact, in the BPS limit, the repulsion is cancelled and the attraction is doubled, because of the long-range nature of the massless Higgs field.

On general grounds, one expects a massless scalar field to produce an attraction between any particles which couple to it. In this sense, scalar interactions are like gravity, which is mediated by a tensor field. (In quantum field theory, the exchange of particles of even spin – spin 0 for a scalar, spin 2 for a graviton – leads to attractive forces.) Our calculation shows that monopoles and antimonopoles experience a scalar attraction of strength π/s^2. It is consistent to say that the monopole and antimonopole both have scalar charges of strength 2π, in addition to their magnetic charges. This charge can be read off from the coefficient of the $1/r$ term in the asymptotic expansion of $|\Phi|$.

Nahm has given a physical reason for the scalar interaction between monopoles [312]. Recall that the Higgs field of one monopole has the hedgehog form (8.56) with the asymptotic behaviour (8.80). When a second monopole or antimonopole is superposed into this field at a separation s from the first, it experiences a Higgs field reduced in magnitude by $1/2s$ from the usual vacuum value. The mass of a Prasad-Sommerfield monopole is normally proportional to the Higgs vacuum value. The second monopole therefore has effectively a reduced mass (and a larger size). This dependence of mass on position leads to forces of the magnitude we have calculated. In fact, the only Lorentz invariant way a particle may interact with a scalar field is through a modification of its mass.

A very interesting observation is that the magnetic and scalar charges determine the interaction of monopoles even when the monopoles are moving. For two or more Prasad-Sommerfield monopoles, the forces do not precisely cancel when the monopoles are in relative motion. We shall show how to calculate these forces in Section 8.12.

Dyons can also be included. By considering the asymptotic Higgs field of a dyon (8.107), one sees that a dyon of magnetic charge $g = -2\pi$ and electric charge q has a scalar charge $(g^2 + q^2)^{1/2}$. Generally, there is a Coulomb force between two dyons. However, two dyons of equal magnetic charge and equal electric charge, and separation s, experience a net force

$$\frac{g^2}{4\pi s^2} + \frac{q^2}{4\pi s^2} - \frac{g^2 + q^2}{4\pi s^2} = 0 \qquad (8.292)$$

where the last term is the scalar contribution, and this is consistent with the existence of exact static solutions for dyons with these charges.

The next two sections of this chapter will be concerned with the motion of several monopoles or dyons, but no antimonopoles, in the BPS limit. We conclude here with some remarks about the dynamics in the monopole-antimonopole sector, where the topological charge is zero, and about monopoles away from the BPS limit.

We have seen that there is always a long-range attraction between a monopole and antimonopole. It would not be surprising if a monopole and antimonopole, released at rest, always annihilated into radiation. Although no simulations of this process have been carried out, this is almost certainly what happens, for generic initial data. But Taubes has proved [398] that there exists at least one static solution of the field equations in this sector. This solution, representing a monopole-antimonopole pair with a special phase relationship, is unstable. It is a saddle point of the energy functional, with one unstable mode. We shall discuss this solution further in Chapter 11.

A scalar charge can be assigned to a monopole away from the BPS limit. Here the Higgs field has a positive mass m_H, and the asymptotic Higgs field of a monopole has a Yukawa behaviour. The coefficient of the Yukawa term $e^{-m_H r}/r$ determines the scalar charge. There is a Yukawa contribution to the force between two monopoles, proportional to the square of this charge, but it has exponential decay, so at sufficiently large monopole separations the purely magnetic Coulomb forces (repulsive for two monopoles, attractive for a monopole-antimonopole pair) dominate. This is true for both quasi-static processes, and in monopole dynamics.

8.11 Moduli spaces and geodesic motion

Multi-monopole dynamics for $SU(2)$ monopoles in the BPS limit was the first example of soliton dynamics to be modelled in terms of geodesics on a moduli space [279]. Subsequently there has been much progress in understanding monopole dynamics this way. The method gives reliable results for the slow motion of monopoles with no restriction on whether they are far apart or close together.

We recall from Chapter 4 the basic principle. The N-monopole solutions of the Bogomolny equation are the minimal energy static fields in the sector of the theory with topological charge N. The monopole fields are identified under based gauge transformations, and are therefore framed. The set of gauge inequivalent, framed monopoles is a $4N$-dimensional manifold, known as the N-monopole moduli space, and denoted $\mathcal{M}_N$. It is natural to include the single framing parameter as one of the moduli space coordinates, as its variation with time has the physical effect of producing net electric charge, and increasing the kinetic energy. The dynamical field equations can be interpreted as Lagrangian motion on the infinite-dimensional field configuration space $\mathcal{C}_N$ for fields of topological charge N. This space has a Riemannian metric and potential energy function, derived from the kinetic and potential parts of the Lagrangian. If the initial motion is tangent to the moduli space, or close to this, with modest kinetic energy, then, for energetic reasons, the subsequent motion remains close to the moduli space. The moduli space has a metric, obtained by restricting the metric on $\mathcal{C}_N$ to its submanifold $\mathcal{M}_N$. For motion on or close to $\mathcal{M}_N$ the metric dominates, as the potential simply constrains the motion to $\mathcal{M}_N$, and is constant when restricted to $\mathcal{M}_N$. The field dynamics is therefore well approximated by geodesic motion on $\mathcal{M}_N$.

Stuart [386] has reformulated this geodesic approximation in a more precise way, and has rigorously proved the validity of the approximation, subject to certain limitations. Stuart's results are rather technical, and we just indicate them here. Stuart argues that any field configuration close to the moduli space has a unique orthogonal projection onto the moduli space. It is therefore characterized by a point in the moduli space plus a deformation vector orthogonal to it. The deformation vector is a superposition of the field modes which can be interpreted as radiation modes in the background of an N-monopole solution of the Bogomolny equation. Stuart assumes that the initial data are a field in $\mathcal{C}_N$ close to the moduli space, with the initial time derivative of the field – the field velocity – being a tangent vector to C_N having at most a small projection orthogonal to the moduli space. That is, most of the kinetic energy is associated with the projection of the field velocity tangent to the moduli space. Stuart supposes that the field velocity is $O(\varepsilon)$, where ε is small, and then proves that for a time of $O(1/\varepsilon)$, the projection of the true field evolution on to the moduli space $\mathcal{M}_N$ is well approximated by the geodesic motion, with errors of order ε. Implied by this is that the orthogonal motion, representing radiation, remains small during this time. A field velocity of $O(\varepsilon)$ corresponds to monopole speeds of the same order (the speed of light is 1).

Note that in a time of order $1/\varepsilon$, the distance travelled in the moduli space is of order 1. Reducing the initial velocity can increase the accuracy of the geodesic approximation, but the approximation has not yet been proved valid over an infinite time. This is not surprising. Two basic types of geodesic motion have been observed on the moduli space. One is a scattering of monopoles, where the time that they are close together is of order $1/v$ (where v is a typical initial monopole speed), and the distance they travel while close together is of order 1. Here we expect only a small amount of radiation to be produced, and the predictions of the geodesic approximation for scattering trajectories should be accurate. If Stuart's results can be extended to infinite time, for motions in which the monopoles are far apart except for a time of order 1, then a proof of the accuracy of the geodesic approximation for monopole scattering could be obtained. Another type of geodesic is a closed or bounded orbit of monopoles. Here the geodesic approximation should fail after a sufficiently long time. There is steady radiation from a bound orbit, so the monopoles slowly lose their energy. The backreaction of the radiation might also destabilize the orbit, so that the monopole trajectories for large times might be very far from the orbit predicted by the geodesic approximation. There are also intermediate geodesics, of the scattering type, but where the monopoles remain close together for arbitrary long times. Here again the geodesic approximation is suspect.

The mathematical theory of the Riemannian geometry of the moduli spaces $\mathcal{M}_N$ is treated in great detail in the book by Atiyah and Hitchin [17]. In particular, they discuss general results that hold for all N. They show that the metric on $\mathcal{M}_1$ is flat, and they present the explicit metric on $\mathcal{M}_2$ and some of its fascinating geodesics. We shall summarize these results but mostly without proofs, and present some more recent results, in particular, some special geodesics for $N > 2$. Some of the latter describe surprising monopole scattering processes.

The basic properties of $\mathcal{M}_N$ and its metric that hold for all N are as follows:

(i) $\mathcal{M}_N$ is a connected and complete Riemannian manifold of dimension $4N$.

(ii) The metric on $\mathcal{M}_N$ is hyperkähler.

(iii) $\mathcal{M}_N$ has a metric decomposition

$$\mathcal{M}_N \simeq \mathbb{R}^3 \times \frac{S^1 \times \widetilde{\mathcal{M}}_N^0}{\mathbb{Z}_N} \qquad (8.293)$$

where the factor $\mathbb{R}^3 \times S^1$ is flat and decouples from $\widetilde{\mathcal{M}}_N^0$, and where $\widetilde{\mathcal{M}}_N^0$ is simply connected and admits an $SO(3)$ isometry group.

Let us make some comments on these. As a differentiable manifold, $\mathcal{M}_N$ is the space of (Donaldson) rational maps. From this, all the topological properties follow. $\mathcal{M}_N$ is naturally complex, with coordinates the coefficients of the polynomials in the numerator and denominator of the rational map. $\mathcal{M}_N$ is topologically complicated, because the coefficients are constrained so that numerator and denominator have no common roots. Thus $\mathcal{M}_N$ is the complement in a linear space of the variety defined by the equation Res $= 0$, where Res is the resultant of the numerator and denominator, given by Eq. (6.15). However, when roots approach coincidence, then at least one monopole moves off to infinity, and this is an infinite distance metrically. So $\mathcal{M}_N$ is geodesically complete. The metric has no singularities while the monopoles are at finite separations.

It can be shown that the metric on $\mathcal{M}_N$ is Kähler with respect to the complex structure associated with the Donaldson maps. It follows that there are three independent complex structures, associated with the three orthogonal directions in $\mathbb{R}^3$. Together, these combine to give a hyperkähler metric on $\mathcal{M}_N$. (It is not known explicitly how the second and third complex structures act on Donaldson maps.)

We have indicated earlier how the decomposition (8.293) arises for Donaldson maps. $\widetilde{\mathcal{M}}_N^0$ is the space of the strongly centred monopoles. $\mathbb{R}^3$ parametrizes the centre of mass coordinate $\mathbf{X}$, and S^1 the total phase χ (with range $0 \leq \chi \leq 2\pi$). The $\mathbb{Z}_N$ quotient occurs because if we take the space $S^1 \times \widetilde{M}_N^0$ to be the space of rational maps

$$e^{i\alpha} R_0(z) \qquad (8.294)$$

where $R_0(z)$ is strongly centred, then such maps are strongly centred whenever α is an integer multiple of $2\pi/N$.

It is not obvious that $\mathcal{M}_N$ should be a metric product compatible with this decomposition. For example, the total inertia (which is the coefficient multiplying the standard metric on $\mathbb{R}^3$) could depend on the relative positions of the monopoles (parametrized by a point in $\widetilde{\mathcal{M}}_N^0$). In fact this does not occur, essentially because of the hyperkähler structure of $\mathcal{M}_N$. A proof is given in [17]. The $\mathbb{R}^3 \times S^1$ factor of $\mathcal{M}_N$ has a fixed, flat metric

$$ds^2 = d\mathbf{X} \cdot d\mathbf{X} + d\chi^2 \tag{8.295}$$

and the metric on $\widetilde{\mathcal{M}}_N^0$ is hyperkähler, orthogonal to the $\mathbb{R}^3 \times S^1$ factor, and independent of $\mathbf{X}$ and χ. The $SO(3)$ action on $\widetilde{\mathcal{M}}_N^0$ is the action of the rotation group in $\mathbb{R}^3$ keeping the centre fixed. $\mathcal{M}_N^0 \cong \widetilde{\mathcal{M}}_N^0 / \mathbb{Z}_N$ is called the reduced N-monopole moduli space, and its first homotopy group $\pi_1(\mathcal{M}_N^0)$ is $\mathbb{Z}_N$. Except in the trivial case $N = 1$, where it is a point, the reduced moduli space has a non-trivial metric. Only for $N = 2$ is the metric known explicitly, though Hitchin [185] has obtained a rather general implicit formula for the Kähler potential (associated to any one of the complex structures) for all N in terms of the Riemann Θ-functions of the spectral curves of the monopoles.

For $N = 1$ monopoles, the flat factor (8.295) is the complete metric on the moduli space. This implies that in the moduli space approximation, the Lagrangian of a single monopole, which is purely kinetic, is

$$L = \pi \dot{\mathbf{X}} \cdot \dot{\mathbf{X}} + \pi \dot{\chi}^2 \,. \tag{8.296}$$

Since the monopole has mass 2π, the first term represents the usual non-relativistic kinetic energy of a particle with velocity $\dot{\mathbf{X}}$. The final term is related to electric charge. If we perform the gauge transformation

$$g(\mathbf{x}) = \exp(\chi \Phi(\mathbf{x})) \tag{8.297}$$

on the Prasad-Sommerfield monopole, we change the framing at infinity by the phase angle χ. If χ is time dependent, and we keep $A_0 = 0$, we obtain a dyon with electric charge $q = -2\pi\dot{\chi}$. (Since $E_i = \dot{A}_i = -\dot{\chi}D_i\Phi = \dot{\chi}B_i$, the electric charge is $\dot{\chi}$ times the magnetic charge.) The expression (8.296) implies that a non-moving dyon has total energy, or mass

$$2\pi + \frac{q^2}{4\pi} \,. \tag{8.298}$$

This is the correct expression if we take the exact formula for the dyon mass $((2\pi)^2 + q^2)^{1/2}$, and expand to quadratic order in q. The geodesic motion for the Lagrangian (8.296) is with $\dot{\mathbf{X}}$ and $\dot{\chi}$ constant. This simply describes a dyon of constant electric charge in uniform motion.

The factor $\mathbb{R}^3 \times S^1$ has a similar interpretation for charge N monopoles. There is a centre of mass $\mathbf{X}$ and an overall phase χ. $\dot{\mathbf{X}}$ is the centre of mass velocity and $\dot{\chi}$ determines the total electric charge. Both of these are conserved. The geodesic motion on $\mathcal{M}_N^0$ decouples from this, and

describes the relative motion of the monopoles and the time evolution
of their relative electric charges, although the latter only have a precise
meaning when the monopoles are well separated.

Note that the geodesic approximation is only correct up to quadratic
order in velocities. It ignores the changes in the monopole shape due to
Lorentz contraction and to the acquisition of electric charge. This suggests
that the geodesic approximation is a non-relativistic one, requiring both
$v \ll 1$ and $q \ll g$. This is true, but the monopole velocities can be a
substantial fraction of the speed of light, up to of order one half, before
radiative effects become substantial. Evidence for this is presented in
ref. [293].

The metric on the reduced two-monopole moduli space $\mathcal{M}_2^0$ was determined by Atiyah and Hitchin, as follows. They argued that because of
the $SO(3)$ symmetry, and because of the reflection symmetries of two-
monopole solutions, the metric has the form

$$ds^2 = f^2(r)\, dr^2 + a^2(r)\, \sigma_1^2 + b^2(r)\, \sigma_2^2 + c^2(r)\, \sigma_3^2. \tag{8.299}$$

σ_1, σ_2 and σ_3 are standard 1-forms on $SO(3)$. In terms of Euler angles,

$$\sigma_1 = -\sin\psi\, d\theta + \cos\psi \sin\theta\, d\varphi \tag{8.300}$$
$$\sigma_2 = \cos\psi\, d\theta + \sin\psi \sin\theta\, d\varphi \tag{8.301}$$
$$\sigma_3 = d\psi + \cos\theta\, d\varphi. \tag{8.302}$$

These angles determine the orientation of the monopole pair, and lie in
the ranges $0 \le \theta \le \pi$, $0 \le \varphi \le 2\pi$, $0 \le \psi \le 2\pi$. r is a measure of
the separation of the monopoles. There is considerable freedom in the
choice of this radial coordinate, and this can be used to fix $f(r)$. Atiyah
and Hitchin [17] chose $f = abc$. Following Gibbons and Manton [148],
we choose $f = -b/r$, as this gives a more intuitive version of the metric
when the monopoles are well separated.

Atiyah and Hitchin next showed that because of the hyperkähler property of the metric, one has

$$\frac{2bc}{f}\frac{da}{dr} = (b - c)^2 - a^2 \tag{8.303}$$

$$\frac{2ca}{f}\frac{db}{dr} = (c - a)^2 - b^2 \tag{8.304}$$

$$\frac{2ab}{f}\frac{dc}{dr} = (a - b)^2 - c^2. \tag{8.305}$$

Remarkably, these equations can be solved explicitly in terms of elliptic
integrals. Let K_k denote the complete elliptic integral of the first kind, as
defined by (8.164), and write the elliptic modulus as $k = \sin(\gamma/2)$. Next

relate r to γ by the definition $r = K_k$, with r taking values in the range $\frac{1}{2}\pi \leq r < \infty$, so that $\gamma \in [0, \pi)$. Then, with our choice of f, the solution is given by

$$a = \sqrt{w_2 w_3 / w_1}, \quad b = \sqrt{w_3 w_1 / w_2}, \quad c = -\sqrt{w_1 w_2 / w_3} \qquad (8.306)$$

where

$$w_1 = -\frac{dr}{d\gamma} r \sin\gamma - \frac{1}{2} r^2 (1 + \cos\gamma) \qquad (8.307)$$

$$w_2 = -\frac{dr}{d\gamma} r \sin\gamma \qquad (8.308)$$

$$w_3 = -\frac{dr}{d\gamma} r \sin\gamma + \frac{1}{2} r^2 (1 - \cos\gamma). \qquad (8.309)$$

The functions $a(r), b(r), -c(r)$ are plotted in Fig. 8.10. The crucial fact in proving that the above is indeed a solution is the differential equation satisfied by K_k, which reads

$$\frac{d}{dk}\left(k(1 - k^2)\frac{dK_k}{dk}\right) = k K_k. \qquad (8.310)$$

The geodesic motion on the Atiyah-Hitchin manifold is not integrable; nevertheless quite a lot is known about it. Before discussing this, it helps

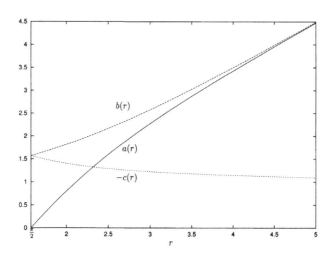

Fig. 8.10. The functions $a(r), b(r), -c(r)$ which arise in the $N = 2$ monopole metric.

to have a picture of the geometrical meaning of quantities occurring in the metric. Figure 8.11 shows a typical centred two-monopole configuration.

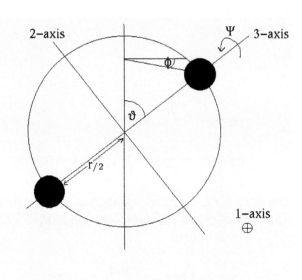

Fig. 8.11. A sketch of a centred $N = 2$ monopole configuration indicating the meaning of the parameters r, θ, φ, ψ that occur in the metric.

It is characterized by three orthogonal, unoriented lines passing through the centre. The configuration is invariant under 180° rotations about the three axes. r determines the monopole separation. When r is large, the individual monopoles are approximately spherical, and have separation r. When $r = \frac{1}{2}\pi$, the minimal value, the monopoles coalesce into a toroidal configuration, with the 1-axis being the axis of symmetry. The 1-, 2- and 3-axes can be thought of as body-fixed axes. They are also the principal axes of inertia, and $a^2(r), b^2(r)$ and $c^2(r)$ are, respectively, the moments of inertia for rotations about these axes. Our choice of Euler angles is such that (θ, φ) specifies the direction in space of the 3-axis, and ψ specifies the orientation of the 1- and 2-axes in the orthogonal plane.

When r is large, $a^2 \sim b^2 \sim r^2$. These are the moments of inertia we expect for the motion of point-like monopoles. Rotation about the 3-axis, for large r, does not move the monopole locations, although there is kinetic energy, since $c^2 \sim 1$. The interpretation is that this motion changes the relative phase of the monopoles, so that they become dyons with opposite electric charges. When $r = \frac{1}{2}\pi$, the monopole configuration is a torus, invariant about the 1-axis; the moments of inertia b^2 and c^2 are equal, because of the extra symmetry, and $a^2 = 0$ because a rotation about the 1-axis has no effect on the fields (at most, it produces a gauge transformation, but there is no associated kinetic energy).

The generic orbit of $SO(3)$ in the Atiyah-Hitchin manifold is given by $SO(3)/(\mathbb{Z}_2 \times \mathbb{Z}_2)$, which has quite a large first homotopy group. But note that there is no singularity at $r = \frac{1}{2}\pi$. Here $a = 0$, and $c = -b$, and the orbit of $SO(3)$ is two-dimensional. The submanifold of $\mathcal{M}_2^0$ at $r = \frac{1}{2}\pi$ is $S^2/\mathbb{Z}^2$, i.e. a copy of $\mathbb{RP}_2$, and $\pi_1(\mathcal{M}_2^0)$ is this $\mathbb{Z}_2$. $\mathbb{RP}_2$ is the space of (strongly centred) axisymmetric 2-monopoles, being the manifold parametrizing the directions of an unoriented axis of symmetry. (The structure of $\mathcal{M}_2^0$ in the neighbourhood of $r = \frac{1}{2}\pi$ is best seen after a change of coordinates; see [148].) Rotations by 180° about each of the principal axes are candidates for the single non-trivial element of $\pi_1(\mathcal{M}_2^0)$. However, one of these rotations (the rotation about the 1-axis) has no effect as $r \to \frac{1}{2}\pi$, and is therefore a contractible loop. The other two rotations become topologically equivalent because of this, and are non-contractible.

As $r \to \infty$, the monopoles become well separated and the Atiyah-Hitchin metric simplifies. Ignoring terms that decay exponentially with r, the asymptotic form of the metric is

$$ds^2 = \left(1 + \frac{m}{r}\right)(dr^2 + r^2\,d\theta^2 + r^2\sin^2\theta\,d\varphi^2) + \left(1 + \frac{m}{r}\right)^{-1}(d\psi + \cos\theta\,d\varphi)^2 \tag{8.311}$$

where $m = -1$. This is a version of the Taub-NUT metric but with a negative value for the mass parameter m. For positive m the metric (8.311) is regular everywhere, including $r = 0$, whereas for $m = -1$ it has singularities, and even changes signature at $r = 1$. However, the singularities are irrelevant in this application, as only the region $r \gg 1$ of this Taub-NUT metric has anything to do with monopoles.

The metric (8.311) has an additional $SO(2)$ symmetry, not possessed by the Atiyah-Hitchin metric. This implies that the geodesic motion of well separated monopoles has, asymptotically, an additional constant of motion, which is not even approximately conserved in a close collision of monopoles. This constant is the difference between the electric charges of the monopoles.

The geodesic motion on the modified Taub-NUT space (8.311) is integrable, and described in detail in ref. [148]. The equations of motion are a variant of the Coulomb problem, and geodesics correspond to trajectories in $\mathbb{R}^3$ describing the relative motion of point-like monopoles or dyons. There are both bound orbits (ellipses) and unbound orbits (hyperbolae). These conics are generally in planes that do not pass through the origin.

Therefore, unlike two electric charges in the Coulomb problem, two dyons move on conics which are not in the same plane. Note that this motion is different from that for pure dyons, which just have electric and magnetic charges. There, each orbit is on a cone but not in any fixed plane. The difference can be ascribed to the scalar part of the interaction between BPS dyons.

To investigate general geodesic motion on the Atiyah-Hitchin manifold, it is convenient, as for Eulerian rigid body motion, to introduce the body-fixed angular velocity components l_m, corresponding to the 1-forms σ_m,

$$l_1 = -\sin\psi\,\dot\theta + \cos\psi\sin\theta\,\dot\varphi \tag{8.312}$$

etc. The Lagrangian for motion on $\mathcal{M}_2^0$ becomes

$$L = \frac{1}{2}\pi\left(f^2(r)\dot r^2 + a^2(r)l_1^2 + b^2(r)l_2^2 + c^2(r)l_3^2\right), \tag{8.313}$$

whose equations of motion give geodesic motion at constant speed on $\mathcal{M}_2^0$. Let us introduce the (scaled) body-fixed angular momenta

$$M_1 = a^2 l_1, \quad M_2 = b^2 l_2, \quad M_3 = c^2 l_3. \tag{8.314}$$

The variational equations obtained from L are

$$\frac{dM_1}{dt} = \left(\frac{1}{b^2} - \frac{1}{c^2}\right)M_2 M_3 \tag{8.315}$$

$$\frac{dM_2}{dt} = \left(\frac{1}{c^2} - \frac{1}{a^2}\right)M_3 M_1 \tag{8.316}$$

$$\frac{dM_3}{dt} = \left(\frac{1}{a^2} - \frac{1}{b^2}\right)M_1 M_2 \tag{8.317}$$

$$f\frac{d}{dt}\left(f\frac{dr}{dt}\right) = \frac{1}{a^3}\frac{da}{dr}M_1^2 + \frac{1}{b^3}\frac{db}{dr}M_2^2 + \frac{1}{c^3}\frac{dc}{dr}M_3^2. \tag{8.318}$$

As for a rigid body, one should first solve these generalized Euler equations, and subsequently find the motion in space (solving for θ, φ and ψ in terms of r, M_1, M_2 and M_3). The only obvious constants of motion are the energy

$$E = \frac{1}{2}\pi\left(f^2\dot r^2 + \frac{M_1^2}{a^2} + \frac{M_2^2}{b^2} + \frac{M_3^2}{c^2}\right) \tag{8.319}$$

and the total angular momentum

$$M_{\text{tot}}^2 = M_1^2 + M_2^2 + M_3^2. \tag{8.320}$$

Equations (8.315)–(8.318) are not integrable. Temple-Raston [399] made a numerical study of solutions with a range of initial data, and, by using a Poincaré return map on a constant energy hypersurface, found evidence for chaotic behaviour. Wojtkowski [430] treated the equations for large r as a small perturbation of the integrable dynamics on Taub-NUT, and showed, using KAM theory, that although most of the tori of the Taub-NUT dynamics would be destroyed, many bounded orbits would remain. Quite what the correct global picture of the dynamics of these equations is, remains to be clarified. Since an energy hypersurface is three-dimensional, any surviving torus may separate two parts of this surface, preventing Arnold diffusion.

While the general geodesic motion is rather complicated, Atiyah and Hitchin showed that very interesting motion occurs when one considers simpler solutions. As for a rigid body, the equations simplify if the motion is with one of the principal axes fixed. In such a motion, two of the quantities M_1, M_2 and M_3 are zero, and the third is constant. Clearly, equations (8.315)–(8.317) are satisfied, and (8.318) is solved by quadrature. Even the case $M_1 = M_2 = M_3 = 0$ is interesting.

The example with M_1 constant and $M_2 = M_3 = 0$ is the simplest case. Suppose the 1-axis is fixed to point along the x^1-axis in space. Then the monopoles move in the (x^2, x^3) plane, and they repel if $M_1 \neq 0$ because da/dr is positive. Geometrically, the geodesic motion is restricted to a two-dimensional submanifold in $\mathcal{M}_2^0$, the Atiyah-Hitchin rounded cone, sketched in Fig. 8.12, which is a surface of revolution with metric

$$ds^2 = f^2(r)dr^2 + a^2(r)d\widetilde{\psi}^2 \, . \tag{8.321}$$

Here $\widetilde{\psi}$ is a new Euler angle, which measures the rotations about the 1-axis. A crucial point is that the range of $\widetilde{\psi}$ is $0 \leq \widetilde{\psi} \leq \pi$, because a rotation about the 1-axis by 180° brings the field configuration back to its starting configuration, and this is a contractible loop.

The vertex of the Atiyah-Hitchin cone, at $r = \frac{1}{2}\pi$, is a smooth point, corresponding to the monopole configuration with axial symmetry about the x^1-axis. It can be verified that the metric near this point is of the form

$$d\widetilde{r}^2 + 4\widetilde{r}^2 d\widetilde{\psi}^2 \tag{8.322}$$

where $\widetilde{r}$ is the proper distance from the vertex. The factor 4 compensates for the range of $\widetilde{\psi}$ being π.

The simplest motion of all on the Atiyah-Hitchin cone is with $M_1 = 0$, so there is no rotation at all, and only r varies with time. This describes a head-on collision of monopoles. The geodesic passes straight over the top of the cone, as shown in Fig. 8.12.

Note that there is a sudden jump in $\widetilde{\psi}$ as the geodesic passes through the vertex (analogous to the jump in polar coordinates for a straight line motion in $\mathbb{R}^3$ which passes through the origin). Since the full range of $\widetilde{\psi}$

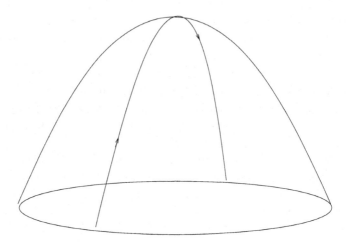

Fig. 8.12.   The Atiyah-Hitchin cone and a geodesic associated with right-angle scattering of two monopoles.

is π, the jump in $\tilde{\psi}$ is $\frac{1}{2}\pi$. So the outgoing monopoles are moving along a line at right angles to the line of the incoming monopoles. This right-angle scattering of monopoles, first predicted by Atiyah and Hitchin, is displayed in Fig. 8.13. It is a direct consequence of the geometry of the Atiyah-Hitchin manifold. A second geodesic surface of revolution is found by holding the 3-axis fixed (say along the x^3-axis in space), so M_3 is constant and $M_1 = M_2 = 0$. The metric is

$$ds^2 = f^2(r)dr^2 + c^2(r)d\psi^2 \qquad (8.323)$$

where ψ is the Euler angle already defined, with the range $0 \le \psi \le \pi$. The surface defined by (8.323) is approximately a cylinder, since $|c|$ has only a small variation with r. But notice that this surface is geodesically incomplete, because it has a boundary at $r = \frac{1}{2}\pi$, consisting of a circle of circumference $\frac{1}{2}\pi^2$ (since $|c(\frac{1}{2}\pi)| = \frac{1}{2}\pi$). This apparent problem is resolved, as Atiyah and Hitchin pointed out, by considering the third surface, in which the 2-axis is fixed in space (also along the x^3-axis). This surface has metric

$$ds^2 = f^2(r)dr^2 + b^2(r)d\psi^2 . \qquad (8.324)$$

It is a surface of revolution which broadens out to infinite width as $r \to \infty$, but it is also geodesically incomplete, having a boundary circle of circumference $\frac{1}{2}\pi^2$ too (since $b(\frac{1}{2}\pi) = \frac{1}{2}\pi$). These two surfaces in fact smoothly join together to form a single surface of revolution inside $\mathcal{M}_2^0$

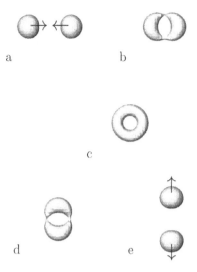

Fig. 8.13. Energy density isosurfaces illustrating the right-angle scattering of two monopoles.

which is a geodesic submanifold and geodesically complete. This surface is known as the Atiyah-Hitchin trumpet, and is sketched in Fig. 8.14.

To clarify how this surface arises, consider the field configurations which occur during the right-angle scattering of monopoles from the x^3-axis to the x^2-axis. (The principal body-axes are always along the spatial Cartesian axes, but before the scattering it is the 3-axis that is along the x^3-axis, and afterwards it is the 2-axis.) Now consider all the configurations obtained by rotating these about the x^3-axis in space. This gives the surface of revolution discussed above, with ψ the rotation angle. It contains as generating curves all the geodesics where monopoles approach along the x^3-axis and scatter at right angles into the (x^1, x^2) plane. The surface contains a circle of axisymmetric monopole configurations, where the axis of symmetry lies in the (x^1, x^2) plane.

Geodesics are of two kinds; they either pass through the trumpet from one end to the other, or they begin and end at the diverging end of the trumpet. The former behaviour occurs if $|M_2| \leq \sqrt{2E/\pi}$, the latter if $|M_2| > \sqrt{2E/\pi}$. Let us describe the physical interpretation of these two kinds of geodesic, assuming in both cases that the geodesic enters from the wider end of the trumpet. Asymptotically, we have two monopoles moving in the (x^1, x^2) plane and oriented so that the body 2-axis is along the spatial x^3-axis. If their speeds are v, and the impact parameter is h,

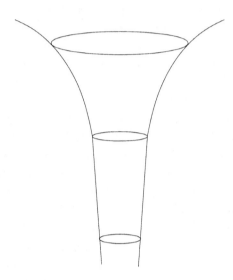

Fig. 8.14.   The Atiyah-Hitchin trumpet.

then $E = 2\pi v^2$ and $M_2 = 2hv$. If $h > 1$, then the monopoles approach but they can not pass through the trumpet, and they emerge, with some scattering angle, in the (x^1, x^2) plane. If $h < 1$, then the geodesic passes through the trumpet. The monopoles emerge along the spatial x^3-axis, and the rotation about this axis implies that they have become dyons with opposite electric charges. The dyons are moving back-to-back but the angular momentum of the initial motion has been conserved, since the dyon pair possesses angular momentum associated with the electric-magnetic interaction. This is a remarkable, truly three-dimensional soliton motion.

Note that if $|M_2|$ is slightly greater than $\sqrt{2E/\pi}$ then the monopoles almost turn into a dyon pair. The monopoles in the (x^1, x^2) plane turn into dyons moving along the x^3-axis, but these dyons attract, so they eventually turn round, and convert back to monopoles in the (x^1, x^2) plane. The scattering process therefore can take an arbitrarily long time, and the scattering angle becomes infinitely sensitive to the precise value of M_2 as the critical value is approached.

The Atiyah-Hitchin cone and trumpet have a rather simple description in terms of the Donaldson rational maps, although the metric information is hidden. Consider the rational maps

$$R(z) = \frac{1}{z^2 - \alpha}. \tag{8.325}$$

These are strongly centred, and invariant under the cyclic group C_2, generated by $z \mapsto -z$. A subset of the monopole moduli space defined by imposing a symmetry is automatically a geodesic submanifold, so the set of monopoles with the rational maps (8.325) is a geodesic surface, parametrized by the complex number α. The maps (8.325) correspond to monopoles lying on an Atiyah-Hitchin cone. The 1-axis is lined up with the x^3-axis in space here. We see this because when $\alpha = 0$, which corresponds to the vertex of the cone, the configuration is axially symmetric about the x^3-axis. $|\alpha|$ is a measure of the monopole separation ($2|\alpha|^{1/2}$ is in fact the separation of the spectral lines in the x^3 direction), and arg α can be identified with $2\widetilde{\psi}$.

Imposing the reflection symmetry $x^2 \mapsto -x^2$ forces α to be real, and defines a geodesic on the cone. This geodesic is the right-angle scattering of monopoles in a head-on collision. When α is large and positive, we see, considering the zeros of $z^2 - \alpha$, that the monopoles lie on the x^1-axis; when α is large and negative, they are on the x^2-axis.

The other C_2-symmetric, strongly centred 2-monopoles have rational maps

$$R(z) = \frac{z/\sqrt{\beta}}{z^2 - \beta}. \tag{8.326}$$

Here β may not vanish. The range of arg β may be taken to be 2π. The change of sign of R as arg β increases by 2π is just a reframing of the monopole. (One could make the range 4π by going to $\widetilde{\mathcal{M}}_2^0$, the double cover of $\mathcal{M}_2^0$.) For large β, there are two monopoles in the (x^1, x^2) plane; as $\beta \to 0$, there are two monopoles on the x^3-axis, as we explained in Section 8.8. The critical circle of values $|\beta| = \frac{1}{16}\pi^2$ is when the configuration is axially symmetric about some axis in the (x^1, x^2) plane. (This axial symmetry is not obvious from (8.326), but we know it from the spectral curve of the axially symmetric 2-monopole (8.206), which tells us that the pair of spectral lines in any direction orthogonal to the symmetry axis is separated by $\frac{1}{2}\pi$.) The rational maps (8.326) therefore parametrize monopole configurations lying on an Atiyah-Hitchin trumpet, the narrow end corresponding to small β. The orientation is the same as we considered before, with the 3-axis (or 2-axis) along the spatial x^3-axis.

One other special geodesic is known on $\mathcal{M}_2^0$. It is a bounded geodesic, and was discovered by Bates and Montgomery [38]. Note from Fig. 8.10 that at one value of r ($\simeq 2.3$), $|a| = |c|$. If r is independent of time, and has this special value, then M_2 is constant, and $M_1^2 + M_3^2$ is constant too. In fact, from (8.319), $M_1 = M_{\text{tot}} \cos \omega t$ and $M_3 = M_{\text{tot}} \sin \omega t$, where ω is related to M_2. Now it can also be verified that for suitable M_{tot}, the right-hand side of (8.318) vanishes, so it is consistent for r to be constant. In this way we obtain a periodic solution of (8.315)–(8.318). It is a solution in which the configuration precesses steadily in space about a fixed line. If the monopoles are thought of as well separated (actually they are not) they are moving on two circles, parallel to a fixed plane, as shown in Fig. 8.15.

Fig. 8.15.   The bound orbit configuration of two monopoles, with the axis of rotation.

There may be other bounded geodesics on $\mathcal{M}_2^0$. This is suggested by the work of Temple-Raston and Wojtkowski, but the situation needs to be clarified. Note that the $\mathbb{RP}_2$ of axisymmetric monopoles is not a geodesic submanifold of $\mathcal{M}_2^0$, so a great circle motion on this $\mathbb{RP}_2$ is *not* a geodesic.

The geodesic motion of more than two monopoles is not understood in any generality, as no explicit expressions for the metric on the moduli spaces $\mathcal{M}_N$ are known, for $N > 2$. However, there are various special kinds of monopole motion, often of a rather symmetric kind, which are understood.

A class of examples comes by considering the Donaldson rational maps for N-monopoles, with cyclic symmetry C_N about the x^3-axis [187]. These rational maps, being defined by imposing a symmetry, give a geodesic submanifold of $\mathcal{M}_N$. Recall from (8.249) that such maps are of the form

$$R(z) = \frac{\alpha z^l}{z^N - \beta} \tag{8.327}$$

with l any fixed integer between 0 and $N - 1$.

We can restrict to maps that are strongly centred, and get a geodesic submanifold of $\widetilde{\mathcal{M}}_N$. Strong centring determines α in terms of β. In this way, we find a totally geodesic surface in $\widetilde{\mathcal{M}}_N$, which we denote Σ_N^l. It is a surface of revolution, with β as a complex coordinate on it. Rotations about the x^3-axis rotate the argument of β.

The first non-trivial cases are with $N = 2$. As we mentioned above, we obtain the Atiyah-Hitchin rounded cone and the Atiyah-Hitchin trumpet by picking $l = 0$ and $l = 1$, respectively. For $N > 2$ and $l = 0$, we obtain surfaces Σ_N^0 which are analogues of the Atiyah-Hitchin cone. The point $\beta = 0$ is included in each of these surfaces, so they are simply connected. For $N > 2$ and $l \neq 0$, we obtain surfaces which are analogous to the Atiyah-Hitchin trumpet. Here $\beta = 0$ is excluded (because there would otherwise be a common factor z^l in numerator and denominator of R), so the surfaces are not simply connected.

A generating geodesic on each of these surfaces is obtained by imposing a reflection symmetry, $x^2 \mapsto -x^2$, which restricts β to be real. On Σ_N^0, there is a geodesic where β runs along the real axis from ∞ to $-\infty$. This corresponds to π/N scattering of monopoles in the (x^1, x^2) plane. N unit charge monopoles approach each other on the vertices of a contracting N-gon, coalesce instantaneously into a toroidal configuration (when $\beta = 0$), and recede on the vertices of an expanding N-gon rotated by π/N relative to the first.

On Σ_N^l, for $l \neq 0$, there is a geodesic where β runs along the real axis from ∞ to 0. The incoming monopoles are on the vertices of an N-gon in the (x^1, x^2) plane as before, but their relative phases are different, and depend on l. As we explained in Section 8.8, the outgoing configuration (as $\beta \to 0$) is of two approximately axisymmetric monopole clusters, of charges l and $N - l$ respectively, moving in opposite directions along the x^3-axis. This truly three-dimensional motion of monopoles has no analogue in any two-dimensional soliton scattering process. Replacing l by $N - l$ makes essentially no difference, being equivalent to the reflection $x^3 \mapsto -x^3$. So for $N = 4$, for example, this analysis establishes three types of C_4-symmetric monopole scattering. In each case the incoming monopoles are on the vertices of a contracting square. There is planar scattering for $l = 0$. For $l = 1$ there is an outgoing toroidal 3-monopole and an outgoing single monopole. Finally, for $l = 2$ there is reflection symmetry under $x^3 \mapsto -x^3$, and there are two outgoing toroidal 2-monopoles, moving oppositely along the x^3-axis. This is displayed in Fig. 8.16.

It has not yet been possible to compute the actual metrics on most of the surfaces of revolution Σ_N^l. The precise speed at which the monopole scattering processes we have just described occur, is not known. Without knowing the metric, it is also not possible to describe the general geodesics on Σ_N^l quantitatively. We may state, however, that for motion on Σ_N^0, the incoming and outgoing monopoles are not electrically charged. A general geodesic on Σ_N^l (for $l \neq 0$), which passes from one end of the trumpet to the other, will have incoming monopoles with net orbital angular momentum but no electric charge (the phases are non-trivial but they are time independent), whereas the outgoing toroidal clusters will be

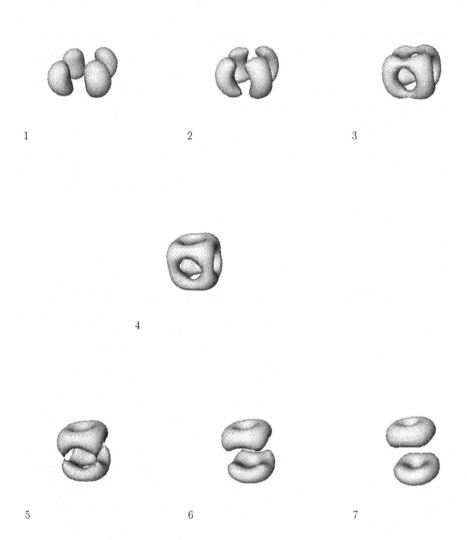

Fig. 8.16.   Energy density isosurfaces for a family of $N = 4$ monopoles with cyclic C_4 symmetry.

oppositely electrically charged (the phase of β rotates steadily as $\beta \to 0$). Orbital angular momentum is converted into the electromagnetic angular momentum of the dyonic clusters which are moving back-to-back.

The spectral curves associated with each of the submanifolds Σ_N^l, when reduced by the action of the cyclic symmetry, lead to curves which have genus greater than 1 (for $N > 2$) so the Nahm data can not be obtained in terms of elliptic functions. The Nahm equation in this case is equivalent [392] to the equations of the periodic Toda chain (a well known integrable system). Formally, the solution can be expressed in terms of a theta function but explicit information about the associated period matrix is not known and this makes it impossible to impose the required boundary conditions. This formal solution is therefore not useful and gives no information about the associated monopoles. However, it is possible [393] to obtain a good approximation to this Nahm data (in terms of elementary functions) which is sufficiently accurate to be used in a numerical computation of the energy density. This is how Fig. 8.9 was produced, which we can now interpret as three-monopole scattering via the geodesic approximation.

There is another family of geodesics that has been found by considering cyclically symmetric Donaldson maps [196]. These maps are of the form

$$R(z) = \frac{\alpha z^l + 1}{z^N} \qquad (8.328)$$

where $N/2 < l \le N - 1$. They are invariant under the combination of a rotation by π/l about the x^3-axis followed by the reflection $x^3 \mapsto -x^3$. Note that this symmetry implies a cyclic C_l symmetry, whose generator is a double application of the above transformation. As before, α parametrizes a geodesic surface of revolution, and the simplest geodesic has α real and running from $-\infty$ to ∞. Because of the nature of the symmetry, this is called twisted line scattering of monopoles. All geodesics in this class describe monopoles which scatter along a line. The initial configuration is of two charge $(N - l)$ monopoles symmetrically approaching a charge $(2l - N)$ monopole at the origin along the positive and negative x^3-axis. All the monopoles merge into the axially symmetric charge N monopole when $\alpha = 0$, and the outgoing configuration is obtained from the incoming one by a rotation by π/l about the x^3-axis. For example, when $N = 3$, the only possibility is $l = 2$, so the incoming and outgoing configuration is of two monopoles moving along the x^3-axis, with a single monopole at the origin. The monopoles coalesce, instantaneously forming the axially symmetric 3-monopole when $\alpha = 0$ (see Fig. 8.17). Note that the geodesic passes twice through tetrahedrally symmetric 3-monopole configurations, with one tetrahedron being the dual of the other. The Nahm data are explicitly known for this family of $N = 3$ monopoles [196].

All the known Platonic monopoles can similarly be found as members of twisted line scatterings [196], but only for $N = 3$ are the Nahm data known for the whole 1-parameter family.

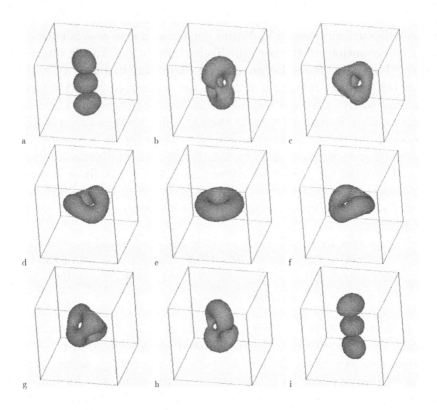

Fig. 8.17. Energy density isosurfaces for a family of $N = 3$ monopoles with twisted line symmetry.

It is possible to find further examples of geodesic motion of monopoles by using the Jarvis rational maps. For example, we explained in Section 8.8 that the maps (8.257) describe 4-monopole configurations with tetrahedral symmetry – in fact, all of them, for the given orientation. This set is therefore a geodesic submanifold of $\mathcal{M}_4$. We can fix the phase by requiring the parameter c to be real, and deduce that if c runs from 0 to ∞ then the corresponding monopole motion is a geodesic motion. Figure 8.6 shows how the motion proceeds.

In this example, the metric along the geodesic is precisely known [66], so the time evolution of the fields along the geodesic is determined if we

specify the initial velocity. The metric has not been computed either
from the rational maps, or from the monopole fields corresponding to
these maps. The calculation is based on Nahm data. As we mentioned
at the end of Section 8.6, the Nahm data for tetrahedrally symmetric 4-
monopoles are known in terms of elliptic functions. The Nahm data also
depend on the parameter c, so let us write them as $T_i(s; c)$. The natural
formula for the metric on the space of Nahm data here simplifies to the
one-dimensional form $g(c) \, dc^2$ where

$$g(c) = -\Omega \int_0^2 \sum_{i=1}^3 \text{Tr} \left(\frac{dT_i}{dc} \frac{dT_i}{dc} \right) ds \qquad (8.329)$$

and Ω is a constant normalization factor. That this is the correct metric relies on the remarkable theorem of Nakajima [314], who has proved
generally that the natural metric on the moduli space of Nahm data,
for $SU(2)$ N-monopoles, is isometric with the metric on $\mathcal{M}_N$ which we
defined earlier, and have been using.

In [66], the integral (8.329) is explicitly performed, leading to an expression for $g(c)$ in terms of elliptic integrals. This expression is rather complicated, but it is smooth, and it has a simple asymptotic form. Changing
to the physical coordinates representing the monopole positions is helpful.
If the monopole positions are $\frac{1}{\sqrt{8}}(\pm r, \pm r, \pm r)$ with an even (alternatively,
odd) number of + signs, so the separation of each pair is r, and if r is
large, then the metric is

$$3 \left(1 - \frac{2}{r} - 288 r e^{-2r} \right) dr^2 \qquad (8.330)$$

with higher order exponentially small corrections.

Another interesting submanifold of $\mathcal{M}_N^0$ (for all $N > 2$) on which the
metric is explicitly known is the four-dimensional Atiyah-Hitchin submanifold [53, 197]. The corresponding Nahm data were discussed earlier, and
are given by (8.168). Recall that they involve the same functions as in the
$N = 2$ case, and for this reason the metric is just a constant multiple of
the Atiyah-Hitchin one. As we mentioned earlier, this submanifold is associated with a string of N monopoles equally spaced along a line. There
is an $SO(3)$ action on the submanifold, and the one remaining non-trivial
parameter is the inter-monopole distance.

8.12 Well separated monopoles

The complete metric on the N-monopole moduli space $\mathcal{M}_N$ can not yet
be calculated in detail. Only a few geodesic submanifolds have been

identified. However, the metric simplifies if all N monopoles are well separated in space.

We saw in Section 8.11 how the Atiyah-Hitchin metric on the two-monopole moduli space simplifies to Taub-NUT as the separation increases. The Taub-NUT metric was rederived by Manton [281] by treating the monopoles (or rather, dyons) as point particles, interacting through their magnetic, electric and scalar charges. This calculation was generalized by Gibbons and Manton to N monopoles [149]. The minimal separation between any pair of monopoles was assumed to be much greater than unity. The result is a hyperkähler metric on a $4N$-dimensional manifold. Gibbons and Manton were unable to prove rigorously that this metric is the asymptotic form of the metric on $\mathcal{M}_N$. However, physically it is very plausible. More recently, Bielawski [54] has investigated the Nahm data for N well separated monopoles, and defined new Nahm data which satisfy the Nahm equation with modified boundary conditions. The metric on the moduli space of this modified Nahm data is automatically hyperkähler, and is precisely the Gibbons-Manton metric. Since the new data differ from the true data for N-monopoles by an amount that goes exponentially fast to zero as the monopoles separate, Bielawski concludes that the Gibbons-Manton metric is the asymptotic form of the metric on $\mathcal{M}_N$, and differs from it by an amount which is exponentially small in the monopole separations.

We now present the calculation of the asymptotic metric. We recall that a BPS dyon can be regarded as a particle with a magnetic charge g, an electric charge q, and a (positive) scalar charge $(g^2 + q^2)^{1/2}$. Its mass M is also $(g^2 + q^2)^{1/2}$. In the normalizations we have chosen for the $SU(2)$ gauge theory we have $g = -2\pi$.

We shall assume that dyon velocities are non-relativistic, and that their electric charges are small relative to $|g|$. We suppose that each dyon is a source for magnetic and electric fields $\mathbf{b}$ and $\mathbf{e}$ which obey the Maxwell equations, and a Lorentz scalar field ϕ obeying the linear, massless wave equation. The magnetic and electric fields are related to the Yang-Mills field strength, and the scalar field is related to the difference between $|\Phi|$ and its vacuum value 1.

The magnetic and electric fields can be expressed locally in terms of a covariant vector and scalar* potential in the usual way

$$-\mathbf{\nabla} \times \mathbf{a} = \mathbf{b} \qquad (8.331)$$

$$-\mathbf{\nabla} a_0 + \dot{\mathbf{a}} = \mathbf{e} \qquad (8.332)$$

because of the Maxwell equations $\mathbf{\nabla} \cdot \mathbf{b} = 0$ and $\mathbf{\nabla} \times \mathbf{e} + \dot{\mathbf{b}} = \mathbf{0}$. Away from the point sources, it follows from the other Maxwell equations, $\mathbf{\nabla} \cdot \mathbf{e} = 0$

---

* scalar here is the conventional terminology but means the time component of a Lorentz 4-vector.

and $\nabla \times \mathbf{b} - \dot{\mathbf{e}} = \mathbf{0}$, that one can introduce dual vector and scalar potentials $\tilde{\mathbf{a}}, \tilde{a}_0$ such that

$$\nabla \times \tilde{\mathbf{a}} = \mathbf{e} \tag{8.333}$$

$$-\nabla \tilde{a}_0 + \dot{\tilde{\mathbf{a}}} = \mathbf{b}. \tag{8.334}$$

These dual potentials $\tilde{\mathbf{a}}$ and $\tilde{a}_0$ are only defined up to a gauge transformation, like $\mathbf{a}$ and a_0.

Suppose now that one of the dyons has trajectory $\mathbf{x}(t)$, and small velocity $\mathbf{v}(t) = \dot{\mathbf{x}}(t)$. The Lorentz scalar field at $\mathbf{x}'$ due to the dyon is

$$\phi = \frac{(g^2 + q^2)^{1/2}}{4\pi s}(1 - v^2)^{1/2} \tag{8.335}$$

where $s = (r^2 - |\mathbf{r} \times \mathbf{v}|^2 + O(v^2))^{1/2}$ and $\mathbf{r} = \mathbf{x}' - \mathbf{x}$. This is the scalar version of a Liénard-Wiechert potential.

Now it will turn out that the leading term in ϕ, namely $|g|/4\pi s$, has no effect. This is because there are no forces between static monopoles. It is therefore a sufficiently good approximation to replace s by r in (8.335) and then to expand to quadratic order in $\mathbf{v}$ and q, which gives

$$\phi = \frac{|g|}{4\pi r}\left(1 + \frac{q^2}{2g^2} - \frac{v^2}{2}\right). \tag{8.336}$$

To write down the vector and scalar potentials, and the dual potentials, we introduce a local vector potential $\mathbf{w}(\mathbf{y})$ for a point Dirac monopole at rest, satisfying

$$-\nabla \times \mathbf{w} = \frac{1}{y^2}\hat{\mathbf{y}} \tag{8.337}$$

and $\mathbf{w}(\mathbf{y}) = \mathbf{w}(-\mathbf{y})$. We do not need to specify the gauge precisely, nor worry about the singularities of $\mathbf{w}(\mathbf{y})$ as we only need to work locally to understand the interactions of the dyons. A dyon at rest would produce the usual electric Coulomb potential, and a Dirac monopole potential. For a dyon in motion, with trajectory $\mathbf{x}(t)$ as before, the potentials and dual potentials at $\mathbf{x}'$, to the accuracy we need, are

$$\mathbf{a} = -\frac{q}{4\pi r}\mathbf{v} + \frac{g}{4\pi}\mathbf{w}$$

$$a_0 = \frac{q}{4\pi r} - \frac{g}{4\pi}\mathbf{v} \cdot \mathbf{w}$$

$$\tilde{\mathbf{a}} = -\frac{g}{4\pi r}\mathbf{v} - \frac{q}{4\pi}\mathbf{w}$$

$$\tilde{a}_0 = \frac{g}{4\pi r} + \frac{q}{4\pi}\mathbf{v} \cdot \mathbf{w}. \tag{8.338}$$

Here $r = |\mathbf{x} - \mathbf{x}'|$ as before, and $\mathbf{w} = \mathbf{w}(\mathbf{x}' - \mathbf{x})$.

Suppose now that a second dyon, with electric charge q' and mass M', moves along the trajectory $\mathbf{x}'(t)$. Its interaction with the first dyon is via the potentials at $\mathbf{x}'(t)$ due to the first, and is described by the Lagrangian

$$L = \left(-M' + (g^2 + q'^2)^{1/2}\phi\right)(1 - v'^2)^{1/2}$$
$$- q'\mathbf{v}' \cdot \mathbf{a} - q'a_0 - g\mathbf{v}' \cdot \tilde{\mathbf{a}} - g\tilde{a}_0 . \tag{8.339}$$

Note that the electric charge is coupled to the usual potentials and the magnetic charge to the dual potentials. This is an ansatz that leads to a generalized Lorentz force law on dyons. The effect of the scalar field ϕ is to modify the effective rest mass of the second dyon, the coupling being proportional to the scalar charge of the second dyon. This is the way that Lorentz scalar fields act on point particles.

If we now substitute the expressions (8.336) and (8.338) for the potentials, set $M' = (g^2 + q'^2)^{1/2}$, and expand out, keeping terms of order v^2, qv and q^2, then L simplifies to

$$L = -M' + \frac{1}{2}M_0 v'^2 - \frac{g^2}{8\pi r}(\mathbf{v}' - \mathbf{v})^2$$
$$- \frac{g}{4\pi}(q' - q)(\mathbf{v}' - \mathbf{v}) \cdot \mathbf{w} + \frac{1}{8\pi r}(q' - q)^2 \tag{8.340}$$

where $M_0 = |g|$ is the monopole mass. The constant term, $-M'$, can now be dropped, as it has no effect on the dynamics.

Note that the interaction terms are symmetric in $\mathbf{v}, q$ and $\mathbf{v}', q'$, so if one just adds the kinetic term $\frac{1}{2}M_0 v^2$, then L becomes a suitable Lagrangian for the dynamics of both the first and second dyon. Extending this to N dyons, of charges $q_1, \ldots, q_N$, with trajectories $\mathbf{x}_i(t)$ and velocities $\mathbf{v}_i = \dot{\mathbf{x}}_i$, the Lagrangian is

$$L = \sum_{i=1}^{N} \frac{1}{2}M_0 \mathbf{v}_i^2 - \frac{g^2}{8\pi} \sum_{1 \leq i < j \leq N} \frac{(\mathbf{v}_j - \mathbf{v}_i)^2}{r_{ji}} \tag{8.341}$$

$$- \frac{g}{4\pi} \sum_{1 \leq i < j \leq N} (q_j - q_i)(\mathbf{v}_j - \mathbf{v}_i) \cdot \mathbf{w}_{ji} + \frac{1}{8\pi} \sum_{1 \leq i < j \leq N} \frac{(q_j - q_i)^2}{r_{ji}} .$$

Here, $r_{ij} = |\mathbf{x}_i - \mathbf{x}_j|$ is the separation between dyons i and j and $\mathbf{w}_{ij} = \mathbf{w}(\mathbf{x}_j - \mathbf{x}_i)$ is the (static) Dirac potential at $\mathbf{x}_j$ due to a source at $\mathbf{x}_i$. The last term is a Coulomb-type potential, but notice that it depends only on the electric charge differences between the dyons, so that if all dyons have the same electric charge, and are at rest, then there is no interaction between them. Note also that the kinetic term can be rewritten as

$$\sum_{i=1}^{N} \frac{1}{2}M_0 \mathbf{v}_i^2 = \frac{1}{2N}M_0(\mathbf{v}_1 + \cdots + \mathbf{v}_N)^2 + \sum_{1 \leq i < j \leq N} \frac{1}{2N}M_0(\mathbf{v}_j - \mathbf{v}_i)^2 \tag{8.342}$$

and since the interaction terms only involve velocity differences, the sum of the dyon velocities decouples. The quantity

$$\mathbf{V} = \frac{1}{N}(\mathbf{v}_1 + \cdots + \mathbf{v}_N) \tag{8.343}$$

is the centre of mass velocity and is conserved.

The Lagrangian (8.341) is defined on the $3N$-dimensional configuration space of the N dyon positions. It is not purely quadratic in velocities, that is, purely kinetic, because of the terms linear in velocity (the electric-magnetic coupling) and because of the Coulomb terms. However, if it were possible to interpret each electric charge as the velocity in an additional, internal one-dimensional space associated with each monopole, then the Lagrangian would be purely kinetic. From the rational map description of well separated monopoles, we have learnt that there is a phase angle associated with each monopole and the time derivative of this phase can in fact be identified with electric charge.

So, we now consider a $4N$-dimensional manifold E_N which is a $\mathbb{T}^N$ (N-torus) bundle over the configuration space of dyon positions, having local coordinates $\{\mathbf{x}_i, \psi_i\}$. ψ_i is an abstract phase angle in the range $0 \leq \psi_i \leq 2\pi$ associated with the ith monopole (which we do not attempt to directly relate either to the rational maps or to the monopole framings). We suppose that E_N is endowed with a $\mathbb{T}^N$-invariant metric. The purely kinetic Lagrangian for motion on E_N then possesses N independent constants of motion which we shall identify with the electric charges of the dyons. We shall also require the remaining equations of motion to be the same as those obtained from the Lagrangian (8.341).

An appropriate ansatz for the Lagrangian on E_N is

$$\mathcal{L} = \frac{1}{2}g_{ij}\mathbf{v}_i \cdot \mathbf{v}_j + \frac{1}{2}h_{ij}(\dot{\psi}_i - \mathbf{W}_{ik} \cdot \mathbf{v}_k)(\dot{\psi}_j - \mathbf{W}_{jl} \cdot \mathbf{v}_l) \tag{8.344}$$

where g_{ij}, h_{ij} and $\mathbf{W}_{ij}$ depend only on the $3N$ coordinates $\{\mathbf{x}_i\}$, and g_{ij} and h_{ij} are symmetric and invertible. Varying with respect to the phases ψ_i, we obtain the N constants of motion

$$q_i = -\kappa h_{ij}(\dot{\psi}_j - \mathbf{W}_{jl} \cdot \mathbf{v}_l) \tag{8.345}$$

which, for a suitable choice of the constant κ, may be identified with the electric charges. Using these constants, we may eliminate the angles from the Lagrangian $\mathcal{L}$ to obtain an effective Lagrangian

$$\mathcal{L}_{\text{eff}} = \frac{1}{2}g_{ij}\mathbf{v}_i \cdot \mathbf{v}_j + \frac{1}{\kappa}q_i\mathbf{W}_{ij} \cdot \mathbf{v}_j - \frac{1}{2\kappa^2}h^{ij}q_iq_j \tag{8.346}$$

where h^{ij} is the inverse of h_{ij}, and q_i are now treated as constant parameters. Note that $\mathcal{L}_{\text{eff}}$ is not simply $\mathcal{L}$ with the $\dot{\psi}_j$ terms replaced by q_j; nevertheless the equations of motion derived from $\mathcal{L}$ and $\mathcal{L}_{\text{eff}}$ are the same.

We now determine g_{ij}, h_{ij} and $\mathbf{W}_{ij}$ by requiring the equations of motion obtained from $\mathcal{L}_{\text{eff}}$ to be the same as those obtained from L (8.341). This means that $\mathcal{L}_{\text{eff}}$ and L should be the same, except possibly for additive or multiplicative constants. Therefore the matrix g_{ij} must be chosen to have components

$$g_{jj} = M_0 - \frac{g^2}{4\pi} \sum_{i \neq j} \frac{1}{r_{ij}} \quad \text{(no sum over } j\text{)} \tag{8.347}$$

$$g_{ij} = \frac{g^2}{4\pi} \frac{1}{r_{ij}} \quad (i \neq j) \tag{8.348}$$

and $\mathbf{W}_{ij}$ must have components

$$\mathbf{W}_{jj} = -\frac{g\kappa}{4\pi} \sum_{i \neq j} \mathbf{w}_{ij} \quad \text{(no sum over } j\text{)} \tag{8.349}$$

$$\mathbf{W}_{ij} = \frac{g\kappa}{4\pi} \mathbf{w}_{ij} \quad (i \neq j). \tag{8.350}$$

The symmetry properties of the Dirac potentials imply that $\mathbf{W}_{ij}$ is a symmetric matrix. Simply identifying $\mathcal{L}_{\text{eff}}$ and L would give a matrix h^{ij} with no inverse. But we may add a constant matrix to h^{ij}, and taking advantage of this, we see that a satisfactory choice is $h^{ij} = \frac{\kappa^2}{g^2} g_{ij}$.

We next fix κ so that the Dirac string singularities in the potential $\mathbf{w}$ do not produce physical singularities in the Lagrangian $\mathcal{L}_{\text{eff}}$. The required value is $\kappa = \frac{4\pi}{g}$. Now, using the values $M_0 = |g| = 2\pi$, and dropping an overall factor of $\frac{1}{2}\pi$, we find that the Lagrangian $\mathcal{L}$ can be interpreted as the purely kinetic Lagrangian for motion on E_N, where the metric is

$$ds^2 = g_{ij} d\mathbf{x}_i \cdot d\mathbf{x}_j + g_{ij}^{-1} (d\psi_i - \mathbf{W}_{ik} \cdot d\mathbf{x}_k)(d\psi_j - \mathbf{W}_{jl} \cdot d\mathbf{x}_l) \tag{8.351}$$

with the slightly modified matrix

$$g_{jj} = 2 - \sum_{i \neq j} \frac{1}{r_{ij}} \quad \text{(no sum over } j\text{)}$$

$$g_{ij} = \frac{1}{r_{ij}} \quad (i \neq j) \tag{8.352}$$

and $\mathbf{W}_{ij}$ as before. This is the Gibbons–Manton metric.

It can easily be verified that the matrices g_{ij} (as immediately above) and $\mathbf{W}_{ij}$ satisfy the equations

$$\frac{\partial}{\partial x_i^a} W_{jk}^b - \frac{\partial}{\partial x_j^b} W_{ik}^a = \varepsilon^{abc} \frac{\partial}{\partial x_i^c} g_{jk} \tag{8.353}$$

$$\frac{\partial}{\partial x_i^a} g_{jk} = \frac{\partial}{\partial x_j^a} g_{ik} \tag{8.354}$$

where the upper indices a, b, c denote the Cartesian components. These conditions were shown by Pedersen and Poon [327], and Papadopoulos and Townsend [323], following earlier work by Hitchin *et al.* [186], to be the conditions for the metric (8.351) to be hyperkähler. The matrices (8.352) give in fact a rather simple and symmetric, non-trivial solution of these conditions.

In summary, we have shown using a physical argument that the dynamics of N well separated BPS dyons is described by geodesic motion on a $4N$-dimensional manifold E_N with metric given by (8.351). This metric is the asymptotic form of the true metric on $\mathcal{M}_N$. The metric is hyperkähler and $\mathbb{T}^N$-invariant. However, it is not complete as there are singularities at finite positive values of r_{ij}, although for well separated monopoles the singularities are not encountered.

In the case of two monopoles, (8.351) is the Taub-NUT metric (times a flat factor), and is the asymptotic form of the Atiyah-Hitchin metric. The difference between the Atiyah-Hitchin and Taub-NUT metrics is exponentially small in the monopole separation. That suggests that for N monopoles the metric (8.351) should agree with the true metric up to exponentially small terms. This is what Bielawski [54] has proved. It is an interesting challenge to understand explicitly the nature of the leading exponential corrections. Physically, they are produced by the short-range, massive gauge fields that are present, predominantly in the monopole cores. Rather remarkably, they can be calculated using ideas from supersymmetric quantum field theory [113].

Since the metric (8.351) is hyperkähler, one might expect to derive it by the hyperkähler quotient construction. This has been done by Gibbons, Rychenkova and Goto [150], starting from a flat $8N$-dimensional space, and imposing a $\mathbb{T}^N$-invariance.

8.13 $SU(m)$ monopoles

So far we have dealt only with the case of $SU(2)$ monopoles. The kind of analysis we have described in this chapter can, of course, be extended to more general gauge groups, where things usually become more complicated. In this section we sketch how the ideas and results are modified for $SU(m)$ gauge groups and discuss some special situations in which the problem simplifies. We only consider monopoles satisfying the Bogomolny equation.

Recall from our general discussion that in a gauge theory where the non-abelian gauge group G is spontaneously broken by the Higgs field Φ

to a residual symmetry group H, then the Higgs field on the 2-sphere at infinity lies in the coset space G/H, so the monopoles have a topological classification determined by the elements of $\pi_2(G/H)$.

For $G = SU(m)$, the boundary conditions at spatial infinity imply that Φ takes values in the gauge orbit of some matrix

$$\Phi_0 = -i \, \mathrm{diag}\,(\mu_1, \mu_2, \ldots, \mu_m)\,. \tag{8.355}$$

By convention, it is assumed that $\mu_1 \leq \mu_2 \leq \cdots \leq \mu_m$ and since Φ is traceless, $\mu_1 + \mu_2 + \cdots + \mu_m = 0$. This Φ_0 is the vacuum expectation value for Φ and the residual symmetry group H is the invariance group of Φ_0 under gauge transformations. Thus, for example, if all the μ_p are distinct then the residual symmetry group is the maximal torus $U(1)^{m-1}$ and this is known as maximal symmetry breaking. In this case, as discussed in Chapter 3,

$$\pi_2 \left(\frac{SU(m)}{U(1)^{m-1}} \right) = \pi_1(U(1)^{m-1}) = \mathbb{Z}^{m-1} \tag{8.356}$$

so the monopoles are topologically characterized by $m - 1$ integers.

In contrast, the minimal symmetry breaking case is where all but the first of the μ_p are identical, so the residual symmetry group is $U(m-1)$. Recall from Chapter 3 that

$$\pi_2 \left(\frac{SU(m)}{U(m-1)} \right) = \mathbb{Z} \tag{8.357}$$

so there is only one topological integer characterizing a monopole. Despite this, a given solution has $m - 1$ integers associated with it, which arise in the following way.

A careful analysis of the boundary conditions [156, 423] indicates that there is a choice of gauge such that the Higgs field for large r, in a given direction, is given by

$$\Phi(r) = -i \, \mathrm{diag}\,(\mu_1, \mu_2, \ldots, \mu_m) - \frac{i}{2r} \, \mathrm{diag}\,(k_1, k_2, \ldots, k_m) + O(r^{-2})\,. \tag{8.358}$$

In the maximal symmetry breaking case the topological charges are given by

$$n_p = \sum_{q=1}^{p} k_q\,, \tag{8.359}$$

for $1 \leq p \leq m - 1$. In the case of minimal symmetry breaking only the first of these numbers, n_1, is a topological charge. Nonetheless, the remaining n_p constitute an integer characterization of a solution, which is gauge invariant up to reordering of the integers k_p. The n_p are known as magnetic weights, with the matrix diag $(k_1, k_2, \ldots, k_m)$ often called the charge matrix and diag $(\mu_1, \mu_2, \ldots, \mu_m)$ the mass matrix.

There are some obvious ways of embedding $su(2)$ in $su(m)$, for example,

$$
\begin{pmatrix} \alpha & \beta \\ -\bar{\beta} & -\alpha \end{pmatrix} \hookrightarrow
\begin{pmatrix} \ddots & & & \\ & \alpha & \cdots & \beta & \\ & \vdots & \ddots & \vdots & \\ & -\bar{\beta} & \cdots & -\alpha & \\ & & & & \ddots \end{pmatrix}.
\tag{8.360}
$$

Important $SU(m)$ monopoles can be produced by embedding the $SU(2)$ charge 1 monopole fields, which are known $su(2)$-valued fields, in $su(m)$. Some care must be taken in constructing these embedded monopoles to ensure that the asymptotic behaviour is correct. The $SU(2)$ monopole may need to be scaled and it may be necessary to add a constant diagonal Higgs field beyond the plain embedding described by (8.360); details can be found in refs. [423, 415]. Obviously there is an embedding of the form (8.360) for each choice of two columns in the target matrix. The embedded 1-monopoles have a single integer $k_p = 1$ and another integer $k_{p'} = -1$, the rest are zero. The choice of columns for the embedding dictates the values for p, p', so there are $m - 1$ different types of fundamental monopole with one of the magnetic weights n_p being unity, and the rest zero, corresponding to the choice $p' = p + 1$.

Recall that in the case of minimal symmetry breaking the choice of order of the k_p is a gauge choice. In fact, in the case of minimal symmetry breaking, the embedded 1-monopole is unique up to position and gauge transformation. Solutions with $n_1 = N$ have N times the energy of this basic solution and so it is reasonable to call these N-monopoles. There are of course different types of such N-monopoles corresponding to different magnetic weights.

For intermediate cases of symmetry breaking the residual symmetry group is $H = U(1)^l \times K$, where K is a rank $m - l - 1$ semi-simple Lie group, the exact form of which depends on how the entries in the mass matrix coincide with each other. Such monopoles have l topological charges.

The various mathematical methods we have described earlier in this chapter can be formulated for the case of general gauge groups. Ward [412] has constructed some explicit $SU(3)$ monopoles via the splitting of appropriate patching matrices over mini-twistor space. The spectral curve approach for maximal symmetry breaking has been formulated by

Hurtubise and Murray [201] and consists of a specification of rank(G) alge-braic curves in mini-twistor space, satisfying reality and non-singularity conditions. For higher rank gauge groups the Donaldson rational map correspondence has been extended by Murray [309] to maps into flag manifolds and a similar extension exists for the rational maps of Jarvis [225].

The Nahm transform for general G is outlined in the original work of Nahm [313] and is discussed further in ref. [201]. Briefly, for $G = SU(m)$, the Nahm data are triples of antihermitian matrix functions (T_1, T_2, T_3) of s over the $m-1$ intervals (μ_p, μ_{p+1}). The size of the matrices depends on the corresponding values of n_p; the matrices (T_1, T_2, T_3) are $n_p \times n_p$ matrices in the interval (μ_p, μ_{p+1}). They are required to be non-singular and to satisfy the Nahm equation (8.113) inside each interval, but there are complicated boundary conditions at the ends of each of the intervals. These boundary conditions are designed so that the Weyl equation (8.142) has the number of solutions required to yield the correct type of monopole fields.

The simplest case is maximal symmetry breaking in an $SU(3)$ theory. There are then two types of fundamental monopole, and the charge is a two-component vector (n_1, n_2). The simplest multi-monopole is of charge $(1,1)$, so $(k_1, k_2, k_3) = (1, 0, -1)$, and its Nahm data were studied by Con-nell [91]. Since there is only one of each type of monopole, the Nahm data are one-dimensional over each of the two intervals, so the Nahm equation is trivially satisfied by constants over each of the two intervals. These two triples of constants determine the positions of the two constituent monopoles and the matching condition at the common boundary of the two intervals determines the relative phase.

The moduli space of these monopoles is eight-dimensional but, as in the $SU(2)$ case, there is an isometric splitting to factor out the position of the centre of mass and the overall phase. The relative moduli space, $\widetilde{\mathcal{M}}^0_{(1,1)}$, is thus four-dimensional. By computing the metric on the space of Nahm data and using a uniqueness argument, Connell was able to show that the metric on $\widetilde{\mathcal{M}}^0_{(1,1)}$ is the Taub-NUT metric with a positive mass parameter. This result was rediscovered some years later [253, 145]. The metric has no singularity and is exact. Recall that asymptotically, the Atiyah-Hitchin metric is also Taub-NUT, but with a negative mass parameter, so the asymptotic metric has a singularity outside its region of validity. This difference in sign results from the fact that in the $SU(3)$ case the two monopoles are electrically charged with respect to different $U(1)$ factors in the residual symmetry group. There is thus conservation of the individual electric charge of each monopole, providing a $U(1)$ symmetry in the metric which is absent in the Atiyah-Hitchin metric, since charge exchange occurs between $SU(2)$ monopoles. This results in a simplified dynamics of charge $(1, 1)$ monopoles, which bounce back off each other in a head-on collision in comparison with the right-angle scattering of $SU(2)$ monopoles.

Similar simplifications can be expected in all cases where there is at most a single monopole of each type. Thus the $4(m-2)$-dimensional relative moduli space $\widetilde{\mathcal{M}}^0_{(1,1,\ldots,1)}$ of charge $(1, 1, \ldots, 1)$ monopoles in an $SU(m)$ theory should be tractable. Indeed, Lee, Weinberg and Yi [254] have computed the asymptotic metric, which is a generalization of Taub-NUT, and conjectured that it is the exact metric. This is supported by a computation of the metric on the space of Nahm data by Murray [310], which gives the same result. Note that this last calculation is not quite a proof, since although it is believed that the transformation between the monopole moduli space metric and the metric on Nahm data is an isometry for all gauge groups and symmetry breaking, it has only been proved for $SU(2)$ monopoles [314] and for special cases of minimally broken $SU(m)$ [395]. These and other monopole metrics have also been obtained by Gibbons, Rychenkova and Goto [150] using the hyperkähler quotient construction.

There is a method which can be used to give a local construction of hyperkähler metrics known as the generalized Legendre transform [186]. This can be used, for example, to give yet another derivation [212] of the Atiyah-Hitchin metric. Using this method, Chalmers [80] was able to rederive the Lee-Weinberg-Yi metric.

In order to examine if there are any other special choices of gauge group, symmetry breaking and monopole charges for which there may be a simplification we need to review a few more details of the Nahm data boundary conditions.

For ease of notation we shall only describe the case where all the integers k_p except for k_1 are negative, and hence $n_p < n_{p-1}$ for $p > 1$, since this will be the case of interest in what follows. Define the function

$$k(s) = \sum_{p=1}^{m} k_p\, \theta(s - \mu_p) \tag{8.361}$$

where $\theta(s)$ is the usual step function. In the interval (μ_p, μ_{p+1}), $k(s) = n_p$, so the graph of $k(s)$ is a rectilinear skyline whose shape depends on the charge matrix of the corresponding monopole. If the graph near μ_p (for $1 < p < m$) is

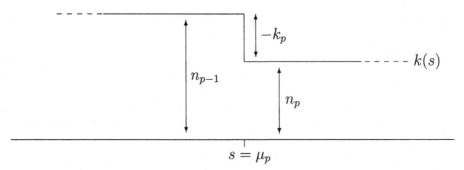

$$s = \mu_p$$

then as s approaches μ_p from below it is required that

$$T_i(s) = \left(\begin{array}{c|c} \frac{1}{t}R_i + O(1) & O(t^{(|k_p|-1)/2}) \\ \hline O(t^{(|k_p|-1)/2}) & T_i' + O(t) \end{array} \right) \begin{array}{c} |k_p| \\ n_p \end{array} \qquad (8.362)$$

where $t = s - \mu_p$ and where T_i' is the limiting value of $T_i(s)$ as s approaches μ_p from above. It follows from the Nahm equation (8.113) that the $|k_p| \times |k_p|$ residue matrices (R_1, R_2, R_3) in (8.362) form a representation of $su(2)$. The boundary conditions require that this representation is the unique irreducible $|k_p|$-dimensional representation of $su(2)$.

In summary, at the boundary between two intervals, if the Nahm matrices are $n_{p-1} \times n_{p-1}$ on the left and $n_p \times n_p$ on the right, an $n_p \times n_p$ block continues through the boundary and there is an $(n_{p-1} - n_p) \times (n_{p-1} - n_p)$ block with a simple pole whose residues form an irreducible representation of $su(2)$.

These conditions now suggest a simplifying case, since if $k_p = -1$ for all $p > 1$ then $k(s)$ is a staircase with each step down of unit height. We shall refer to this situation as the countdown case since for this situation the magnetic weights are given by $(m - 1, m - 2, \ldots, 2, 1)$. Thus, since all the one-dimensional representations of $su(2)$ are trivial, the Nahm data have only one pole, which is at $s = \mu_1$. Taking the limiting case of minimal symmetry breaking, by setting $\mu_1 = -(m - 1)$ and $\mu_2 = \cdots = \mu_m = 1$, we find that the Nahm data are defined on a single interval $[-m + 1, 1]$ with the only pole occurring at the left-hand end of the interval. This is very similar to the Nahm data for $SU(2)$ monopoles, except that the pole at the right-hand end of the interval is lost. This allows a construction of Nahm data for charge $m - 1$ monopoles in a minimally broken $SU(m)$ theory in terms of rescaled Nahm data for $SU(2)$ monopoles, where the rescaling moves the second pole in the Nahm data outside the interval. For convenience we now shift s so that the Nahm data are defined over the interval $[-1, m - 1]$, to agree with our earlier $SU(2)$ notation.

As an illustration, we present the Nahm data for an $SU(m)$ spherically symmetric monopole of charge $m-1$. They are given by $T_i(s) = -\frac{1}{2(s+1)}\rho_i$ where ρ_1, ρ_2, ρ_3 form the standard irreducible representation of $su(2)$ of dimension $m - 1$. The associated spectral curves are simply $\eta^{m-1} = 0$. Spherically symmetric $SU(m)$ monopoles were first studied by Bais and Wilkinson [27], Leznov and Saveliev [264], and Ganoulis, Goddard and Olive [140], using a radial ansatz to reduce the Bogomolny equation to a Toda equation for the radial profile functions.

The simplest countdown example to consider further is the class of $SU(3)$ monopoles of charge 2 with minimal symmetry breaking. For $k_1 = 2$ there are two distinct types corresponding to magnetic weights $(2, 0)$ and $(2, 1)$. (The cases $(2, 2)$ and $(2, 0)$ are equivalent by a reordering of k_2 and k_3.) For weights $(2, 0)$ the monopoles are all embeddings of $SU(2)$ 2-monopoles and this case is not interesting as an example of $SU(3)$ 2-monopoles. For weights $(2, 1)$ this is a countdown case and was first studied by Dancer [99, 98]. Given the comments above, it is fairly clear that the appropriate Nahm data are similar to the $SU(2)$ two-monopole Nahm data (8.155). The functions f_1, f_2, f_3 are almost the same as in the $SU(2)$ case (8.163), except that the scale factor associated with the complete elliptic integral K_k, whose value was chosen to place the second pole at $s = 1$, is now replaced by a parameter D, whose range is such that no second pole occurs in the interval, i.e. $D < \frac{2}{3}K_k$. Explicitly, the Nahm data are

$$T_1(s) = -\frac{iD\,\mathrm{dn}_k(Ds)}{2\,\mathrm{sn}_k(Ds)}\tau_1, \ T_2(s) = -\frac{iD}{2\,\mathrm{sn}_k(Ds)}\tau_2, \ T_3(s) = \frac{iD\,\mathrm{cn}_k(Ds)}{2\,\mathrm{sn}_k(Ds)}\tau_3.$$
$$(8.363)$$

The moduli space of such monopoles has dimension 12, so after centring we are left with an eight-dimensional relative moduli space $\widetilde{\mathcal{M}}^0_{(2,1)}$. There is an isometric $SO(3) \times SU(2)/\mathbb{Z}_2$ action on $\widetilde{\mathcal{M}}^0_{(2,1)}$. The $SU(2)/\mathbb{Z}_2$ action is a gauge transformation on the Nahm matrices, equal to the identity at $s = -1$, while the $SO(3)$ action both rotates the three Nahm matrices as a vector and gauge transforms them [99]. Taking the quotient of $\widetilde{\mathcal{M}}^0_{(2,1)}$ by the $SU(2)/\mathbb{Z}_2$ action gives a five-dimensional manifold D_5 which admits an $SO(3)$ action, since the $SU(2)/\mathbb{Z}_2$ and $SO(3)$ actions on $\widetilde{\mathcal{M}}^0_{(2,1)}$ commute. The Nahm data for D_5 are precisely the orbit under $SO(3)$ of the 2-parameter family of Nahm data (8.363). Using this, Dancer [99] computed an explicit expression for the metric on D_5 and an implicit form for the metric on the whole of $\widetilde{\mathcal{M}}_{(2,1)}$. A more explicit form for the metric on $\widetilde{\mathcal{M}}_{(2,1)}$, in terms of invariant 1-forms corresponding to the two group actions, together with a study of the corresponding asymptotic monopole fields, has been given by Irwin [209].

A totally geodesic two-dimensional submanifold Y of $\widetilde{\mathcal{M}}_{(2,1)}$ is obtained by imposition of a $\mathbb{Z}_2 \times \mathbb{Z}_2$ symmetry, representing monopoles which are symmetric under reflection in all three Cartesian axes. In fact Y consists of six copies of the space $D_5/SO(3)$. This submanifold was introduced by Dancer and Leese and the geodesics and corresponding monopole dynamics investigated [100, 101]. There are two interesting new phenomena which occur. The first is that there can be double scatterings, where the two monopoles scatter at right angles in two orthogonal planes. The second is that there are unusual geodesics where two monopoles approach from infinity but stick together, with the motion taking the configuration asymptotically towards an embedded $SU(2)$ field, which is on the boundary of the $SU(3)$ monopole moduli space and metrically at infinity. This kind of behaviour is still not completely understood but the interpretation is that there is a non-abelian cloud [255, 209], whose radius is related to the parameter D in the Nahm data (8.363). It is the motion of this cloud which carries off the kinetic energy when the monopoles stick. Lee, Weinberg and Yi [255] interpret this cloud as the limit of a charge $(2, 1)$ monopole in a maximally broken theory, in which the mass of the $(\ ,1)$ monopole tends to zero, thereby losing its identity and becoming the cloud.

For the case of charge $(2, 1)$ monopoles in the maximally broken $SU(3)$ theory, Chalmers has conjectured an implicit form for the metric [80]. This uses the generalized Legendre transform technique, modifying the similar construction of the Atiyah-Hitchin metric [212].

Nahm data for other $SU(m)$ countdown examples can be obtained by a modification of $SU(2)$ Nahm data. For example, Platonic $SU(m)$ monopoles can be studied using the $SU(2)$ Nahm data discussed earlier. Again, exotic phenomena are found such as double scatterings and pathological geodesics where the monopoles never separate [198].

8.14 Hyperbolic monopoles

In this brief final section we mention another generalization of monopoles, namely, monopoles in three-dimensional hyperbolic space, $\mathbb{H}^3$.

Hyperbolic space has a constant negative curvature, which we denote by $-\kappa^2$. Perhaps the most familiar description of $\mathbb{H}^3$ is as the interior of the unit 3-ball. In terms of angular coordinates $z, \bar{z}$ and a radial coordinate $\rho \in [0, 1)$ the metric is

$$ds^2 = \frac{4}{\kappa^2(1-\rho^2)^2}\left(d\rho^2 + \rho^2\frac{4dzd\bar{z}}{(1+|z|^2)^2}\right) = dr^2 + \frac{\sinh^2(\kappa r)}{\kappa^2}\frac{4dzd\bar{z}}{(1+|z|^2)^2}$$

$$\text{(8.364)}$$

where we have introduced r, the hyperbolic distance from the origin, through the relation $\rho = \tanh(\kappa r/2)$. From the final expression in (8.364) it is clear that in the zero curvature limit $\kappa \to 0$, the metric becomes that of flat Euclidean space $\mathbb{R}^3$, with $z, \bar{z}$ the standard angular coordinates and r the usual radial coordinate.

The fact that monopoles on hyperbolic space are interesting was first recognized by Atiyah [15], who noted that for special values of the curvature, $-\kappa^2$, hyperbolic monopoles may be interpreted as four-dimensional instantons with a circle symmetry; this is discussed further in Chapter 10.

Given any three-dimensional Riemannian manifold with metric h_{ij}, the Bogomolny equation on this manifold is

$$D_i \Phi = \frac{1}{2} \sqrt{\det h} \, \varepsilon_{ijk} h^{jl} h^{km} F_{lm} \, . \tag{8.365}$$

The Euclidean case, of course, corresponds to $h_{ij} = \delta_{ij}$. The requirement that Eq. (8.365) is integrable becomes the geometrical condition that the manifold has constant curvature. This leaves three possibilities, namely, flat Euclidean space, hyperbolic space, or the 3-sphere. However, the last possibility is not compatible with smooth fields carrying magnetic charge, so hyperbolic space remains as the only natural generalization of the Euclidean case where there are monopole solutions.

There is a correspondence between hyperbolic monopoles and Jarvis rational maps, which closely mirrors the Euclidean case. The rational map is again constructed as the scattering data of Hitchin's equation along radial lines out from the origin, and the only difference is that the metric along this line has now changed, but this is of no real importance. In terms of the spherical coordinates $r, z, \bar{z}$ on $\mathbb{H}^3$, given in (8.364), the Bogomolny equation on hyperbolic space is equivalent to the equation

$$\partial_r \left(H^{-1} \partial_r H \right) + \frac{\kappa^2 (1 + |z|^2)^2}{\sinh^2(\kappa r)} \partial_{\bar{z}} \left(H^{-1} \partial_z H \right) = 0 \tag{8.366}$$

where the equations for the monopole fields in terms of H are still given by the Euclidean expressions (8.260), but with r now being hyperbolic distance. Each solution of this equation is determined by a rational map through the same boundary condition (8.271) as in the Euclidean case,

Some time ago, Atiyah conjectured that in the limit as the curvature of hyperbolic space tends to zero, Euclidean monopoles are recovered, but only recently has this been rigorously established [226]. The proof relies on Eq. (8.366) and the observation that in the zero curvature limit, $\kappa \to 0$, the Euclidean equation (8.261) is recovered.

The simplest way to present the fields of the spherically symmetric hyperbolic 1-monopole is via this rational map formalism. The rational map is again $R = z$ and the solution of (8.366) is

$$H = \exp\left\{ \frac{-w(r)}{2(1+|R|^2)} \begin{pmatrix} |R|^2 - 1 & -2\bar{R} \\ -2R & 1 - |R|^2 \end{pmatrix} \right\} \qquad (8.367)$$

where

$$w(r) = 2\log \frac{(2+\kappa)\sinh(\kappa r)}{\kappa \sinh((2+\kappa)r)} . \qquad (8.368)$$

The solution (8.368) reduces to the Euclidean solution (8.273) as $\kappa \to 0$.

9

Skyrmions

9.1 The Skyrme model

The Skyrme model [377, 379] is a nonlinear theory of pions in three spatial dimensions, with the Skyrme field, $U(t, \mathbf{x})$, being an $SU(2)$-valued scalar. Although not involving quarks, it can be regarded as an approximate, low energy effective theory of QCD, becoming exact as the number of quark colours becomes large [428]. Remarkably, and this was Skyrme's main motivation for constructing and studying this model, it has topological soliton solutions that can be interpreted as baryons. These solitons are called Skyrmions.

The model is defined by the Lagrangian

$$
L = \int \left\{ \frac{F_\pi^2}{16} \mathrm{Tr}(\partial_\mu U \partial^\mu U^\dagger) \right.
$$

$$
\left. + \frac{1}{32e^2} \mathrm{Tr}([\partial_\mu U U^\dagger, \partial_\nu U U^\dagger][\partial^\mu U U^\dagger, \partial^\nu U U^\dagger]) \right\} d^3x, \quad (9.1)
$$

where F_π and e are parameters, whose values are fixed by comparison with experimental data. These parameters can be scaled away by using energy and length units of $F_\pi/4e$ and $2/eF_\pi$ respectively, which we adopt from now on. In terms of these standard units the Skyrme Lagrangian can be written as

$$
L = \int \left\{ -\frac{1}{2} \mathrm{Tr}(R_\mu R^\mu) + \frac{1}{16} \mathrm{Tr}([R_\mu, R_\nu][R^\mu, R^\mu]) \right\} d^3x, \quad (9.2)
$$

where we have introduced the $su(2)$-valued current $R_\mu = (\partial_\mu U)U^\dagger$. The Euler-Lagrange equation which follows from (9.2) is the Skyrme field equation

$$
\partial_\mu \left(R^\mu + \frac{1}{4}[R^\nu, [R_\nu, R^\mu]] \right) = 0, \quad (9.3)
$$

which is a nonlinear wave equation for $U(t, \mathbf{x})$. An interesting feature of (9.3) is that it is in the form of a current conservation equation $\partial_\mu \tilde{R}^\mu = 0$, where $\tilde{R}^\mu = R^\mu + \frac{1}{4}[R^\nu, [R_\nu, R^\mu]]$.

One imposes the boundary condition $U(\mathbf{x}) \rightarrow 1_2$ as $|\mathbf{x}| \rightarrow \infty$. The vacuum, the unique field of minimal energy, is then $U(\mathbf{x}) = 1_2$ for all $\mathbf{x}$.

The Skyrme Lagrangian has an $(SU(2) \times SU(2))/\mathbb{Z}_2 \cong SO(4)$ chiral symmetry corresponding to the transformations $U \mapsto \mathcal{O}_1 U \mathcal{O}_2$, where $\mathcal{O}_1$ and $\mathcal{O}_2$ are constant elements of $SU(2)$. However, the boundary condition $U(\infty) = 1_2$ spontaneously breaks this chiral symmetry to an $SO(3)$ isospin symmetry given by the conjugation

$$U \mapsto \mathcal{O} U \mathcal{O}^\dagger, \qquad \mathcal{O} \in SU(2). \tag{9.4}$$

In order to make explicit the nonlinear pion theory, we write

$$U = \sigma + i\boldsymbol{\pi} \cdot \boldsymbol{\tau}, \tag{9.5}$$

where $\boldsymbol{\tau}$ denotes the triplet of Pauli matrices, $\boldsymbol{\pi} = (\pi_1, \pi_2, \pi_3)$ is the triplet of pion fields and σ is an additional field determined by the pion fields through the constraint $\sigma^2 + \boldsymbol{\pi} \cdot \boldsymbol{\pi} = 1$, which is required since $U \in SU(2)$. Not only the magnitude, but also the sign of σ may be regarded as determined by the requirement of continuity of the field, and the boundary conditions $\boldsymbol{\pi}(\infty) = 0$, $\sigma(\infty) = 1$. In terms of the pion fields, an isospin transformation is $\boldsymbol{\pi} \mapsto M\boldsymbol{\pi}$, where M is the $SO(3)$ matrix corresponding to the $SU(2)$ matrix $\mathcal{O}$,

$$M_{ij} = \frac{1}{2}\text{Tr}(\tau_i \mathcal{O} \tau_j \mathcal{O}^\dagger). \tag{9.6}$$

Pion particles arise from the quantization of small fluctuations of the pion field around the vacuum $\boldsymbol{\pi} = \mathbf{0}$, $\sigma = 1$. Note that substituting (9.5) into the Lagrangian (9.2) reveals that the pions are massless. They are the Goldstone bosons of the spontaneously broken chiral symmetry. An additional term

$$L_{\text{mass}} = m_\pi^2 \int \text{Tr}(U - 1_2) \, d^3x \tag{9.7}$$

can be included in the Lagrangian of the Skyrme model and gives the pions a (tree-level) mass m_π. As most of our discussion is independent of this extra term we do not include it at this stage, but in Section 9.9 we address the modifications that it generates.

If one restricts to static fields, $U(\mathbf{x})$, then the Skyrme energy functional derived from the Lagrangian (9.2) is

$$E = \frac{1}{12\pi^2} \int \left\{ -\frac{1}{2}\text{Tr}(R_i R_i) - \frac{1}{16}\text{Tr}([R_i, R_j][R_i, R_j]) \right\} d^3x, \tag{9.8}$$

where we have introduced the additional factor of $1/12\pi^2$ for later convenience. Static solutions of the Skyrme field equation (9.3) are therefore critical points (either minima or saddle points) of this energy.

At first sight U, at a fixed time, is a map from $\mathbb{R}^3$ into S^3, the group manifold of $SU(2)$. However, the boundary condition implies a one-point compactification of space, so that topologically $U : S^3 \mapsto S^3$, where the domain S^3 is to be identified with $\mathbb{R}^3 \cup \{\infty\}$. As discussed in Chapter 3 the homotopy group $\pi_3(S^3)$ is $\mathbb{Z}$, which implies that maps between 3-spheres fall into homotopy classes indexed by an integer, which we denote by B. This integer is also the degree of the map U and has the explicit representation

$$B = -\frac{1}{24\pi^2} \int \varepsilon_{ijk} \mathrm{Tr}\,(R_i R_j R_k)\,d^3x\,, \qquad (9.9)$$

where $R_i = (\partial_i U)U^\dagger$, as before. As B is a topological invariant, it is conserved under continuous deformations of the field, including time evolution. It is this conserved topological charge which Skyrme identified with baryon number. B is the principal property of a Skyrmion.

The presence of a topological charge is, by itself, not sufficient to ensure the existence of stable topological solitons since we also need to evade Derrick's theorem [107]. But note that the static Skyrme energy decomposes into two components, $E = E_2 + E_4$, corresponding to the terms which are quadratic and quartic in spatial derivatives of the Skyrme field. Under a rescaling of the spatial coordinates $\mathbf{x} \mapsto \mu\mathbf{x}$, the energy becomes

$$e(\mu) = \frac{1}{\mu} E_2 + \mu E_4\,. \qquad (9.10)$$

The two terms therefore scale in opposite ways, leading to a minimal value of $e(\mu)$ for a finite $\mu \neq 0$. This implies that any soliton will have a well defined scale and will neither expand to cover all of space nor contract to be localized at a single point. Note that for any static solution, and in particular for a Skyrmion, which is the minimal energy configuration in a given topological sector, $e(\mu)$ must take its minimal value when $\mu = 1$, so the energy contributions from the quadratic and quartic terms are exactly equal. From this discussion it is clear why the sigma model (the Lagrangian consisting of only the first term in (9.2)) does not support stable solitons. This problem is cured by the addition of the second term in (9.2), known as the Skyrme term. Clearly any term which is of degree 4 or higher in the spatial derivatives would do equally well in this respect, but the Skyrme term is the unique expression of degree 4 which is Lorentz invariant and for which the resulting field equation remains second order in the time derivative.

A more geometrical description of the static Skyrme energy exists [282], which is useful in several contexts. As in nonlinear elasticity theory, the energy density of a Skyrme field depends on the local stretching associated with the map $U : \mathbb{R}^3 \mapsto S^3$. For this formulation, let us introduce the strain tensor D_{ij}, defined at each point $\mathbf{x} \in \mathbb{R}^3$ by

$$D_{ij} = -\frac{1}{2}\text{Tr}(R_i R_j)\,, \qquad (9.11)$$

which is a symmetric, positive definite 3×3 matrix, and which can be thought of as quantifying the deformation induced by the map U. The image under U of an infinitesimal sphere of radius ε and centre $\mathbf{x}$ in $\mathbb{R}^3$, to leading order in ε, is an ellipsoid with principal axes $\varepsilon\lambda_1, \varepsilon\lambda_2, \varepsilon\lambda_3$, where $\lambda_1^2, \lambda_2^2, \lambda_3^2$ are the three non-negative eigenvalues of the matrix D_{ij}. The signs of λ_1, λ_2 and λ_3 are chosen so that $\lambda_1 \lambda_2 \lambda_3$ is positive (negative) if U is locally orientation preserving (reversing). In terms of these eigenvalues, the static energy E, and baryon number B, can be computed as integrals over $\mathbb{R}^3$ of the corresponding densities $\mathcal{E}$ and $\mathcal{B}$ given by

$$\mathcal{E} = \frac{1}{12\pi^2}(\lambda_1^2 + \lambda_2^2 + \lambda_3^2 + \lambda_1^2\lambda_2^2 + \lambda_2^2\lambda_3^2 + \lambda_3^2\lambda_1^2)\,, \quad \mathcal{B} = \frac{1}{2\pi^2}\lambda_1\lambda_2\lambda_3\,. \quad (9.12)$$

From the simple inequality

$$(\lambda_1 \pm \lambda_2\lambda_3)^2 + (\lambda_2 \pm \lambda_3\lambda_1)^2 + (\lambda_3 \pm \lambda_1\lambda_2)^2 \geq 0\,, \qquad (9.13)$$

it follows from the formulae (9.12) that $\mathcal{E} \geq |\mathcal{B}|$ and therefore the Skyrme energy satisfies the Faddeev-Bogomolny lower bound [126]

$$E \geq |B|\,. \qquad (9.14)$$

In contrast to monopoles and vortices, this bound can not be saturated for any non-trivial (i.e. $B \neq 0$) finite energy configuration. This is because the bound is attained only when all the eigenvalues of the strain tensor have modulus 1 at all points in space – an isometry – and this is obviously not possible since $\mathbb{R}^3$ is not isometric to S^3. Note that the bound can be attained if the spatial domain is taken to be the 3-sphere of unit radius; we discuss this further in Section 9.9.

After the baryon number and energy, the most significant characteristic of a static solution of the Skyrme equation is its asymptotic field, which satisfies the linearized form of the equation. To leading order, the three components of the pion field $\boldsymbol{\pi}$ each obey Laplace's equation, and σ can be taken to be unity. More precisely, $\boldsymbol{\pi}$ has a multipole expansion, in which each term is an inverse power of $r = |\mathbf{x}|$, say $r^{-(l+1)}$, times a triplet of angular functions. The leading term, with the smallest l, obeys Laplace's equation, whereas subleading terms may not, because of the nonlinear aspect of the Skyrme equation. For the leading term, therefore, the angular functions are a triplet of linear combinations of the spherical harmonics $Y_{l,m}(\theta, \varphi)$, with m taking integer values in the range $-l \leq m \leq l$. These spherical harmonics can also be expressed in Cartesian coordinates, which often gives more convenient and elegant formulae for the asymptotic fields.

One of the few precise results concerning the Skyrme equation (9.3) is that this multipole expansion can not lead with a monopole term, with $l = 0$. The leading term is a dipole or higher multipole. The proof is as follows [286]. For a static field, the equation implies that the spatial current

$$\tilde{R}_i = R_i + \frac{1}{4}[R_j, [R_j, R_i]] \tag{9.15}$$

has zero divergence and no singularity. Therefore the flux of $\tilde{R}_i$ through a large sphere of radius R (centred at the origin) vanishes, that is,

$$\int_{S_R^2} \tilde{R}_i n^i \, dS = 0, \tag{9.16}$$

where n^i is the unit outward normal. Now, in the asymptotic region, $\tilde{R}_i$ can be replaced by R_i, which in turn simplifies to $i(\partial_i \boldsymbol{\pi}) \cdot \boldsymbol{\tau}$. For a monopole asymptotic field,

$$\boldsymbol{\pi} = \frac{\mathbf{c}}{r} \tag{9.17}$$

where $\mathbf{c}$ is a constant vector, so $\tilde{R}_i$ has the leading asymptotic behaviour $-i\mathbf{c} \cdot \boldsymbol{\tau} x_i / r^3$. Then $\tilde{R}_i n^i = -i\mathbf{c} \cdot \boldsymbol{\tau} / r^2$, so the flux through the sphere is $-4\pi i \mathbf{c} \cdot \boldsymbol{\tau}$. This vanishes only if $\mathbf{c} = \mathbf{0}$.

Recently, it has been rigorously proved [294] that for any non-vacuum solution of the Skyrme equation, the multipole expansion is non-trivial. In other words, the pion field does not vanish to all orders in l, and the leading term is a multipole satisfying the Laplace equation.

9.2 Hedgehogs

Esteban [123] has proved the existence of a $B = 1$ Skyrmion, that is, a minimizer of the energy functional (9.8) within the charge 1 sector, following earlier work of Kapitansky and Ladyzenskaia [230] in which it was proved that a minimizer exists within the family of spherically symmetric charge 1 Skyrme fields. It is believed to be true, though not yet proven, that these two minimizers are the same, that is, the minimal energy Skyrmion in the $B = 1$ sector is spherically symmetric. Here, spherically symmetric does not mean that the Skyrme field is just a function of the radial coordinate r, since it is easily seen that such a field must have $B = 0$. When we refer to a spatial symmetry of a Skyrmion, such as spherical symmetry, we mean that the field has the equivariance property that the effect of a spatial rotation can be compensated by an isospin transformation (9.4). This implies that both the energy density $\mathcal{E}$, and baryon density $\mathcal{B}$, are strictly invariant under the symmetry.

The spherically symmetric $B = 1$ Skyrmion was presented in the original work of Skyrme and takes the hedgehog form (cf. Section 4.3)

$$U(\mathbf{x}) = \exp\{if(r)\hat{\mathbf{x}} \cdot \boldsymbol{\tau}\} \equiv U_H(\mathbf{x}). \tag{9.18}$$

In terms of $\boldsymbol{\pi}$ and σ fields,

$$\boldsymbol{\pi} = \sin f(r)\,\hat{\mathbf{x}}, \quad \sigma = \cos f(r). \tag{9.19}$$

The name hedgehog derives from the fact that the pion fields of this configuration point radially outward from the origin at all points in space, so $\hat{\boldsymbol{\pi}} = \hat{\mathbf{x}}$. f is a real radial profile function with the boundary conditions $f(0) = \pi$ and $f(\infty) = 0$. The latter condition ensures that $U(\infty) = 1_2$, while the former guarantees that $U(0)$ is well defined and that $B = 1$. The value of B is confirmed by substituting the hedgehog ansatz into the expression (9.9) for the baryon number, giving

$$B = -\frac{2}{\pi}\int_0^\infty f' \sin^2 f \, dr = \frac{1}{\pi}f(0) = 1. \tag{9.20}$$

Alternatively, we can easily verify that if f monotonically decreases, then each point of the target space $SU(2)$ (except $U = 1_2$) has exactly one preimage in $\mathbb{R}^3$, with positive Jacobian.

Substituting the hedgehog ansatz (9.18) into the static Skyrme equation yields the second order nonlinear ordinary differential equation

$$(r^2 + 2\sin^2 f)f'' + 2rf' + \sin 2f\left(f'^2 - 1 - \frac{\sin^2 f}{r^2}\right) = 0. \tag{9.21}$$

The solution of this equation, satisfying the boundary conditions, can not be obtained in closed form but it is a simple task to compute it numerically using a shooting method. The numerical solution is presented in Fig. 9.1.

The energy, given by

$$E = \frac{1}{3\pi}\int_0^\infty \left\{ r^2 f'^2 + 2\sin^2 f\,(1 + f'^2) + \frac{\sin^4 f}{r^2} \right\} dr, \tag{9.22}$$

is calculated to be $E = 1.232$, to three decimal places, and so the $B = 1$ Skyrmion exceeds the Faddeev-Bogomolny bound by approximately 23%.

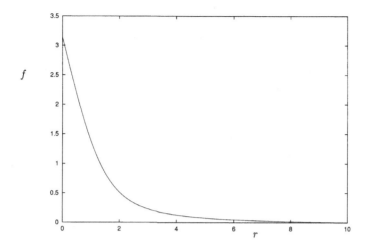

Fig. 9.1. The profile function $f(r)$ for the $B = 1$ Skyrmion.

The Skyrmion described by (9.18) is located at the origin, but it can be positioned at any point in space and given any orientation by acting with the translation and rotation groups of $\mathbb{R}^3$. The moduli space of charge 1 Skyrmions is therefore six-dimensional. In general, it is to be expected that the moduli space of a charge B Skyrmion is nine-dimensional, since in addition to translations and rotations there is also the action of the three-dimensional isospin group (9.4). However, for $B = 1$, an isospin transformation is equivalent to a spatial rotation, which is of course why the Skyrmion is spherically symmetric, so three moduli are lost.

A linearization of Eq. (9.21) reveals the large r asymptotic behaviour of the profile function, $f \sim C/r^2$, for some constant C, which numerically is found to be $C = 2.16$. Therefore, the leading order asymptotic fields are

$$\boldsymbol{\pi} = \frac{C}{r^2}\hat{\mathbf{x}}, \quad \sigma = 1. \tag{9.23}$$

In other words, from far away a single Skyrmion resembles a triplet of orthogonal pion dipoles, with dipole strength $4\pi C$. In Section 9.3 we discuss the asymptotic interactions of well separated Skyrmions, and their interpretation in terms of dipole-dipole forces.

There are further solutions involving the hedgehog ansatz (9.18). Note that U is well defined provided $f(0) = k\pi$, where $k \in \mathbb{Z}$, and a glance at Eq. (9.20) shows that the field in this case describes a spherically symmetric configuration with $B = k$. The pion field still points radially, but inwards or outwards. There appear to be solutions of the equation for the profile function for all values of k [379, 220]. Solutions have been constructed numerically for several values. The $k = -1$ solution is the antiSkyrmion, whose profile function is obtained from that of the Skyrmion by the replacement $f \mapsto -f$. For $|k| > 1$ these hedgehog solutions do not represent the minimal energy Skyrmions with $B = k$, and in fact these solutions are not even bound against break-up into $|k|$ well separated Skyrmions (or antiSkyrmions if $k < 0$). For example, the $k = 2$ hedgehog has an energy $E = 3.67 > 1.232 \times 2$, and has been shown to have six unstable modes. The hedgehog solutions, for $|k| > 1$, are therefore almost certainly all unstable, saddle points of the energy.

A rigorous proof of the existence of charge B, minimal energy Skyrmions with $|B| > 1$ appears to be difficult, and has not yet been found. Their existence has been established by Esteban [123], but only under the assumption that

$$E_B < E_{B'} + E_{B-B'} , \tag{9.24}$$

for all $B' \in \mathbb{Z} - \{0, B\}$, where $E_{B'}$ denotes the infimum of the energy (9.8) within the space of Skyrme fields with baryon number B'. Esteban [123] was able to prove the weaker inequality

$$E_B \leq E_{B'} + E_{B-B'} , \tag{9.25}$$

but the strict inequality is not yet proved in general. The strict inequality would prevent the break-up of a charge B field into infinitely separated clusters of charge B' and $B - B'$, and would imply that the energy E_B was attained by a Skyrmion solution. In the following section we present a physical perspective on these inequalities, in terms of the forces between well separated Skyrmions. Later, we will also describe the solutions that have been discovered numerically, that are believed to be the minimal energy Skyrmions.

9.3 Asymptotic interactions

As noted above, the asymptotic field of a single Skyrmion is that of a triplet of orthogonal dipoles and we can make use of this interpretation to calculate the asymptotic forces between two well separated Skyrmions by computing the interaction energy of the pair of dipole triplets. It is convenient to rewrite (9.23) in the form

$$\pi_j = \frac{C}{r^2} \hat{x}^j = \frac{\mathbf{p}_j \cdot \mathbf{x}}{4\pi r^3} , \tag{9.26}$$

where we have introduced the three orthogonal dipole moments

$$\mathbf{p}_j = 4\pi C \mathbf{e}_j , \tag{9.27}$$

with $\{\mathbf{e}_j\}$ being the standard basis vectors of $\mathbb{R}^3$. More generally, the frame of dipoles may be rotated, but their magnitudes are unchanged. With the energy normalized as in (9.8) the interaction energy of two individual dipoles, with moments $\mathbf{p}, \mathbf{q}$ and relative position vector $\mathbf{X}$, is given by

$$E_{\text{dip}} = \frac{1}{24\pi^3}(\mathbf{p} \cdot \tilde{\partial})(\mathbf{q} \cdot \tilde{\partial})\frac{1}{|\mathbf{X}|}, \tag{9.28}$$

where $\tilde{\partial}_i = \frac{\partial}{\partial X^i}$. This is similar to the formula for the interaction energy of two electric dipoles, but has the opposite sign, because the pion field is a scalar, so like charges attract.

We can use the translation and isospin symmetries to position the first Skyrmion at the origin in standard orientation, and the second Skyrmion at the point $\mathbf{X} \in \mathbb{R}^3$, with $X = |\mathbf{X}| \gg 1$, and with an orientation determined by the $SU(2)$ matrix $\mathcal{O}$. The dipole moments of the second Skyrmion are then $\mathbf{q}_j = M\mathbf{p}_j$, where M is the $SO(3)$ matrix corresponding to $\mathcal{O}$, as given in (9.6). There is a dipole interaction between $\mathbf{p}_j$ and $\mathbf{q}_k$ only if $j = k$, so summing the interactions of the three pairs and using (9.28) we obtain the total interaction energy

$$E_{\text{int}} = \frac{2C^2}{3\pi}(\tilde{\partial} \cdot M\tilde{\partial})\frac{1}{X}. \tag{9.29}$$

To get a better understanding of this, we can reexpress the matrix M in terms of a rotation through an angle ψ about an axis $\hat{\mathbf{n}}$,

$$M_{ij} = \cos\psi\, \delta_{ij} + (1 - \cos\psi)\hat{n}_i\hat{n}_j + \sin\psi\, \varepsilon_{ijk}\hat{n}_k. \tag{9.30}$$

The interaction energy (9.29) then takes the form

$$E_{\text{int}} = -\frac{2C^2}{3\pi}(1 - \cos\psi)\frac{1 - 3(\hat{\mathbf{X}} \cdot \hat{\mathbf{n}})^2}{X^3}. \tag{9.31}$$

Clearly, by a suitable choice of the axis $\hat{\mathbf{n}}$, the two Skyrmions can be made to either repel or attract, corresponding to a positive or negative interaction energy respectively. The attraction is maximal (that is, the interaction energy is minimal) if $\hat{\mathbf{X}} \cdot \hat{\mathbf{n}} = 0$ and $\psi = \pi$, in other words, one Skyrmion is rotated relative to the other through an angle of $180°$ about a line perpendicular to the line joining them. This is known as the attractive channel. Note that in making this statement we are using the fact that an isospin rotation of a single Skyrmion is equivalent to a spatial rotation, so we may think in terms of the latter.

In Section 9.8, where we discuss Skyrmion dynamics, we return to formula (9.31) in relation to setting up initial conditions for several well separated Skyrmions such that they mutually attract.

The dipole calculation described above can not serve as a rigorous derivation of the asymptotic interaction energy of two Skyrmions since it assumes that a Skyrmion whose field is asymptotically of the dipole triplet form also reacts to an external field like a dipole triplet. Below we present a more formal calculation of the interaction energy, closely following the presentation in [365], which confirms the result obtained from the dipole picture.

In Eq. (9.8) we have expressed the static energy in terms of the right currents $R_i = \partial_i U U^\dagger$, but we could also have chosen to write it in terms of the left currents $L_i = U^\dagger \partial_i U$, giving an identical expression after the replacement of R_k by L_k. These two equivalent formulations are useful in what follows, as are the quantities $\tilde{R}_i$ and $\tilde{L}_i$ defined as

$$\tilde{R}_i = R_i - \frac{1}{4}[R_j, [R_j, R_i]], \quad \tilde{L}_i = L_i - \frac{1}{4}[L_j, [L_j, L_i]]. \tag{9.32}$$

It follows from the Skyrme field equation (9.3) that for a static solution, both these currents are divergenceless, that is,

$$\partial_i \tilde{R}_i = \partial_i \tilde{L}_i = 0. \tag{9.33}$$

To calculate the interaction energy of two well separated Skyrmions we use the product ansatz of two hedgehog fields

$$U = U^{(1)}U^{(2)}, \quad U^{(1)} = U_{\mathrm{H}}(\mathbf{x}), \quad U^{(2)} = \mathcal{O}U_{\mathrm{H}}(\mathbf{x} - \mathbf{X})\mathcal{O}^\dagger. \tag{9.34}$$

In computing the energy of the product field (9.34) it is helpful to note the following relation

$$L_i = U^\dagger \partial_i U = U^{(2)\dagger}(L_i^{(1)} + R_i^{(2)})U^{(2)}, \tag{9.35}$$

where $L_i^{(1)}$ denotes the left current constructed from the field $U^{(1)}$, and so on. Substituting this expression into the Skyrme energy gives a term involving only $L_i^{(1)}$, one involving only $R_i^{(2)}$ and a cross term. The first two terms each contribute precisely the energy of a single Skyrmion and the cross term gives the interaction energy which, neglecting terms that are quadratic in both $L_i^{(1)}$ and $R_i^{(2)}$, has the leading order contribution

$$E_{\mathrm{int}} \sim -\frac{1}{12\pi^2}\int_{\mathbb{R}^3} \mathrm{Tr}(L_i^{(1)}\tilde{R}_i^{(2)} + \tilde{L}_i^{(1)}R_i^{(2)} - L_i^{(1)}R_i^{(2)})\,d^3x. \tag{9.36}$$

In order to evaluate this integral for large X, we divide $\mathbb{R}^3$ into three regions, I, II and III, given by $I = \{\mathbf{x} : |\mathbf{x}| < \rho\}$, $II = \{\mathbf{x} : |\mathbf{x} - \mathbf{X}| < \rho\}$ and $III = \mathbb{R}^3 - I - II$, with $2\rho < X$. For large X we choose ρ large enough so that outside region I we can apply the asymptotic expression

$$L_i^{(1)} \sim l_i^{(1)} \equiv iC\partial_i\left(\frac{\mathbf{x} \cdot \boldsymbol{\tau}}{|\mathbf{x}|^3}\right), \tag{9.37}$$

and similarly outside region II

$$R_i^{(2)} \sim r_i^{(2)} \equiv iC\partial_i \left(\frac{\mathcal{O}(\mathbf{x} - \mathbf{X}) \cdot \tau \mathcal{O}^\dagger}{|\mathbf{x} - \mathbf{X}|^3} \right). \tag{9.38}$$

Note that since $\tilde{L}_i^{(1)}$ differs from $L_i^{(1)}$ only by a triple product of $L_i^{(1)}$'s (and similarly for $\tilde{R}_i^{(2)}$) then in the above limits we also have that

$$\tilde{L}_i^{(1)} \sim l_i^{(1)} \quad \text{and} \quad \tilde{R}_i^{(2)} \sim r_i^{(2)}. \tag{9.39}$$

Furthermore, we also require that ρ is small enough that $l_i^{(1)}$ may be taken to be constant over region II and $r_i^{(2)}$ constant over region I. This is achieved by letting $\rho \to \infty$ as $X \to \infty$ in such a way that $\rho/X \to 0$. Substituting these approximations into (9.36) we arrive at

$$
\begin{aligned}
E_{\text{int}} \quad \sim \quad & -\frac{1}{12\pi^2} \text{Tr}\Big\{ r_i^{(2)}\Big|_{\mathbf{x}=0} \int_I (\tilde{L}_i^{(1)} - l_i^{(1)})\, d^3x \\
& + \; l_i^{(1)}\Big|_{\mathbf{x}=\mathbf{X}} \int_{II} (\tilde{R}_i^{(2)} - r_i^{(2)})\, d^3x + \int_{\mathbb{R}^3} l_i^{(1)} r_i^{(2)}\, d^3x \Big\}. \tag{9.40}
\end{aligned}
$$

Expanding $\tilde{L}_i^{(1)}$ in terms of Pauli matrices as

$$\tilde{L}_i^{(1)} = i\tilde{L}_{im}\tau_m, \tag{9.41}$$

we see from Eq. (9.33) that for each $m = 1, 2, 3$, $\tilde{L}_{im}$ are the components of a divergenceless vector field, which implies that there exists a potential Z_{km} such that

$$\tilde{L}_{im} = \varepsilon_{ijk}\partial_j Z_{km}. \tag{9.42}$$

Explicitly, it can be checked that this potential is given by

$$Z_{km} = \left(\sin^2 f - \frac{\sin^4 f}{r^2} \right) \hat{x}_k \hat{x}_m + \frac{rf'}{2}\left(1 + 2\frac{\sin^2 f}{r^2} \right) \varepsilon_{kmn}\hat{x}_n. \tag{9.43}$$

Thus

$$
\begin{aligned}
\int_I \tilde{L}_i^{(1)}\, d^3x \quad = \quad & i\tau_m \int_{\partial I} \varepsilon_{ijk} Z_{km} \hat{x}_j\, dS \\
= \quad & i\tau_m \int_{\partial I} (\delta_{im} - \hat{x}_i \hat{x}_m) \frac{rf'}{2}\left(1 + 2\frac{\sin^2 f}{r^2} \right) dS \\
\sim \quad & -\frac{8\pi iC}{3}\tau_i, \tag{9.44}
\end{aligned}
$$

where the final line is obtained by making use of the asymptotic expression $f(\rho) \sim C/\rho^2$ and keeping only leading order terms in ρ. Next, we have that

$$
\begin{aligned}
\int_I l_i^{(1)} \, d^3x &= iC \int_I \partial_i \left(\frac{\mathbf{x} \cdot \boldsymbol{\tau}}{|\mathbf{x}|^3} \right) d^3x = -iC\tau_m \int_I \partial_i \partial_m \frac{1}{|\mathbf{x}|} \, d^3x \\
&= -\frac{iC}{3} \tau_i \int_I \nabla^2 \frac{1}{|\mathbf{x}|} \, d^3x = \frac{4\pi iC}{3} \tau_i \,,
\end{aligned}
\tag{9.45}
$$

where the final expression is obtained by using the identity $\nabla^2 \frac{1}{|\mathbf{x}|} = -4\pi\delta(\mathbf{x})$. Combining these two results we have that

$$
\int_I (\tilde{L}_i^{(1)} - l_i^{(1)}) \, d^3x \sim -4\pi iC\tau_i \,.
\tag{9.46}
$$

From (9.38) we see that

$$
r_i^{(2)} \Big|_{\mathbf{x}=0} = iC\mathcal{O}\tau_n \mathcal{O}^\dagger \frac{(\delta_{in} - 3\hat{X}_i\hat{X}_n)}{X^3} \,,
\tag{9.47}
$$

so the first term in (9.40) has been calculated to be

$$
-\frac{1}{12\pi^2} \mathrm{Tr}\left\{ r_i^{(2)} \Big|_{\mathbf{x}=0} \int_I (\tilde{L}_i^{(1)} - l_i^{(1)}) \, d^3x \right\}
$$

$$
\sim -\frac{C^2}{3\pi X^3} \mathrm{Tr}(\mathcal{O}\tau_n \mathcal{O}^\dagger \tau_n - 3\hat{X}_i\hat{X}_n \mathcal{O}\tau_n \mathcal{O}^\dagger \tau_i)
$$

$$
= -\frac{2C^2(\mathrm{Tr}M - 3\hat{\mathbf{X}} \cdot M\hat{\mathbf{X}})}{3\pi X^3} \,,
\tag{9.48}
$$

where M is the $SO(3)$ matrix corresponding to $\mathcal{O}$ as given in (9.6).

A similar calculation for the second term in (9.40) yields the same result

$$
-\frac{1}{12\pi^2} \mathrm{Tr}\left\{ l_i^{(1)} \Big|_{\mathbf{x}=\mathbf{X}} \int_{II} (\tilde{R}_i^{(2)} - r_i^{(2)}) \, d^3x \right\} \sim -\frac{2C^2(\mathrm{Tr}M - 3\hat{\mathbf{X}} \cdot M\hat{\mathbf{X}})}{3\pi X^3} \,.
\tag{9.49}
$$

The final term in (9.40) is relatively simple to calculate using an integration by parts and the relation $\nabla^2 \frac{1}{|\mathbf{x}|} = -4\pi\delta(\mathbf{x})$,

$$
-\frac{1}{12\pi^2} \mathrm{Tr}\left\{ \int_{\mathbb{R}^3} l_i^{(1)} r_i^{(2)} \, d^3x \right\} \sim \frac{C^2}{6\pi^2} M_{kj} \int_{\mathbb{R}^3} \partial_i \partial_j \frac{1}{|\mathbf{x} - \mathbf{X}|} \partial_i \partial_k \frac{1}{|\mathbf{x}|} \, d^3x
$$

$$
= \frac{2C^2(\mathrm{Tr}M - 3\hat{\mathbf{X}} \cdot M\hat{\mathbf{X}})}{3\pi X^3} \,.
\tag{9.50}
$$

Adding together the three terms in (9.40) we arrive at the final answer

$$
E_{\mathrm{int}} \sim -\frac{2C^2(\mathrm{Tr}M - 3\hat{\mathbf{X}} \cdot M\hat{\mathbf{X}})}{3\pi X^3} = \frac{2C^2}{3\pi} (\tilde{\partial} \cdot M\tilde{\partial}) \frac{1}{X} \,,
\tag{9.51}
$$

which agrees with the earlier result obtained from the asymptotic dipole calculation.

The general form of the interaction energy of two charge 1 Skyrmions was originally presented by Skyrme [379] and verified by Jackson *et al.* [218] and Vinh Mau *et al.* [408]. Castillejo and Kugler [76] noted that if the asymptotic interaction energy of two well separated clusters of Skyrmions, of any charge, is positive, then it can be made negative by performing an appropriate isospin transformation on one of the clusters. We have already explicitly seen that this is true in the case of two charge 1 Skyrmions, as illustrated by Eq. (9.31). It may appear that this result constitutes a proof of the strict inequality (9.24), and hence that Skyrmions exist for any baryon number, since it is always possible to arrange that two clusters have a negative interaction energy, and hence a total energy which is lower than the sum of their individual energies. However, the flaw in Castillejo and Kugler's argument is that, to lowest order, the asymptotic interaction energy may vanish. In this case, the lowest order contribution to the asymptotic interaction energy can not be made negative by an isospin rotation and the calculation must be performed to higher order. A similar caveat obviously applies at each order and so it is not possible to conclude that the interaction energy is negative, only that it is non-positive. This is another manifestation of the fact that the weaker energy inequality (9.25) has been proved, but the strict inequality (9.24), required for the proof of existence of arbitrary charge Skyrmions, remains unproven at present.

However, further progress on this problem has recently been made. Now that it has been established that any Skyrmion has a leading multipole [294], it can be shown that in most cases a pair of well separated Skyrmions of any baryon number can be oriented and positioned so as to attract. Unfortunately, the argument breaks down because of the non-linear terms if the leading multipole of one of the Skyrmions is of high order, or more precisely, if the orders of the multipoles differ by more than two. Nevertheless, as Schroers has shown [367], some rigorous conclusions about the existence of Skyrmions of higher baryon number are possible.

9.4 Low charge Skyrmions

In this section we discuss the properties of minimal energy Skyrmions with charges $1 \leq B \leq 8$, constructed using numerical methods. Details of the numerical codes used to compute these solutions can be found in the papers cited below, and a detailed discussion appears in [45], to which we refer the interested reader.

All known solutions appear to be isolated and their only moduli are the obvious ones associated with the nine-dimensional symmetry group of the

Skyrme model. Generic solutions therefore have nine moduli, although solutions with axial or spherical symmetry have, respectively, one or three fewer.

As we have already noted, for charges $B > 1$ the minimal energy Skyrmion is not spherically symmetric. For $B = 2$, it turns out that it has an axial symmetry [244, 283, 406]. The energy density has a similar toroidal structure to that of the charge 2 axisymmetric monopole solution discussed in the previous chapter, despite the fact that the fields of the two models are very different. In displaying Skyrmions it is conventional to plot surfaces of constant baryon density $\mathcal{B}$ (baryon density isosurfaces), where $\mathcal{B}$ is the integrand in Eq. (9.9), although energy density isosurfaces are qualitatively very similar. In Fig. 9.2 we display baryon density iso-surfaces for the minimal energy Skyrmions of charges $1 \le B \le 8$.

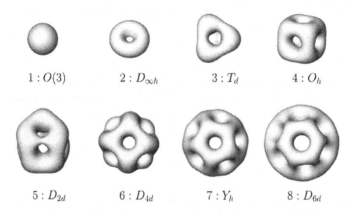

Fig. 9.2.   Baryon density isosurfaces for $1 \le B \le 8$. The baryon number and symmetry of each solution is shown.

There are axially symmetric solutions of the Skyrme equation for $B > 2$ [244], but these are not the minimal energy solutions, and in fact for $B > 4$ they are not even sufficiently bound to prevent break-up into B single Skyrmions, so they correspond to saddle points.

The Skyrmions presented in Fig. 9.2 have only discrete symmetries for $B > 2$. The $B = 3$ and $B = 4$ Skyrmions have tetrahedral symmetry T_d and cubic symmetry O_h, respectively [65], and again are very similar to particular monopoles of the same charge, which we have already discussed. The associated polyhedra, where the baryon density is concentrated, are a tetrahedron and cube, as the figure shows. It is perhaps of interest to point out that these Skyrmion solutions were computed before the existence of the corresponding monopoles was known. At the time it was therefore very surprising to find these highly symmetric Platonic

Skyrmions emerging from asymmetric initial conditions. Their existence was a major motivation for the search for Platonic monopole solutions, and although a deep connection between Skyrmions and monopoles is still lacking, a link between these two kinds of soliton has been found, via rational maps, and has led to an improved understanding of the structure of Skyrmions, as we discuss in detail in the next section.

The $B = 5$ Skyrmion has a relatively small symmetry, namely D_{2d}. The associated polyhedron comprises four squares and four pentagons, the top and bottom of the structure being related by a relative rotation of $90°$. In case the reader is not familiar with extended dihedral symmetries we briefly recount them here. The dihedral group D_n is obtained from C_n, the cyclic group of order n, by the addition of a C_2 axis which is orthogonal to the main C_n symmetry axis. The group D_n can be extended by the addition of a reflection symmetry in two ways: by including a reflection in the plane perpendicular to the main C_n axis, which produces the group D_{nh} or, alternatively, a reflection in a plane which contains the main symmetry axis and bisects a pair of the C_2 axes obtained by applying the C_n symmetry to the generating C_2 axis, which produces the group D_{nd}.

Recall that a charge 5 monopole exists with octahedral symmetry, so given the similarity between monopoles and Skyrmions it may seem a little curious that the $B = 5$ Skyrmion has relatively little symmetry. In fact, as we discuss further in the next section, there is an octahedrally symmetric charge 5 solution, but it is a saddle point whose energy is a little higher than the less symmetric D_{2d} minimum.

The $B = 6$ and $B = 8$ Skyrmions also have extended dihedral symmetries, this time D_{4d} and D_{6d} respectively. The $B = 7$ Skyrmion is icosahedrally symmetric [41], its symmetry group Y_h being an extension of D_{5d}. The baryon density of the $B = 7$ Skyrmion is localized around the edges of a dodecahedron, the structure closely resembling the icosahedrally symmetric charge 7 monopole.

The polyhedron associated with the $B = 6$ Skyrmion consists of two halves, each formed from a square with pentagons hanging down from all four sides. To join these two halves, the two squares must be parallel, with one rotated by $45°$ relative to the other. The $B = 8$ Skyrmion has a similar structure, except that the squares are replaced by hexagons, and each half has six pentagons hanging down. The top hexagon is parallel to the bottom hexagon but rotated by $30°$. The halves of the $B = 7$ Skyrmion have pentagons hanging from a pentagon, hence the larger symmetry.

In Fig. 9.3 we display models (not to scale) of the polyhedra associated with the Skyrmions of charge $5, 6, 7, 8$, and in Table 9.1 we present, for charges 1 to 8, the symmetries and energies per baryon, E/B, of the

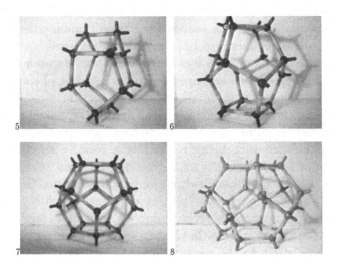

Fig. 9.3. Polyhedral models (not to scale) representing the Skyrmions with $B = 5, 6, 7, 8$.

Skyrmions, computed from the numerical solutions of the field equation [45].

In ref. [41] a phenomenological rule for the structure of the minimal energy Skyrmions was proposed, called the Geometric Energy Minimization (GEM) rule. This states that, for $B > 2$, the polyhedron associated with the charge B Skyrmion is composed of almost regular polygons meeting at $4(B - 2)$ trivalent vertices, and the baryon density is concentrated along the edges of the polygons. Note that there are several equivalent ways in which the GEM rule can be stated, since, by using the trivalent property together with Euler's formula, any one of the three parameters of the structure, the number of vertices V, faces F, or edges E, determines the other two. Explicitly, $V = 4(B - 2)$, $F = 2(B - 1)$, $E = 6(B - 2)$. Since the baryon density isosurface has a hole in the centre of each face, the GEM rule is consistent with the observation of ref. [65] that the isosurface contains $2(B - 1)$ holes. For $3 \leq B \leq 8$ we have already described the Skyrmions, and it is a simple task to confirm that the GEM rule is obeyed in these cases. However, as B increases, the number of possible structures satisfying the GEM rule grows rapidly, so that by no means does it uniquely predict the structure.

For $B \geq 7$ it is possible to satisfy the GEM rule with a trivalent polyhedron formed from 12 pentagons and $2B - 14$ hexagons. We will refer to such structures as fullerene-like and to the conjecture that the Skyrmion's

Table 9.1.   The symmetry, K, and energy per baryon, E/B, for the numerically computed minimal energy Skyrmions of charge $1 \leq B \leq 8$.

B	K	E/B
1	$O(3)$	1.2322
2	$D_{\infty h}$	1.1791
3	T_d	1.1462
4	O_h	1.1201
5	D_{2d}	1.1172
6	D_{4d}	1.1079
7	Y_h	1.0947
8	D_{6d}	1.0960

baryon density isosurface has this form as the fullerene hypothesis since precisely the same fullerene (a shortening of Buckminsterfullerene) structures arise in carbon chemistry, where carbon atoms sit at the vertices of such polyhedra [137]. It is then plausible [41] that the minimal energy Skyrmion of charge B has the same symmetry as a fullerene from the family $C_{4(B-2)}$. For low charges ($B = 7$, $B = 8$) this leads to a unique prediction for the structures, which are those we have already encountered, but as the charge increases the number of possible structures again increases. In particular, for $B = 9$ there are two possibilities with D_2 and T_d symmetries respectively, for $B = 10$ there are six, for $B = 11$ there are 15, with a rapid increase for $B > 11$. However, there is a unique icosahedrally symmetric configuration with $B = 17$ corresponding to the famous fullerene structure of C_{60}, and given its high symmetry it is not surprising that the minimal energy $B = 17$ configuration has this structure. In Section 9.6 we discuss Skyrmions of higher charge, up to $B = 22$, and find that the fullerene hypothesis is valid for all but two charges, where interesting caveats apply. In the next section we discuss an approximate analytic description of Skyrmions and see that within this approach at least one aspect of the GEM rule, namely, that the number of faces is $2(B - 1)$, can be understood.

9.5 The rational map ansatz

The observed similarities between Skyrmions and monopoles leads naturally to the question whether there is an approximate construction of Skyrmions from monopoles. Of course, it is not expected that an exact correspondence exists, since the Yang-Mills-Higgs and Skyrme models

have a number of very different properties, but for certain monopole solutions a mapping does exist which generates a good approximation to a
related exact Skyrmion solution. As yet, there is no known direct transformation between the fields of a monopole and those of a Skyrmion, but
as we describe in this section, there is an indirect transformation via rational maps between Riemann spheres. Recall from Chapter 8 that there
is a precise correspondence between charge N monopoles and degree N
rational maps (we have in mind here the Jarvis maps); thus a Skyrme
field constructed from a rational map is indirectly constructed from a
monopole.

One needs an ansatz for a Skyrme field in terms of a rational map, and
the shell-like fullerene structures of the numerically computed Skyrmions
suggest how to proceed. Rational maps are maps from $S^2 \mapsto S^2$, whereas
Skyrmions are maps from $\mathbb{R}^3 \mapsto S^3$. The main idea behind the rational
map ansatz, introduced in [193], is to identify the domain S^2 of the rational map with concentric spheres in $\mathbb{R}^3$, and the target S^2 with spheres
of latitude on S^3. It is convenient to use 3-vector notation to present the
ansatz explicitly. Recall that via stereographic projection, the complex
coordinate z on a sphere can be identified with conventional polar coordinates by $z = \tan(\theta/2)e^{i\varphi}$. Equivalently, the point z corresponds to the
unit vector

$$\widehat{\mathbf{n}}_z = \frac{1}{1 + |z|^2}(z + \bar{z}, \, i(\bar{z} - z), \, 1 - |z|^2). \qquad (9.52)$$

Similarly the value of the rational map $R(z)$ is associated with the unit
vector

$$\widehat{\mathbf{n}}_R = \frac{1}{1 + |R|^2}(R + \bar{R}, \, i(\bar{R} - R), \, 1 - |R|^2). \qquad (9.53)$$

Let us denote a point in $\mathbb{R}^3$ by its coordinates (r, z), where r is the radial
distance from the origin and z specifies the direction from the origin. The
ansatz for the Skyrme field, depending on a rational map $R(z)$ and a
radial profile function $f(r)$, is

$$U(r, z) = \exp(i f(r) \, \widehat{\mathbf{n}}_{R(z)} \cdot \boldsymbol{\tau}), \qquad (9.54)$$

where, as usual, $\boldsymbol{\tau} = (\tau_1, \tau_2, \tau_3)$ denotes the triplet of Pauli matrices. For
this to be well defined at the origin, $f(0) = k\pi$ for some integer k. We
take $k = 1$ in what follows. The boundary condition $U = 1_2$ at $r = \infty$ is
satisfied by setting $f(\infty) = 0$. It is straightforward to verify (see below)
that the baryon number of this field configuration is $B = N$, where N is
the degree of R.

Mathematically, this construction of a map from compactified $\mathbb{R}^3$ to
S^3, out of a map from S^2 to S^2, is a suspension; the suspension points on
the domain are the origin and the point at infinity, and on the target the

points $U = -1_2$ and $U = 1_2$. Suspension is an isomorphism between the homotopy groups $\pi_2(S^2)$ and $\pi_3(S^3)$, which explains why $B = N$.

An $SU(2)$ Möbius transformation on the domain S^2 of the rational map corresponds to a spatial rotation, whereas an $SU(2)$ Möbius transformation on the target S^2 corresponds to a rotation of $\hat{\mathbf{n}}_R$, and hence to an isospin rotation of the Skyrme field. Thus if a rational map $R : S^2 \mapsto S^2$ is symmetric in the sense defined in Chapter 6 (i.e. a rotation of the domain can be compensated by a rotation of the target), then the resulting Skyrme field is symmetric in the sense defined in Section 9.2 (i.e. a spatial rotation can be compensated by an isospin rotation).

Note that if we introduce the Hermitian projector

$$P = \frac{1}{1 + |R|^2} \begin{pmatrix} 1 & \bar{R} \\ R & |R|^2 \end{pmatrix}, \tag{9.55}$$

satisfying $P^2 = P = P^\dagger$, then the ansatz (9.54) can be written as

$$U = \exp(if(2P - 1_2)), \tag{9.56}$$

which is similar to the expression (8.262), describing the asymptotic form of the solution of the Jarvis equation corresponding to the monopole with rational map R.

The simplest degree 1 rational map is $R = z$, which is spherically symmetric. The ansatz (9.54) then reduces to Skyrme's hedgehog field (9.18) with $f(r)$ being the usual profile function. In this case the ansatz is compatible with the static Skyrme equation but in general it is not, so it can not produce exact solutions, only low energy approximations to these.

An attractive feature of the ansatz is that it leads to a simple energy expression which can be minimized with respect to the rational map R and the profile function f to obtain close approximations to the numerical, exact Skyrmion solutions. To calculate the energy we exploit the geometrical formulation of the Skyrme model presented in Section 9.1. For the ansatz (9.54), the strain in the radial direction is orthogonal to the strain in the angular directions. Moreover, because $R(z)$ is conformal, the angular strains are isotropic. If we identify λ_1^2 with the radial strain and λ_2^2 and λ_3^2 with the angular strains, we can easily compute that

$$\lambda_1 = -f'(r), \quad \lambda_2 = \lambda_3 = \frac{\sin f}{r} \frac{1 + |z|^2}{1 + |R|^2} \left| \frac{dR}{dz} \right|. \tag{9.57}$$

From Eq. (9.12), the baryon number is

$$B = -\int \frac{f'}{2\pi^2} \left(\frac{\sin f}{r} \frac{1 + |z|^2}{1 + |R|^2} \left| \frac{dR}{dz} \right| \right)^2 \frac{2i \, dz d\bar{z}}{(1 + |z|^2)^2} r^2 \, dr, \tag{9.58}$$

where $2i\,dzd\bar{z}/(1+|z|^2)^2$ is equivalent to the usual area element on a 2-sphere $\sin\theta\,d\theta d\varphi$. Now the part of the integrand

$$\left(\frac{1+|z|^2}{1+|R|^2}\left|\frac{dR}{dz}\right|\right)^2\frac{2i\,dzd\bar{z}}{(1+|z|^2)^2} \tag{9.59}$$

is precisely the pull-back of the area form $2i\,dRd\bar{R}/(1+|R|^2)^2$ on the target sphere of the rational map R; therefore its integral is 4π times the degree N of R. So (9.58) simplifies to

$$B = -\frac{2N}{\pi}\int_0^\infty f'\sin^2 f\,dr = N\,, \tag{9.60}$$

where we have used the boundary conditions $f(0)=\pi$, $f(\infty)=0$. This verifies again that the baryon number of the Skyrme field generated from the ansatz is equal to the degree of the rational map.

Substituting the strains (9.57) into the expression (9.12) for the energy density yields the energy

$$E = \frac{1}{12\pi^2}\int\left\{f'^2 \;+\; 2\frac{\sin^2 f}{r^2}(f'^2+1)\left(\frac{1+|z|^2}{1+|R|^2}\left|\frac{dR}{dz}\right|\right)^2\right. \tag{9.61}$$

$$\left.+\;\frac{\sin^4 f}{r^4}\left(\frac{1+|z|^2}{1+|R|^2}\left|\frac{dR}{dz}\right|\right)^4\right\}\frac{2i\,dzd\bar{z}}{(1+|z|^2)^2}\,r^2\,dr\,,$$

which can be simplified, using the above remarks about baryon number, to

$$E = \frac{1}{3\pi}\int_0^\infty\left(r^2f'^2 + 2B\sin^2 f(f'^2+1)+\mathcal{I}\frac{\sin^4 f}{r^2}\right)dr\,. \tag{9.62}$$

$\mathcal{I}$ denotes the purely angular integral

$$\mathcal{I} = \frac{1}{4\pi}\int\left(\frac{1+|z|^2}{1+|R|^2}\left|\frac{dR}{dz}\right|\right)^4\frac{2i\,dzd\bar{z}}{(1+|z|^2)^2}\,, \tag{9.63}$$

which only depends on the rational map R.

Note the following pair of inequalities associated with the expression (9.62) for the energy E. The elementary inequality

$$\left(\int 1\,dS\right)\left(\int\left(\frac{1+|z|^2}{1+|R|^2}\left|\frac{dR}{dz}\right|\right)^4 dS\right) \geq \left(\int\left(\frac{1+|z|^2}{1+|R|^2}\left|\frac{dR}{dz}\right|\right)^2 dS\right)^2\,, \tag{9.64}$$

where $dS = 2i\,dzd\bar{z}/(1+|z|^2)^2$, implies that $\mathcal{I}\geq B^2$. Next, by using a Bogomolny-type argument, we see that

$$E = \frac{1}{3\pi}\int_0^\infty\left\{\left(rf' + \sqrt{\mathcal{I}}\frac{\sin^2 f}{r}\right)^2 \;+\; 2B\sin^2 f(f'+1)^2\right. \tag{9.65}$$

$$\left. -\; 2(2B+\sqrt{\mathcal{I}})f'\sin^2 f\right\}dr$$

so

$$E \geq \frac{1}{3\pi}(2B + \sqrt{\mathcal{I}}) \int_0^\infty (-2f' \sin^2 f) \, dr = \frac{1}{3\pi}(2B + \sqrt{\mathcal{I}}) \left[-f + \frac{1}{2} \sin 2f \right]_0^\infty \tag{9.66}$$

and so

$$E \geq \frac{1}{3}(2B + \sqrt{\mathcal{I}}). \tag{9.67}$$

Combined with the earlier inequality for $\mathcal{I}$, we recover the usual Fadeev-Bogomolny bound $E \geq B$. The bound (9.67) is stronger than this for fields given by the rational map ansatz, but there is no reason to think that true solutions of the Skyrme equation are constrained by this bound.

To minimize E one should first minimize $\mathcal{I}$ over all maps of degree B. The profile function f minimizing the energy (9.62) may then be found by a simple gradient flow algorithm with B and $\mathcal{I}$ as fixed parameters. In Section 9.6 we discuss the results of a numerical search for $\mathcal{I}$-minimizing maps among all rational maps of degree B, but in this section we first consider the simpler problem in which we restrict the map to a given symmetric form, with symmetries corresponding to one of the numerically known Skyrmion solutions. If these maps still contain a few free parameters, $\mathcal{I}$ can be minimized with respect to these. This procedure is appropriate for all baryon numbers up to $B = 8$, for which we know the symmetries of the numerically computed Skyrmions, and there is sufficient symmetry to highly constrain the rational map.

For $B = 2, 3, 4, 7$ the symmetries of the numerically computed Skyrmions are $D_{\infty h}, T_d, O_h, Y_h$ respectively. From the general discussion and specific examples of Chapters 6 and 8, we see that in each of these cases there is a unique rational map with the given symmetry. We recall that they are

$$R = z^2, \ R = \frac{z^3 - \sqrt{3}iz}{\sqrt{3}iz^2 - 1}, \ R = \frac{z^4 + 2\sqrt{3}iz^2 + 1}{z^4 - 2\sqrt{3}iz^2 + 1}, \ R = \frac{z^7 - 7z^5 - 7z^2 - 1}{z^7 + 7z^5 - 7z^2 + 1}. \tag{9.68}$$

Using these maps, and computing the optimal profile functions $f(r)$, one obtains Skyrme fields whose baryon density isosurfaces are indistinguishable from those presented in Fig. 9.2. In Table 9.2 we list the energies per baryon of the approximate solutions obtained using the rational map ansatz, together with the values of $\mathcal{I}$ and $\mathcal{I}/B^2$, in order to compare with the bound $\mathcal{I}/B^2 \geq 1$.

Recall that the Wronskian of a rational map $R(z) = p(z)/q(z)$ of degree B is the polynomial

$$W(z) = p'(z)q(z) - q'(z)p(z) \tag{9.69}$$

of degree $2B - 2$, and observe that the zeros of the Wronskian give interesting information about the shape of the Skyrme field constructed using

Table 9.2. Approximate Skyrmions obtained using the rational map ansatz. For $1 \leq B \leq 8$ we list the symmetry of the rational map, K, the value of $\mathcal{I}$, its comparison with the bound $\mathcal{I}/B^2 \geq 1$, and the energy per baryon E/B obtained after computing the profile function which minimizes the Skyrme energy function.

B	K	$\mathcal{I}$	$\mathcal{I}/B^2$	E/B
1	$O(3)$	1.0	1.000	1.232
2	$D_{\infty h}$	5.8	1.452	1.208
3	T_d	13.6	1.509	1.184
4	O_h	20.7	1.291	1.137
5	D_{2d}	35.8	1.430	1.147
6	D_{4d}	50.8	1.410	1.137
7	Y_h	60.9	1.242	1.107
8	D_{6d}	85.6	1.338	1.118

the ansatz (9.54). Where W is zero, the derivative dR/dz is zero, so the strain eigenvalues in the angular directions, λ_2 and λ_3, vanish. The baryon density, being proportional to $\lambda_1\lambda_2\lambda_3$, therefore vanishes along the entire radial line in the direction specified by any zero of W. The energy density will also be low along such a radial line, since there will only be the contribution λ_1^2 from the radial strain eigenvalue. The ansatz thus makes manifest why the Skyrme field baryon density contours look like polyhedra with holes in the directions given by the zeros of W, and why there are $2B-2$ such holes, precisely the structure seen in all the plots in Fig. 9.2. This explains the GEM rule $F = 2(B-1)$, and although there is no firm rational map explanation of the other aspects of the GEM rule, we will make some further comments on them in the following section.

As an example, consider the icosahedrally symmetric degree 7 map in (9.68). The Wronskian is

$$W(z) = 28z(z^{10} + 11z^5 - 1),\tag{9.70}$$

which is proportional to the Klein polynomial $\mathcal{Y}_v$, and it vanishes at the twelve face centres of a dodecahedron [237]. This explains why the baryon density isosurface of the $B = 7$ Skyrmion displayed in Fig. 9.2 is localized around the edges of a dodecahedron.

For the remaining charges, $B = 5, 6, 8$, the Skyrmions have extended dihedral symmetries, so we need to consider degree B rational maps with dihedral symmetries D_n, and their extensions by reflections to D_{nd} and D_{nh}. Constructing D_n-symmetric maps does not require the general group theory formalism discussed in Chapter 6 since it is simple to explicitly apply the two generators of D_n to a map. In terms of the Riemann sphere coordinate z the generators of the dihedral group D_n may be taken to be $z \mapsto e^{2\pi i/n} z$ and $z \mapsto 1/z$. The reflection required to extend the symmetry to D_{nh} is represented by $z \mapsto 1/\bar{z}$, whereas $z \mapsto e^{\pi i/n} \bar{z}$ results in the symmetry group D_{nd}.

Explicitly, an s-parameter family of D_n-symmetric maps is given by[*]

$$R(z) = \frac{\sum_{j=0}^{s} a_j z^{jn+u}}{\sum_{j=0}^{s} a_{s-j} z^{jn}}, \tag{9.71}$$

where $u = B \bmod n$ and $s = (B - u)/n$. Here $a_s = 1$ and $a_0, \ldots, a_{s-1}$ are arbitrary complex parameters. Clearly, these maps satisfy the conditions for symmetry under D_n,

$$R(e^{2\pi i/n} z) = e^{2\pi i u/n} R(z), \quad R(1/z) = 1/R(z), \tag{9.72}$$

and imposing a reflection symmetry constrains each complex coefficient a_j to be either real, or pure imaginary. In the case of D_{nh} symmetry, all coefficients a_j are real, whereas for D_{nd} symmetry a_j is real or imaginary depending on whether $(s - j) \bmod 2$ is, respectively, 0 or 1.

Consider now the $B = 5$ maps with D_{2d} symmetry. Setting $B = 5$ and $n = 2$ in the above gives $u = 1$ and $s = 2$, so there is a family of degree 5 maps with two real parameters,

$$R(z) = \frac{z(a + ibz^2 + z^4)}{1 + ibz^2 + az^4}, \tag{9.73}$$

with a and b real. Additional symmetry occurs if $b = 0$; $R(z)$ then has D_{4h} symmetry, the symmetry of a square. There is octahedral symmetry if, in addition, $a = -5$. This value ensures the $120°$ rotational symmetry

$$R\left(\frac{iz + 1}{-iz + 1}\right) = \frac{iR(z) + 1}{-iR(z) + 1}. \tag{9.74}$$

The octahedral map $R(z) = z(z^4 - 5)/(-5z^4 + 1)$ has Wronskian

$$W(z) = -5(z^8 + 14z^4 + 1), \tag{9.75}$$

which is proportional to $\mathcal{O}_f$, the face polynomial of an octahedron. Using (9.73) in the rational map ansatz for the Skyrme field gives a structure which is a polyhedron with eight faces. In the special case $a = -5$, $b = 0$, this polyhedron is an octahedron, and the angular integral is $\mathcal{I} = 52.1$; however, a numerical search over the parameters a and b shows that $\mathcal{I}$

---

[*] There are other D_n-symmetric families of maps, but we will not need these.

is minimized when $a = -3.07$, $b = 3.94$, taking the value $\mathcal{I} = 35.8$. The approximate Skyrmion generated from the map with these parameter values has a baryon density isosurface which is virtually identical to that of the numerically computed solution displayed in Fig. 9.2. From this analysis we therefore understand that there is an octahedrally symmetric $B = 5$ solution, but that it is a saddle point with an energy higher than that of the less symmetric D_{2d} Skyrmion. There is a further, higher saddle point at $a = b = 0$, where the map (9.73) simplifies to $R(z) = z^5$, and gives a toroidal Skyrme field. Although many minimal energy Skyrmions are highly symmetric, symmetry is not the most important factor in determining the structure of the minimal energy solution, and less symmetric configurations sometimes have lowest energy.

Another example of a symmetric saddle point is the $B = 7$ configuration with cubic symmetry. The relevant rational map is given by $R(z) = (7z^4 + 1)/(z^7 + 7z^3)$ and has Wronskian $W(z) = -21z^2(z^4 - 1)^2$. Each root of this Wronskian is a double root (including the one at infinity) and they lie at the face centres of a cube. A baryon density isosurface for this saddle point configuration is therefore qualitatively similar to that of the minimal energy $B = 4$ Skyrmion. This cubic $B = 7$ saddle point will play a role in a scattering process discussed in Chapter 10.

The analysis of the relevant dihedrally symmetric $B = 6$ and $B = 8$ maps is similar to the $B = 5$ case, the only difference being that just one real parameter appears, so the energy minimization is easier. These maps can be found in ref. [193].

Given the rational map describing a Skyrmion it is possible to infer information regarding its asymptotic fields. For a Skyrmion which is symmetric under a group K, its pion fields will be invariant under combinations of rotations by elements of K and isospin rotations given by some (not necessarily irreducible) real three-dimensional representation of K, which we denote by ρ. Now the dipole fields of a single Skyrmion, being spherically symmetric, are also K-symmetric by restriction, and the corresponding representation ρ is the defining representation of K, regarded as a subgroup of $SO(3)$, which we denote by $\hat{\rho}$, so $\hat{\rho}(k) = k$. By comparing ρ and $\hat{\rho}$ it is possible to determine whether a given Skyrmion looks from far away like a single Skyrmion or antiSkyrmion, that is, like a triplet of orthogonal dipoles. This information is important in understanding the interaction between Skyrmion solutions and will be used in Section 9.8 when we discuss Skyrmion dynamics and scattering.

As an example, consider the tetrahedrally symmetric $B = 3$ Skyrmion described by the map $R(z) = (z^3 - \sqrt{3}iz)/(\sqrt{3}iz^2 - 1)$. A straightforward calculation reveals that $\rho = \hat{\rho} = F$, that is, the pion fields transform via the same three-dimensional irreducible representation of the tetrahedral group as the hedgehog fields of a single Skyrmion or antiSkyrmion. In order to distinguish between these last two possibilities we can compute the value of the rational map along the three Cartesian directions, finding $R(0) = 0$, $R(1) = -1$, $R(i) = i$, which demonstrates that the asymptotic dipole fields are those of an antiSkyrmion, since the pion fields are obtained from those of a Skyrmion by the reflection $\pi_2 \mapsto -\pi_2$.

The fact that the $B = 3$ Skyrmion is asymptotically like an anti-Skyrmion can be understood more naively by a simple addition of the dipole moments of its constituent single Skyrmions. First consider two single Skyrmions brought together along the x^1-axis. They are in the attractive channel if the first is in standard orientation and the second is rotated by 180° around the x^3-axis. This gives triplets of dipole moments $\mathbf{p} = 4\pi C(\mathbf{e}_1, \mathbf{e}_2, \mathbf{e}_3)$ and $\mathbf{q} = 4\pi C(-\mathbf{e}_1, -\mathbf{e}_2, \mathbf{e}_3)$. Their sum is $4\pi C(\mathbf{0}, \mathbf{0}, 2\mathbf{e}_3)$, implying that the toroidal $B = 2$ Skyrmion has only a single dipole, with roughly twice the usual strength. Now bring in a third Skyrmion, along the x^3-axis, and rotated by 180° around the x^1-axis, giving the dipole moments $\mathbf{r} = 4\pi C(\mathbf{e}_1, -\mathbf{e}_2, -\mathbf{e}_3)$. The total of the dipoles is $\mathbf{p} + \mathbf{q} + \mathbf{r} = 4\pi C(\mathbf{e}_1, -\mathbf{e}_2, \mathbf{e}_3)$, precisely those of an antiSkyrmion.

A similar analysis suggests that the $B = 4$ cubic Skyrmion will have no dipoles, since it can be constructed from two $B = 2$ tori. These have a single dipole each, which by an appropriate relative isospin rotation can be made to cancel. The symmetry of the degree 4 cubic map (9.68) is consistent with this result, since the representation ρ is the sum of a one- and two-dimensional irreducible representation of O, whereas $\hat{\rho}$ is a three-dimensional irreducible representation. The fact that the $B = 4$ Skyrmion has no dipole fields explains why it is so tightly bound, and why it interacts only weakly with other Skyrmions.

For the dodecahedral $B = 7$ Skyrmion the naive dipole picture appears to fail, since the combination of the $B = 4$ cubic Skyrmion, with no dipole fields, and the $B = 3$ tetrahedral Skyrmion, with antiSkyrmion dipole fields, suggests that the $B = 7$ Skyrmion has the dipole fields of an antiSkyrmion. However, the representation analysis of the degree 7 dodecahedral map (9.68) reveals that although both ρ and $\hat{\rho}$ are three-dimensional irreducible representations of Y, they are not the same (one is F_1 and the other is F_2). Hence the asymptotic fields can not be those of a single antiSkyrmion (or Skyrmion). In fact, there are no dipole moments at all. The reason why the simple dipole picture fails in this case is not yet understood.

9.6 Higher charge Skyrmions

In the preceding section, for each charge $B \leq 8$, the map R was selected so that the symmetry of the resulting Skyrme field matched that of the numerically computed Skyrmion. Recently, an alternative approach to

Table 9.3.   Results from the simulated annealing of rational maps of degree B. For $9 \leq B \leq 22$ we list the symmetry of the rational map, K, the minimal value of $\mathcal{I}$, the value of $\mathcal{I}/B^2$ (which is bounded below by 1), and the energy per baryon E/B obtained after computing the profile function which minimizes the Skyrme energy functional.

B	K	$\mathcal{I}$	$\mathcal{I}/B^2$	E/B
9	D_{4d}	109.3	1.349	1.116
10	D_{4d}	132.6	1.326	1.110
11	D_{3h}	161.1	1.331	1.109
12	T_d	186.6	1.296	1.102
13	O	216.7	1.282	1.098
14	D_2	258.5	1.319	1.103
15	T	296.3	1.317	1.103
16	D_3	332.9	1.300	1.098
17	Y_h	363.4	1.257	1.092
18	D_2	418.7	1.292	1.095
19	D_3	467.9	1.296	1.095
20	D_{6d}	519.7	1.299	1.095
21	T	569.9	1.292	1.094
22	D_{5d}	621.6	1.284	1.092

constructing the appropriate rational map R, based on energy minimization rather than symmetry, has been applied for all charges $B \leq 22$ [45]. In this approach, no assumption is made as to the possible symmetry of the minimal energy Skyrmion, which has the advantage that full numerical simulations of the Skyrme equation need not first be performed (although it is obviously useful to have these results for comparison, as we discuss later). The main task is to search for the rational map of degree B that minimizes $\mathcal{I}$, which may be viewed as an interesting energy function on the space of rational maps. This is still quite difficult numerically but has been performed using a simulated annealing algorithm, a Monte-Carlo based minimization method which has a major advantage over other conventional minimization techniques in that changes which increase the energy are allowed, enabling the algorithm to escape from local minima that are not the global minimum.

For $B \leq 8$ the simulated annealing algorithm reproduces the rational maps discussed previously (whose properties are listed in Table 9.2), providing a nice numerical check on both the minimizing rational map strategy and also the full field simulations – since these produce very similar configurations.

The results of the simulated annealing algorithm applied to a general rational map of degree $9 \leq B \leq 22$ are presented in Table 9.3. In each case, we tabulate the identified symmetry group K, the minimum value of $\mathcal{I}$, the quantity $\mathcal{I}/B^2$ (which is strikingly uniform at around 1.25–1.35), and the value of E/B for the profile function which minimizes the energy functional (9.62) for the particular map.

By minimizing within certain symmetric families of maps, where the symmetries are *not* shared by the minimal energy map, it is possible to find other critical points of $\mathcal{I}$. In Table 9.4 we present the results of an extensive search for such minimal energy maps with particular symmetries, usually dihedral groups or those groups suggested by the extensive tables of fullerenes presented in ref. [137], which lends further weight to the conclusion that the maps presented in Table 9.3 are in fact the global minima for the energy functional $\mathcal{I}$. These results do, however, turn up the possibility that in certain cases the $\mathcal{I}$-minimizing map may not necessarily be the one which represents the true Skyrmion, since some of the $\mathcal{I}$ values in Tables 9.4 and 9.3 are very close. For the moment we will denote the maps in Table 9.4 by $*$, and conclude at least that they are not global minima of $\mathcal{I}$, but represent other critical points.

For most charges there is a sufficient gap between the minimal value of $\mathcal{I}$ and that of any other critical point to be confident that the minimal map corresponds to the Skyrmion. However, for charges $B = 10, 16, 22$ a glance at Tables 9.3 and 9.4 reveals that there are different maps (with different symmetries) whose associated Skyrme fields have energies which differ by less than 0.1%. Given that the rational map ansatz is an approximation which tends to overestimate the energy by around 1%, it is not clear which of these maps will best describe the angular form of the minimal energy Skyrmion. This question has been addressed using full field simulations [45] in which various initial conditions, consisting of a number of well separated Skyrmion clusters, are relaxed. Although it is difficult to make definitive statements, the results suggest that for these three charges the maps presented in Table 9.4, rather than in Table 9.3, represent the minimal energy Skyrmions. The case $B = 14$ is anomalous, in that the rational map describing the Skyrmion, which again is not the $\mathcal{I}$-minimizing map, is not currently known. The solution obtained from full field simulations (believed to be the minimal energy configuration) is rather elongated, so the rational map approximation to this configuration probably has a substantially higher energy, since it assumes a spherical shape. This explains why it is likely that the rational map which describes the more spherical version of this Skyrmion is not the $\mathcal{I}$-minimizing map. There is a technical reason why we are unable to compute this map,

Table 9.4. Same as for Table 9.3, but for other critical points of $\mathcal{I}$. Notice that the $\mathcal{I}$ values for the $B = 10$ configurations with D_3 and D_{3d} symmetry, for $B = 13$ with D_{4d}, $B = 16$ with D_2, and $B = 22$ with D_3 are extremely close to the corresponding values in Table 9.3, suggesting the possibility of local minima or low-lying saddle points.

B	K	$\mathcal{I}$	$\mathcal{I}/B^2$	E/B
9*	T_d	112.8	1.393	1.123
10*	D_3	132.8	1.328	1.110
10*	D_{3d}	133.5	1.335	1.111
10*	D_{3h}	143.2	1.432	1.126
13*	D_{4d}	216.8	1.283	1.098
13*	O_h	265.1	1.568	1.140
15*	T_d	313.7	1.394	1.113
16*	D_2	333.4	1.302	1.098
17*	O_h	367.2	1.271	1.093
19*	T_h	469.8	1.301	1.096
22*	D_3	623.4	1.288	1.092

which is that the associated Skyrmion has very little symmetry, in fact only C_2, and this is already contained within the symmetry group of the $\mathcal{I}$-minimizing map, which is D_2.

Taking into account the above comments, we present, in Table 9.5, the symmetry K, and energy per baryon E/B, for all minimal energy Skyrmions with $B \leq 22$. These values were computed by relaxation of the full Skyrme energy function with initial conditions created from the corresponding rational map (see ref. [45] for further details). We also list the energy E, the ionization energy $I = E_{B-1} + E_1 - E_B$, which is the energy required to remove a single Skyrmion, and the binding energy per baryon given by $\Delta E/B = E_1 - (E/B)$, which is the energy required to separate the solution into single Skyrmions divided by the total baryon number.

In Fig. 9.4 we plot baryon density isosurfaces (to scale) for each of the Skyrmions with $7 \leq B \leq 22$, and also display models (not to scale) of the associated polyhedra. For all charges except $B = 9$ and $B = 13$ (which we discuss below) the Skyrmions are fullerene-like, and the associated polyhedra can be found in the classification of fullerenes [137].

A particularly interesting example is the $B = 17$ Skyrmion, which has the icosahedrally symmetric structure of the famous C_{60} Buckyball, as

Table 9.5. A summary of the symmetries and energies of the Skyrmion configurations which have been identified as the energy minima. Included also are the ionization energy I – that required to remove one Skyrmion – and the binding energy per Skyrmion $\Delta E/B$ – the energy required to split the charge B Skyrmion into B charge 1 Skyrmions divided by the total number of Skyrmions. (*) This symbol indicates Skyrmions whose angular form differs from that of the minimal energy solutions within the rational map ansatz. (**) The values quoted for $B = 14$ are computed using an initial configuration with D_2 symmetry.

B	K	E/B	E	I	$\Delta E/B$
1	$O(3)$	1.2322	1.2322	0.0000	0.0000
2	$D_{\infty h}$	1.1791	2.3582	0.1062	0.0531
3	T_d	1.1462	3.4386	0.1518	0.0860
4	O_h	1.1201	4.4804	0.1904	0.1121
5	D_{2d}	1.1172	5.5860	0.1266	0.1150
6	D_{4d}	1.1079	6.6474	0.1708	0.1243
7	Y_h	1.0947	7.6629	0.2167	0.1375
8	D_{6d}	1.0960	8.7680	0.1271	0.1362
9	D_{4d}	1.0936	9.8424	0.1578	0.1386
10*	D_3	1.0904	10.9040	0.1706	0.1418
11	D_{3h}	1.0889	11.9779	0.1583	0.1433
12	T_d	1.0856	13.0272	0.1829	0.1466
13	O	1.0834	14.0842	0.1752	0.1488
14**	C_2	1.0842	15.1788	0.1376	0.1480
15	T	1.0825	16.2375	0.1735	0.1497
16*	D_2	1.0809	17.2944	0.1753	0.1513
17	Y_h	1.0774	18.3158	0.2108	0.1548
18	D_2	1.0788	19.4184	0.1296	0.1534
19	D_3	1.0786	20.4934	0.1572	0.1536
20	D_{6d}	1.0779	21.5580	0.1676	0.1543
21	T_d	1.0780	22.6380	0.1522	0.1542
22*	D_3	1.0766	23.6852	0.1850	0.1556

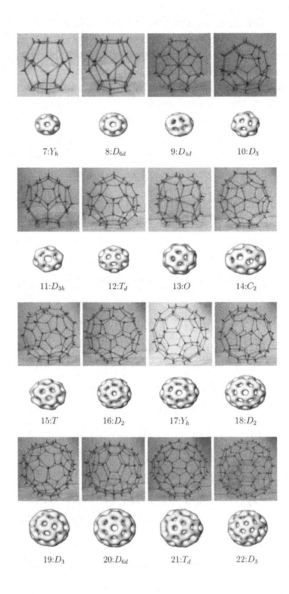

Fig. 9.4. Baryon density isosurfaces for $7 \leq B \leq 22$, and the associated symmetry groups and polyhedral models (not to scale).

indicated earlier. It is formed from 12 pentagons and 20 hexagons and is the structure with isolated pentagons having the least number of vertices. The decomposition which determines the relevant rational map is

$$\underline{18}|_Y = E_2' \oplus G' \oplus 2I', \qquad (9.76)$$

whose single two-dimensional component E_2' demonstrates that there is a unique Y-symmetric degree 17 map. In fact, the map is [193]

$$R(z) = \frac{17z^{15} - 187z^{10} + 119z^5 - 1}{z^2(z^{15} + 119z^{10} + 187z^5 + 17)}, \qquad (9.77)$$

and it is Y_h-symmetric.

In general, even in highly symmetric cases there will still be a few parameters in the family of symmetric maps of interest. For example, the decomposition

$$\underline{6n+4}|_T = nE' \oplus (n+1)E_1' \oplus (n+1)E_2', \qquad (9.78)$$

valid for any non-negative integer n, shows that there is an n-parameter family of tetrahedral maps of degree $B = 6n + 3$, corresponding to the middle component in the above. For $n = 0, 2, 3$ $(B = 3, 15, 21)$ this family includes the minimal energy map, and for $n = 1$ $(B = 9)$ it includes a map which is very close to minimal. Thus it seems possible that other members of this family will be minimal maps, for example, for $B = 27$. The explicit form of all the relevant rational maps for $B \leq 22$ can be found in ref. [45].

The charge $B = 9$ and $B = 13$ Skyrmions are not fullerene-like. Their symmetry groups, D_{4d} and O, both contain C_4 subgroups, and this is incompatible with the trivalent vertex structure of a fullerene. As can clearly be seen in Fig. 9.4, these Skyrmions both contain tetravalent vertices, which can be obtained from fullerenes by a process known as symmetry enhancement (see Fig. 9.5).

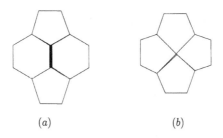

(a)   (b)

Fig. 9.5. An illustration of symmetry enhancement.

Consider part of a fullerene with the form shown in Fig. 9.5(a), consisting of two pentagons and two hexagons with a C_2 symmetry. The symmetry enhancement process shrinks the edge common to the two hexagons (the thick line) to zero length, resulting in the coalescence of two vertices. The object formed is shown in Fig. 9.5(b). It has a tetravalent vertex connecting four pentagons and the symmetry is enhanced to C_4. We find, empirically, that pairs of symmetry enhancement processes occur on antipodal edges of a fullerene structure.

There is a C_{28} fullerene with D_2 symmetry (denoted 28:1 in ref. [137]) that contains two of the structures shown in Fig. 9.5(a). If symmetry enhancement is performed on both, then the resulting object is precisely the D_{4d} configuration of the $B = 9$ Skyrmion described earlier. There are also D_2-symmetric C_{44} fullerenes (denoted 44:75 and 44:89 in ref. [137]) with an equal number of pentagons and hexagons (12 of each), and a very symmetric configuration can be obtained by symmetry enhancement at all six possible vertices, which results in the cubic $B = 13$ Skyrmion.

In the context of fullerenes it is, of course, impossible for vertices to coalesce since they correspond to the positions of the carbon atoms, but for Skyrmions the vertices represent concentrations of the baryon density and they need not be distinct; it just appears that in most cases it is energetically favourable to have distinct vertices. Note that, by an examination of the baryon density isosurface by eye, it can often be difficult to identify whether a given vertex is trivalent or tetravalent, since the edge length which must be zero for symmetry enhancement could be small, but non-zero.

Although we do not have a general global characterization of the vertices of the polyhedron associated with a rational map (as we do for the face centres, via the Wronskian) it is possible, by a local analysis of the rational map, to check whether a given point is a local baryon density maximum and to obtain its valency. By using the freedom to perform rotations of both the domain and target 2-spheres it is always possible to choose the given point to be $z = 0$ and the rational map to have a local expansion

$$R(z) = \alpha(z + \beta z^{p+1} + O(z^{p+2})), \qquad (9.79)$$

where α and β are real positive constants. (The derivative of the map is non-zero at $z = 0$, since the baryon density is assumed to be non-zero there.)

Substituting the expansion (9.79) into the expression for the angular contribution to the baryon density (9.59) we obtain the following result. If $p = 1$ then $z = 0$ is not a vertex. If $p > 2$ and $\alpha > 1$ then $z = 0$ is a p-valent vertex, with the baryon density being a local maximum there. The remaining case of $p = 2$ is a little more subtle. In many cases, all the local maxima of the baryon density correspond to vertices of the polyhedron. However, in some cases (the lowest charge example being $B = 5$) some of the maxima are at edge midpoints. Such edges may consequently appear thicker than others. The rational map description of such a bivalent maximum is the $p = 2$ case, and a local maximum requires that $\alpha > \sqrt{1 + 3\beta}$.

To illustrate this analysis, consider the cubic $B = 13$ Skyrmion. For computing O-symmetric degree 13 maps the relevant decomposition is

$$\underline{14}|_O = 2E_2' \oplus E_1' \oplus 2G' . \tag{9.80}$$

From the $2E_2'$ component there is a 1-parameter family of maps, with the explicit form

$$R(z) = \frac{z(a + (6a - 39)z^4 - (7a + 26)z^8 + z^{12})}{1 - (7a + 26)z^4 + (6a - 39)z^8 + az^{12}} , \tag{9.81}$$

whose minimal value of $\mathcal{I}$ occurs at $a = 0.40 + 5.18i$. This gives a Skyrme field whose baryon density is virtually identical to the one shown in Fig. 9.4. The associated polyhedron is similar to a cube, each face of which consists of four pentagons with a tetravalent bond. In order for them to fit together, with all the other bonds being trivalent, each of the six faces must be rotated slightly relative to the one diametrically opposite, which removes the possibility of the cube having reflection symmetries and symmetry group O_h. The polyhedron has 24 pentagonal faces, as opposed to the 12 pentagons and 12 hexagons that would have been expected of a fullerene structure. Expanding the map (9.81) about $z = 0$ gives

$$R(z) = az + z^5(7a^2 - 32a - 39) + \cdots , \tag{9.82}$$

and since $|a| > 1$, a comparison with Eq. (9.79) confirms that the point $z = 0$ is a tetravalent vertex. The $B = 9$ minimizing map also contains tetravalent vertices (this time two of them) and this can be checked in a similar way.

A more global characterization of the vertices would be useful. Usually they correspond to local maxima of the integrand defining $\mathcal{I}$ in Eq. (9.63). This density depends on the modulus of the rational map and its derivative, but there is generally no simple formula for finding its maxima. However, in particularly symmetric cases the vertices can be identified with the zeros of the Hessian. Explicitly, the Hessian is the polynomial

$$H(z) = (2B - 2)W(z)W''(z) - (2B - 3)W'(z)^2 , \tag{9.83}$$

where $W(z)$ is the Wronskian. It has degree $4(B - 2)$, which is consistent with the GEM rule for the number of vertices. For example, for the icosahedral rational map describing the minimal energy $B = 7$ Skyrmion,

$$R(z) = \frac{z^7 - 7z^5 - 7z^2 - 1}{z^7 + 7z^5 - 7z^2 + 1},\qquad (9.84)$$

the Hessian is

$$H(z) = -8624(z^{20} - 228z^{15} + 494z^{10} + 228z^5 + 1),\qquad (9.85)$$

which is proportional to the Klein polynomial $\mathcal{Y}_f$ associated with the vertices of a dodecahedron [237].

9.7 Lattices, crystals and shells

So far we have only discussed Skyrmions with a finite baryon number, but in fact the lowest known value for the energy per baryon, E/B, occurs for an infinite crystal of Skyrmions. As we have seen, for certain relative orientations, well separated Skyrmions attract. At high density it is expected that the Skyrmions will form a crystal, though a crystal structure has not yet been seen dynamically for a finite baryon number, probably due to the fact that so far only simulations up to $B = 22$ have been performed.

To study Skyrmion crystals one imposes periodic boundary conditions on the Skyrme field and works within a unit cell (equivalently, 3-torus) $\mathbb{T}^3$. The first attempted construction of a crystal was by Klebanov [235], using a simple cubic lattice of Skyrmions whose symmetries maximize the attraction between nearest neighbours. After relaxation, Klebanov's crystal has an energy 1.08 per baryon. Other symmetries were proposed which lead to slightly lower, but not minimal, energy crystals [160, 221]. Following the work of Castillejo *et al.* [75] and Kugler and Shtrikman [248], it is now understood that it is best to arrange the Skyrmions initially as a face-centred cubic lattice, with their orientations chosen symmetrically to give maximal attraction between all nearest neighbours. Explicitly, the Skyrme field is strictly periodic after translation by $2L$ in the x^1, x^2 or x^3 directions. A unit cell is a cube of side length $2L$, with Skyrmions in standard orientation on the vertices, and further Skyrmions at the face centres, each rotated by $180°$ about the axis which is normal to the face. With this set-up each Skyrmion has twelve nearest neighbours which are all in the attractive channel. Inside one unit cell, the total baryon number is $B = 4$. If we fix the origin at the centre of one of the unrotated Skyrmions, this configuration has the combined spatial plus isospin symmetries generated by

$$(x^1, x^2, x^3) \mapsto (-x^1, x^2, x^3),\ (\sigma, \pi_1, \pi_2, \pi_3) \mapsto (\sigma, -\pi_1, \pi_2, \pi_3);\qquad (9.86)$$

$$(x^1, x^2, x^3) \mapsto (x^2, x^3, x^1), \; (\sigma, \pi_1, \pi_2, \pi_3) \mapsto (\sigma, \pi_2, \pi_3, \pi_1); \qquad (9.87)$$

$$(x^1, x^2, x^3) \mapsto (x^1, x^3, -x^2), \; (\sigma, \pi_1, \pi_2, \pi_3) \mapsto (\sigma, \pi_1, \pi_3, -\pi_2); \qquad (9.88)$$

$$(x^1, x^2, x^3) \mapsto (x^1 + L, x^2 + L, x^3), \; (\sigma, \pi_1, \pi_2, \pi_3) \mapsto (\sigma, -\pi_1, -\pi_2, \pi_3). \qquad (9.89)$$

Symmetry (9.86) is a reflection in a face of the cube, (9.87) is a rotation around a three-fold axis along a diagonal, (9.88) is a four-fold rotation around an axis through opposite face centres, and (9.89) is a translation from the corner of the cube to a face centre.

At low densities (large L), the Skyrmions are localized around their lattice positions, each having an almost spherical isosurface where $\sigma = 0$, separating the core of the Skyrmion ($\sigma < 0$) from its tail ($\sigma > 0$). Since the Skyrmions are well separated, the average value of σ in a unit cell, $\langle \sigma \rangle$, is close to one.

As the density is increased (that is, L reduced) the energy decreases and there is a phase transition to a crystal of half-Skyrmions. At this point the symmetry is increased by the addition of the generator

$$(x^1, x^2, x^3) \mapsto (x^1 + L, x^2, x^3), \quad (\sigma, \pi_1, \pi_2, \pi_3) \mapsto (-\sigma, -\pi_1, \pi_2, \pi_3), \qquad (9.90)$$

a translation half-way along the cube edge. Note that this symmetry involves a chiral $SO(4)$ rotation, rather than just an $SO(3)$ isospin transformation as before. The previous translational symmetry (9.89) can be obtained by applying this new generator, together with this generator rotated by 90°.

This phase is where the minimal energy Skyrme crystal occurs. The $\sigma < 0$ and $\sigma > 0$ regions are perfect cubes of side length L, with $\sigma = 0$ on all the faces. Each cube has identical pion field distributions and baryon number $\frac{1}{2}$. For this configuration, $\langle \sigma \rangle = 0$, and there is a restoration of chiral symmetry. The minimum of the energy occurs at $L \approx 4.7$. A variational method, based on a truncated Fourier series expansion of the fields, approximates the energy per baryon to be $E/B = 1.038$, and a recent numerical calculation [42], using far larger grids than previous studies [75], gives a very similar value of $E/B = 1.036$. In Fig. 9.6 we plot a baryon density isosurface for the Skyrme crystal. Each lump represents a half-Skyrmion and the total baryon number shown is 4. The fields obtained either numerically, or by optimizing the Fourier series, are very well approximated by the analytic formulae [75]

$$\sigma = c_1 c_2 c_3, \qquad (9.91)$$

$$\pi_1 = -s_1 \sqrt{1 - \frac{s_2^2}{2} - \frac{s_3^2}{2} + \frac{s_2^2 s_3^2}{3}} \qquad \text{and cyclic}, \qquad (9.92)$$

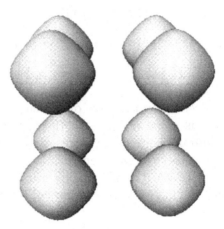

Fig. 9.6.   A baryon density isosurface for a portion of the Skyrme crystal.

where $s_i = \sin(\pi x^i/L)$ and $c_i = \cos(\pi x^i/L)$. This approximation to the Skyrme crystal field has the right symmetries and is motivated by an exact solution for a crystal in the two-dimensional $O(3)$ sigma model, which has a similar form but with the trigonometric functions replaced by Jacobi elliptic functions.

Table 9.5 shows that the energy per baryon of the shell-like Skyrmions is decreasing as B increases, but is still some way above that of the Skyrme crystal. The asymptotic value of E/B for the shell-like structures for large B can be compared with the value for the crystal by studying yet another periodic arrangement of Skyrmions, a two-dimensional lattice, rather than a three-dimensional crystal.

In very large fullerenes, where hexagons are dominant, the twelve pentagons may be viewed as defects inserted into a flat structure, to generate the curvature necessary to close the shell. Energetically, the optimum infinite structure is a hexagonal lattice, that is, a graphite sheet – the most stable form of carbon thermodynamically. The reason that closed shells are preferred for a finite number of carbon atoms is that the penalty for introducing the pentagonal defects is not as severe as that incurred by having dangling bonds at the edges of a truncated graphite sheet. A prediction of the fullerene approach to Skyrmions is the existence of a Skyrme field analogous to a graphite sheet. This configuration would have infinite energy, since it has infinite extent in two directions, but its energy per baryon should be lower than that of any of the known finite energy Skyrmions, and will be the asymptotic value approached by large fullerene-like Skyrmions.

Such a hexagonal Skyrme lattice can be constructed using the ansatz of ref. [42]

$$U(x^1, x^2, x^3) = \exp\left(\frac{if}{1 + |R|^2}(R\tau_- + \bar{R}\tau_+ + (1 - |R|^2)\tau_3)\right), \quad (9.93)$$

a variant of the rational map ansatz. Here $\tau_\pm = \tau_1 \pm i\tau_2$, R is a meromorphic, periodic function of $z = x^1 + ix^2$, and f is a real function of x^3 chosen so that the Skyrme lattice physically occupies the (x^1, x^2) plane. The direction of the vector of pion fields is determined by $R(z)$, whereas the magnitude of the vector also depends on the profile function f, and hence on the height above or below the lattice. If Ω_1 and Ω_2 are the fundamental periods of $R(z)$, then

$$U(z + n\Omega_1 + m\Omega_2, x^3) = U(z, x^3) \quad \forall\, n, m \in \mathbb{Z}. \quad (9.94)$$

Let $\mathbb{T}^2$ denote the associated torus, the parallelogram in the complex plane with vertices $0, \Omega_1, \Omega_2, \Omega_1 + \Omega_2$ and opposite edges identified.

To understand the boundary conditions on f we need to recall our motivation. The lattice is being thought of as an infinite limit of the shell-like Skyrmions containing pentagons and hexagons. Thus, below the lattice is the outside of the shell, where $U \to 1_2$. Above the lattice is the inside of the shell, where the Skyrme field is approaching the value associated with the centre of the Skyrmion, so $U \to -1_2$. We therefore require

$$f(-\infty) = 0, \; f(\infty) = \pi. \quad (9.95)$$

This implies that the Skyrme lattice is a novel domain wall, separating differing vacua.

To compute the baryon number and energy of the Skyrme field (9.93) it is again convenient to use the geometrical strain formulation. The strain in the direction normal to the lattice is orthogonal to the two strains tangential to the lattice, which are equal. λ_i may be interpreted as the strain in the x^i direction, and it is easy to show that

$$\lambda_1 = \lambda_2 = 2J \sin f, \quad \lambda_3 = f', \quad (9.96)$$

where

$$J = \frac{1}{1 + |R|^2}\left|\frac{dR}{dz}\right|. \quad (9.97)$$

Therefore, the energy and baryon densities (9.12) are

$$\mathcal{E} = \frac{1}{12\pi^2}\left(f'^2 + 8J^2(f'^2 + \sin^2 f) + 16J^4 \sin^4 f\right), \quad (9.98)$$

$$\mathcal{B} = \frac{2}{\pi^2}J^2 f' \sin^2 f. \quad (9.99)$$

We now compute the baryon number B in a fundamental region of the lattice, $x^3 \in (-\infty, \infty)$ and $(x^1, x^2) \in \mathbb{T}^2$. Since R is a map from $\mathbb{T}^2$ to S^2, its degree, k, is the integral over $\mathbb{T}^2$ of the pull-back of the area 2-form on S^2, $dRd\bar{R}/(1 + |R|^2)^2$, that is

$$k = \frac{1}{\pi} \int_{\mathbb{T}^2} J^2 \, dx^1 dx^2, \qquad (9.100)$$

since R is a holomorphic function of z. Using (9.99) it is now easy to see that the baryon number is equal to the degree k, since

$$B = \frac{2}{\pi^2} \int_{-\infty}^{\infty} f' \sin^2 f \, dx^3 \int_{\mathbb{T}^2} J^2 \, dx^1 dx^2 = \frac{k}{\pi} \left[f - \frac{1}{2} \sin 2f \right]_{-\infty}^{\infty} = k, \qquad (9.101)$$

using (9.100) and the boundary conditions (9.95).

To calculate the energy E in the fundamental region it is useful to introduce a scale parameter μ, write $u = x^3/\mu$ and set $f(x^3) = g(u)$. Then, if A is the area of the fundamental torus $\mathbb{T}^2$, integrating the density (9.98) gives

$$E = \int_{-\infty}^{\infty} dx^3 \int_{\mathbb{T}^2} \mathcal{E} \, dx^1 dx^2 = \frac{A}{\mu} E_1 + \frac{1}{\mu} E_2 + \mu E_3 + \frac{\mu}{A} E_4, \qquad (9.102)$$

where

$$E_1 = \frac{1}{12\pi^2} \int_{-\infty}^{\infty} g'^2 \, du, \quad E_2 = \frac{2k}{3\pi} \int_{-\infty}^{\infty} g'^2 \sin^2 g \, du,$$

$$E_3 = \frac{2k}{3\pi} \int_{-\infty}^{\infty} \sin^2 g \, du, \quad E_4 = \frac{4\mathcal{I}}{3\pi^2} \int_{-\infty}^{\infty} \sin^4 g \, du. \qquad (9.103)$$

E depends on the map R only through the quantity

$$\mathcal{I} = A \int_{\mathbb{T}^2} J^4 \, dx^1 dx^2, \qquad (9.104)$$

a combination independent of A. The scale μ and area A are fixed in terms of the E_i's, by minimizing (9.102). Requiring $\frac{\partial E}{\partial \mu} = \frac{\partial E}{\partial A} = 0$ gives

$$\mu = \sqrt{E_2/E_3}, \quad A = \sqrt{E_2 E_4/E_1 E_3}, \qquad (9.105)$$

and hence the minimized energy is

$$E = 2(\sqrt{E_1 E_4} + \sqrt{E_2 E_3}). \qquad (9.106)$$

To proceed further we choose $R(z)$ to be an elliptic function with a hexagonal period lattice. The simplest is the Weierstrass function $\wp(z)$ satisfying

$$\wp'^2 = 4(\wp^3 - 1), \qquad (9.107)$$

which has periods $\Omega_1 = \Gamma(\frac{1}{6})\Gamma(\frac{1}{3})/(2\sqrt{3\pi})$ and $\Omega_2 = \Omega_1 \exp(\pi i/3)$. Obviously we can scale both the elliptic function and its argument and still have a hexagonal period lattice; hence we take

$$R(z) = c\wp(z/\alpha),\tag{9.108}$$

where c and α are arbitrary real constants. For computational purposes it is actually more convenient to work with a rectangular fundamental torus, $(x^1, x^2) \in [0, \alpha\Omega_1] \times [0, \alpha\sqrt{3}\Omega_1]$, whose area is $A = \sqrt{3}\alpha^2\Omega_1^2$. As this torus contains two fundamental parallelograms and the $\wp$-function has one double pole in each, then by counting preimages, we see that the degree of the map from the rectangular torus to the sphere is $k = 4$.

E is minimized by choosing c so as to minimize $\mathcal{I}$. The minimal value is $\mathcal{I} \approx 193$, when $c \approx 0.7$. (Recall that $\mathcal{I}$ is independent of α.)

We now make the simplifying ansatz that $g(u)$ is the sine-Gordon kink profile function

$$g(u) = 2\tan^{-1} e^u,\tag{9.109}$$

which is a reasonably good choice, and has the advantage that all the integrals in (9.103) can be performed exactly. The results are

$$E_1 = \frac{1}{6\pi^2}, \quad E_2 = \frac{32}{9\pi}, \quad E_3 = \frac{16}{3\pi}, \quad E_4 = \frac{16\,\mathcal{I}}{9\pi^2},\tag{9.110}$$

from which we find that the scale and area are

$$\mu = \sqrt{\frac{2}{3}}, \quad A = \frac{8}{3}\sqrt{\mathcal{I}},\tag{9.111}$$

and using (9.106), that the energy is

$$E = \frac{4}{3\pi^2}\sqrt{\frac{2}{3}}(\sqrt{\mathcal{I}} + 8\pi).\tag{9.112}$$

Recalling the numerical value of $\mathcal{I}$, and that $B = k = 4$, we thus find an energy per baryon

$$E/B = 1.076.\tag{9.113}$$

The true lattice has been determined by numerical relaxation, using the ansatz above, involving the Weierstrass function and sine-Gordon kink profile, to give a starting approximation [42]. Its energy is found to be

$$E/B = 1.061.\tag{9.114}$$

In Fig. 9.7 we display a surface of constant baryon density for this hexagonal Skyrme lattice. The structure is clearly visible, the baryon density

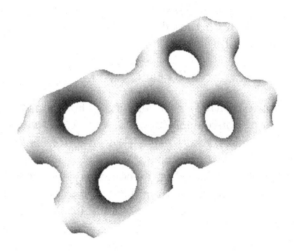

Fig. 9.7. A baryon density isosurface for a portion of the Skyrme lattice.

having a hole in the centre of each of the hexagonal faces. Note that the displayed region contains exactly eight full hexagons and has baryon number 4, so each hexagon may be thought of as having baryon number $\frac{1}{2}$. This is the expected limit of the polyhedron structures discussed earlier, where a charge B Skyrmion has $2(B-1)$ faces. Other lattices, such as a tetravalent square lattice, can be created by choosing a Weierstrass function different from (9.107), but these have energies which are slightly higher than the trivalent hexagonal lattice.

Since the energy per baryon of the Skyrme lattice exceeds that of the Skyrme crystal it is reasonable to expect that above some critical charge, the minimal energy Skyrmion will resemble a portion of the crystal rather than a shell constructed from the planar lattice by inserting pentagonal defects. As the crystal is basically a stack of $B = 4$ cubes, $B = 32$ is the first charge at which any sizeable, symmetric chunk of the crystal can emerge. Attempts have been made [35] to construct Skyrme fields by cutting out a portion of the crystal and interpolating its surface fields to the vacuum, but these all have rather high energies.

An alternative to either a single-shell or crystal structure is a two-shell structure. This has been investigated [290] using yet another variant of the rational map ansatz,

$$U(r, z) = \exp(\theta(r_0 - r)if_1(r)\hat{\mathbf{n}}_{R_1(z)} \cdot \boldsymbol{\tau} + \theta(r - r_0)if_2(r)\hat{\mathbf{n}}_{R_2(z)} \cdot \boldsymbol{\tau}), \quad (9.115)$$

where $\theta(r)$ is the Heaviside step function and r_0 is a radius where the two shells meet. The two profile functions, f_1 and f_2, satisfy the boundary conditions $f_1(0) = 2\pi$, $f_1(r_0) = f_2(r_0) = \pi$, $f_2(\infty) = 0$, and the angular distributions of the fields on the two shells are determined by two rational maps R_1 and R_2, with degrees k_1 and k_2 respectively. The baryon number of this configuration is $B = k_1 + k_2$. The multi-shell generalization is obvious.

Some two-shell and three-shell configurations for $B = 12, 13, 14$ have been studied, and also used as initial configurations in a numerical relaxation of the full Skyrme energy. In most cases they relax to a single-shell structure, with energy a bit higher than that described in Section 9.6, so they probably describe saddle points. Note that two-shell configurations have $U = 1_2$ at the origin, so can not relax to the minimal energy single-shell Skyrmions discussed in Section 9.6, for which $U = -1_2$ there.

The two-shell ansatz with baryon number $k_1 + k_2$ has an interpretation in terms of $k_1 + k_2$ individual Skyrmions on a single shell, which is often the end point of a numerical relaxation. To see this, consider $U(r, z)$ for a given value of z, and compare the values of U at the two radii where $f_1(r) = \frac{3}{2}\pi$ and $f_2(r) = \frac{1}{2}\pi$. If these values are close, the field configuration along this radial line can be relaxed to be approximately constant, but if they are antipodal then the radial gradient energy is large and may be interpreted as due to a single Skyrmion at $r = r_0$, with angular location z. The condition that the values of U are antipodal is that $R_1(z) = R_2(z)$, since the rational maps then have the same value but the profile functions have opposite sign, that is, $\sin f_1 = -1$, $\sin f_2 = 1$. If $R_1 = p_1/q_1$ and $R_2 = p_2/q_2$ then the antipodal condition is

$$p_1(z)q_2(z) - p_2(z)q_1(z) = 0, \qquad (9.116)$$

which is a polynomial equation of degree $k_1 + k_2$. The $k_1 + k_2$ roots determine the angular locations of the Skyrmions on the shell $r = r_0$.

In summary, there are a number of alternatives to a single-shell structure for Skyrmions and what is remarkable is that none of these alternatives appears to give minimal energy Skyrmions for $B \leq 22$. However, single-shells can not be the whole story for large enough baryon number.

9.8 Skyrmion dynamics

In the preceding sections we have been concerned with static Skyrmions, but in this section we turn to Skyrmion dynamics and scattering. To begin with, we describe how some of the static, symmetric, minimal energy Skyrmions can be formed from the collision of well separated single Skyrmions [40].

The time dependent Skyrme field equation is solved using a finite difference method (see ref. [45] for a detailed discussion), which is most conveniently implemented using a nonlinear sigma model formulation. Explicitly, the Skyrme field is parametrized by the unit 4-vector $\phi = (\sigma, \pi_1, \pi_2, \pi_3)$, in terms of which the Lagrangian density becomes

$$\mathcal{L} = \partial_\mu \phi \cdot \partial^\mu \phi - \tfrac{1}{2}(\partial_\mu \phi \cdot \partial^\mu \phi)^2 + \tfrac{1}{2}(\partial_\mu \phi \cdot \partial_\nu \phi)(\partial^\mu \phi \cdot \partial^\nu \phi) + \lambda(\phi \cdot \phi - 1), \quad (9.117)$$

with the Lagrange multiplier λ introduced in order to enforce the constraint $\phi \cdot \phi = 1$.

The Euler-Lagrange equation is

$$\begin{aligned}(1 - \partial_\mu \phi \cdot \partial^\mu \phi)\partial_\alpha \partial^\alpha \phi \quad &- \quad (\partial^\nu \phi \cdot \partial_\mu \partial_\nu \phi - \partial_\mu \phi \cdot \partial_\alpha \partial^\alpha \phi)\partial^\mu \phi \\ &+ \quad (\partial^\mu \phi \cdot \partial^\nu \phi)\partial_\mu \partial_\nu \phi - \lambda \phi = 0, \quad (9.118)\end{aligned}$$

where λ can be calculated by contracting (9.118) with ϕ and using the second derivative of the constraint, giving

$$\lambda = -(\partial_\mu \phi \cdot \partial_\nu \phi)(\partial^\mu \phi \cdot \partial^\nu \phi) - (1 - \partial_\mu \phi \cdot \partial^\mu \phi)\partial_\nu \phi \cdot \partial^\nu \phi. \quad (9.119)$$

The simplest possible scattering event involves the head-on collision of two Skyrmions in the attractive channel. As discussed in Section 9.3, an initial configuration can be constructed using the product ansatz $U = U^{(1)}U^{(2)}$ for well separated Skyrmions, each of which may also be independently Lorentz boosted. An example that has been calculated has an initial configuration consisting of two Skyrmions with positions

$$\mathbf{X}_1 = (0, 0, a), \quad \mathbf{X}_2 = (0, 0, -a), \quad (9.120)$$

where $a = 1.5$; the second Skyrmion is rotated relative to the first by a 180° rotation around the x^2-axis, and each Skyrmion is Lorentz boosted towards the other with a velocity $v = 0.3$, in order to speed up the interaction.

Figure 9.8 shows an isosurface plot of the baryon density at regular time intervals. We see that the initially well separated Skyrmions deform as they come together, before coalescing into a toroidal configuration very close to the exact minimal energy $B = 2$ Skyrmion. The torus then breaks up, with the result that the Skyrmions scatter at right angles. This right-angle scattering was predicted analytically [283] and is a familiar property of two-soliton scattering; for example, we have already seen that monopoles and vortices exhibit this behaviour. The Skyrmions then attract once more and pass through the torus again. This almost elastic process repeats itself a number of times, with a little energy being radiated each time, eventually settling down to the exact static solution.

In order to discuss attractive configurations of $B > 2$ Skyrmions we first introduce some notation. Take the positions of the single Skyrmions to be $\mathbf{X}_i$ for $i = 1, \ldots, B$, and define the relative position vectors $\mathbf{X}_{ij} = \mathbf{X}_i - \mathbf{X}_j$.

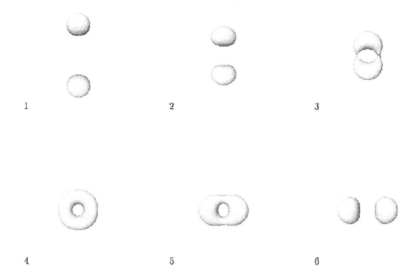

Fig. 9.8. Baryon density isosurfaces at increasing times during the head-on collision of two Skyrmions.

Suppose the orientation of the Skyrmion at $\mathbf{X}_i$ relative to that at $\mathbf{X}_j$ is obtained by a rotation by 180° about an axis with unit vector $\mathbf{n}_{ij}$. Then all pairs will mutually, maximally attract if $\mathbf{X}_{ij} \cdot \mathbf{n}_{ij} = 0$ (no sum) for all $i \neq j$.

Three Skyrmions can scatter close to the tetrahedral $B = 3$ Skyrmion. In choosing Skyrmion initial configurations, the analogous monopole scattering is a good guide. Recall from Chapter 8 that the tetrahedral 3-monopole is formed during the C_3-symmetric scattering in which three monopoles are initially on the vertices of a large contracting equilateral triangle. We therefore take three well separated Skyrmions in such a configuration, with

$$\mathbf{X}_1 = (-a, -a, -a), \quad \mathbf{X}_2 = (-a, a, a), \quad \mathbf{X}_3 = (a, -a, a). \qquad (9.121)$$

The first Skyrmion is in standard orientation, and the orientations of the second and third are fixed by taking

$$\mathbf{n}_{12} = (1, 0, 0), \quad \mathbf{n}_{13} = (0, 1, 0). \qquad (9.122)$$

This implies that $\mathbf{n}_{23} = (0, 0, 1)$, so all pairs are in the attractive channel, since $\mathbf{X}_{ij} \cdot \mathbf{n}_{ij} = 0$ for all $i \neq j$.

Again we choose $a = 1.5$, and this time each Skyrmion is boosted to have an initial velocity of $v = 0.17$ towards the centre of the triangle. The evolution of this configuration is shown in Fig. 9.9.

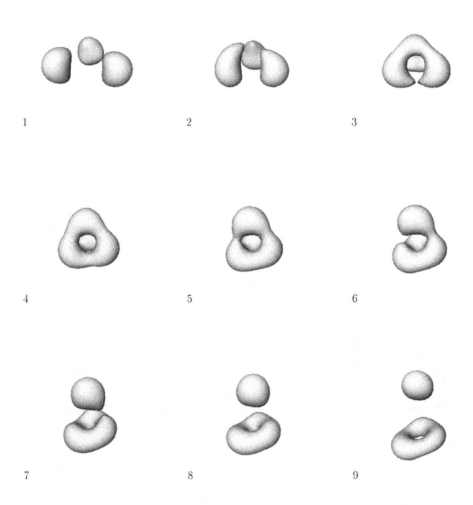

Fig. 9.9.   Baryon density isosurfaces at increasing times during the scattering of three Skyrmions with approximate C_3 symmetry.

We should point out that the C_3 symmetry is slightly broken by the product ansatz implementation of the initial data, $U = U^{(1)}U^{(2)}U^{(3)}$, which is clearly asymmetric under permutations of the indices. If a were larger, the product ansatz would be closer to having exact cyclic symmetry.

The Skyrmions deform as they coalesce, and each behaves slightly differently. The dynamics is, nonetheless, remarkably similar to the monopole case, except for the influence of the varying potential energy, in that the Skyrmions form an approximately tetrahedral configuration, which then splits into a single Skyrmion and a charge 2 torus.

We have seen a second scattering process passing through the tetrahedral 3-monopole – the twisted line scattering of three collinear monopoles. A similar scattering process also occurs for three collinear Skyrmions with appropriate orientations [40].

Recall that four monopoles on the vertices of a contracting regular tetrahedron scatter through the cubic charge 4 solution. There is an analogous four-Skyrmion scattering process. To the $B = 3$ system given by (9.121) and (9.122), we add a fourth Skyrmion at $\mathbf{X}_4 = (a, a, -a)$ with orientation given by $\mathbf{n}_{14} = (0, 0, 1)$. This completes a regular tetrahedron. The additional relative orientations are $\mathbf{n}_{24} = (0, 1, 0)$ and $\mathbf{n}_{34} = (1, 0, 0)$, so still we have $\mathbf{X}_{ij} \cdot \mathbf{n}_{ij} = 0$ for all $i \neq j$, and all Skyrmion pairs maximally attract. Once more we take $a = 1.5$, but this time no initial Lorentz boosts are required, because of the strong attractions.

The evolution of this configuration is displayed in Fig. 9.10. The mutual attractions cause the Skyrmions to coalesce and form a cubic configuration. This then splits up, and the Skyrmions are found on the vertices of a tetrahedron dual to the initial one. Again the product ansatz implementation results in the tetrahedral symmetry being only approximately attained. Aside from this technicality, however, the scattering process is once again a close copy of what happens for monopoles.

Another configuration is four Skyrmions on the corners of the square

$$\mathbf{X}_1 = (a, a, 0), \ \ \mathbf{X}_2 = (a, -a, 0), \ \ \mathbf{X}_3 = (-a, -a, 0), \ \ \mathbf{X}_4 = (-a, a, 0).$$
$$(9.123)$$

If

$$\mathbf{n}_{12} = (1, 0, 0), \quad \mathbf{n}_{13} = (0, 0, 1), \quad \mathbf{n}_{14} = (0, 1, 0), \quad (9.124)$$

then $\mathbf{n}_{23} = (0, 1, 0)$, $\mathbf{n}_{24} = (0, 0, 1)$, $\mathbf{n}_{34} = (1, 0, 0)$ which implies that all pairs mutually attract. The dynamics of this configuration is exhibited in Fig. 9.11 for initial conditions with no Lorentz boost. The initial D_4-symmetric configuration scatters through the $B = 4$ cube and emerges as two $B = 2$ tori; yet another well known monopole process.

Given that N-monopole dynamics at low energy can be well approximated by geodesic motion on the monopole moduli space, a natural ques-

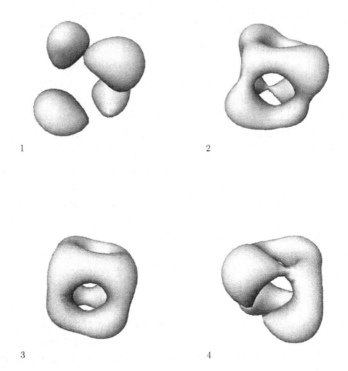

1

2

3   4

Fig. 9.10. Baryon density isosurfaces at increasing times during the scattering of four Skyrmions with approximate tetrahedral symmetry.

tion is whether a similar moduli space approximation exists for Skyrmions. Since there are weak forces between Skyrmions, the moduli space of the exact minimal energy Skyrmion of charge B does not contain adequate degrees of freedom to describe all the required low energy configurations. It is at most nine-dimensional, corresponding to the action of translations, rotations and isospin rotations on the otherwise unique solution. Another manifold $\mathcal{M}_B$, whose coordinates parametrize a suitably larger set of low energy field configurations, is required. Ideally, $\dim \mathcal{M}_B = 6B$, since this is the dimension of the space of B well separated Skyrmions with all possible orientations.

An obvious candidate for $\mathcal{M}_B$ is the parameter space of field configurations obtained using the product ansatz for B Skyrmions. This is certainly $6B$-dimensional and adequately describes well separated Skyrmions, but it is not acceptable since the product ansatz fails near the minimal energy charge B Skyrmion. For example, the product ansatz for two Skyrmions satisfactorily defines $\mathcal{M}_2$ when the Skyrmion separation is large compared to the Skyrmion size, and the energy initially decreases in the attractive

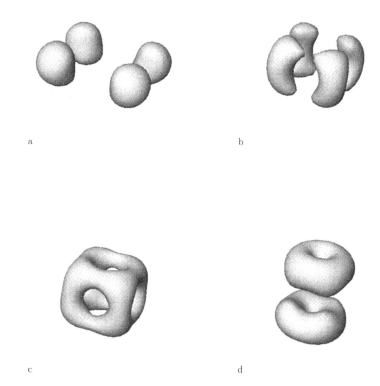

Fig. 9.11. Baryon density isosurfaces at increasing times during the scattering of four Skyrmions with approximate D_4 symmetry.

channel as the separation is reduced, as we have seen from the calculation of the asymptotic interaction energy in Section 9.3. However, as the separation is reduced further the interaction energy obtained from the product ansatz begins to increase [218] and a product of coincident Skyrmions does not resemble the minimal energy $B = 2$ torus.

A more promising definition [284] of $\mathcal{M}_2$ is as the unstable manifold of the spherically symmetric $B = 2$ hedgehog solution, which we discussed in Section 9.2. This saddle point solution may be thought of as two coincident Skyrmions, with one wrapped around the other. (It is also well approximated in the product ansatz by two coincident Skyrmions with the same orientation.) It has six unstable modes, and six zero modes, three translational and three rotational. Of the six unstable modes, three correspond to rotating one of the Skyrmions with respect to the other, while three are associated with separating the Skyrmions. The union of gradient flow curves descending from the $B = 2$ hedgehog in all possible positions and orientations is a 12-dimensional manifold. A generic curve will end at the minimal energy $B = 2$ torus, but a submanifold of curves will end at infinitely separated Skyrmions. Curves close to this submanifold will go out to well separated Skyrmions and then return to the torus. This definition of $\mathcal{M}_2$ will therefore include well separated Skyrmions in all possible orientations, but it will also include the low energy configurations where the Skyrmions are close together.

An attempt has been made to construct $\mathcal{M}_2$ numerically [409] by solving the gradient flow equation

$$R_0 - \frac{1}{4}[R_i, [R_i, R_0]] = \partial_i(R_i - \frac{1}{4}[R_j, [R_j, R_i]]).\tag{9.125}$$

Particularly interesting is one of the steepest and shortest gradient flow trajectories, where the constituent Skyrmions of the hedgehog simultaneously separate a little, and twist, then recombine into the torus. A systematic construction of some two-dimensional submanifolds of $\mathcal{M}_2$ has been carried out, and with the action of the nine-dimensional symmetry group this is effectively a construction of some 11-dimensional submanifolds of $\mathcal{M}_2$. The 10-dimensional attractive channel of two Skyrmions has also been found using the gradient flow, starting with well separated Skyrmions. However, it is difficult numerically to implement gradient flow in regions where the Skyrmions are well separated. As a technical simplification, in this region the product ansatz can be used. In fact for well separated Skyrmions the gradient flow equations within the product ansatz can be solved exactly [211]. In conclusion, the work in [409] and [211] shows that it is feasible, if difficult, to construct $\mathcal{M}_2$ using numerical gradient flow.

Given the manifold $\mathcal{M}_2$ one can now attempt to define a truncated dynamics on it by restriction of the Skyrme Lagrangian. Note that, unlike the moduli space approximation for monopoles, there will be both a non-trivial metric and potential energy function on $\mathcal{M}_2$. These have been partly calculated in ref. [409]. The potential is easy to calculate along any gradient flow curve. The metric coefficient along a gradient flow curve can be inferred from the rate at which the potential energy decreases. Several of the remaining metric coefficients are (spin and isospin) moments of inertia of the configurations generated during the gradient flow. The topography of $\mathcal{M}_2$ is a valley within the infinite-dimensional configuration space of $B = 2$ Skyrme fields, with the attractive channel being an almost flat submanifold of this. The highest point in $\mathcal{M}_2$ is the $B = 2$ hedgehog, whose energy is about one and a half times that of either the torus or well separated Skyrmions. So at really low energies the region near the hedgehog will not be explored, even though this is the solution on which the whole construction of $\mathcal{M}_2$ is based. The fact that the valley is not precisely flat, because of the weak inter-Skyrmion forces, means that the motion can not be assumed to be vanishingly slow. For example, the attraction of two Skyrmions may build up modest speeds even if they start at rest.

In principle, $\mathcal{M}_B$ could be the unstable manifold of the charge B hedgehog solution. The product ansatz suggests that this solution has $6B - 6$ unstable modes and six zero modes. However, a practical implementation is even less feasible.

In the simpler case of a $(2+1)$-dimensional Baby Skyrme model, a rather differently defined moduli space involving both a metric and a potential function has been constructed to study the classical dynamics of two solitons [388], and yields results which are in good agreement with full field simulations. The Baby Skyrme model may be considered as a deformation of the $O(3)$ sigma model, for which a precisely defined moduli space, $\mathcal{M}$, of static Bogomolny lump solutions exists. $\mathcal{M}$, with a deformed metric, is a suitable approximate moduli space for the deformed theory. The potential is approximated by evaluating the energy of sigma model lumps using the Baby Skyrmion potential energy function. Unfortunately the Skyrme model can not be treated in this way as there is no known deformation of the model to a nearby one with Bogomolny equations.

A related aspect of Skyrmion dynamics is of interest, namely, an analysis of the vibrational modes of minimal energy Skyrmions. This leads to a model of the linearization of the moduli space $\mathcal{M}_B$, near the Skyrmion. The low frequency vibrational modes provide a coordinate independent description of the configuration space around the static solution. Calculating the frequencies of the lowest-lying vibrational modes also provides a first step in an attempt to quantize the Skyrmion within a harmonic approximation. We will not discuss the quantization aspect, but we will discuss how the vibrational modes of Skyrmions provide yet another link to monopoles.

A numerical computation of the vibration frequencies, and the classification of degenerate modes into irreducible representations of the symmetry group of the static Skyrmion, has been performed for charges $B = 2$ and $B = 4$ [34], and a qualitative analysis has been given for $B = 7$ [36]. The method employed is to solve a semi-linearized form of the time dependent Skyrme equation, with as initial condition a rather general, slightly perturbed Skyrmion. The frequencies of the normal modes are found by Fourier transforming the fields at a given spatial location with respect to time. The spectrum obtained can be divided into two parts, corresponding to vibration frequencies below and above that of the breather mode, which is the oscillation corresponding to a change in the scale size of the Skyrmion. We are more interested in the lower-lying modes below the breather, since they can be identified with variations of the parameters in the rational map describing the static Skyrmion.

To be specific, let us consider the vibrations of the cubic $B = 4$ Skyrmion, whose modes lie in multiplets transforming under real irreducible representations of the octahedral group O. The computations of ref. [34] reveal that there are nine modes below the breather, which transform under the representations E, A_1, F_2, F_2, in order of increasing frequency.

Recall that the rational map of degree 4 with octahedral symmetry is

$$R_0(z) = \frac{z^4 + 2\sqrt{3}iz^2 + 1}{z^4 - 2\sqrt{3}iz^2 + 1}. \tag{9.126}$$

The general variation of this map, in which we preserve the leading coefficient, 1, of the numerator as a normalization, is

$$R(z) = \frac{z^4 + \alpha z^3 + (2\sqrt{3}i + \beta)z^2 + \gamma z + 1 + \delta}{(1+\lambda)z^4 + \mu z^3 + (-2\sqrt{3}i + \nu)z^2 + \sigma z + 1 + \tau} \tag{9.127}$$

where $\alpha, \beta, \gamma, \delta, \lambda, \mu, \nu, \sigma, \tau$ are small complex numbers. We now calculate the effect of the transformations of the octahedral group. For example, the 90° rotation, represented by the transformation $R(z) \mapsto 1/R(iz)$ leaves R_0 fixed, but transforms the more general map $R(z)$ to

$$\tilde{R}(z) = \frac{(1+\lambda)z^4 - i\mu z^3 + (2\sqrt{3}i - \nu)z^2 + i\sigma z + 1 + \tau}{z^4 - i\alpha z^3 - (2\sqrt{3}i + \beta)z^2 + i\gamma z + 1 + \delta}. \tag{9.128}$$

Normalizing this by dividing top and bottom by $1 + \lambda$, and ignoring quadratic and smaller terms in the small parameters, we get

$$\tilde{R}(z) = \frac{z^4 - i\mu z^3 + (2\sqrt{3}i - \nu - 2\sqrt{3}i\lambda)z^2 + i\sigma z + 1 + \tau - \lambda}{(1-\lambda)z^4 - i\alpha z^3 + (-2\sqrt{3}i - \beta + 2\sqrt{3}i\lambda)z^2 + i\gamma z + 1 + \delta - \lambda}. \tag{9.129}$$

Hence, the transformation acts linearly on the nine parameters $\alpha, \ldots, \tau$ via a complex 9×9 representation matrix that can be read off from this expression. As we want to deal with a real representation, we consider this as a real 18×18 matrix. The only contribution to the trace of this matrix is associated with the replacement of λ by $-\lambda$ in the leading term of the denominator. Since λ has a real and imaginary part, the character of the 90° rotation in this representation is -2.

Similar calculations for elements of each conjugacy class of the octahedral group give the remaining characters and allow us to identify the irreducible content of this representation as $2A_1 \oplus 2E \oplus 2F_1 \oplus 2F_2$.

To determine which of these irreducible representations correspond to true vibrations we need to remove those corresponding to zero modes. To find the zero mode representation associated with isospin rotations of the Skyrme field, we consider the infinitesimal $SU(2)$ Möbius deformations

$$R_0(z) \mapsto \frac{(1 + i\varepsilon)R_0(z) + \varepsilon'}{-\bar{\varepsilon}'R_0(z) + (1 - i\varepsilon)} \tag{9.130}$$

where ε is real, and ε' complex. Under the transformations of the octahedral group a computation of the characters reveals that these variations transform as $A_1 \oplus E$. Similarly, the variations which correspond to translations and rotations transform under the octahedral group as $F_1 \oplus F_1$. From the above 18-dimensional representation we therefore remove $A_1 \oplus E \oplus F_1 \oplus F_1$ to obtain the representation of the true vibrations. This has the irreducible components $A_1 \oplus E \oplus F_2 \oplus F_2$, and is nine-dimensional. These irreducible representations are precisely the ones obtained from the Fourier analysis of the field vibrations, given earlier.

As we saw, a number of scattering events through the symmetric minimal energy Skyrmions have a remarkable similarity to monopole scatterings. These monopole-like, Skyrmion scattering processes correspond precisely to the extension of the low-lying vibrational modes (which we refer to as monopole modes) to large amplitude, splitting the minimal energy Skyrmion into clusters of lower charge. Each monopole mode corresponds to a different cluster decomposition and it is often possible to identify the correspondence by comparing the symmetries of the scattering process and the vibration mode. A more sophisticated approach is to use the irreducible representation of each vibration mode to identify the mode with an explicit rational map deformation. Via the Jarvis correspondence between monopoles and rational maps, the extension of this deformation to large parameter values determines a monopole configuration with well separated clusters. The cluster decomposition of the Skyrmion can thus be identified.

As an example, the one-dimensional A_1 mode in the vibrational spectrum of the $B = 4$ Skyrmion is represented by the 1-parameter family of rational maps

$$R(z) = c \frac{z^4 + 2\sqrt{3}iz^2 + 1}{z^4 - 2\sqrt{3}iz^2 + 1}, \tag{9.131}$$

with c close to 1. Extending c to arbitrary positive values, and using the Jarvis correspondence, we recognize this family of tetrahedrally symmetric maps as describing the dynamics of four monopoles which approach and separate on the vertices of dual tetrahedra and pass through the cubic 4-monopole. Therefore this vibrational mode, extended to large amplitude, will separate the $B = 4$ Skyrmion into four single Skyrmions on the vertices of a tetrahedron, which is one of the attractive channel scatterings that we have already discussed. We denote this process by $1 + 1 + 1 + 1$ to signify the charges of the clusters into which the Skyrmion separates. The other $B = 4$ attractive channel scattering we have considered is the D_4-symmetric scattering, which emerges as $2 + 2$, that is, two $B = 2$ tori. This cluster decomposition corresponds to the two-dimensional vibrational representation E. The two remaining three-dimensional representations correspond to the cluster decompositions $3 + 1$, in which a single Skyrmion collides with the tetrahedral $B = 3$ Skyrmion preserving cyclic C_3 symmetry throughout, and the final decomposition is $2 + 1 + 1$, which is a D_{3d} twisted line scattering in which two single Skyrmions collide symmetrically with a $B = 2$ torus. All these scattering processes have been computed using full field simulations, verifying the above picture.

The Jarvis rational maps of degree B have $4B + 2$ parameters. For general B one therefore expects the minimal energy Skyrmion to have $4B - 7$ monopole vibrational modes below the breather, where the nine zero modes describing translations, rotations and isospin rotations have been subtracted off. As another example, for the $B = 3$ tetrahedral Skyrmion, there are five monopole modes, and a rational map symmetry analysis [193] suggests that they form an irreducible doublet and triplet of the tetrahedral group. The two distinct modes correspond to the two possible cluster decompositions, $2 + 1$ and $1 + 1 + 1$, and the corresponding processes are the C_3-symmetric and D_{2d} twisted line scatterings as seen for monopoles in Chapter 8. The Skyrmion collision for the first of these has already been described earlier in this section and the twisted line scattering is described in ref. [40]. For $B = 2$, the monopole mode separates the two Skyrmions and the corresponding collision process is right-angle scattering.

In summary, we see that there is a strong correlation between the low-lying vibrational modes of a Skyrmion and the zero modes of the associated monopole. An analysis of rational maps clarifies the correlation. Furthermore, an extension of these modes to large amplitude shows a correspondence between monopole dynamics, studied within the geodesic approximation, and attractive channel Skyrmion scattering, which has been confirmed using full field simulations. These results suggest that a $(4B + 2)$-dimensional moduli space of Skyrme fields, which includes the nine exact zero modes of a general Skyrmion, may model low energy Skyrmion dynamics. However, no precise construction of a suitable manifold of Skyrme fields directly from rational maps, or from monopole fields, has yet been achieved.

9.9 Generalizations of the Skyrme model

In arriving at the Skyrme model as a low energy effective theory from QCD in the limit in which the number of colours, N_c, is large, one finds that the Skyrme field takes values in $SU(N_f)$, where N_f is the number of flavours of light quarks. So far we have only considered the case of $N_f = 2$, which is physically the most relevant since the up and down quarks are almost massless, and the $SU(2)$ flavour symmetry between up and down quarks is only weakly broken in nature; but the model with

$SU(3)$ flavour symmetry, to allow for the strange quark, with appropriate additional symmetry breaking terms to take account of the higher strange quark mass, is also a reasonable approximation and allows the possibility to study strange baryons and nuclei within the Skyrme model, and also scattering processes involving ordinary baryons and strange mesons. The basic fields (of the linearized model) now describe pions, kaons, and the eta meson. There is still just one topological charge, identified as baryon number, arising from the homotopy group $\pi_3(SU(3)) = \mathbb{Z}$. In the absence of any symmetry breaking mass terms, the three flavour Skyrme Lagrangian is given by the usual expression (9.2), but with $U \in SU(3)$. There is also a Wess-Zumino term, which we discuss below, but this only plays a role in the quantization of Skyrmions and can be ignored for the present discussion of classical solutions.

Obviously, solutions of the $SU(3)$ model can be obtained by a simple embedding of $SU(2)$ Skyrmions, and current evidence suggests that these are the minimal energy solutions at each charge. However, there are also solutions which do not correspond to $SU(2)$ embeddings, and although they have energies which are slightly higher than the embedded Skyrmions, they are still low energy configurations, and they have symmetries that are very different from the $SU(2)$ solutions and so may be of some interest.

An example of a non-embedded solution is the dibaryon of Balachandran *et al.* [30], which is a spherically symmetric solution with $B = 2$. Explicitly, the Skyrme field is given by

$$U(\mathbf{x}) = \exp\left\{ i f_1(r) \mathbf{\Lambda} \cdot \hat{\mathbf{x}} + i f_2(r) \left((\mathbf{\Lambda} \cdot \hat{\mathbf{x}})^2 - \frac{2}{3} 1_3 \right) \right\} , \qquad (9.132)$$

where $\mathbf{\Lambda}$ is a triplet of $su(3)$ matrices generating $so(3)$ and f_1, f_2 are real profile functions satisfying the boundary conditions $f_1(0) = f_2(0) = \pi$ and $f_1(\infty) = f_2(\infty) = 0$. Substituting this ansatz into the static Skyrme equation leads to two coupled ordinary differential equations for f_1 and f_2. Solving these numerically yields an energy per baryon of $E/B = 1.19$, which is about 1% higher than the energy of the embedded $SU(2)$ torus of charge 2.

Recently, an extension of the rational map ansatz has been proposed [206], to create $SU(N_{\mathrm{f}})$ Skyrme fields from rational maps of the Riemann sphere into $\mathbb{CP}^{N_{\mathrm{f}}-1}$. Explicitly, the ansatz extends the $SU(2)$ projector form (9.56) to

$$U = \exp\left(i f \left(2P - \frac{2}{N_{\mathrm{f}}} 1_{N_{\mathrm{f}}} \right) \right) , \qquad (9.133)$$

where P is now an $N_{\mathrm{f}} \times N_{\mathrm{f}}$ Hermitian projector, constructed from a vector

v with N_f components via

$$P = \frac{\mathbf{v} \otimes \mathbf{v}^\dagger}{|\mathbf{v}|^2}, \tag{9.134}$$

and $f(r)$ is a real radial profile function with the usual boundary conditions. The vector $\mathbf{v}(z) : S^2 \mapsto \mathbb{CP}^{N_f-1}$ appears to be a rational map from the Riemann sphere into $\mathbb{C}^{N_f}$, but it is only defined projectively due to the relation (9.134). In fact, we can use this projective property to take $\mathbf{v}$ to be a vector in which all components are polynomials in z, and the degree of this projector, which is equal to the baryon number of the resulting Skyrme field, is given by the highest degree of the component polynomials. When $N_f = 2$ this ansatz coincides with the usual $SU(2)$ ansatz after the identification $\mathbf{v} = (q, p)^t$, where $R = p/q$ is the usual rational map and we have made use of the equivalence $\mathbb{CP}^1 \cong S^2$.

Although there are some difficulties with this ansatz [394], it can be used to produce some low energy field configurations and to understand the existence of certain symmetric Skyrme fields, which do not exist at the same charge in the $SU(2)$ model.

The $SU(N_f)$ Skyrme model has a global $SU(N_f)/\mathbb{Z}_{N_f}$ symmetry corresponding to the conjugation $U \mapsto \mathcal{O}U\mathcal{O}^\dagger$, where $\mathcal{O} \in SU(N_f)$. In terms of the ansatz (9.133) this symmetry is represented by the target space transformation

$$\mathbf{v} \mapsto \mathcal{O}\mathbf{v}. \tag{9.135}$$

The identification of K-symmetric maps (and hence K-symmetric Skyrme fields) is analogous to the $SU(2)$ case. The set of target space rotations accompanying spatial rotations needs to form an N_f-dimensional representation of K, so the simplest situation in which a degree B symmetric map exists is when

$$\underline{B+1}|_K = X_{N_f} \oplus \cdots, \tag{9.136}$$

where $\underline{B+1}|_K$ is the restriction of the $(B+1)$-dimensional irreducible representation of $SU(2)$ to the subgroup K, and X_{N_f} denotes any N_f-dimensional irreducible representation of K. In this case a basis for X_{N_f} consists of N_f polynomials in z of degree B, which can be taken to be the N_f components of the vector **v**.

To illustrate these ideas let us consider $B = 6$ Skyrme fields with icosahedral symmetry in the $SU(3)$ model. The relevant decomposition is

$$\underline{7}|_Y = F_2 \oplus G. \tag{9.137}$$

The presence of the three-dimensional F_2 shows that there is an icosahedrally symmetric degree 6 map from $\mathbb{CP}^1$ into $\mathbb{CP}^2$. Explicitly, this map is given by

$$\mathbf{v}(z) = (z^6 + 3z, \, 1 - 3z^5, \, \sqrt{50}z^3)^t \tag{9.138}$$

and is Y_h-symmetric. Thus there is an icosahedrally symmetric $B = 6$ Skyrme field in the $SU(3)$ model, whereas, as we have seen earlier, the lowest charge for which there is an icosahedrally symmetric $SU(2)$ Skyrmion is $B = 7$.

Substituting the ansatz (9.133) into the Skyrme Lagrangian leads to an energy function on the space of rational maps into $\mathbb{CP}^{N_f-1}$, and an essentially independent energy function for the profile function. In the case of $N_f = 3$ and $B = 6$ a numerical search for the minimizing map produces the map above [206], suggesting that the minimal energy non-embedded $SU(3)$ Skyrmion of charge 6 may be Y_h-symmetric. The profile function is also easily determined numerically. Numerical investigations of the full $SU(3)$ Skyrme model need to be performed to find the precise solutions of lowest energy, but this has yet to be done.

We now turn to a different generalization, the Skyrme model on a 3-sphere, in which the domain $\mathbb{R}^3$ is replaced by S_L^3, the 3-sphere of radius L, but the Skyrme field is still a map to the target space $SU(2)$. The baryon number is the degree of U. This generalization has been studied in ref. [291], and in a more geometrical context in ref. [282], where it was also shown that the geometrical strain formulation discussed earlier can be used to define a Skyrme energy functional for a map between any three-dimensional Riemannian manifolds. By taking the limit $L \to \infty$ the Euclidean model is recovered, but it is possible to gain some additional understanding of Skyrmions by first considering finite values of L.

Let μ, z be coordinates on S_L^3, with μ the polar angle (the co-latitude) and z the Riemann sphere coordinate on the 2-sphere at polar angle μ. Take f, R to be similar coordinates on the unit 3-sphere S_1^3, which we identify with the target manifold $SU(2)$.

In general, a static field is given by functions $f(\mu, z, \bar{z})$ and $R(\mu, z, \bar{z})$, but various simplifications are possible. To find the $B = 1$ Skyrmion we consider an analogue of the hedgehog field, an $SO(3)$-symmetric map of the form

$$f = f(\mu), \quad R = z, \tag{9.139}$$

whose energy is

$$E = \frac{1}{3\pi} \int_0^\pi \left\{ L \sin^2 \mu \left(f'^2 + \frac{2 \sin^2 f}{\sin^2 \mu} \right) + \frac{\sin^2 f}{L} \left(\frac{\sin^2 f}{\sin^2 \mu} + 2 f'^2 \right) \right\} d\mu. \tag{9.140}$$

Among these maps there is the 1-parameter family of degree 1 conformal maps

$$\tan \frac{f}{2} = e^a \tan \frac{\mu}{2}, \tag{9.141}$$

where a is a real constant. These may be pictured as a stereographic projection from S_L^3 to $\mathbb{R}^3$, followed by a rescaling by e^a, and then an inverse stereographic projection from $\mathbb{R}^3$ to S_1^3. Substituting the expression (9.141) into the energy (9.140), and performing the integral gives

$$E = \frac{L}{1 + \cosh a} + \frac{\cosh a}{2L} \,. \tag{9.142}$$

If $a = 0$ then (9.141) is the identity map with energy

$$E = \frac{1}{2}\left(L + \frac{1}{L}\right). \tag{9.143}$$

Note that if $L = 1$ then $E = 1$, so the Faddeev-Bogomolny bound is attained. We can therefore be certain that, in this case, the $B = 1$ Skyrmion is given by the identity map. We mentioned earlier that the bound could only be attained by a mapping which is an isometry, and this occurs when $L = 1$, the domain then being isometric to the target space.

Computing a to minimize the energy (9.142), for a fixed, general value of L, results in

$$\cosh a = \sqrt{2}L - 1\,. \tag{9.144}$$

For $L < \sqrt{2}$ this is clearly unattainable, and in fact the minimum occurs at $a = 0$. This shows that, for $L < \sqrt{2}$, the identity map is stable with respect to conformal transformations, though actually a stronger result, that the identity map is stable against any deformation for $L < \sqrt{2}$, is true [282]. The identity map is thus very likely the Skyrmion. The energy density of the identity map is distributed evenly over the 3-sphere, so no point of either the domain or target spheres is singled out as special. The unbroken symmetry group is the diagonal $SO(4)$ subgroup of the full symmetry group, which may be interpreted either as spatial or chiral $SO(4)$ rotations.

For $L > \sqrt{2}$ there are two roots of equation (9.144), related by the symmetry $a \mapsto -a$, but they give geometrically equivalent solutions since this sign change can be undone by making the replacement $\mu \mapsto \pi - \mu$, which exchanges poles on S_L^3. The energy is

$$E = \sqrt{2} - \frac{1}{2L}\,, \tag{9.145}$$

which is clearly less than (9.143). If a is positive, there is a preferred point in S_L^3, which corresponds to the point at infinity in $\mathbb{R}^3$, where the energy density is minimal, and the image of this point is a preferred point in S_1^3. The unbroken symmetry is therefore $SO(3)$ isospin symmetry, as in the Euclidean case, and chiral symmetry is broken. The energy density is maximal at the antipodal point. These conformal maps are not the exact Skyrmion solutions for $L > \sqrt{2}$, but they are expected to be close, and have the same symmetry. In the Euclidean limit $L \to \infty$ the radial variable should be identified as the combination $r = L\mu$, in which case the expression for the energy (9.140) reproduces the result for the hedgehog profile function (9.22). In the limit, the conformal map with $e^a \sim \sqrt{8}L$, that is, $f(r) = 2\tan^{-1}(\sqrt{2}r)$, has energy $E = \sqrt{2}$, which is higher than the value $E = 1.232$ of the minimizing hedgehog profile function, but the Skyrme field is qualitatively similar.

In summary, we see that on a small 3-sphere the energy density of a $B = 1$ Skyrmion is uniformly distributed over S_L^3 and the unbroken symmetry group is $SO(4)$, but as the radius of the 3-sphere is increased beyond the critical value $L = \sqrt{2}$ there is a bifurcation to a Skyrmion localized around a point and chiral symmetry is broken. Thus a phase transition occurs, as in the Skyrme crystal, when one moves from conditions of high to low baryon density, with a corresponding breaking of chiral symmetry. This may have relevance to the physical issue of whether quark confinement occurs at the same time as chiral symmetry breaking as very dense quark matter becomes less dense.

For charge $B > 1$ the rational map ansatz can again be applied to produce low energy Skyrme fields which approximate the minimal energy Skyrmions on S_L^3 [246], by taking $R(z)$ to be a degree B rational map and $f(\mu)$ the associated energy minimizing profile function. This produces fields which tend to those of the Euclidean model as $L \to \infty$ and for all cases except $B = 2$, this ansatz produces the lowest energy configurations yet discovered. The energy is particularly low if one chooses the optimal value of L, which depends on B. For $B = 2$ an exact solution is known [219] which has lower energy than the $O(2)$ symmetric field obtained from the rational map ansatz with $R = z^2$. This solution has a doubly axially symmetric form with the larger symmetry $O(2) \times O(2)$, a subgroup of the $O(4)$ symmetry group of the 3-sphere Skyrme model that is lost in the Euclidean limit.

Finally, in introducing the Skyrme model in Section 9.1 we already mentioned that a possible modification of the model is the addition of the pion mass term (9.7). The qualitative results of our previous discussions are unchanged by its inclusion, but here we briefly mention the small quantitative differences it generates. The most important effect is that the Skyrmion becomes exponentially localized, in contrast to the algebraic asymptotic behaviour of the Skyrme field in the massless pion model. This is because the modified equation for the hedgehog profile function,

$$(r^2 + 2\sin^2 f)f'' + 2rf' + \sin 2f\left(f'^2 - 1 - \frac{\sin^2 f}{r^2}\right) - m_\pi^2 r^2 \sin f = 0,$$

$$(9.146)$$

has the asymptotic Yukawa-type solution

$$f(r) \sim \frac{A}{r} e^{-m_\pi r} . \tag{9.147}$$

Clearly the energy of a single Skyrmion with $m_\pi > 0$ will be slightly higher than with $m_\pi = 0$, because the pion mass term is positive for all fields. For higher charge Skyrmions, the rational map approach works as before, but the profile function will again be slightly modified, leading to slightly higher energies.

9.10 Quantization of Skyrmions

Quantization is a vital issue for Skyrmions, more so than for the other solitons we have discussed, because Skyrmions are supposed to model physical baryons and nuclei, and a single baryon is a spin half fermion. We consider here both the $SU(2)$ and $SU(N_{\mathrm{f}})$ Skyrme models in $\mathbb{R}^3$.

We first briefly discuss the Wess-Zumino term [424], which is an additional contribution to the action of the $SU(N_{\mathrm{f}})$ Skyrme model given by

$$S_{\mathrm{WZ}} = -\frac{iN_{\mathrm{c}}}{240\pi^2} \int \varepsilon^{\mu\nu\alpha\beta\gamma} \mathrm{Tr}(R_\mu R_\nu R_\alpha R_\beta R_\gamma) \, d^5x , \tag{9.148}$$

where the integration is performed over a five-dimensional region whose boundary is four-dimensional space-time. The Wess-Zumino term does not contribute to the classical energy, but it plays an important role in the quantum theory. Its introduction breaks the time reversal and parity symmetries of the model down to the combined symmetry operation

$$t \mapsto -t, \quad \mathbf{x} \mapsto -\mathbf{x}, \quad U \mapsto U^\dagger, \tag{9.149}$$

which appears to be realized in nature, unlike these individual symmetry operations. A topological argument shows that N_{c} must be an integer, and Witten [428] argued that it should be identified with the number of quark colours, based on considerations of flavour anomalies in the quark and Skyrme models.

To determine whether a Skyrmion should be quantized as a fermion we can compare the amplitudes for the processes in which a Skyrmion remains at rest for some long time T, and in which the Skyrmion is slowly rotated through an angle 2π during this time. The sigma model and Skyrme terms in the action do not distinguish between these two processes, since they involve two or more time derivatives, but the Wess-Zumino term is only linear in time derivatives and so can distinguish them. In fact it results in the amplitudes for these two processes differing by a factor $(-1)^{N_c}$, which shows that the Skyrmion should be quantized as a fermion when N_c is odd, and in particular, in the physical case $N_c = 3$ [428].

For $N_f = 2$ the above analysis does not apply, since the Wess-Zumino term vanishes for an $SU(2)$-valued field. To determine the appropriate quantization of an $SU(2)$ Skyrmion one may follow the approach of Finkelstein and Rubinstein [132], who showed that it is possible to quantize a soliton as a fermion by lifting the classical configuration space to its simply connected covering space. In the $SU(2)$ Skyrme model, this is a double cover for any value of B. To treat a single soliton as a fermion, states should be multiplied by a factor of -1 when acted upon by any operation corresponding to a circuit around a non-contractable loop in the configuration space. Equivalently, the wavefunction has opposite signs on the two points of the covering space that cover one point in the configuration space. These authors also showed that the exchange of two $B = 1$ Skyrmions is a loop which is homotopic to a 2π rotation of a Skyrmion, in agreement with the spin-statistics result. It was verified by Williams [426] that a 2π rotation of a single Skyrmion is a non-contractible loop, thus requiring the Skyrmion to be quantized as a fermion. This result was generalized by Giulini [154], who showed that a 2π rotation of a charge B Skyrmion is a non-contractible loop if B is odd and contractible if B is even.

A practical, approximate quantum theory of Skyrmions is achieved by a rigid body quantization of the spin and isospin rotations. Vibrational modes whose excited states usually have considerably higher energy are ignored. For the $B = 1$ Skyrmion, this quantization was carried out by Adkins, Nappi and Witten [7], who showed that the lowest energy states (compatible with the Finkelstein-Rubinstein constraints) have spin half and isospin half, and may be identified with states of a proton or neutron.

The quantization of the $B = 2$ Skyrmion was discussed by Braaten and Carson [64], using a rigid body quantization. Their analysis was extended by Leese, Manton and Schroers [261], who also allowed the toroidal Skyrmion to break up in the direction of the lowest vibrational mode, which corresponds to the attractive channel. Both calculations find that the lowest energy quantum state has isospin zero and spin 1, as expected for the deuteron. The second calculation gets closer to the usual physical picture of the deuteron as a rather loose proton-neutron bound state.

For higher charge Skyrmions symmetric under a discrete group K, the moduli space of zero modes is $(SO(3) \times SO(3))/K$, where in this case K really denotes the group and not its double cover. K can be replaced by its double cover, K', if the $SO(3)$ factors in the above quotient space are promoted to $SU(2)$. A quantization of the zero modes can be performed by quantizing on this quotient space, but there are a number of inequivalent ways to do this labelled by the irreducible representations of K'. It is most convenient to define the wavefunctions on $SU(2) \times SU(2)$, and require them to be eigenstates of the operations corresponding to the elements of K'. The Finkelstein-Rubinstein constraints are imposed by requiring the eigenvalues to be ± 1 depending on whether the particular element of K' corresponds to a contractible or non-contractible loop. This has been performed [64, 74, 411, 210] for charges $B \leq 8$, and gives the correct quantum numbers (spin, isospin and parity) for the experimentally observed ground states of nuclei in all cases except $B = 5$ and $B = 7$. A further study, making use of the topological properties of the space of rational maps, has allowed an extension of this analysis up to $B = 22$ [247]. The fact that some results do not agree with the experimental data is probably due to the restricted zero mode quantization, which does not allow any vibrational or deformation modes, and assumes a rigid rotor approximation so that the symmetry of the static solution is maintained even in the presence of spin.

9.11 The Skyrme-Faddeev model

Some time ago, Faddeev [125] suggested that stable closed strings may exist as topological solitons in a modified $O(3)$ sigma model which includes a fourth order derivative term, with the topology arising due to the twisting of a planar soliton along the length of the string. Each slice normal to the string carries the localized planar soliton. The Skyrme-Faddeev model, which realizes this idea, involves a map $\mathbf{n} : \mathbb{R}^3 \mapsto S^2$, and can be obtained from the Skyrme model simply by restricting the field values to an equatorial 2-sphere of the usual $SU(2)$ target space. Explicitly, the field of the model is a real three-component vector $\mathbf{n} = (n_1, n_2, n_3)$, with unit length, $\mathbf{n} \cdot \mathbf{n} = 1$. The associated restricted Skyrme field is $U = i\mathbf{n} \cdot \boldsymbol{\tau}$. Substituting this into the Skyrme Lagrangian (9.2) results in

$$L = \int \left\{ \partial_\mu \mathbf{n} \cdot \partial^\mu \mathbf{n} - \frac{1}{2} (\partial_\mu \mathbf{n} \times \partial_\nu \mathbf{n}) \cdot (\partial^\mu \mathbf{n} \times \partial^\nu \mathbf{n}) \right\} d^3 x, \qquad (9.150)$$

which is the Skyrme-Faddeev Lagrangian. Its first term is that of the $O(3)$ sigma model and the higher order derivative Skyrme term is, as usual, required to give the possibility of configurations which are stable under a spatial rescaling.

In order for a field configuration to have finite energy the vector $\mathbf{n}$ must tend to a constant value at spatial infinity, which we may take to be the vector $\mathbf{n}^\infty = (0, 0, 1)$. Finite energy field configurations have a topological classification, but the novel aspect of this model is that the topological charge is not a topological degree, as it is for the solitons we have considered so far in this book, such as vortices, monopoles or Skyrmions.

The boundary condition again compactifies space to S^3, so that at any given time the field is a map $\mathbf{n} : S^3 \mapsto S^2$. Since $\pi_3(S^2) = \mathbb{Z}$, there is an associated integer topological charge N, the Hopf charge, which gives the soliton number. This charge can not be the degree of the mapping, since the domain and target spaces have different dimensions. Instead, one definition is the following. Let ω denote the area 2-form on the target S^2 and let $f = \mathbf{n}^*\omega$ be its pull-back under $\mathbf{n}$ to the domain S^3. Since ω is closed, f is closed. Then, due to the triviality of the second cohomology group of the 3-sphere, $H^2(S^3) = 0$, this pull-back must be an exact 2-form, say $f = da$. The Hopf charge is constructed by integrating the Chern-Simons 3-form over S^3,

$$N = \frac{1}{4\pi^2} \int_{S^3} f \wedge a \,. \tag{9.151}$$

This integral is independent of the choice of a, because if $a \mapsto a + d\alpha$, then the change of N is

$$\Delta N = \frac{1}{4\pi^2} \int_{S^3} f \wedge d\alpha = \frac{1}{4\pi^2} \int_{S^3} (d(f\alpha) - (df)\alpha) = 0 \tag{9.152}$$

because $df = 0$, and by Stokes' theorem the integral of $d(f\alpha)$ vanishes over a closed 3-manifold.

An important point to note is that the Hopf charge can not be written as the integral of any density which is local in the field $\mathbf{n}$. For this reason it is useful to consider an alternative interpretation of N. Generically, the preimage of a point on the target S^2 is a closed loop in S^3. Now if a field has Hopf number N then the two loops consisting of the preimages of any two distinct points on the target S^2 will be linked exactly N times. In Fig. 9.12 we schematically represent the preimages of two points for a configuration with $N = 1$.

Solitons have been found in the Skyrme-Faddeev model for a range of values of N. They are string-like, but not all of a simple shape. Recall that the position of a lump or Skyrmion is sometimes defined to be the point in space at which the field takes the value antipodal to the vacuum value. Here, the position of a soliton is the curve in space which is the preimage of the vector $-\mathbf{n}^\infty = (0, 0, -1)$. Displaying this closed string is a useful way to represent the solution. Alternatively, a thickened version of the soliton may be represented by the preimage of the circle of vectors with $n_3 = \text{const}$. The Skyrme-Faddeev model has a global $O(3)$ symmetry, but

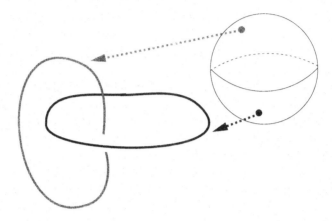

Fig. 9.12. A sketch showing two loops corresponding to the preimages of two points on the target 2-sphere. The loops are linked exactly once, indicating that the configuration has Hopf charge $N = 1$.

the choice of a vacuum value $\mathbf{n}^\infty$ breaks this to an $O(2)$ symmetry, which rotates the (n_1, n_2) components. As usual, when we refer to a symmetry of a configuration we mean that the effect of a spatial transformation can be undone by acting with an element of the unbroken global symmetry group of the theory, in this case $O(2)$. This implies that both the n_3 component (which determines the position of the soliton) and the energy density are strictly invariant under the symmetry operation.

Not only is there a topological Hopf charge in this model, but there is also a lower bound on the energy in terms of the charge N [405, 249]. Explicitly,

$$E > c|N|^{3/4} \qquad\qquad (9.153)$$

where $c = 16\pi^2 3^{3/8} \approx 238$. This energy bound is rather unusual in that a fractional power of the topological charge occurs, reflecting the fact that this bound is not obtained from the usual Bogomolny-type argument, but relies on a sophisticated use of Sobolev inequalities for its derivation. As such, the above value for the constant c may not be very tight. We will comment further on this shortly.

As pointed out in ref. [405], spherically symmetric fields automatically have zero Hopf charge, so it is not immediately obvious how to write down even the simplest field configurations which have non-zero values of N. However, a toroidal field can be constructed for any N, based on Faddeev's original idea. One may think of this field as a two-dimensional Baby Skyrmion which is embedded in the normal slice to a circle in space and has its internal phase rotated through an angle $2\pi N$ as it travels around the circle once. The construction can be implemented in toroidal coordinates if the size of the circle is fixed in advance, and was the method used in the numerical investigations [127, 155], which established the existence of axially symmetric solitons with charges $N = 1$ and $N = 2$, but it is rather cumbersome. A more elegant approach to constructing field configurations with non-zero Hopf charge makes use of the observation [302] that a field with Hopf charge N can be obtained by applying the standard Hopf projection $H : S^3 \mapsto S^2$ to a map U between 3-spheres with winding number N – in other words, a Skyrme field. Precisely, let $U(\mathbf{x})$ be a Skyrme field, that is, any smooth map from $\mathbb{R}^3$ into $SU(2)$ which satisfies the boundary condition that U tends to the identity as $|\mathbf{x}| \to \infty$. Let U have baryon number (degree) B. By writing the matrix entries of U in terms of complex numbers Z_0 and Z_1 as

$$U = \begin{pmatrix} Z_0 & -\bar{Z}_1 \\ Z_1 & \bar{Z}_0 \end{pmatrix}, \qquad (9.154)$$

where $|Z_0|^2 + |Z_1|^2 = 1$, the image of the Hopf map H can be written in terms of the column vector $Z = (Z_0, Z_1)^{\mathrm{t}}$ as

$$\mathbf{n} = Z^\dagger \boldsymbol{\tau} Z . \qquad (9.155)$$

It is easy to see that $\mathbf{n}$ is a real 3-vector of unit length and satisfies the boundary condition $\mathbf{n}(\infty) = \mathbf{n}^\infty$. Furthermore, it can be shown that the Hopf charge of the configuration constructed in this way is equal to the baryon number of the Skyrme field U, that is, $N = B$.

A useful supply of Skyrme fields for this purpose can be obtained using the rational map ansatz, as described in Section 9.5. Recall that this involves a rational map $R(z)$ and profile function $f(r)$. In particular, choosing the map $R(z) = z^N$ gives an axially symmetric field $\mathbf{n}$ of Hopf charge N, which has the same qualitative properties as those constructed by hand using toroidal coordinates. Note that in the case $N = 1$ the Skyrme field is spherically symmetric, but the Hopf projection breaks this, so that the vector $\mathbf{n}$ has only an axial symmetry. To determine the position of any approximate soliton constructed using this method we need to calculate the points in $\mathbb{R}^3$ at which $\mathbf{n} = (0, 0, -1)$. Equation (9.155) shows that this is equivalent to finding where $Z_0 = 0$. In the rational map ansatz, $Z_0 = 0$ if $f(r) = \frac{1}{2}\pi$ and also $|R(z)| = 1$. For the family of maps $R = z^N$ the second condition gives $|z| = 1$, the equatorial circle on the Riemann sphere. Therefore the position of the soliton is a circle in the (x^1, x^2) plane, whose radius is determined by the first condition.

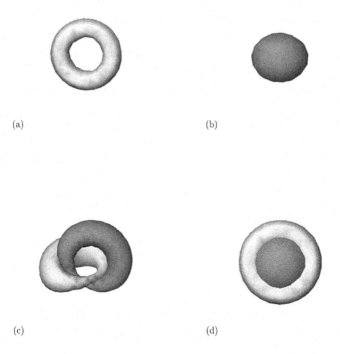

Fig. 9.13.   Isosurface plots for the $N = 1$ soliton displaying (a) the thickened locus of the position, (b) the energy density, (c) linking structure between two independent points on the target 2-sphere, and (d) a comparison between the position and energy density. Notice that the linking number is indeed 1 and that the energy density is not toroidal, but rather its maximum occurs at a point inside the locus of the position.

Using these axially symmetric configurations as initial data in a full three-dimensional numerical relaxation [43], it is found that for $N = 1$ and $N = 2$ the minimal energy soliton fields are very close to the initial data. In Fig. 9.13 and Fig. 9.14 we present, for the $N = 1$ and $N = 2$ solitons respectively, the position, the energy density, the linking number (by plotting the preimages of the points $\mathbf{n} = (-1, 0, 0)$ and $\mathbf{n} = (0, -1, 0)$), and the position and energy density isosurface together for comparison.

The energy of the $N = 1$ soliton has been computed several times [155, 43, 179, 419], using a variety of numerical schemes, and within the accuracy of the numerical calculations it is $E \approx 545$. Note that this is more than double the bound (9.153) with the quoted value of c, in agreement with our earlier remark that this value is probably not optimal. Ward [418] has argued (but it has not yet been proven) for the stronger value

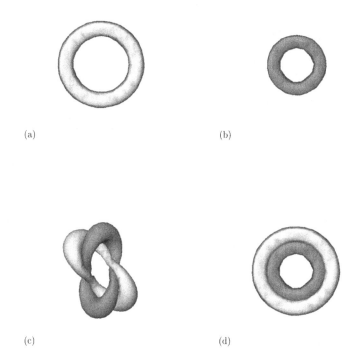

Fig. 9.14. The same quantities as in Fig. 9.13, but for the $N = 2$ soliton. Notice that the locus of the position and the energy density are both toroidal, but that the energy density is peaked inside the position.

$c = 32\pi^2\sqrt{2} \approx 447$. This is better from the point of view of the $N = 1$ soliton, since its energy would then only exceed the bound by roughly 20%, as would the energies of the higher charge solitons [43]. Ward's value is arrived at by considering the Skyrme-Faddeev model on S_L^3 rather than $\mathbb{R}^3$, in analogy with the discussion of Skyrmions on a 3-sphere. As in the Skyrme model, there is a special radius of the sphere, in this case[†] $L = \sqrt{2}$, for which an exact solution can be obtained, which corresponds to the identity map from S_L^3 to S_1^3 followed by the standard Hopf projection. The energy of this solution, which is possibly an absolute minimum for a soliton of unit charge, is precisely the value of c proposed by Ward. Thus if the Skyrme-Faddeev model mimics the result in the Skyrme model, where the topological energy bound is attained at the special radius, and is otherwise exceeded, then this energy is a natural candidate for the optimal constant c. Other aspects of the Skyrme model on a 3-sphere also

---

[†] The fact that the special radius is not $L = 1$ is simply due to our choice of coefficients in front of the two terms in the Lagrangian (9.150).

find parallels in the Skyrme-Faddeev model. For example, the identity map followed by the Hopf projection is an unstable solution if the radius L exceeds a critical value, which in the normalization we have chosen is $L > 2$.

Returning to solitons of the Skyrme-Faddeev model in flat space, for $N > 2$ the results of the numerical relaxation show that the minimal energy solution does not have the axially symmetric form described above. For example, the position of the $N = 3$ soliton has the structure of a twisted loop; this is displayed in the first plot of Fig. 9.15. Faddeev

a   b   c

Fig. 9.15.   The position of the soliton for (a) $N = 3$, (b) $N = 6$, (c) $N = 7$.

and Niemi [127] conjectured that the string-like solitons in this model would form knotted configurations for large enough values of N. This was verified numerically in ref. [43] (and later in ref. [179]) where both links and knots were found as the minimal energy solutions at various Hopf charges. The second and third plots of Fig. 9.15 show the position of the soliton for $N = 6$ and $N = 7$. The $N = 7$ soliton has the form of a trefoil knot, while the $N = 6$ soliton is composed of two linked loops which each resemble the $N = 2$ soliton. The total Hopf charge is here $N = 6$ because there is an additional two units of charge associated with the double counting of the linking number of two preimages, when the preimage of a single point itself has disconnected, linked components. The fact that the linking number is not simply additive, as this example demonstrates, is probably the physical reason why the energy bound (9.153) grows slowly, as a fractional power of the Hopf charge N.

As with Skyrmions, it is expected that the configuration space of the Skyrme-Faddeev model is very complicated, leading to many solutions which are local energy minima but not global minima, in addition to saddle point solutions. In fact, because of the string-like nature of the solutions, it is very likely that the difficulties associated with finding the global minimum at each charge will be much worse than in the Skyrme model. It has already been demonstrated [419] that even the space of $N = 2$ field configurations has quite a complicated structure.

Further numerical and analytical studies are required to fully investigate the soliton solutions which are expected to exist for higher Hopf charge, and to determine whether more complicated knots and links arise as the minimal energy solutions. There is physical motivation for this, since it has been proposed that the Skyrme-Faddeev model arises as a dual description of strongly coupled $SU(2)$ Yang-Mills theory [128], with the solitonic strings possibly representing glueballs.

Finally, we note that in the model with Lagrangian

$$L = \int \{(\partial_\mu \mathbf{n} \times \partial_\nu \mathbf{n}) \cdot (\partial^\mu \mathbf{n} \times \partial^\nu \mathbf{n})\}^{3/4} \, d^3x \,, \tag{9.156}$$

exact solutions describing axially symmetric Hopf solitons can be found explicitly [12]. This rather strange model, involving a fractional power in the Lagrangian density, is scale invariant. The solitons are therefore similar to lumps in the $O(3)$ sigma model, in that they have a zero mode associated with changes in the scale of the soliton, which might lead to soliton collapse in a finite time in dynamical situations.

10

Instantons

10.1 Self-dual Yang-Mills fields

This chapter is concerned with instantons, which are topological solitons of pure Yang-Mills theory defined in four-dimensional Euclidean space-time. If we regard instantons as static solitons in four space dimensions then they are the same kind of soliton that we have been discussing throughout this book, and in particular there are a number of similarities with static sigma model lumps in two space dimensions, which are often regarded as lower-dimensional analogues of Yang-Mills instantons. Instantons would be dynamical, particle-like solitons in a (4+1)-dimensional Yang-Mills theory, but we will not pursue this interpretation.

The physical motivation for considering four-dimensional Euclidean space is that in quantum field theory in (3+1)-dimensional Minkowski space-time one is led to the computation of path integrals which need to be analytically continued in order to be well defined. This continuation, known as a Wick rotation, is implemented by the replacement of the time coordinate $t \mapsto it$, which converts the Minkowski metric to the Euclidean one. We will denote Euclidean time by the coordinate x^4, and regard it as a fourth space coordinate in a static theory. The reason classical solutions are important is that they dominate the path integral, and in particular the instanton solutions generate non-perturbative quantum effects.

Let us consider an $SU(2)$ gauge theory with $su(2)$-valued gauge potential A_μ, $\mu = 1, \ldots, 4$, and associated field tensor

$$F_{\mu\nu} = \partial_\mu A_\nu - \partial_\nu A_\mu + [A_\mu, A_\nu]. \qquad (10.1)$$

Pure Yang-Mills theory is defined by the action

$$S = -\frac{1}{8} \int \mathrm{Tr}(F_{\mu\nu} F_{\mu\nu}) \, d^4x. \qquad (10.2)$$

Here, and in the remainder of this chapter, we use the Euclidean metric with signature $(+,+,+,+)$, and repeated indices are summed over with the naive summation convention. Note that the action S is non-negative, and so is equivalent to an energy for a static field.

Variation of this action produces the Yang-Mills field equation for the stationary points,

$$D_\mu F_{\mu\nu} = 0. \tag{10.3}$$

Let $F = \frac{1}{2}F_{\mu\nu}dx^\mu \wedge dx^\nu$ be the 2-form field strength. Then, because F is defined on a four-dimensional manifold, its Hodge dual $^\star F$ is also a 2-form. In components the dual is defined by

$$^\star F_{\mu\nu} = \frac{1}{2}\varepsilon_{\mu\nu\alpha\beta}F_{\alpha\beta}, \tag{10.4}$$

where $\varepsilon_{\mu\nu\alpha\beta}$ is the alternating tensor, and we use the convention that $\varepsilon_{1234} = -1$. Using the fact that $\text{Tr}(F_{\mu\nu}F_{\mu\nu}) = \text{Tr}(^\star F_{\mu\nu} {}^\star F_{\mu\nu})$, the action (10.2) can be rewritten as

$$S = -\frac{1}{16}\int \left\{ \text{Tr}((F_{\mu\nu} \mp {}^\star F_{\mu\nu})(F_{\mu\nu} \mp {}^\star F_{\mu\nu})) \pm 2\,\text{Tr}(F_{\mu\nu} {}^\star F_{\mu\nu}) \right\} d^4x. \tag{10.5}$$

The first term is a total square, and hence non-negative, so we see that this rearrangement leads to the lower bound

$$S \geq \pi^2|N|, \tag{10.6}$$

where we have defined the quantity

$$N = -\frac{1}{8\pi^2}\int \text{Tr}(F_{\mu\nu} {}^\star F_{\mu\nu})\, d^4x. \tag{10.7}$$

Recall from Section 3.5 that the second Chern number of an $SU(2)$ gauge field in $\mathbb{R}^4$ is

$$c_2 = \frac{1}{8\pi^2}\int_{\mathbb{R}^4} \text{Tr}(F \wedge F). \tag{10.8}$$

N is just the second Chern number expressed in terms of the components of the field tensor.

We are concerned with finite action fields, which means that the field strength must tend to zero as $|x| \to \infty$. This implies that as $|x| \to \infty$ the gauge potential tends to a pure gauge, that is

$$A_\mu = -\partial_\mu g^\infty(g^\infty)^{-1} \tag{10.9}$$

for some $g^\infty(x) \in SU(2)$, defined on the 3-sphere at spatial infinity. As we explained in Section 3.5, N is an integer in this case, and is equal to the degree of the map $g^\infty : S^3_\infty \mapsto SU(2)$.

One may also regard the gauge potential as a connection on an $SU(2)$ bundle over S^4, with field strength F. The fact that we can equally well regard the action as defined on S^4 or $\mathbb{R}^4$ is because it is conformally invariant, so the field equation is the same in either case. (Recall a similar situation in Chapter 6 where we considered the conformally invariant $O(3)$ sigma model both in the plane and on the unit 2-sphere.) The integer N is then the second Chern number of the bundle. The map g^∞ between 3-spheres arises as the transition function specifying how the connection defined over almost the whole of S^4 is glued on to a connection defined over a small neighbourhood of the point on S^4 representing the point at infinity in $\mathbb{R}^4$, and its degree is again N.

From (10.5) it is immediately clear that the bound (10.6) is attained only by fields which are either self-dual or anti-self-dual

$$F_{\mu\nu} = \pm {}^\star F_{\mu\nu} \, . \tag{10.10}$$

Finite action solutions of the (anti-)self-dual Yang-Mills equation are called (anti-)instantons and are global minima of the action within the space of charge N fields. Such fields automatically satisfy the second order Yang-Mills equation (10.3), because of the Bianchi identity. N is positive for non-trivial self-dual fields and is interpreted as the number of instantons. Generically a solution will have an action density which is localized around N points in $\mathbb{R}^4$. ($|N|$ is the number of anti-instantons if $N < 0$.)

The general instanton solution for $N = 1$ was first found by Belavin *et al.* [47]. This, and some particular multi-instanton solutions were later constructed by 't Hooft [402], using an ansatz that had been proposed previously [93, 425] to simplify the Yang-Mills equations. To present this ansatz we introduce the antisymmetric tensor $\sigma_{\mu\nu}$, with 2×2 Pauli matrix values, defined by

$$\sigma_{i4} = \tau_i, \quad \sigma_{ij} = \varepsilon_{ijk}\tau_k, \quad i, j \in \{1, 2, 3\}, \tag{10.11}$$

which has the property that it is anti-self-dual, $\frac{1}{2}\varepsilon_{\mu\nu\alpha\beta}\sigma_{\alpha\beta} = -\sigma_{\mu\nu}$. The self-dual gauge potential is constructed from a real scalar field ρ via

$$A_\mu = \frac{i}{2}\sigma_{\mu\nu}\partial_\nu \log \rho \, . \tag{10.12}$$

(With the sign of σ_{i4} reversed, $\sigma_{\mu\nu}$ would be self-dual and one would obtain an anti-self-dual gauge potential.) Substituting this ansatz into the self-dual Yang-Mills equation (10.10) leads to the Laplace equation in $\mathbb{R}^4$

$$\partial_\mu\partial_\mu\rho = 0 \, . \tag{10.13}$$

The 1-instanton is generated by the one-pole solution

$$\rho(x) = 1 + \frac{\lambda^2}{|x - a|^2}, \tag{10.14}$$

where $a \in \mathbb{R}^4$ is an arbitrary constant 4-vector and $|x|^2 = x^\mu x^\mu$ denotes the standard Euclidean norm. The positive real constant λ is arbitrary and represents the width of the instanton, in the sense that the action density is maximal at the point $x = a$ and decays algebraically with the distance from this point in such a way that the action inside the 4-ball $|x - a| \leq \lambda$ is $\frac{1}{2}\pi^2$, precisely half the total. Note that it may appear that the gauge potential obtained from this ansatz is singular at a but in fact this singularity is merely a gauge artifact and can be removed by a suitable gauge transformation.

The five real parameters a, λ are easily understood as a consequence of the translational and conformal symmetries of the self-duality equation, in analogy with our discussion of sigma model lumps in Chapter 6.

As in our definition of the monopole moduli space, where we included an additional $U(1)$ factor corresponding to an overall phase, it is convenient to include a constant $SU(2)$ gauge transformation within the definition of the instanton moduli space, leading to an additional three real parameters specifying the $SU(2)$ orientation of the instanton. Thus, with this addition, the moduli space $\mathcal{M}_1$ is eight-dimensional.

Although there are no dynamical aspects associated with instantons it is still of interest to study the instanton moduli spaces and their metrics [111, 112]. For instantons on $\mathbb{R}^4$ these moduli spaces are concrete examples of hyperkähler manifolds [186]. The metric on moduli space is defined by restricting the natural metric on the configuration space of the gauge theory to the submanifold of instantons. Let A_μ be an instanton and let X_μ and Y_μ be two tangent vectors to the space of instantons. Explicitly, X_μ (and similarly Y_μ) must be a solution of the linearized self-dual equation, that is, $D_\mu^A X_\nu - D_\nu^A X_\mu$ must be self-dual. X_μ must also be orthogonal to the gauge orbits, which is equivalent to the requirement that it satisfies the background gauge condition

$$D_\mu^A X_\mu = 0. \tag{10.15}$$

Then the metric is defined by

$$g(X, Y) = -\int_{\mathbb{R}^4} \text{Tr}(X_\mu Y_\mu) \, d^4 x. \tag{10.16}$$

The moduli space $\mathcal{M}_1$ is $\mathbb{R}^4 \times (\mathbb{R}^4)^*$, where $(\mathbb{R}^4)^*$ is $\mathbb{R}^4$ with the origin removed. The group $SU(2)$, parametrizing the global orientation of the instanton, acts by rotations on the factor $(\mathbb{R}^4)^*$, and the quotient is the upper half space in $\mathbb{R}^5$, with coordinates $a \in \mathbb{R}^4$ and $\lambda > 0$. The metric on $\mathcal{M}_1$ is the flat metric of $\mathbb{R}^8$, and is incomplete at the origin because the point-like instanton with $\lambda = 0$ is not a true solution of charge 1.

Note that this is the metric for one instanton on $\mathbb{R}^4$. Although the self-dual condition is conformally invariant, the metric on the instanton moduli space is not, essentially because the metric is related to instanton dynamics which would take place in (4+1)-dimensions. The moduli space metric for one instanton on S^4 has been calculated [166], and is relatively complicated.

By an index theorem calculation it was shown by Atiyah, Hitchin and Singer [19] and by Schwarz [368] that the N-instanton moduli space, $\mathcal{M}_N$, has dimension $8N$. When all the instantons are far apart, the $8N$ parameters may be interpreted as eight parameters for each of the N instantons, giving the positions, scales and $SU(2)$ orientations of each. For instantons on $\mathbb{R}^4$, the metric on $\mathcal{M}_N$ is hyperkähler, and Maciocia [271] has shown that (up to a constant factor) the hyperkähler potential $\mathcal{K}$ has the simple form

$$\mathcal{K} = \int_{\mathbb{R}^4} |x|^2 \mathrm{Tr}(F_{\mu\nu} F_{\mu\nu}) \, d^4x \,. \tag{10.17}$$

Again, $\mathcal{M}_N$ is incomplete because one or more instantons can collapse to zero size.

The 't Hooft ansatz can be used to construct some charge N instantons by taking the solution of the Laplace equation (10.13) to have N distinct poles,

$$\rho(x) = 1 + \sum_{j=1}^{N} \frac{\lambda_j^2}{|x - a_j|^2} \,, \tag{10.18}$$

with arbitrary widths and positions, producing a $(5N + 3)$-dimensional family of charge N instantons when the overall $SU(2)$ orientation is included. Only for $N = 1$ does this ansatz capture the most general instanton solution. For example, for $N = 2$ there are only 13 parameters here instead of the 16 in the general solution. Note that the gauge potential produced by the 't Hooft ansatz tends rapidly to zero at spatial infinity; in addition there are N singularities which can each be removed by a gauge transformation $g : S^3 \mapsto SU(2)$ of unit degree, defined on a small 3-sphere surrounding the singularity. The topological charge is therefore N.

The 't Hooft ansatz was generalized by Jackiw, Nohl and Rebbi [215], who noticed that (10.18) could be extended by acting with the conformal group to produce the JNR ansatz

$$\rho(x) = \sum_{j=0}^{N} \frac{\lambda_j^2}{|x - a_j|^2} \,, \tag{10.19}$$

in which the number of poles is one greater than the number of instantons. The topological charge is still N, because each singularity contributes one unit to the charge, but one unit is subtracted because of the behaviour of the gauge potential at infinity. The 't Hooft form of ρ (10.18) can be recovered from (10.19) in the limit in which $\lambda_0 = |a_0|$ and the location of the first pole, a_0, is sent to infinity. Although the JNR ansatz appears to yield a $(5N+8)$-dimensional family of instantons, note that the logarithmic derivative in (10.12) means that the multiplication of ρ by a constant does not alter the gauge potential, so only the ratios of the weights λ_j/λ_0, $j = 1, \ldots, N$, are relevant, reducing the parameter count to $5N + 7$.

If $N = 1$ then all the additional parameters of the JNR extension are redundant, since the 't Hooft ansatz together with global gauge rotations already produces the general 8-parameter instanton. Explicitly, the two-pole JNR ansatz with

$$\rho(x) = \frac{\lambda_0^2}{|x - a_0|^2} + \frac{\lambda_1^2}{|x - a_1|^2} \tag{10.20}$$

produces an instanton which is gauge equivalent to a 't Hooft instanton with its position and scale given by

$$a = \frac{\lambda_0^2 a_1 + \lambda_1^2 a_0}{\lambda_0^2 + \lambda_1^2}, \qquad \lambda = \frac{\lambda_0 \lambda_1}{\lambda_0^2 + \lambda_1^2} |a_0 - a_1|, \tag{10.21}$$

and its $SU(2)$ orientation defined by the direction of the line through the poles a_0 and a_1.

Similarly, for the special case of $N = 2$, where the parameter count of $5N + 7$ appears to exceed the 16 dimensions of the moduli space $\mathcal{M}_2$, it turns out that one of the degrees of freedom corresponds to a gauge transformation, which makes one parameter redundant, leaving precisely the required number for the general solution. This is most easily understood using a nice geometrical description of the two-instanton moduli space, due to Hartshorne [173]. In this description each charge 2 instanton is uniquely associated with a circle in $\mathbb{R}^4$ together with an ellipse, which is in the same plane as the circle and interior to it, and furthermore satisfies the Poncelet condition that there exists a triangle with vertices on the circle and sides tangent to the ellipse. This pair of conics, an example of which is shown in Fig. 10.1, is the gauge invariant two-instanton data. The conics and the associated triangle are related to the JNR parameters as follows. The three poles a_0, a_1 and a_2 are the vertices of the triangle. Let b_0, b_1 and b_2 be the points on the sides of this triangle at which the sides are tangent to the interior ellipse. Then the ratios of the weights are given by the formula

$$\frac{\lambda_i^2}{\lambda_j^2} = \frac{|a_i - b_k|}{|a_j - b_k|},\qquad (10.22)$$

where i, j, k are three distinct elements of the set $\{0, 1, 2\}$.

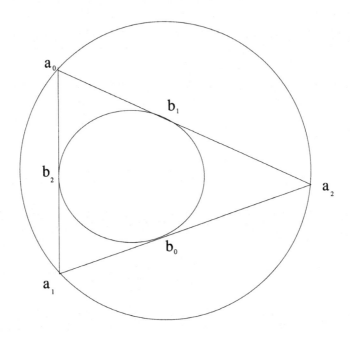

Fig. 10.1.   The circle and ellipse associated with a 2-instanton, and one of the family of triangles with vertices on the circle and tangent to the ellipse.

Given the two conics, Poncelet's theorem states that there is a whole 1-parameter family of triangles with vertices on the circle and sides tangent to the ellipse. Given that the conics are the gauge invariant data, this means that JNR data associated with two triangles in the same family yield gauge equivalent instantons. The infinitesimal motion of the triangle within the family corresponds to moving each pole a_i around the circle by an angle proportional to λ_i^2. It is this freedom to move the poles around the circle which accounts for the one redundant parameter in the JNR two-instanton data.

For instantons on S^4 the Hartshorne picture is similar. One should regard S^4 as embedded in $\mathbb{R}^5$. The circle and ellipse are coplanar in $\mathbb{R}^5$ and still satisfy the Poncelet condition. The circle lies on S^4 whereas the ellipse is in the *interior*. For an instanton in $\mathbb{R}^4$ whose JNR poles are collinear, the Hartshorne circle degenerates to a line and the ellipse is not really visible in $\mathbb{R}^4$; however, this is just the case where the circle on S^4 passes through the point corresponding to the point at infinity in $\mathbb{R}^4$. If one of the JNR poles is also at infinity, then this corresponds to the 't Hooft ansatz.

Let us now return to the situation where the Hartshorne circle in $\mathbb{R}^4$ does not degenerate to a line. The two instantons being well separated corresponds to the ellipse having a high eccentricity. For example, suppose the ellipse almost touches the circle at the points a_1 and a_2, in which case $\lambda_i \ll \lambda_0$ for $i = 1, 2$. Let us normalize the weights so that $\lambda_0 = 1$. Then this configuration describes a superposition of two charge 1 instantons with approximate positions a_1 and a_2 and widths $\lambda_1|a_1 - a_0|$ and $\lambda_2|a_2 - a_0|$ respectively. The $SU(2)$ orientation of the ith instanton is associated with the direction of the line through the poles a_i and a_0, for $i = 1, 2$.

A particularly symmetric 2-instanton arises if the three JNR poles have equal weights and are located at the vertices of an equilateral triangle. In this case the instanton is $SO(2)$-symmetric, because the two conics are a pair of concentric circles with the ratio of their radii equal to 2. From this description the symmetry is obvious. In terms of the JNR data a rotation simply maps one member of the Poncelet family of triangles into another, hence producing only a gauge transformation.

A generalization of this symmetric placement of JNR poles, with equal weights, can be made to produce instantons of higher charge with discrete symmetries, such as those of the Platonic solids. For example, for $N = 3$ there are four poles and if these are taken to have equal weights and to be located at the vertices of a regular tetrahedron in a subspace $\mathbb{R}^3 \subset \mathbb{R}^4$, such as the slice $x^4 = 0$, then clearly such an arrangement has tetrahedral symmetry and involves an arbitrary scale, giving the size of the tetrahedron. The generalization to other Platonic symmetry groups is obvious; for example, the first JNR instanton with cubic symmetry occurs for $N = 5$, when the six poles are placed at the vertices of an octahedron. However, for the case of the cubic group there are symmetric instantons of lower charge, in fact $N = 4$, but these are not of the JNR type, since it is not possible to place five distinct points in $\mathbb{R}^3$ or $\mathbb{R}^4$ with cubic symmetry. Later in this chapter we discuss how general symmetric instantons can be obtained, including those which are not of the JNR type, such as an icosahedrally symmetric 7-instanton and the above-mentioned cubic 4-instanton.

In summary, for $N = 1$ and $N = 2$ the JNR ansatz produces the most general charge N instanton but for $N > 2$ it only generates a $(5N + 7)$-dimensional submanifold of the $8N$-dimensional moduli space. The missing moduli, whose number for large N grows like $3N$, may be thought of as independent $SU(2)$ orientations for each instanton.

Historically, the first multi-instanton solutions were found by Witten [427] before the discovery of the 't Hooft solutions. Witten's approach involves searching for instantons which have a four-dimensional cylindrical symmetry, that is, an $SO(3)$ rotational symmetry about the x^4-axis. The metric of $\mathbb{R}^4 - \mathbb{R}^1$ in cylindrical coordinates,

$$ds^2 = (dx^4)^2 + dr^2 + r^2(d\theta^2 + \sin^2\theta \, d\varphi^2) \tag{10.23}$$

with $r = |\mathbf{x}| > 0$, becomes that of $\mathbb{H}^2 \times S^2$ by dividing by $\frac{1}{2}r^2$. x^4 and r are coordinates on $\mathbb{H}^2$, the hyperbolic plane with curvature $-\frac{1}{2}$ in the upper half plane model, and $SO(3)$ acts on θ and φ in the standard way. As a result of this conformal equivalence

$$\mathbb{R}^4 - \mathbb{R}^1 \sim \mathbb{H}^2 \times S^2 \,, \tag{10.24}$$

the symmetry reduction of four-dimensional Yang-Mills theory leads to an abelian Higgs model in $\mathbb{H}^2$.

Explicitly, in the Witten ansatz, the Cartesian components of the $SU(2)$ gauge potential are

$$A_i = \frac{1}{2}\left(\frac{\phi_2 + 1}{r^2}\varepsilon^{iak}x^k + \frac{\phi_1}{r^3}(\delta^{ia}r^2 - x^ix^a) + a_r\frac{x^ix^a}{r^2}\right)t^a$$

$$A_4 = a_4\frac{x^a}{2r}t^a \,. \tag{10.25}$$

Here, $t^a = i\tau^a$ and ϕ_1, ϕ_2, a_4 and a_r are arbitrary functions of x^4 and r. Witten argued that the ansatz is the most general one for fields invariant under combined rotations and rigid gauge transformations, since it uses all the available tensors. Witten's insight was later verified by an analysis of symmetric gauge fields along the lines described in Section 4.3 [136]. This analysis clarifies that a_4 and a_r are the components of a gauge potential on $\mathbb{H}^2$, the space of $SO(3)$ orbits, and that (ϕ_1, ϕ_2) can be combined as $\phi = \phi_1 + i\phi_2$, a complex Higgs field on $\mathbb{H}^2$. It also clarifies why the reduced gauge group is $U(1)$.

Using the ansatz, the self-dual Yang-Mills equation reduces to the Bogomolny equations for vortices in the abelian Higgs model on $\mathbb{H}^2$, which we discussed in Section 7.14.3. These equations are

$$D_4\phi + iD_r\phi = 0$$

$$B - \frac{1}{r^2}(1 - \bar{\phi}\phi) = 0 \tag{10.26}$$

where $B = \partial_4 a_r - \partial_r a_4$. The equations can again be solved by reducing them to the Liouville equation. The charge N instanton that Witten found is obtained from a degree $N + 1$ rational map in the complex coordinate $y = x^4 + ir$. There is a reality condition which relates the poles and zeros of the rational map, together with a residual gauge invariance, leading to $2N$ real degrees of freedom in the solution. In addition there are three degrees of freedom for the gauge orientation. These instantons are in fact a special case of 't Hooft instantons in which the pole positions are restricted to lie on the x^4-axis [277]. The $2N$ degrees of freedom are then the widths of the N unit charge instantons and their locations along the axis.

In the following section we describe a construction which, in principle, can be used to obtain the general charge N instanton by solving a purely algebraic system.

10.2 The ADHM construction

The integrability of the self-dual Yang-Mills equation was first recognized by Ward [412], who demonstrated that the twistor transform of Penrose could be used to provide a correspondence between instantons and certain holomorphic vector bundles over the twistor space $\mathbb{CP}^3$ [420]. There are two alternative methods for constructing the appropriate bundles; the first involves obtaining the bundle as an extension of line bundles and leads to the Atiyah-Ward construction [23, 95], whereas the second is the method of monads which was applied by Atiyah, Drinfeld, Hitchin and Manin [18] to yield the ADHM construction.

The ADHM construction was formulated in terms of local data by physicists [94, 84], where it becomes a prescription that generates the gauge potential of the general charge N instanton from matrices satisfying certain algebraic, but nonlinear, constraints. It is this form of the ADHM construction which we now present.

The ADHM data for an $SU(2)$ N-instanton combine into a matrix

$$\widehat{M} = \begin{pmatrix} L \\ M \end{pmatrix} \tag{10.27}$$

where L is a row of N quaternions and M is a symmetric $N \times N$ matrix of quaternions. In other words, each element of the matrix $\widehat{M}$ is a quaternion of the form $q = \sum_{\mu=1}^{4} q_\mu e_\mu$, where $q_\mu \in \mathbb{R}$, $e_4 = 1$ denotes the quaternionic identity element and e_1, e_2, e_3 satisfy the quaternion relations $e_1^2 = e_2^2 = e_3^2 = -1$ and $e_1 e_2 = e_3 = -e_2 e_1$, etc. We use a 2×2 Pauli matrix representation of the quaternions in which e_4 is the identity matrix and $e_j = -i\tau_j$ for $j = 1, 2, 3$. With this choice it is clear that a pure quaternion, that is, a quaternion q for which $q_4 = 0$, can be identified with an element of $su(2)$, which is an important aspect of the ADHM

construction. On the other hand, a real quaternion q_4e_4 is identified with
the real number q_4.

To be valid ADHM data the matrix $\widehat{M}$ must satisfy the nonlinear reality
constraint

$$\widehat{M}^\dagger \widehat{M} = R_0 , \tag{10.28}$$

where † denotes the quaternionic conjugate transpose and R_0 is any real
non-singular $N \times N$ matrix. (The operation † transposes the matrix
and replaces each entry $q = q_1e_1 + q_2e_2 + q_3e_3 + q_4e_4$ by its conjugate
$q^\dagger = -q_1e_1 - q_2e_2 - q_3e_3 + q_4e_4$.)

The first step in constructing the instanton from the ADHM data is to
form the matrix

$$\Delta(x) = \begin{pmatrix} L \\ M - x1_N \end{pmatrix} , \tag{10.29}$$

where 1_N denotes the $N \times N$ identity matrix and x is the quaternion corresponding to a point in $\mathbb{R}^4$ via $x = x^\mu e_\mu$. The second step is then to find
the $(N+1)$-component column vector $\Psi(x)$ of unit length, $\Psi(x)^\dagger \Psi(x) = 1$,
which solves the equation

$$\Psi(x)^\dagger \Delta(x) = 0 . \tag{10.30}$$

The final step is to compute the gauge potential $A_\mu(x)$ from $\Psi(x)$ using
the formula

$$A_\mu(x) = \Psi(x)^\dagger \partial_\mu \Psi(x) . \tag{10.31}$$

This defines a pure quaternion which can be regarded as an element of
$su(2)$ as discussed above.

In order for all these steps to be valid, the ADHM data must satisfy an
additional invertibility condition, which is that the columns of $\Delta(x)$ span
an N-dimensional quaternionic space for all x. In other words,

$$\Delta(x)^\dagger \Delta(x) = R(x) \tag{10.32}$$

where $R(x)$ is a real $N \times N$ invertible matrix for every x.

There is a freedom in choosing $\Psi(x)$ given by $\Psi(x) \mapsto \Psi(x)q(x)$, where
$q(x)$ is a unit quaternion, satisfying $q^\dagger q = 1$. The unit quaternions can
be identified with $SU(2)$ and from Eq. (10.31) we see that this freedom
corresponds to a gauge transformation.

It is relatively straightforward to verify that the above procedure yields
a self-dual gauge field, as we now show.

From the definition (10.31) it follows that the gauge field tensor is given
by

$$F_{\mu\nu} = \partial_\mu \Psi^\dagger \partial_\nu \Psi + \Psi^\dagger \partial_\mu \Psi \Psi^\dagger \partial_\nu \Psi - \partial_\nu \Psi^\dagger \partial_\mu \Psi - \Psi^\dagger \partial_\nu \Psi \Psi^\dagger \partial_\mu \Psi , \tag{10.33}$$

which can be rearranged, using the fact that Ψ has unit length, as

$$F_{\mu\nu} = \partial_\mu \Psi^\dagger (1_N - \Psi\Psi^\dagger)\partial_\nu \Psi - \partial_\nu \Psi^\dagger (1_N - \Psi\Psi^\dagger)\partial_\mu \Psi \,. \tag{10.34}$$

Note that the operator $1_N - \Psi\Psi^\dagger$ projects onto the quaternionic subspace orthogonal to Ψ. Using the definition of R in Eq. (10.32) and the orthogonality property (10.30) this projector can be rewritten as

$$1_N - \Psi\Psi^\dagger = \Delta R^{-1}\Delta^\dagger \,, \tag{10.35}$$

so that the field tensor becomes

$$F_{\mu\nu} = \partial_\mu \Psi^\dagger \Delta R^{-1}\Delta^\dagger \partial_\nu \Psi - \partial_\nu \Psi^\dagger \Delta R^{-1}\Delta^\dagger \partial_\mu \Psi \,. \tag{10.36}$$

Differentiating Eq. (10.30) provides the identity $\partial_\mu \Psi^\dagger \Delta = -\Psi^\dagger \partial_\mu \Delta$, whose application, together with its conjugate, to the above expression leads to

$$F_{\mu\nu} = \Psi^\dagger \partial_\mu \Delta R^{-1} \partial_\nu \Delta^\dagger \Psi - \Psi^\dagger \partial_\nu \Delta R^{-1} \partial_\mu \Delta^\dagger \Psi \,. \tag{10.37}$$

Now

$$\partial_\mu \Delta = -e_\mu \tilde{1}_N \,, \tag{10.38}$$

where $\tilde{1}_N$ is the constant real $(N+1) \times N$ matrix whose first row is zero and whose remaining $N \times N$ block is the identity matrix. We therefore arrive at the final expression

$$F_{\mu\nu} = -\Psi^\dagger \tilde{1}_N R^{-1}(e_\mu e_\nu^\dagger - e_\nu e_\mu^\dagger)\tilde{1}_N^\dagger \Psi \,, \tag{10.39}$$

where we have used the fact that R^{-1} is a real matrix and hence commutes with the quaternion e_μ. The purely tensorial part of this expression is the combination

$$\eta_{\mu\nu} = e_\mu e_\nu^\dagger - e_\nu e_\mu^\dagger \,, \tag{10.40}$$

which, as is easily checked using the quaternion or Pauli matrix algebra, is self-dual, $^\star\eta_{\mu\nu} = \eta_{\mu\nu}$, and therefore the construction yields a self-dual gauge field as stated.

Let us now check that the number of real parameters in the ADHM data is $8N$, as required to produce the general N-instanton solution. In the ADHM matrix $\widehat{M}$ there are $4N$ real parameters in the row vector L and $2N(N+1)$ real parameters in the symmetric matrix M. The constraint (10.28) removes $\frac{3}{2}N(N-1)$ of these, three for each of the upper triangular entries of the matrix R as a consequence of setting the pure quaternion part to zero. There is a further redundancy in the ADHM data corresponding to the transformation

$$\Delta(x) \mapsto \begin{pmatrix} q & 0 \\ 0 & \mathcal{O} \end{pmatrix} \Delta(x)\mathcal{O}^{-1} \,, \tag{10.41}$$

where $\mathcal{O}$ is a constant real orthogonal $N \times N$ matrix, q is a constant unit quaternion and the decomposition into blocks is as in Eq. (10.29). The transformation rotates the components of the vector Ψ, as can be seen from its definition (10.30), but this does not change the gauge potential derived from the formula (10.31). There are $\frac{1}{2}N(N-1)$ parameters in the matrix $\mathcal{O} \in O(N)$, and three in the unit quaternion q, but we do not subtract out these last three since they are balanced by the three which occur in the overall $SU(2)$ orientation. The final tally reads

$$4N + 2N(N+1) - \frac{3}{2}N(N-1) - \frac{1}{2}N(N-1) = 8N, \qquad (10.42)$$

as required.

For the simple case of $N = 1$ the ADHM data may be taken to have the form

$$\widehat{M} = \begin{pmatrix} \lambda \\ a \end{pmatrix}, \qquad (10.43)$$

where λ is real and positive and a is an arbitrary quaternion. The ADHM constraint (10.28) is then trivially satisfied and this generates an instanton with width λ and position in $\mathbb{R}^4$ corresponding to a. For $N = 2$ and $N = 3$ the general solution of the ADHM constraints can also be found [84, 245], but for $N > 3$ it is difficult to obtain explicit solutions and the general solution is not known. In the following section we describe how some explicit ADHM data can be found by searching for particularly symmetric instantons.

As a final point, the tangent vectors required to compute the moduli space metric can also be determined within the ADHM formalism, allowing the metric to be calculated on the space of ADHM data. This approach is similar to that described in Chapter 8, where the monopole moduli space metric was in some examples calculated as a metric on the space of Nahm data. Stated more formally, the natural metric on the space of $(N+1) \times N$ matrices satisfying the ADHM constraints, modulo the $O(N)$ action, is isometric to the natural metric on $\mathcal{M}_N$, the space of N-instantons modulo gauge transformations [271].

10.3 Symmetric instantons

Symmetric instantons within the ADHM formulation are described in detail in ref. [376]. We will describe the main aspects here, and give a few examples. We are interested in instantons which are symmetric under the action of a finite rotation group $K \subset SO(3)$ acting on the coordinates (x^1, x^2, x^3) of $\mathbb{R}^3 \subset \mathbb{R}^4$ and leaving x^4 alone. As in our previous discussion of symmetric Skyrmions, it is convenient to work with the double group of K, which we continue to denote by K. Now we can exploit the equivalence of $SU(2)$ and the group of unit quaternions to represent an element of K by a unit quaternion, which fits with the quaternionic representation of a point $x \in \mathbb{R}^4$ used in the ADHM construction. Explicitly, a spatial rotation acts by conjugation by a unit quaternion k

$$x \mapsto kxk^{-1}, \qquad (10.44)$$

which clearly fixes the x^4 (real) component and transforms the pure part by the $SO(3)$ rotation corresponding to the $SU(2)$ element represented by k. The ADHM data of an N-instanton are K-symmetric if for every $k \in K$ the spatial rotation (10.44) leads to gauge equivalent ADHM data. Recalling the redundancy (10.41), the requirement is that for every k

$$\begin{pmatrix} L \\ M - kxk^{-1}1_N \end{pmatrix} = \begin{pmatrix} q & 0 \\ 0 & \mathcal{O}k \end{pmatrix} \begin{pmatrix} L \\ M - x1_N \end{pmatrix} k^{-1}\mathcal{O}^{-1}, \qquad (10.45)$$

where, as earlier, $\mathcal{O} \in O(N)$ and q is a unit quaternion, both being k-dependent. The set of matrices $\mathcal{O}(k)$, as k runs over all the elements of K, forms a real N-dimensional representation of K, and similarly the set of quaternions $q(k)$ forms a complex two-dimensional representation. The procedure to calculate K-symmetric ADHM data is therefore first to choose a real N-dimensional representation and a complex two-dimensional representation of K and then to find the most general matrices L and M compatible with Eq. (10.45). Hopefully, these matrices then contain just a few free parameters to make the ADHM constraint (10.28) tractable, yet non-trivial.

Although we have already pointed out that the simplest example of a tetrahedrally symmetric 3-instanton can be obtained easily within the JNR approach, it is instructive to see how the more general symmetric ADHM scheme works in this simple case [191, 376].

The relevant representations for the T-symmetric 3-instanton are (following our earlier notation) the three-dimensional representation F and the two-dimensional representation E'. The tetrahedral group is generated by a 180° rotation about the x^3-axis, which in the double group becomes the unit quaternion $k_1 = e_3$, together with a 120° rotation about the line $x^1 = x^2 = x^3$, which becomes the unit quaternion $k_2 = \frac{1}{2}(1 - e_1 - e_2 - e_3)$. In F these two elements are represented by

$$\mathcal{O}_1 = \begin{pmatrix} -1 & 0 & 0 \\ 0 & -1 & 0 \\ 0 & 0 & 1 \end{pmatrix}, \quad \mathcal{O}_2 = \begin{pmatrix} 0 & 1 & 0 \\ 0 & 0 & 1 \\ 1 & 0 & 0 \end{pmatrix}, \qquad (10.46)$$

respectively, whereas the two-dimensional representation E' is the restriction $\underline{2}|_T$ so that the two generators are simply represented by $q_1 = k_1$ and

$q_2 = k_2$. It is then a simple matter to verify that the ADHM matrix

$$\widehat{M} = \begin{pmatrix} e_1 & e_2 & e_3 \\ 0 & e_3 & e_2 \\ e_3 & 0 & e_1 \\ e_2 & e_1 & 0 \end{pmatrix} \tag{10.47}$$

satisfies the constraint (10.28) and the symmetry condition (10.45) for both these generators of the tetrahedral group, with the explicit matrices given above. An arbitrary scale can be introduced by multiplying $\widehat{M}$ by a constant real number, and this 1-parameter family precisely corresponds to the family of instantons generated using the JNR ansatz by placing the four poles on the vertices of a tetrahedron centred at the origin, as described earlier.

A more complicated example is the ADHM data of a 7-instanton with icosahedral symmetry [376]. In this case, the appropriate representations of Y are the real, reducible seven-dimensional representation $F_2 \oplus G$ and the complex two-dimensional representation E_2'. The icosahedral (double) group is generated by the three elements in the group of unit quaternions

$$k_1 = e_2, \quad k_2 = -\frac{1}{2}(e_1 + \tau e_2 - \tau^{-1} e_3), \quad k_3 = e_1, \tag{10.48}$$

where $\tau = \frac{1}{2}(\sqrt{5}+1)$ is the golden mean.

In E_2' the three generators are represented by

$$q_1 = e_2, \quad q_2 = -\frac{1}{2}(e_1 - \tau^{-1} e_2 + \tau e_3), \quad q_3 = e_1, \tag{10.49}$$

(note the replacement $\tau \mapsto -\tau^{-1}$). In F_2 and G they are represented by

$$\mathcal{O}_1 = \begin{pmatrix} -1 & 0 & 0 \\ 0 & 1 & 0 \\ 0 & 0 & -1 \end{pmatrix}, \quad \mathcal{O}_2 = -\frac{1}{2}\begin{pmatrix} 1 & \tau^{-1} & -\tau \\ \tau^{-1} & \tau & 1 \\ -\tau & 1 & -\tau^{-1} \end{pmatrix},$$

$$\mathcal{O}_3 = \begin{pmatrix} 1 & 0 & 0 \\ 0 & -1 & 0 \\ 0 & 0 & -1 \end{pmatrix}, \tag{10.50}$$

and

$$\mathcal{O}_1' = \begin{pmatrix} 1 & 0 & 0 & 0 \\ 0 & -1 & 0 & 0 \\ 0 & 0 & 1 & 0 \\ 0 & 0 & 0 & -1 \end{pmatrix}, \quad \mathcal{O}_2' = \frac{1}{4}\begin{pmatrix} -1 & \sqrt{5} & -\sqrt{5} & -\sqrt{5} \\ \sqrt{5} & 3 & 1 & 1 \\ -\sqrt{5} & 1 & -1 & 3 \\ -\sqrt{5} & 1 & 3 & -1 \end{pmatrix},$$

$$\mathcal{O}'_3 = \begin{pmatrix} 1 & 0 & 0 & 0 \\ 0 & 1 & 0 & 0 \\ 0 & 0 & -1 & 0 \\ 0 & 0 & 0 & -1 \end{pmatrix},$$

(10.51)

respectively. The reader may then verify that the constraint and symmetry conditions with these matrices are satisfied by the ADHM data

$$\widehat{M} = \begin{pmatrix} 1 & e_1 & e_2 & e_3 & 0 & 0 & 0 \\ 0 & 0 & 0 & 0 & e_1 & e_2 & e_3 \\ 0 & 0 & 0 & 0 & 0 & \tau e_3 & \tau^{-1}e_2 \\ 0 & 0 & 0 & 0 & \tau^{-1}e_3 & 0 & \tau e_1 \\ 0 & 0 & 0 & 0 & \tau e_2 & \tau^{-1}e_1 & 0 \\ e_1 & 0 & \tau^{-1}e_3 & \tau e_2 & 0 & 0 & 0 \\ e_2 & \tau e_3 & 0 & \tau^{-1}e_1 & 0 & 0 & 0 \\ e_3 & \tau^{-1}e_2 & \tau e_1 & 0 & 0 & 0 & 0 \end{pmatrix}.$$

(10.52)

We have again centred the instanton at the origin in $\mathbb{R}^4$, and still have the freedom to multiply $\widehat{M}$ by an arbitrary real scale factor.

Other symmetric ADHM data can be constructed in a similar fashion; for example, the ADHM data of a cubic 4-instanton are presented in [260]. It would be amusing to investigate instantons in $\mathbb{R}^4$ that are symmetric under one of the Platonic symmetry groups of $\mathbb{R}^4$, especially the symmetries of one of the exotic polytopes (the 24-cell, the 120-cell or the 600-cell) that have no $\mathbb{R}^3$ analogues.

In fact, the original motivation for searching for the symmetric instantons described above stems from a connection with Skyrmions, to which we now turn.

10.4 Skyrme fields from instantons

In Chapter 9 we discussed two methods of constructing approximate charge B Skyrmions, the product ansatz and the rational map ansatz, but both of these approximations have disadvantages.

The problem with the product ansatz is that it is only a good description of B well separated, unit charge Skyrmions; it can also be used to approximate the hedgehog, saddle point solution, but not to get close to the minimal energy Skyrmion. The rational map ansatz suffers from the opposite deficiency, in that it provides a good approximation to Skyrmions of minimal energy, and also to some low energy saddle point solutions, but does not contain any degrees of freedom to allow the individual Skyrmions to separate. (This is perhaps just a defect of the ansatz as we have presented it. There may be a cleverer ansatz which lets the Skyrmions separate as the rational map parameters vary, just as monopoles separate when the parameters of a Jarvis rational map vary.) In this section we describe a third method which can be used to construct charge B Skyrme fields, including well separated Skyrmions in arbitrary positions and orientations, and good approximations to the minimal energy solutions. This approach uses Yang-Mills instantons.

The Skyrme fields from instantons scheme was first proposed in [21] and involves computing the holonomy of $SU(2)$ instantons in Euclidean $\mathbb{R}^4$ along lines parallel to the x^4-axis. Explicitly, the prescription for the Skyrme field is to take

$$U(\mathbf{x}) = \mathcal{P} \exp \left(\int_{-\infty}^{\infty} A_4(\mathbf{x}, x^4) \, dx^4 \right) \tag{10.53}$$

where $\mathcal{P}$ denotes path ordering and A_μ is the gauge potential of a Yang-Mills instanton in $\mathbb{R}^4$, and where $\mathbf{x} = (x^1, x^2, x^3)$. Since A_4 takes values in the Lie algebra $su(2)$, its exponential is valued in the group $SU(2)$, so that $U(\mathbf{x}) : \mathbb{R}^3 \mapsto SU(2)$, as required for a static Skyrme field.

More precisely, the end points $-\infty$ and ∞ should both refer to the single point on S^4 corresponding to the point at infinity in $\mathbb{R}^4$. The holonomy is then along a closed loop in S^4, and is almost gauge invariant. The only effect of a gauge transformation $g(x)$ is to conjugate $U(\mathbf{x})$ by a fixed element $g(\infty)$. This corresponds to an isospin rotation of the Skyrme field. Also, the boundary condition $U \to 1_2$ as $|\mathbf{x}| \to \infty$ is satisfied in this scheme, because the loop on S^4 tends to zero size in this limit. In practice, ensuring the holonomy is along a closed loop means that sometimes an additional factor should be included in the formula (10.53), namely the transition function that connects ∞ back to $-\infty$. For an instanton given by the 't Hooft ansatz, the formula (10.53) is complete as it stands, but for the JNR ansatz, an additional factor of -1 is required. In an axial gauge, where $A_4 = 0$, the holonomy would be entirely contained in the transition function at infinity.

There is no real need for the Yang-Mills field to be an instanton. However, by restricting the scheme to instantons one obtains a quite large, but still finite-dimensional family of interesting Skyrme fields. Such fields are never exact solutions of the Skyrme equation but some can be good approximations to minimal energy Skyrmions and other important field configurations.

If A_μ is a self-dual Yang-Mills field with instanton number (second Chern number) N then it follows from general topological considerations that the resulting Skyrme field has baryon number $B = N$. This can also be verified using specific examples, and the general result follows by continuity. The construction yields an $(8N - 1)$-dimensional family of Skyrme fields from the $8N$-dimensional moduli space of charge N instantons; one parameter is lost since a translation of the instanton in the x^4

direction does not change the Skyrme field, due to the integration over the x^4 coordinate.

As the basic example, the charge 1 instanton, given by the one-pole 't Hooft ansatz (10.14) with width λ and position $a = 0$, generates a Skyrme field of the hedgehog form (9.18) with a profile function given by

$$f(r) = \pi \left[1 - \left(1 + \frac{\lambda^2}{r^2} \right)^{-1/2} \right]. \tag{10.54}$$

Instantons are scale invariant, so the parameter λ is arbitrary and can be chosen to minimize the energy of the resulting Skyrme field. The appropriate value of the scale is $\lambda^2 = 2.11$, and then the energy is $E = 1.243$, which is only 1% above that of the true Skyrmion solution.

The main difficulty with this construction is the computation of the holonomy, due to the fact that the integration is path ordered. To compute the path ordered exponential (10.53) one must introduce the quantity $\widetilde{U}(\mathbf{x}, x^4)$ and solve the (matrix) ordinary differential equation

$$\frac{\partial \widetilde{U}}{\partial x^4} = A_4 \widetilde{U} \tag{10.55}$$

along the real line $x^4 \in (-\infty, \infty)$, with $\mathbf{x}$ regarded as a parameter, and with the initial condition $\widetilde{U}(\mathbf{x}, -\infty) = 1_2$. The Skyrme field $U(\mathbf{x})$ is then obtained as the end point of the x^4-flow, that is, $U(\mathbf{x}) = \widetilde{U}(\mathbf{x}, \infty)$. In general, the direction of A_4 in the $su(2)$ algebra will vary with x^4, so Eq. (10.55) represents a complicated set of coupled equations whose solution can not be obtained in closed form, even if the instanton gauge potential is given explicitly, which it sometimes is, but not always. However, a simplification arises if A_4 is proportional to a fixed element of the algebra for all x^4, since the holonomy is then essentially abelian, with the result that the solution of (10.55) can be obtained by elementary integration. This is the case for the 1-instanton, leading to the explicit hedgehog profile expression (10.54). For more general instantons, although the integration can not be performed analytically it can be done numerically, by solving an ODE at each spatial point where the Skyrme field is required.

In this scheme, the charge 2 instantons generate a 15-dimensional family of Skyrme fields which includes configurations with two well separated Skyrmions with arbitrary positions, orientations and scales (which can be fixed by minimization of the energy as in the charge 1 sector). This accounts for fourteen of the instanton parameters and the final one, which corresponds to the x^4-separation of the two instantons, has little effect. As described earlier, a particularly symmetric 2-instanton arises if the three JNR poles have equal weights and are located at the vertices of an equilateral triangle in the spatial slice $x^4 = 0$. In this case, the instanton and resulting Skyrme field is axially symmetric. With a suitable scale for the equilateral triangle this instanton produces a good approximation to the minimal energy axially symmetric $B = 2$ Skyrmion. An approximation to the spherically symmetric $B = 2$ hedgehog solution can also be obtained, this time by placing the three poles on the x^4-axis, and minimizing the energy within this class. These examples show that the 15-dimensional space of Skyrme fields generated by instantons are a good candidate for a finite-dimensional truncation of the charge 2 sector of the Skyrme model. Furthermore, this 15-dimensional manifold appears to contain a 12-dimensional submanifold obtained as the unstable manifold of the hedgehog, and this is likely to be qualitatively very similar to the 12-dimensional manifold constructed using the gradient flow in the full field configuration space starting with the exact $B = 2$ hedgehog solution, as described in Chapter 9. For further details see ref. [22]. A 10-dimensional submanifold of the instanton-generated Skyrme fields, corresponding to two Skyrmions in the attractive channel, was actually used to study the quantization of the $B = 2$ Skyrmion in [261], producing a reasonable model of the deuteron. The relevant instantons have a Hartshorne circle and ellipse in a spatial plane, with coincident centres, and hence a triplet of reflection symmetries.

Recall from Chapter 9 that for $B > 2$ the minimal energy Skyrmions have discrete symmetries. This naturally leads to the question whether suitable symmetric instantons exist to generate Skyrme fields with these symmetries, and motivated the original study of symmetric instantons. The explicit ADHM data presented in the previous section provide the answer for the examples of the tetrahedrally symmetric 3-Skyrmion and the icosahedrally symmetric 7-Skyrmion. The holonomies of these symmetric instantons yield good approximations to the Skyrmions, provided the scales are chosen optimally [260, 376].

By considering infinitesimal variations of the ADHM data in the neighbourhood of the tetrahedrally symmetric 3-instanton it is possible to classify many of the vibrational modes of the associated minimal energy 3-Skyrmion (in fact, 23 modes in total, including nine zero modes and the breather mode), in a similar way as described in Chapter 9 using the rational map ansatz. This calculation [191] reproduces the results of the rational map approach, which provided the lowest-lying vibrational modes below the breather, and provides further modes which match the next lowest set of vibrational modes computed from full field simulations.

As discussed in Chapter 9, there is a close analogy between Skyrmion and monopole scattering through symmetric configurations. As a particular example, this analogy suggests that there should be a T_d-symmetric scattering of seven Skyrmions, in which the field passes through the icosahedrally symmetric 7-Skyrmion twice (in different orientations) and the cubic charge 7 saddle point once. Such a scattering can be approximated using instanton holonomies created from a 1-parameter family of T_d-symmetric 7-instantons, where variation of the parameter is interpreted as time evolution, in a similar manner to the geodesic description of monopole scattering. The ADHM data of an appropriate family of tetrahedral 7-instantons were derived in [376], which of course contain the icosahedral ADHM data (10.52) as a special case, and also contain a cubic instanton which is equivalent to one obtained by placing eight equal weight JNR poles on the vertices of a cube. The associated dynamics of the generated Skyrme field is displayed in Fig. 10.2 as a sequence of baryon density isosurfaces.

At the start (picture 1) there are clearly six unit charge Skyrmions on the Cartesian axes and a Skyrmion at the origin. As the Skyrmions approach (picture 2), the one at the origin shrinks until it disappears completely (picture 3). The Skyrmions then merge until the $B = 7$ dodecahedron is formed (picture 5), after which the configuration deforms until it turns into a cube (picture 8). This process is then reversed, rotated by 90° around the x^1-axis, so that the dual dodecahedron is formed (picture 11), and the Skyrmions finally separate again along the Cartesian axes (picture 15). The true dynamical evolution depends upon the initial speeds of the incoming Skyrmions, which affects the amount of energy lost through radiation as the process evolves. If the incoming speeds are great enough then the whole scattering process displayed in Fig. 10.2 should take place. Radiation effects will mean that for most low speeds the incoming Skyrmions will eventually get trapped at one of the dodecahedra, and perhaps if the Skyrmions are initially static then only the first portion of the scattering process will occur and the cube may never be formed. Full field simulations with initial conditions given by the instanton-generated Skyrme field verify that the true dynamical evolution does follow the sequence described above, so the instanton-generated Skyrme fields provide an accurate approximation to the Skyrmion scattering process.

The Skyrmions from instantons scheme also gives an approximation to the Skyrme crystal, as the holonomy of an instanton on $\mathbb{T}^4$ [296], but unfortunately there is no known explicit expression for the relevant periodic instanton or Skyrme field.

Recall that in earlier chapters we commented that sine-Gordon kinks represent a toy model for Skyrmions in two dimensions lower, and planar sigma model lumps are similarly an analogue of Yang-Mills instantons. It is amusing that these two analogies can be linked by the instanton holonomy construction, in that good approximations to sine-Gordon kinks can be obtained by computing the holonomy of lumps in the $O(3)$ sigma

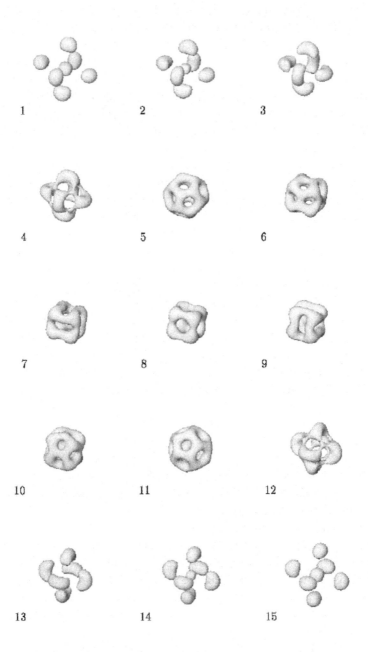

Fig. 10.2. Baryon density isosurfaces for a family of $B = 7$ Skyrme fields obtained from a family of $N = 7$ instantons.

model when formulated as a $U(1)$ gauge theory [389].

10.5 Monopoles as self-dual gauge fields

In the previous section we have described an approximate connection between Skyrmions and instantons, but in fact there is an exact link between monopoles and anti-self-dual gauge fields which has been known for longer [77, 276].

Consider a gauge potential A_μ in $\mathbb{R}^4$ which has a translational symmetry, so that it is independent of the coordinate x^4, and rename the component along the direction of symmetry $A_4 = \Phi$. The result is a gauge potential A_i and scalar field Φ defined in $\mathbb{R}^3$, with Φ transforming as an adjoint Higgs field under x^4-independent gauge transformations. Moreover, with this dimensional reduction the anti-self-dual Yang-Mills equation (10.10) becomes the Bogomolny equation (8.86) for monopoles in $\mathbb{R}^3$. Note that no anti-instanton can correspond to a monopole via this identification since the property of finite action required for an anti-instanton prevents a translationally invariant gauge potential, which trivially has infinite action. However, rather remarkably, the spherically symmetric 1-monopole can be derived from the anti-self-dual version of the 't Hooft ansatz (10.12) by taking the solution of the Laplace equation (10.13) to be

$$\rho = \frac{\sinh 2r}{r} e^{2ix^4},$$ (10.56)

where r is the three-dimensional radial coordinate. There are a couple of remarks to make about this construction. The first is that although ρ depends on x^4, the gauge potential defined by the logarithmic derivative (10.12) is independent of x^4, as required here. The second remark is that in the construction of anti-instantons the solution of the Laplace equation is required to be real, which is not the case here. However, it turns out that the fields generated by the solution (10.56), although complex, can be made real via a complex gauge transformation, and then the fields are precisely those of the standard 1-monopole. Unfortunately, no real multi-monopoles can be constructed from the 't Hooft ansatz.

An alternative approach to constructing the 1-monopole in $\mathbb{R}^3$ involves an infinite chain of anti-instantons in $\mathbb{R}^4$ [78], with particular locations and scales so that the infinite sum can be calculated in closed form and yields a monopole in a certain limit. The apparent non-trivial periodicity in x^4 disappears in the limit, and the field becomes independent of x^4. This also has a lower-dimensional analogue, with an exact sine-Gordon kink produced by an infinite chain of sigma model lumps [390].

There is a close relationship between finite action anti-instantons and monopoles in hyperbolic space, as pointed out by Atiyah [15]. In the above we considered anti-self-dual gauge fields with a translational symmetry but Atiyah's observation is based on anti-self-dual gauge fields with a rotational symmetry. To be explicit, consider a gauge potential in $\mathbb{R}^4$ symmetric under a circle action, say, rotations in the (x^3, x^4) plane. The fixed set of the circle action is the (x^1, x^2) plane. If this is removed, then the circle action is free, and there is a conformal equivalence between $(\mathbb{R}^4 - \mathbb{R}^2)/S^1$ and hyperbolic 3-space $\mathbb{H}^3$, which can be understood in terms of coordinates as follows. Write the Euclidean metric on $\mathbb{R}^4$ in the form

$$
\begin{aligned}
ds^2 &= (dx^1)^2 + (dx^2)^2 + (dx^3)^2 + (dx^4)^2 \\
&= \frac{r^2}{R^2} \left[R^2 \left(\frac{(dx^1)^2 + (dx^2)^2 + dr^2}{r^2} \right) + R^2 d\theta^2 \right], \quad (10.57)
\end{aligned}
$$

where R is a positive constant parameter, and we have introduced polar coordinates r, θ in the (x^3, x^4) plane, with $r > 0$ and $0 \leq \theta < 2\pi$. Now drop the conformal factor r^2/R^2. The first term in the square brackets is the metric on hyperbolic 3-space of curvature $-1/R^2$, in terms of its standard description as the upper half space in $\mathbb{R}^3$, and the remaining term is the metric on the circle of radius R, which can be removed by quotienting by the circle action. Note that $r = 0$ is the boundary (at infinity) of $\mathbb{H}^3$, which is why this plane must be deleted in the conformal identification.

In analogy with the dimensional reduction by a translation symmetry, one may here perform the dimensional reduction by the circle symmetry, generated by the vector field ∂_θ. Since the Yang-Mills equation is conformally invariant, the dropping of the conformal factor r^2/R^2 has no effect. After defining $A_\theta = \Phi$, the anti-self-dual Yang-Mills equation (10.10) becomes the Bogomolny equation for monopoles in $\mathbb{H}^3$ (8.365). There do exist circularly symmetric finite action anti-instantons and these yield hyperbolic monopoles. Now the anti-instanton extends smoothly to the (removed) (x^1, x^2) plane, and here a rotation by α can be compensated by a gauge rotation by $p\alpha$ for some integer p. For consistency, the curvature of the hyperbolic space must be $-1/p^2$. This procedure has been performed explicitly [79, 315] to yield the spherically symmetric 1-monopole in hyperbolic space, but it is not a practical method for constructing multi-monopoles.

Instead, the ADHM construction restricted to circularly symmetric anti-instantons can be interpreted as a set of difference equations [62] whose solutions can be used to obtain hyperbolic monopoles, in a similar way that solutions of the Nahm equation give Euclidean monopoles. In fact, this set of difference equations is an integrable, lattice discretization of the Nahm equation [311], and the continuum limit in which the lattice spacing tends to zero corresponds to the zero curvature limit, so the appearance of the Nahm equation on one side of the transform and Euclidean monopoles on the other is consistent.

The connection between circularly symmetric Yang-Mills fields in $\mathbb{R}^4$ and monopoles in $\mathbb{H}^3$ has suggested a method to prove the existence of solutions of the second order Yang-Mills field equation which are neither self-dual nor anti-self-dual [375]. In parallel with the above description of S^1-invariant anti-self-dual gauge fields as Bogomolny hyperbolic monopoles, there is a similar correspondence at the level of the second order field equations. Adapting the methods of Taubes, mentioned in Chapter 8 and to be described in more detail in Section 11.4, to prove the existence of non-Bogomolny solutions of the Yang-Mills-Higgs equations in Euclidean $\mathbb{R}^3$, Sibner, Sibner and Uhlenbeck [375] were able to prove a similar result in hyperbolic space $\mathbb{H}^3$, and hence prove the existence of circularly symmetric solutions of the second order Yang-Mills equation in $\mathbb{R}^4$ which are not (anti-)self-dual. These unstable solutions are expected to be composed of instanton-anti-instanton pairs, though no explicit solutions are available, even numerically, to investigate their detailed properties.

An intermediate object between an anti-instanton and a monopole is a non-trivially periodic anti-instanton, or caloron [172, 116]. This is an anti-self-dual gauge field which is periodic in one spatial direction, in other words an anti-instanton on $\mathbb{R}^3 \times S^1$. The name caloron refers to the relevance of Yang-Mills fields on $\mathbb{R}^3 \times S^1$ to the quantum theory at finite temperature. As the period tends to infinity an anti-instanton on $\mathbb{R}^4$ can be recovered, and as the period tends to zero a monopole is obtained. The Nahm transform applies equally well to calorons as to monopoles and recently progress has been made in explicitly constructing calorons and understanding their structure in terms of monopole constituents [141].

As we have seen, a simple dimensional reduction of the integrable anti-self-dual Yang-Mills equation leads to the integrable Bogomolny equation for monopoles. The Nahm equation, which is also integrable, can be interpreted in a similar fashion as the dimensional reduction of the self-dual Yang-Mills equation under three translational symmetries. Ward [417] pointed out that other well known integrable systems, such as the sine-Gordon equation, can also be obtained as dimensional and algebraic reductions of self-dual Yang-Mills, and suggested that the self-dual Yang-Mills equation might be a master integrable system from which many (if not all) integrable equations could be derived. Since this original suggestion, a plethora of known integrable equations (and some new ones) have been proved to fit into this scheme – see for example the book on this topic by Mason and Woodhouse [297] – though some higher-dimensional integrable systems such as the KP equation do not appear to fit in a natural way. In this context it is often necessary to begin with the self-duality equation in (2+2)-dimensions, which is still real, rather than in (4+0)-dimensions. The intermediate case of (3+1)-dimensions is not tractable since the Hodge dual introduces a factor i, which means there can be no real self-dual gauge fields in ordinary Minkowski space.

10.6 Higher rank gauge groups

In this chapter we have restricted our discussion to $SU(2)$ instantons, but this gauge group can obviously be replaced by one of higher rank, such as $SU(m)$, $SO(m)$ or $Sp(m)$, and not only do instantons still exist, classified by a single integer, the second Chern number, but the ADHM construction can be applied in essentially the same manner. In the case of $SU(m)$, the moduli space of N-instantons has dimension $4mN$ and is always a hyperkähler manifold. If $0 < N < \frac{1}{2}m$ then all N-instantons can be obtained by a simple embedding of instantons of a smaller gauge group, but if $N \geq \frac{1}{2}m$ this is no longer the case and there are non-embedded instantons. Similar results apply for other gauge groups.

For $SU(m)$ N-instantons on the 4-torus $\mathbb{T}^4$ there is a duality, because the Nahm transformation maps this space of instantons to the space of instantons on the dual torus (with periods replaced by their inverses) where the gauge group and instanton number are also swapped, that is, the mapping is to $SU(N)$ m-instantons [63].

The instanton holonomy construction has also been applied to $SU(m)$ instantons [144, 204] to obtain approximations to known spherically symmetric $SU(m)$ Skyrmions, which have some amusing properties [163, 205].

11

Saddle points – sphalerons

11.1 Mountain passes

For much of this book we have been seeking and studying stable soliton solutions in various field theories. Occasionally we have found unstable solutions of the field equations, for example, the hedgehog solutions of the Skyrme equation for $|B| > 1$. These were usually minima of the energy within a subclass of fields with a certain symmetry, but saddle points of the energy in the space of all field configurations. One might say that these saddle points were found accidentally. In this chapter we shall describe a more systematic, topological, approach to saddle point solutions. We shall also comment on their interpretation and physical significance.

The basic idea is the following [267]. Suppose on a connected, compact manifold X, there is defined a bounded, twice differentiable potential energy function V. Think of V as the height on X. Suppose that V has two isolated local minima, at $\mathbf{x}_0$ and $\mathbf{x}_1$. Then there is also a saddle point of V at some point of X. This saddle point is the "mountain pass" between $\mathbf{x}_0$ and $\mathbf{x}_1$. One can show its existence by considering all the paths from $\mathbf{x}_0$ to $\mathbf{x}_1$. Along each path c, V has a maximal value V_c attained at some point $\mathbf{x}_c$. Let V_* be the infimum over all paths c of the values V_c. Then there is a sequence of paths $c_1, c_2, \ldots$ such that $\lim_{n\to\infty} V_{c_n} = V_*$, and among these is a subsequence, such that $\lim_{n\to\infty} \mathbf{x}_{c_n} = \mathbf{x}_*$ exists. At the limiting point $\mathbf{x}_*$, which is the mountain pass, $V = V_*$. $\mathbf{x}_*$ is the highest point along a path from $\mathbf{x}_0$ to $\mathbf{x}_1$, but it is the lowest possible such highest point. See Fig. 11.1 for a sketch of this situation. An important extension of this result is the following. Suppose $\pi_1(X)$ is non-trivial. Let $\mathbf{x}_0$ be the isolated point on X where V has its minimum. Consider all the paths on X, beginning and ending at $\mathbf{x}_0$, in some particular homotopy class of $\pi_1(X)$ that is not the identity class. Thus the loops are non-contractible.

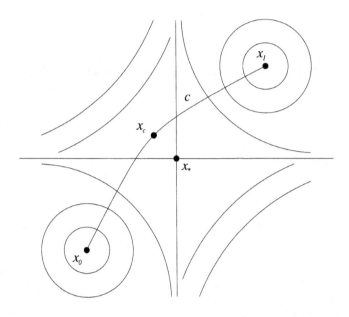

Fig. 11.1.   Sketch of a contour plot of a function V, showing two local minima at $\mathbf{x}_0$ and $\mathbf{x}_1$, and a saddle point at $\mathbf{x}_*$. The maximum of V along the path c is at $\mathbf{x}_c$.

On each path c, again find the point where V attains its maximum V_c. Then find the infimum V_* of the values V_c. Since X is compact, this infimum is attained. There is a path (not unique) from $\mathbf{x}_0$ to $\mathbf{x}_0$, along which the maximal value of V is V_* at a point $\mathbf{x}_*$. $\mathbf{x}_*$ is a saddle point of V on X.

Note that $\mathbf{x}_*$ can not be either $\mathbf{x}_0$ or $\mathbf{x}_1$ in the first example. If it were, then there would be a path from $\mathbf{x}_0$ to $\mathbf{x}_1$ along which V is constant (since the maximal and minimal values of V along the path would be equal). So the minima of V at $\mathbf{x}_0$ and $\mathbf{x}_1$ would not be isolated. Similarly, $\mathbf{x}_*$ can not be $\mathbf{x}_0$ in the second example. If it were, there would be a non-contractible loop in X, based at $\mathbf{x}_0$, along which V was constant. Again, the minimum of V at $\mathbf{x}_0$ would not be isolated.

The arguments leading to these saddle points of V each suggest that at $\mathbf{x}_*$, V has just one unstable direction. Indeed, consider the (symmetric) matrix of second derivatives of V at $\mathbf{x}_*$. Provided its eigenvalues are all non-zero, then there is precisely one negative eigenvalue. If there were two or more negative eigenvalues, then a path through $\mathbf{x}_*$ could be deformed so that the potential energy along the path would be everywhere lower than V_*, contradicting the definition of $\mathbf{x}_*$.

Of course, V could have unstable saddle points where the second derivative matrix has more than one negative eigenvalue. But such saddle points will not be found by considering a set of non-contractible loops (or paths between two points). One may find them by considering a set of non-contractible spheres S^n, all belonging to a given class of $\pi_n(X)$. However, the analysis relating the existence of the saddle points to the homotopy groups is trickier.

The argument, as presented so far, is for a function V defined on a compact manifold X. One would like to extend the argument to the potential energy function E of some field theory. This was first achieved by Taubes [398]. The manifold X is replaced by the field configuration space $\mathcal{C}$. It is much trickier to prove rigorously that saddle points of E exist, using the mountain pass idea. However, one can try the method, and see what saddle points are suggested. In this way, genuine saddle point solutions of certain field theories have been discovered.

Most of the required ingredients are present in field theory. Although $\mathcal{C}$ is not generally connected, the connected components are labelled by elements of $\pi_0(\mathcal{C})$, or equivalently by the topological charge(s) of the fields, and we can apply our techniques to one of these components. Although each component of the configuration space $\mathcal{C}$ is infinite-dimensional, the potential energy E is formally differentiable. The derivative is the first variation of E (the left-hand side of the Euler-Lagrange field equation) and it vanishes at a stationary point. The second variation operator, or Hessian, is well defined there, and has a finite number of negative eigenvalues. However, there are some possible problems. $\mathcal{C}$ is generally not compact. As a result, we may not be able to find a saddle point by taking a limit of a sequence of field configurations. As the limit is taken, the field energy may split into two or more clusters moving away to infinite separation. This is a real problem in certain field theories, but not others. Each case needs to be considered carefully. Another problem is that the minima and saddle points of E may not be isolated. This can happen if the theory has a large symmetry group. Saddle point solutions may still exist, but one needs to use more refined topological arguments to establish their existence. In particular, in gauge theories, one needs to avoid the infinite-dimensional degeneracy associated with the topologically complicated group of gauge transformations. Gauge fixing helps, but must be carried out in a continuous way as one varies the field. This is not always possible.

We shall now consider a number of examples of field theories where non-contractible loops of field configurations can be constructed, in some cases leading to the existence of saddle point solutions of the field equations. Such saddle points in field theory are called "sphalerons" [241] –

from the ancient Greek $\sigma\phi\alpha\lambda\varepsilon\rho os$ (sphaleros), meaning "unstable", or "ready to fall". They are static, but unstable, finite energy solutions. Sphalerons, like solitons, have a localized and smooth energy density. Following Taubes [398], we shall only use ideas from homotopy theory in the following discussion. However, there is also the possibility of using homology ideas and Morse theory [303] to investigate minima and saddle points. See refs. [20, 61] for the application of homology ideas to the study of instanton moduli spaces, and ref. [16], where the Yang-Mills action is used as a Morse function in the context of Yang-Mills theory defined over a Riemann surface.

11.2 Sphalerons on a circle

This example [292, 24] is one of the simplest to understand, though perhaps not as physically interesting as the examples we consider later.

Let $\phi(x)$ be a real scalar field defined in one space dimension, with "space" a circle of length $2\pi L$. We take x to lie in the range $0 \leq x \leq 2\pi L$, and impose the periodic boundary condition $\phi(2\pi L) = \phi(0)$. The possible time dependence of ϕ is unimportant here, and is suppressed. Consider the ϕ^4 theory, whose potential energy function is

$$E = \frac{1}{2} \int_0^{2\pi L} \left((\partial_x \phi)^2 + (1 - \phi^2)^2 \right) dx, \tag{11.1}$$

and whose static field equation is

$$\frac{d^2\phi}{dx^2} + 2(1 - \phi^2)\phi = 0. \tag{11.2}$$

The configuration space $\mathcal{C}$ consists of all field configurations $\phi(x)$ defined on the circle, with finite energy. There are two distinct classical vacua, the constant fields $\phi = 1$ and $\phi = -1$, which we denote by $\mathcal{V}_+$ and $\mathcal{V}_-$. Both have zero energy.

A path $c(\mu)$ in $\mathcal{C}$ connecting $\mathcal{V}_+$ and $\mathcal{V}_-$ is a 1-parameter, continuous family of finite energy configurations $\{\phi(x;\mu) : 0 \leq \mu \leq 1\}$ with $\phi(x;0) = 1$, $\phi(x;1) = -1$. All such paths lie in one homotopy class. For a given path c we define E_c to be the maximal energy along it,

$$E_c = \max_{\mu\in[0,1]} E(\mu). \tag{11.3}$$

We seek the minimum, among all paths, of E_c.

A candidate path c_0 is the set of constant fields

$$\phi(x;\mu) = 1 - 2\mu, \tag{11.4}$$

with energies

$$E(\mu) = \pi L \left(1 - (1 - 2\mu)^2\right)^2$$
$$= 16\pi L(\mu(1 - \mu))^2 . \tag{11.5}$$

The maximal energy along c_0 is $E_{c_0} = \pi L$, attained at $\mu = \frac{1}{2}$.

For small L, E_{c_0} is the minimal value of E_c for any path c, and the constant field $\phi = 0$ is the sphaleron solution. However, for large L, there is a different type of path which connects the vacua, but climbs to a lesser height. Schematically, the path is as in Fig. 11.2. Out of $\mathcal{V}_+$, a kink-antikink pair is produced; the pair separate and move half-way around the circle in opposite directions, then annihilate, leaving the vacuum $\mathcal{V}_-$.

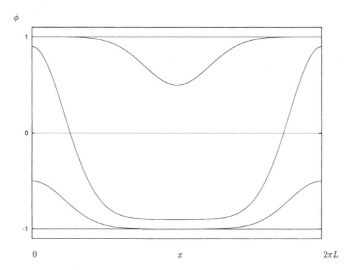

Fig. 11.2. Sketch of a path connecting the two vacua. Out of the vacuum $\mathcal{V}_+$ a kink-antikink pair is produced; the pair separate and move half-way around the circle in opposite directions, then annihilate, leaving the vacuum $\mathcal{V}_-$.

Via this path, the energy barrier between the vacua is approximately $\frac{8}{3}$, twice the energy of a single ϕ^4 kink on the infinite interval. This is independent of L, and less than πL when L is large.

We can find the critical length of the circle where $\phi = 0$ ceases to be the mountain pass between vacua by counting the number of unstable modes. Let $\phi(x) = \eta(x)$ be a small perturbation of the solution $\phi = 0$. To quadratic order in η, the energy is

$$E = \pi L + \frac{1}{2} \int_0^{2\pi L} \left((\partial_x \eta)^2 - 2\eta^2\right) dx$$

$$= \pi L + \frac{1}{2} \int_0^{2\pi L} \eta \left(-\frac{d^2}{dx^2} - 2 \right) \eta \, dx \, . \tag{11.6}$$

The eigenvalue equation for the modes is

$$-\frac{d^2\eta}{dx^2} - 2\eta = \nu\eta \, , \tag{11.7}$$

subject to the boundary condition $\eta(2\pi L) = \eta(0)$. A negative value of ν corresponds to an instability. The modes are $\cos(nx/L)$, $n = 0, 1, \ldots$, and $\sin(nx/L)$, $n = 1, 2, \ldots$, with

$$\nu = \frac{n^2}{L^2} - 2 \, . \tag{11.8}$$

Notice that the lowest mode is singly degenerate, but the higher modes are doubly degenerate. For $L < 1/\sqrt{2}$, only the lowest mode $\eta = \text{const}$ is an unstable mode; for $L > 1/\sqrt{2}$ there are more unstable modes. This shows that $L = 1/\sqrt{2}$ is the critical length.

We can verify this directly by solving the static field equation (11.2) and finding the sphaleron solution. In addition to the three constant solutions $\phi = 1$, $\phi = 0$ and $\phi = -1$, there are non-constant solutions

$$\phi(x) = \sqrt{\frac{2}{1+k^2}} \, k \, \text{sn}_k \left(\sqrt{\frac{2}{1+k^2}} \, x \right) \, , \tag{11.9}$$

which satisfy the periodicity condition if

$$L = \frac{\sqrt{2(1+k^2)} K_k n}{\pi} \tag{11.10}$$

where $n \in \mathbb{Z}$. These solutions can also be translated without changing their energy. The non-constant solutions exist only if $L > 1/\sqrt{2}$, and their number increases as L increases. We plot their energies, together with the energy of the solution $\phi = 0$, in Fig. 11.3. Notice that for $L > 1/\sqrt{2}$ the solution of lowest energy, aside from the vacua $\phi = \pm 1$, is always the solution (11.9) with k determined by Eq. (11.10) with $n = 1$. This is the sphaleron for $L > 1/\sqrt{2}$. It represents a kink and antikink separated by πL, and thus on opposite sides of the circle.

The reason for the existence of the sphaleron of this type is physically clear. A kink and antikink attract and will annihilate if possible. When they are on opposite sides of the circle they are in unstable equilibrium; the energy decreases if they are brought together either on one side of the circle or the other, but since they are exactly opposite there is no tendency to move in either direction. The other non-constant solutions,

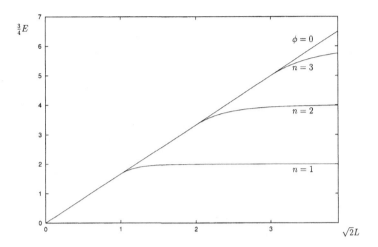

Fig. 11.3. The energies of the $\phi = 0$ and sphaleron solutions with $n = 1, 2, 3$, as a function of the period $2\pi L$, for the ϕ^4 theory on a circle.

for larger values of L, have the interpretation of a chain of alternating kinks and antikinks, equally spaced around the circle. Any perturbation of the relative positions of the kinks and antikinks reduces the energy, so there are several unstable modes.

A more detailed analysis of these various solutions, giving the precise number of unstable modes and some of the corresponding eigenvalues ν, is given in ref. [265].

11.3 The gauged kink

Let $\phi(x)$ be a complex-valued scalar field defined on the whole real line $\mathbb{R}$, with potential energy function

$$E = \frac{1}{2} \int_{-\infty}^{\infty} \left(\partial_x \bar{\phi} \partial_x \phi + (1 - \bar{\phi}\phi)^2 \right) dx . \qquad (11.11)$$

The vacuum manifold is the circle $|\phi| = 1$, so the classification of solutions is rather different than in the real ϕ^4 theory. The vacuum is any constant solution $\phi(x) = c$, with $|c| = 1$. The field equation

$$\frac{d^2\phi}{dx^2} + 2(1 - \bar{\phi}\phi)\phi = 0 \qquad (11.12)$$

has a kink solution $\phi_K(x) = \tanh x$ which connects $\phi = -1$ at $x = -\infty$ to $\phi = 1$ at $x = \infty$. However, unlike in the real theory, there is no reason to think this is topologically stable. The field for $x \ll 0$ can be deformed around the vacuum manifold from the value -1 to 1, keeping the field for $x \gg 0$ unchanged. In this way the kink can be deformed to the vacuum $\phi(x) = 1$. The energy E can be made to monotonically decrease during this process.

Since the kink can be unwound to produce the vacuum in two opposite ways, this seems to imply that the kink solution of the complex ϕ^4 theory is a sphaleron, a mountain pass along a non-contractible loop from the vacuum to the vacuum. To check this, consider small variations away from the kink. Varying $\mathrm{Re}\,\phi$ can only increase the energy (since the kink is stable in the real ϕ^4 theory) so we consider just a variation of $\eta = \mathrm{Im}\,\phi$ (which is in an orthogonal direction in field configuration space). Thus, set $\phi(x) = \phi_{\mathrm{K}}(x) + i\eta(x)$. The energy, to quadratic order in η, is

$$E = \frac{4}{3} + \frac{1}{2}\int_{-\infty}^{\infty} \eta \left(-\frac{d^2}{dx^2} - 2\,\mathrm{sech}^2 x \right) \eta \, dx \,. \tag{11.13}$$

To study the instability of ϕ_{K}, we look at the eigenvalue equation

$$\left(-\frac{d^2}{dx^2} - 2\,\mathrm{sech}^2 x \right) \eta = \nu\eta \,. \tag{11.14}$$

This is a classic example of an integrable stationary Schrödinger equation, and it has the one negative mode

$$\eta(x) = \mathrm{sech}\, x \tag{11.15}$$

with eigenvalue $\nu = -1$. The kink is therefore unstable.

There is still a problem with interpreting the kink as a sphaleron in this theory. The unstable mode is normalizable, so it only deforms the kink near the origin. The effect of lowering the energy towards zero, using this mode, is to produce a field configuration $\phi(x)$ which is close to -1 for $x \ll 0$, whose value slowly winds round the semicircle $|\phi| = 1$, $\mathrm{Im}\,\phi > 0$ (or the semicircle $|\phi| = 1$, $\mathrm{Im}\,\phi < 0$) in some large finite interval of x, and which is close to 1 for $x \gg 0$. Via this mode, the sphaleron does not actually decay to the vacuum solution $\phi = 1$. To reach the vacuum, the field at large negative x must be changed from -1 to 1. However, for this to occur in a finite time, an infinite kinetic energy is needed even though there is no potential barrier to cross. Thus the sphaleron can not really decay to the vacuum.

This difficulty is avoided in the gauged ϕ^4 theory [55], with a complex field ϕ and a $U(1)$ gauge potential a_μ. The full Lagrangian is

$$L = \frac{1}{2}\int_{-\infty}^{\infty} \left(f_{01}^2 + \overline{D_\mu\phi}D^\mu\phi - (1 - \bar\phi\phi)^2 \right) dx \,, \tag{11.16}$$

where $f_{01} = \partial_0 a_1 - \partial_1 a_0$ is the electric field and $D_\mu \phi = \partial_\mu \phi - i a_\mu \phi$ is the covariant derivative of ϕ. The potential energy of static fields is

$$E = \frac{1}{2} \int_{-\infty}^{\infty} \left(\overline{D_1 \phi} D_1 \phi + (1 - \bar{\phi}\phi)^2 \right) dx. \qquad (11.17)$$

In the gauge $a_1 = 0$, the static field equation is as in the ungauged theory, and the solution is $\phi_K(x) = \tanh x$ as before. However, this is now a genuine sphaleron.

To study the mode of instability of this solution, it is convenient to use the gradient flow equations (2.169) and (2.170), with $\kappa = 1$. These can be linearized by setting $\phi = \phi_K + i\eta$ and $a_1 = a$, and ignoring terms quadratic in η and a. The resulting equations are

$$\partial_0 \eta = \partial_1 (\partial_1 \eta - \phi_K a) - (\partial_1 \phi_K) a + 2(1 - \phi_K^2)\eta \qquad (11.18)$$
$$\partial_0 a = \phi_K (\partial_1 \eta - \phi_K a) - (\partial_1 \phi_K)\eta. \qquad (11.19)$$

Using the expressions on the right-hand side of (11.18) and (11.19), one can check that

$$\phi_K \partial_0 \eta - \partial_1 \partial_0 a = 0. \qquad (11.20)$$

This confirms that the flow $(i\partial_0 \eta, \partial_0 a)$ is orthogonal to $(i\phi_K \alpha, \partial_1 \alpha)$, an infinitesimal gauge transformation of ϕ_K with arbitrary parameter $\alpha(x)$.

We now seek an exponentially growing solution of these equations of the form $\eta(t, x) = \eta(x)e^{-\nu t}$, $a(t, x) = a(x)e^{-\nu t}$, with ν negative, which must also satisfy

$$\phi_K \eta - \partial_1 a = 0. \qquad (11.21)$$

The coupling of η to a means that (11.15) is no longer a solution, but by trying a variable power of $\text{sech}\, x$ one finds the solution

$$\eta(x) = \tau (\text{sech}\, x)^\tau, \quad a(x) = -(\text{sech}\, x)^\tau \qquad (11.22)$$

with $\nu = -\tau$, where $\tau = \frac{1}{2}(\sqrt{5} + 1)$ is the golden mean. This appears to be the only unstable mode of the kink sphaleron. The mode deforms the kink into a field configuration which is close to being gauge equivalent to the vacuum. A subsequent change of the phase of ϕ is no longer associated with infinite kinetic energy, because we can choose a_0 so that $D_0 \phi = 0$ even though $\partial_0 \phi$ is non-zero.

An alternative way to consider this example is to change gauge. Require that $\phi = 1$ both at $x = -\infty$ and $x = \infty$, for all field configurations. One vacuum solution is the trivial configuration

$$\phi(x) = 1, \quad a_1 = 0. \qquad (11.23)$$

There is another, topological non-trivial vacuum, with a unit net winding,

$$\phi(x) = e^{i\alpha(x)}, \quad a_1 = \partial_1 \alpha, \qquad (11.24)$$

where $\lim_{x \to -\infty} \alpha(x) = -2\pi$, $\lim_{x \to \infty} \alpha(x) = 0$. (Vacua with multiple windings are obtained by replacing -2π by $-2\pi n$ here.) Although α is not completely determined by these boundary conditions,

$$\int_{-\infty}^{\infty} a_1 \, dx = 2\pi \tag{11.25}$$

in all cases.

The trivial vacuum and the unit winding vacuum both have zero energy. There is no path connecting them which consists only of vacuum configurations. Paths connecting them have to pass over a mountain pass, and this is the kink sphaleron. Note that in the gauged theory, the kink can be presented in a gauge where $\lim_{x \to \pm\infty} \phi(x) = 1$. Since this involves a phase rotation by π at $x = -\infty$, in this gauge

$$\int_{-\infty}^{\infty} a_1 \, dx = \pi \,, \tag{11.26}$$

the value intermediate between the values for the two vacua. Now recall from Section 3.6 that the Chern-Simons number of an abelian gauge field defined on $\mathbb{R}$ is

$$y_1 = \frac{1}{2\pi} \int_{-\infty}^{\infty} a_1 \, dx \,. \tag{11.27}$$

The vacua we have been discussing have $y_1 = 0$ and $y_1 = 1$, respectively. The sphaleron that lies in between has $y_1 = \frac{1}{2}$, and deforming it by the unstable mode either increases or decreases y_1, depending on the direction.

There is one further, important remark. We have been discussing topologically distinct vacua, but need to stress that these are gauge equivalent, by a "large" gauge transformation, and so are physically the same. The path from one vacuum to the neighbouring vacuum, via the sphaleron, is not really a path with distinct end points, but instead a closed, non-contractible loop. The Chern-Simons number of a vacuum field (satisfying the boundary conditions on ϕ) must have an integer value, and a path with topologically distinct vacua at the ends corresponds to a non-contractible loop because it can not be deformed continuously into a path consisting entirely of vacuum fields.

The analogy with a pendulum is perhaps helpful. Suppose $\theta = 0$ is the stable position of the pendulum. Then a continuous path from $\theta = 0$ to $\theta = 2\pi$ is not a path with distinct end points, but rather a non-contractible closed loop that must have passed at least once through the unstable position $\theta = \pi$.

We shall see below that in other gauge theory examples, the sphaleron is the mountain pass between topologically distinct vacua, or really, a mountain pass along a non-contractible loop, and has fractional Chern-Simons number.

11.4 Monopole-antimonopole dipole

In the $SU(2)$ Yang-Mills-Higgs theory with adjoint Higgs field there are, as we saw in Chapter 8, magnetic monopole solutions of positive and negative charge. In the BPS limit, where the forces between monopoles cancel, there are also multi-monopole solutions, which are solutions of the Bogomolny equation $B_i = -D_i\Phi$. So far, the only solution in the zero charge sector that we have presented is the vacuum solution, which is gauge equivalent to $\Phi = i\tau_3$, $A_i = 0$. In pioneering work in 1982, Taubes [398] used a topological argument together with rigorous analysis to show that there is at least one more static solution in this sector, a configuration of a monopole and antimonopole in unstable equilibrium, which satisfies the second order field equations. Similar methods have been applied in a number of other theories to find such unstable solutions, now often referred to as "sphalerons".

Taubes considered the $SU(2)$ Yang-Mills-Higgs theory in the BPS limit, though this is not essential, as the Bogomolny equation plays only a minor (and approximate) role. The true configuration space of the theory is $\mathcal{C} = \mathcal{A}/\mathcal{G}_0$, where $\mathcal{A}$ is the space of all finite energy field configurations, and $\mathcal{G}_0$ is the space of based gauge transformations $g(\mathbf{x})$, satisfying $g(\mathbf{0}) = 1$. This quotient space $\mathcal{C}$ is the same as the space of fields satisfying the radial gauge condition $A_r = 0$. $\mathcal{C}$ is homotopic to the space $\mathrm{Maps}(S^2 \mapsto S^2)$, which can be identified with the space of Higgs fields at infinity. No topological information is carried by the Higgs and gauge field in any bounded region. The group of rigid gauge transformations acts on $\mathcal{C}$, and hence on the target space S^2. One may partly remove this freedom by fixing the Higgs field in one direction. Then $\mathcal{C}$ is homotopically the space of based maps from S^2 to S^2.

As we have discussed before, the connected components of $\mathcal{C}$ are labelled by an integer N, since

$$\pi_0(\mathcal{C}) = \pi_0(\mathrm{Maps}(S^2 \mapsto S^2)) = \pi_2(S^2) = \mathbb{Z}, \qquad (11.28)$$

using the result (3.12). N is the degree of the map, and is the net monopole number.

The basic solution in the component $\mathcal{C}_1$ is the monopole, and in $\mathcal{C}_{-1}$ the antimonopole. If one constructs a field in $\mathcal{C}_0$ which is a superposition of a well separated monopole and antimonopole, then it is always possible (since $\mathcal{C}_0$ is connected) to bring the monopole and antimonopole together and annihilate them.

Note that the loop space of $\mathcal{C}_0$ is topologically non-trivial. Indeed, using (3.12),

$$\pi_1(\mathcal{C}_0) = \pi_1(\mathrm{Maps}_0(S^2 \to S^2)) = \pi_3(S^2) = \mathbb{Z}, \qquad (11.29)$$

where Maps$_0$ denotes the based maps of degree zero, and similarly we have that $\pi_1(\mathcal{C}_N) = \pi_1(\mathrm{Maps}_N(S^2 \to S^2)) = \mathbb{Z}$ for maps of general degree N. One may choose the generator of $\pi_1(\mathcal{C}_0)$ to be a non-contractible loop in $\mathcal{C}_0$, beginning and ending at the vacuum. To construct it, create a monopole-antimonopole pair out of the vacuum, and separate the pair. Then rotate the monopole by 2π keeping the antimonopole fixed, and finally bring the monopole and antimonopole together again until they annihilate. The rotation can also be regarded as changing the relative phase of the monopole and antimonopole.

Now the energy of a monopole-antimonopole pair is dominated by the Coulomb force, and in the BPS limit there is both a magnetic and scalar contribution. At large separation s, the energy of the pair is

$$2M - \frac{2g^2}{4\pi s}, \tag{11.30}$$

where $M = 2\pi$ is the mass of a single monopole and $g = -2\pi$ is its magnetic charge (and 2π its scalar charge). This formula is only an approximation. Nevertheless, Taubes constructed a non-contractible loop in $\mathcal{C}_0$, with precise formulae for the fields along it, along which the energy remains strictly less than $2M$ throughout. The energy starts at zero, increases to approximately $2M - \frac{2g^2}{4\pi s}$ when the pair is separated by s, remains approximately constant at this value as the monopole is rotated, and decreases to zero as the pair annihilates.

Consider the space of all loops in the same homotopy class, beginning and ending at the vacuum. The maximal energy along each loop is well defined, as is the infimum of the maximal energies. The infimum is less than $2M$, because of the explicit example above. It is greater than zero, because only trivial loops can have arbitrarily small energy. Taubes could finally prove that the infimum is attained, by showing that the only way it could fail to be attained is if the monopole and antimonopole drifted away to infinite separation. But that is impossible because it would require energy $2M$.

The conclusion is that there is an unstable static solution in the theory, with one mode of instability, and energy less than $2M$. Although this argument says little about what the solution is like, it is expected to be a monopole-antimonopole pair, with the monopole rotated relative to the antimonopole by π, and the pair relaxed to the smallest separation possible.

Rüber, with Nahm, constructed the solution explicitly by a combination of analytic and numerical methods [355]. They realized that the rotation could be done around the line joining the monopole and antimonopole, and that the entire loop could be constructed using axisymmetric fields. We will not give the ansatz for the fields, but the ansatz for the loop of maps $S^2 \mapsto S^2$ has its first part of the form $\tilde{\theta} = f(\theta)$, $\tilde{\varphi} = \varphi$, where f evolves as in Fig. 11.4 as the monopoles separate. Since at the end of this first part $f(\frac{1}{2}\pi) = \pi$, the equator maps to a point, and in the second part of the loop one can now rotate the Southern hemisphere by 2π ($\tilde{\varphi} = \varphi + \alpha$ for $\theta > \frac{1}{2}\pi$, α runs from 0 to 2π). Then in the final part of the loop the earlier evolution of f can be reversed.

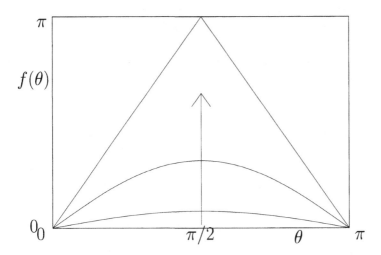

Fig. 11.4. Sketch of the first part of the evolution of the function $f(\theta)$, which occurs in the loop of maps between 2-spheres.

The unstable solution is related to the half-way point along this loop. Its Higgs field at infinity is based on a map similar to

$$\tilde{\theta} = 2\theta, \quad \tilde{\varphi} = \varphi \quad \left(\theta < \frac{1}{2}\pi\right),$$
$$\tilde{\theta} = 2\pi - 2\theta, \quad \tilde{\varphi} = \varphi + \pi \quad \left(\theta > \frac{1}{2}\pi\right) \qquad (11.31)$$

(actually, precisely this map, but the gauge $A_r = 0$ is not chosen). It consists of a monopole-antimonopole pair close together, with neither object much distorted. The solution is axially symmetric and also has a reflection symmetry in the plane separating the monopole from the antimonopole.

Kleihaus and Kunz [236] have repeated the numerical work of Rüber and Nahm and extended it beyond the BPS limit. They find that the monopole-antimonopole static solution persists, and the energy and other properties change continuously as the Higgs mass increases. In the BPS limit, the solution has energy $1.70M$, and the zeros of the Higgs field have separation $s = 2.12$. At this separation, the naive energy, taking into account the Coulomb effects, would be $1.53M$. The actual energy is greater, as there is a short range repulsion balancing the Coulomb attraction. The solution has a net magnetic dipole moment of magnitude

$$p = 14.8 \tag{11.32}$$

as can be determined from the asymptotic, abelian magnetic field. p can be interpreted as the charge $|g| = 2\pi$ times a length $l = 2.36$, which is not much different from the separation of the Higgs zeros.

One could study the instability of this solution by using gradient flow. There is presumably a solution of the gradient flow equations descending from this solution to the vacuum, starting with a relative twist, which allows the monopole-antimonopole pair to annihilate. In the fully dynamical theory, a small perturbation of the solution would initiate motion towards the vacuum, but energy conservation would make the motion oscillatory. Because the relative phase would oscillate, like a pendulum disturbed from its unstable vertical position, the solution would involve an oscillating electric dipole moment, as well as a varying magnetic dipole moment. Energy would ultimately radiate away and the solitons would again annihilate.

The non-triviality of the homotopy groups of $\text{Maps}_N(S^2 \mapsto S^2)$ implies the existence of non-contractible loops and spheres of ever higher dimension in $\mathcal{C}_N$. However, the moduli space of Bogomolny monopoles $\mathcal{M}_N$ also carries non-trivial topology. In fact, because of the Jarvis construction, $\mathcal{M}_N$ is the space of rational maps in $\text{Maps}_N(S^2 \mapsto S^2)$. It has been proved by Segal [370] that for $n \leq N$, the nth homotopy group of the space of rational maps is isomorphic to the nth homotopy group of the space of all maps. So the space of Bogomolny monopoles captures the topology of the space of all finite energy configurations of charge N, for these values of n. As an example, any non-contractible loop in $\mathcal{C}_2$ can be deformed to lie entirely in $\mathcal{M}_2$ (where it corresponds to an end-over-end rotation, one or more times, of the charge 2 toroidal monopole). Thus one has to go to non-contractible spheres in $\mathcal{C}_N$ of dimension N or higher to capture unstable solutions. Such solutions have not been constructed, but one can imagine, for example, an unstable cluster in $\mathcal{C}_1$, consisting of an antimonopole sandwiched between two monopoles.

11.5 The electroweak sphaleron

The standard model of the electromagnetic and weak interactions is a $U(2)$ gauge theory, with a complex doublet of Higgs fields. $U(2)$ is not simple, and the $SU(2)$ and $U(1)$ gauge fields are coupled at different strengths to the Higgs field. The ratio of the strengths is determined by the weak mixing angle Θ_w. The particles of the theory (ignoring fermions) are the charged $W^\pm$ bosons, the neutral Z boson, the photon and the neutral, scalar Higgs particle (which is yet to be discovered).

The classical equations of the electroweak theory have a sphaleron solution [241]. Mathematically, it is easiest to discuss it in the limit $\Theta_w = 0$, where the $U(1)$ field decouples, and the $W^\pm$ and Z bosons have equal masses. The theory then just involves an $SU(2)$ gauge field A_μ and the Higgs field

$$\Phi = \begin{pmatrix} \Phi_1 \\ \Phi_2 \end{pmatrix}. \tag{11.33}$$

The potential energy function is

$$E = \int \left(-\frac{1}{2}\text{Tr}(F_{ij}F_{ij}) + \frac{1}{2}(D_i\Phi)^\dagger D_i\Phi + \frac{\lambda}{4}(1 - \Phi^\dagger\Phi)^2 \right) d^3x, \tag{11.34}$$

where $\lambda > 0$. Note the factor $\frac{1}{2}$ in the Yang-Mills part. U has its minimum where $\Phi^\dagger\Phi = 1$, so the vacuum manifold is a 3-sphere of radius 1. The Higgs boson to W boson mass ratio is $2\sqrt{2\lambda}$.

Since there are no nonlinear constraints on the Higgs field, the topology of a field configuration, or family of them, is captured by the Higgs and gauge field at spatial infinity. As before, we fix the radial gauge condition $A_r = 0$, which completely determines a field configuration aside from a rigid gauge rotation. The Higgs field at infinity defines a map $\Phi^\infty : S^2 \mapsto S^3$. $SU(2)$ acts transitively on S^3 (by left action of $SU(2)$ on itself) so we can fix the gauge even more completely by imposing a base point condition that in the spatial direction $(0, 0, 1)$, say, Φ^∞ always takes the value $\begin{pmatrix} 0 \\ 1 \end{pmatrix}$.

The field configuration space $\mathcal{C}$ is thus topologically equivalent to the space of based maps, $\text{Maps}(S^2 \mapsto S^3)$. This space is connected, and any single map can be deformed to the constant map $\Phi^\infty = \begin{pmatrix} 0 \\ 1 \end{pmatrix}$, as a consequence of $\pi_0(\text{Maps}(S^2 \mapsto S^3)) = \pi_2(S^3) = I$. Therefore, there is no topological charge associated with a field configuration of the electroweak theory, so the theory has no topological solitons. In particular, there are no monopoles.

However, the space $\text{Maps}(S^2 \mapsto S^3)$ is far from topologically trivial. In particular,

$$\pi_1(\text{Maps}(S^2 \mapsto S^3)) = \pi_3(S^3) = \mathbb{Z}, \tag{11.35}$$

so there are non-contractible loops in $\mathcal{C}$. It is easy to construct such a loop [280] – easier than in the case of the monopole-antimonopole pair of the previous section. One just takes the image of S^2 to be a 2-sphere resting on S^3 and slides it over the equator, as in Fig. 11.5. In total, this gives a topologically non-trivial map from S^3 to S^3.

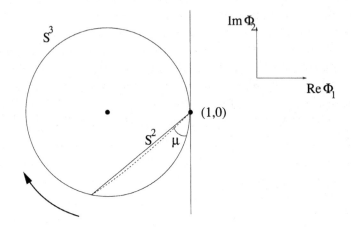

Fig. 11.5.   A loop of maps from S^2 to S^3, parametrized by $\mu \in [0, \pi]$.

Let us use polar coordinates θ, φ on S^2, and let the parameter along the loop be μ, in the range $0 \leq \mu \leq \pi$. A family of maps parametrized by μ, realizing the loop sketched in Fig. 11.5 is

$$\Phi^\infty(\theta, \varphi; \mu) = \begin{pmatrix} \sin \mu \sin \theta \, e^{i\varphi} \\ e^{-i\mu}(\cos \mu + i \sin \mu \cos \theta) \end{pmatrix}. \tag{11.36}$$

This family is a generating loop of the homotopy group $\pi_1(\mathcal{C})$, since it covers the generic point of S^3 exactly once, so the degree of the map $S^3 \mapsto S^3$ is 1.

We now need to smoothly extend Φ^∞ to a set of Higgs and gauge fields defined throughout $\mathbb{R}^3$. We first need a gauge field at infinity, satisfying

$$D_\theta \Phi^\infty = 0, \quad D_\varphi \Phi^\infty = 0 \tag{11.37}$$

to ensure the total energy is finite. Suitable expressions are

$$A_\theta^\infty = -\partial_\theta U^\infty (U^\infty)^{-1}, \quad A_\varphi^\infty = -\partial_\varphi U^\infty (U^\infty)^{-1} \tag{11.38}$$

where U^∞ is the matrix

$$U^\infty = \begin{pmatrix} \bar{\Phi}_2^\infty & \Phi_1^\infty \\ -\bar{\Phi}_1^\infty & \Phi_2^\infty \end{pmatrix} \tag{11.39}$$

with Φ_1^∞ and Φ_2^∞ the two components of (11.36). U^∞ has the property that

$$U^\infty \begin{pmatrix} 0 \\ 1 \end{pmatrix} = \Phi^\infty. \tag{11.40}$$

Then a suitable ansatz for the fields throughout $\mathbb{R}^3$ is

$$
\begin{aligned}
\Phi(r,\theta,\varphi;\mu) &= (1 - h(r))\begin{pmatrix} 0 \\ e^{-i\mu}\cos\mu \end{pmatrix} + h(r)\Phi^\infty(\theta,\varphi;\mu) \\
A_\theta(r,\theta,\varphi;\mu) &= f(r)A_\theta^\infty(\theta,\varphi;\mu) \\
A_\varphi(r,\theta,\varphi;\mu) &= f(r)A_\varphi^\infty(\theta,\varphi;\mu)
\end{aligned}
\tag{11.41}
$$

with $A_r = 0$. The profile functions h and f must satisfy the boundary conditions

$$
\begin{aligned}
h(\infty) &= 1, & h(0) &= 0 \\
f(\infty) &= 1, & f(0) &= 0
\end{aligned}
\tag{11.42}
$$

to obtain the desired field at infinity and to avoid singularities at the origin.

The energy of these fields can be computed as a function of the parameter along the loop, μ. It depends of course on the profile functions. The starting and finishing point of the loop is the vacuum, so the energy is always zero for $\mu = 0$ and $\mu = \pi$. Moreover, for most choices of h and f, the maximal energy along the loop occurs when $\mu = \frac{1}{2}\pi$ [280].

One can now use the ansatz (11.41) to find a candidate sphaleron. One should really find the maximal energy along *all* loops, and seek the minimum of this. We make a restricted minimization, over fields of the form (11.41). We impose $\mu = \frac{1}{2}\pi$ and then minimize the energy over the space of radial profile functions h, f.

The principle of symmetric criticality helps here. The fields (11.41), with $\mu = \frac{1}{2}\pi$, are actually $SO(3)$-symmetric and have an additional reflection symmetry. This is seen most easily by regarding the Higgs field as a quartet of real scalar fields, acted on by a global $SO(4)$ group. The $SU(2)$ gauge group may be regarded as the subgroup $SU(2)_L$ of this $SO(4)$. The functions h and f are the only undetermined quantities after imposing these symmetries. The energy density is spherically symmetric, and the energy reduces to the following expression in terms of h and f,

$$
E = 4\pi \int_0^\infty \mathcal{E}\, dr \,,
$$

where

$$
\mathcal{E} = 4f'^2 + \frac{8}{r^2}f^2(1-f)^2 + \frac{r^2}{2}h'^2 + (1-f)^2 h^2 + \frac{\lambda}{4}r^2(1-h^2)^2. \tag{11.43}
$$

The variational equations of this dimensionally reduced theory, defined on a half-line, are

$$
\begin{aligned}
(r^2 h')' &= 2(1-f)^2 h - \lambda r^2(1-h^2)h \tag{11.44} \\
r^2 f'' &= 2f(1-f)(1-2f) - \frac{r^2}{4}(1-f)h^2 \,. \tag{11.45}
\end{aligned}
$$

They can not be solved analytically, but Burzlaff has established rigorously that a smooth solution satisfying the boundary conditions exists [70]. A unique solution, depending on λ, has also been found numerically. Numerical results for h and f are shown in Fig. 11.6 for $\lambda = \frac{1}{2}$, corresponding to $M_H = 2M_W$. The energy as a function of λ is shown in Fig. 11.7. It increases from $1.52 \times 4\pi$ to $2.70 \times 4\pi$ as λ increases from 0 to ∞, and is $1.98 \times 4\pi$ when $\lambda = \frac{1}{2}$.

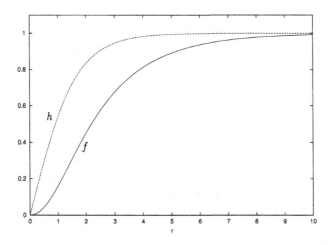

Fig. 11.6.   The electroweak sphaleron profile functions $h(r)$ and $f(r)$ for the value $\lambda = \frac{1}{2}$.

By the principle of symmetric criticality, this method undoubtedly finds a solution of the full field equations, and it was for this solution that the name "sphaleron" was invented. The solution is unstable, by the argument given earlier, because for the obtained profile functions the energy maximum along the loop does occur at $\mu = \frac{1}{2}\pi$. The solution was actually known before its topological significance in the electroweak theory was realized. It was found by Dashen, Hasslacher and Neveu [103] and rediscovered by Boguta [58], in the context of hadronic models. Its instability had also been noted [242, 70].

The Chern-Simons number of the sphaleron is $\frac{1}{2}$. Conceptually, this is for the following reason. Regard the loop parameter μ, suitably rescaled, as a Euclidean time x^4 running from $-\infty$ to ∞, with $x^4 = 0$ corresponding to $\mu = \frac{1}{2}\pi$. The fields along the loop can be regarded as a single field configuration defined in $\mathbb{R}^4$. The gauge field at infinity is a pure gauge $A = -dU^\infty(U^\infty)^{-1}$ (when the component A_4 is included), and the complete gauge field is a vacuum to vacuum transition. The fact that $U^\infty(\theta, \varphi; \mu)$

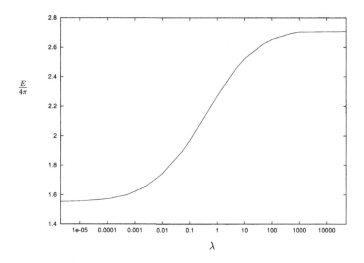

Fig. 11.7. The sphaleron energy, in units of 4π, as a function of the parameter λ, plotted on a logarithmic scale.

covers all of S^3 once means that the second Chern number of the gauge field is $c_2 = 1$, so it is topologically the same as a unit charge instanton. Now recall that c_2 is the change of the Chern-Simons number y_3, so for a suitable gauge choice y_3 increases from 0 to 1 as x^4 increases from $-\infty$ to ∞.

Now the loop is reflection symmetric, with U^∞ at $\mu = \frac{1}{2}\pi$ covering an equatorial S^2 of S^3 and the fields at finite radius respecting the reflection symmetry. Therefore for any profile functions h and f, satisfying the appropriate boundary conditions, the Chern-Simons number of the $\mu = \frac{1}{2}\pi$ field configuration is $y_3 = \frac{1}{2}$. In particular, the sphaleron has $y_3 = \frac{1}{2}$.

The result can be checked [241, 433] using the standard formula for the Chern-Simons number (3.85). However, one must first change gauge so that $A = 0$ at infinity. The required gauge transformation is one defined throughout $\mathbb{R}^3$, and approaching $(U^\infty)^{-1}$ at infinity. Such a gauge transformation is possible preserving the $SO(3)$ symmetry, but breaking the reflection symmetry.

Note that dynamically, there is no instanton in the electroweak theory. This is because Derrick's theorem rules out a stationary point of the action for a Yang-Mills field coupled to a Higgs field in $\mathbb{R}^4$. Conversely, the pure Yang-Mills theory in $\mathbb{R}^4$ has an instanton interpolating between vacua, and the $\mathbb{R}^3$ slice through the middle is similar to the gauge field of the sphaleron, with Chern-Simons number $\frac{1}{2}$, but pure Yang-Mills theory in $\mathbb{R}^3$ has no true sphaleron solution, again by Derrick's theorem. In fact, there is no well defined energy barrier that an instanton traverses. Because an instanton has an arbitrary scale, the energy of its central $\mathbb{R}^3$ slice has an arbitrary value, being small for a large instanton and large for a small instanton.

Let us now consider a further aspect of the sphaleron. One should verify that the constructed solution has *one* unstable mode. The study of the relevant second variation operator has been carried out by Yaffe [433], and also by Kunz and Brihaye [250] and Akiba *et al.* [9]. For $\lambda < 18.1$ there is precisely one unstable mode, but for $\lambda > 18.1$ there are more. The picture is rather as for sphalerons on a circle. The most symmetric solution has an increasing number of negative modes as a parameter (here λ) increases. There is still a sphaleron with one negative mode, for all λ, but it bifurcates off from the symmetric solution at $\lambda = 18.1$. This deformed sphaleron, which was also found by Kunz and Brihaye, and by Yaffe, still has $SO(3)$ symmetry, but it no longer has the reflection symmetry. Consequently, its Chern-Simons number differs from $\frac{1}{2}$. Also, for the deformed sphaleron, Φ does not vanish at the origin, whereas for the original sphaleron it does. The deformed sphaleron is the maximal energy configuration along a non-contractible loop, whose fields are more complicated than those of (11.41) [239].

The discovery of the deformed sphaleron solves a paradox. The theory at $\lambda = \infty$ effectively constrains the Higgs field to the vacuum manifold $\Phi^\dagger \Phi = 1$ (except, perhaps, at isolated points). The $SU(2)$ gauged sigma model, with a nonlinear scalar field satisfying this constraint, is known to have a solution – the weak Skyrmion [119]. The weak Skyrmion has lower energy than the original $\lambda = \infty$ sphaleron, and a different structure. However, it is precisely the limiting form of the deformed sphaleron, with energy $2.54 \times 4\pi$.

Our discussion of the sphaleron and its properties has so far been in terms of dimensionless fields and dimensionless units. Using the more standard formulation of the electroweak Lagrangian, together with the experimental data, one can determine the physical energy and size of the sphaleron. In our formulae, the unit of length is $1/gv$ and the unit of energy is v/g, where v and g are the usual parameters of the electroweak theory. The physical masses of the W and Higgs bosons (ignoring radiative corrections) are $M_W = \frac{1}{2}gv$ and $M_H = \sqrt{2\lambda}gv$. Experimentally, the particle masses are [170]

$$M_W = 80.4\,\text{GeV}\,, \quad M_Z = 91.2\,\text{GeV}\,, \quad M_{\text{photon}} = 0\,, \quad M_H > 114\,\text{GeV}\,, \tag{11.46}$$

and a precision study of radiative effects suggests that $M_H < 200\,\text{GeV}$. By definition, $\sin^2 \Theta_w = 1 - M_W^2/M_Z^2 = 0.223$, and $g^2 \sin^2 \Theta_w/4\pi =$

7.30×10^{-3} is the fine structure constant. Therefore, the energy unit $4\pi v/g$ is 4.91 TeV, and the length unit $(gv)^{-1}$ is $(161 \text{ GeV})^{-1}$, which corresponds to 1.22×10^{-3} fm. We deduce that the sphaleron energy ranges from 7.5 TeV for $\lambda = 0$ up to 13.3 TeV as λ approaches infinity. The deformed sphaleron has energy a few per cent less for large λ.

It is interesting to study how the sphaleron, if it were produced, would decay. This is done by perturbing the sphaleron by its unstable mode and solving the time dependent field equations numerically. The task is simplified because the fields remain $SO(3)$-symmetric, although the reflection symmetry is broken. The sphaleron is converted into a fireball of radiation. By Fourier analysing the outgoing waves, and using semi-classical quantization ideas, it can be estimated that the sphaleron decays into approximately eight Higgs particles and 14 each of W^+, W^- and Z bosons [176]. These would subsequently decay into fermions. Understanding the change in the Chern-Simons number during this process is rather delicate [129].

The inclusion of the $U(1)$ gauge field, required for any non-zero value of Θ_w, complicates the sphaleron solution. It is no longer consistent to impose $SO(3)$ symmetry, but the solution retains an axial symmetry together with a reflection symmetry. For small values of Θ_w one can treat the $U(1)$ effects perturbatively [241, 224]. The energy of the sphaleron is lowered by less than 1%. The sphaleron acquires a small asymptotic $U(1)$ field which can be interpreted as a magnetic dipole field. The dipole moment is

$$p = \frac{2\pi}{3} \frac{1}{g^2 v} \tan \Theta_w \int_0^\infty r^2 h^2(r)(1 - f(r))\, dr\,, \qquad (11.47)$$

whose value, for $\lambda = \frac{1}{2}$, is approximately $p = 47 \tan \Theta_w / g^2 v$. For comparison, the magnetic dipole moment of a W boson is $2 \sin \Theta_w / v$.

This perturbative approach is a very good approximation for Θ_w up to 30° (the experimental value is about 28°). It fails to be a good approximation as Θ_w approaches 90°. In this limit (and keeping λ of order 1), the sphaleron has a prolate (cigar-shaped) axisymmetric form [251]. Being also a magnetic dipole, it is possible to interpret the sphaleron as effectively a monopole-antimonopole pair [182] even though the electroweak theory does not support isolated monopole solutions. The best description of the field is not entirely clear, because the numerics are difficult for Θ_w close to 90°. However, it appears that the monopole and antimonopole are connected by a Z-string.

In summary, the electroweak theory has a sphaleron solution. Provided the Higgs particle exists with a mass less than 200 GeV, the parameter λ is of order 1, so the deformed sphaleron does not come into play. The non-zero value of Θ_w has a rather small effect on the sphaleron and its energy. The best current estimate for the sphaleron energy is that it is in the range 9–10 TeV, the exact value depending on the Higgs mass. It is likely that the theory has several further unstable solutions of higher energy. One such solution, related to a non-contractible 2-sphere in $\mathcal{C}$, and with an energy slightly less than double that of the sphaleron, has been found by Klinkhamer [240].

The physical significance of the electroweak sphaleron is a controversial matter, and goes rather beyond the scope of this book, since fermions play a decisive role. For a review, see [354]. One possibility, rather vaguely formulated, is that the energy 9–10 TeV is a threshold beyond which non-perturbative phenomena become important. Thus in particle collisions just above current accelerator energies (e.g. e^+e^- annihilation at centre of mass energies of order 500 GeV), one expects to produce a small number of W or Z bosons, together with Higgs particles and photons. At higher energies more of these would be produced, and possibly coherently in the form of a sphaleron. This would subsequently decay in a recognizable way, as a rather symmetric fireball. The argument against this is that the small weak coupling constant $g^2/4\pi$ makes production of many W's etc. unlikely, and the probability of producing these particles coherently so that they form the sphaleron, even approximately, is utterly negligible. The alternative argument is that when many particles are produced at high energy, an enormous number of Feynman diagrams are relevant and the usual perturbative rules of quantum field theory are no longer valid. Instead one should rely more on semi-classical methods, and they make classical solutions more significant [345, 122].

While sphaleron production in high energy collisions may be unlikely, there is more consensus that sphalerons can be easily produced in a high temperature situation, such as prevailed in the early universe. Numerical evidence suggests that at high temperature, a field evolves randomly, and approximately classically, subject to Boltzmann statistical mechanics. For a field to be locally excited and pass over the sphaleron barrier does not appear to be unlikely. This is shown by keeping track of the Chern-Simons number in a thermal field simulation, where random jumps by one unit are seen to occur from time to time [165, 11]. At even higher temperatures, there may be no suppression of barrier crossing at all [252, 13].

In the absence of fermions, the excitation and decay of the sphaleron is not terribly significant. However, it has been realized since the pioneering work of 't Hooft on instantons [403], that there is a fermion number "anomaly" [8, 48] associated with vacuum to vacuum transitions in electroweak theory. More precisely, if an electroweak field starts at the vacuum, evolves to the sphaleron and then further evolves to the vacuum "on the other side" – in other words, if the field traverses a non-contractible loop of the theory, with the Chern-Simons number changing by one unit

– then there is a net production of one unit of baryon number B for each of the three generations of quarks, and simultaneously a net production of one unit of lepton number L for each generation.

Thus in the early universe, or even in particle collisions, there is the possibility of B violation and L violation, with $B - L$ conserved. The quantities B and L are conserved according to all perturbative calculations, and they have not been seen to be non-conserved in any experiment. However, the presence of a net B in the universe is a great mystery, especially if the Big Bang produced a universe with $B = 0$. (The net L of the universe is unknown, because the number of neutrinos and antineutrinos can not yet be determined.) Sakharov [361] formulated three conditions for a fundamental resolution of this mystery. There must exist B violating processes; there must be C and CP violation, so that there is the possibility of a drift of B preferentially in one direction, from zero towards a positive value; and there must have been a lack of thermal equilibrium at certain stages of the cooling universe's history, so that some net change of B could have occurred. The electroweak theory satisfies these conditions. B violating transitions can occur through the production and decay of the sphaleron; CP violation is encoded in the Cabbibo-Kobayashi-Maskawa quark mixing matrix, and is experimentally observed; and there could have been a lack of thermal equilibrium during a phase transition at which the Higgs field acquired its vacuum expectation value.

There has been much research on the detailed behaviour of field theories near to phase transitions, and on whether sphaleron production and decay, or topologically related processes, can occur at a significant enough rate to explain the current baryon number of the universe. Unfortunately these investigations are difficult. It appears that because the CP violating effects are so small in the electroweak theory, these processes do not appear powerful enough to explain the observed baryon to photon ratio of $\sim 10^{-10}$ (they give estimates of $\sim 10^{-20}$). More speculative extensions of electroweak theory, with strong CP violation as in supersymmetric models, may do the trick. However, there are many such extensions and no evidence for any one of them being correct.

11.6 Unstable solutions in other theories

There are a number of further field theories where unstable, sphaleron-type solutions are known to exist, and some where they are known not to exist. We summarize some of these results here. For a general discussion, see [134].

The $O(3)$ sigma model in one space dimension, modified by a potential term, has a sphaleron solution which is a lower-dimensional analogue of the electroweak sphaleron [308]. The field is a based map $S^1 \mapsto S^2$, and the theory has non-contractible loops because $\pi_1(\text{Maps}(S^1 \mapsto S^2)) = \pi_2(S^2) = \mathbb{Z}$.

In the gauged GL theory in two space dimensions at critical coupling, discussed in Chapter 7, Jaffe and Taubes [223] have proved that there are no static classical solutions except the multi-vortex solutions satisfying the Bogomolny equation, which have minimal energy and are stable. This result is consistent with the topology. The Higgs field at infinity defines a map $S^1 \mapsto S^1$, and the space of such maps has components labelled by the winding number N. For each winding number, the space is contractible, so there are no non-contractible loops or higher homotopy spheres; thus no saddle point solutions are expected.

In the $\mathbb{CP}^1$ sigma model discussed in Chapter 6, the fields are maps $S^2 \mapsto S^2$, and the minima of the energy are the rational maps, depending only on the complex coordinate z. It has been proved [118, 431] that all finite energy static solutions are of this type, so there are no higher energy saddle point solutions despite the rich topological structure of $\text{Maps}(S^2 \mapsto S^2)$. However, for the $\mathbb{CP}^m$ sigma model with $m > 1$ there are saddle point solutions. Furthermore, unlike in most of the field theories we have discussed in this book, the static second order field equation is integrable (in addition to the first order Bogomolny equation being integrable) and explicit closed form solutions can be obtained for all the saddle points. Recall from Section 6.1 that the energy function of the static $\mathbb{CP}^m$ sigma model in the plane is

$$E = \int \text{Tr}(\partial_i P \partial_i P) \, d^2x \,, \tag{11.48}$$

whose variation yields the static second order equation

$$[\partial_i \partial_i P, P] = 0 \,, \tag{11.49}$$

where P is the $(m+1) \times (m+1)$ hermitian projector which provides coordinates on $\mathbb{CP}^m$. The energy minimizing multi-lump solutions are given by

$$P = P_0 = \frac{\mathbf{f}_0 \, \mathbf{f}_0^\dagger}{|\mathbf{f}_0|^2} \,, \tag{11.50}$$

where $\mathbf{f}_0(z)$ is a holomorphic $(m+1)$-vector. The saddle point solutions are obtained by introducing the operator Δ, whose action on any vector $\mathbf{f}$ is defined by

$$\Delta \mathbf{f} = \partial_z \mathbf{f} - \frac{\mathbf{f} \, (\mathbf{f}^\dagger \, \partial_z \mathbf{f})}{|\mathbf{f}|^2} \tag{11.51}$$

where, as before, ∂_z denotes differentiation with respect to z. Given the holomorphic vector $\mathbf{f}_0$, let $\mathbf{f}_j = \Delta^j \mathbf{f}_0$ $(j = 1, \ldots, m)$ be the sequence of vectors obtained by the repeated application of Δ. Then it can be shown that each of the associated projectors

$$P_j = \frac{\mathbf{f}_j \, \mathbf{f}_j^\dagger}{|\mathbf{f}_j|^2} \tag{11.52}$$

solves Eq. (11.49). It turns out that the vector $\mathbf{f}_m$ (after cancelling any overall factors) is antiholomorphic, so it is again a minimal energy solution. The operator Δ therefore converts a multi-lump solution into a multi-antilump solution by its application m times. Δ can not be applied more than m times, since it gives zero when applied to an antiholomorphic vector.

For $m > 1$ the intermediate solutions are neither holomorphic nor anti-holomorphic and consist of mixtures of lumps and antilumps arranged in unstable equilibria. For a more detailed description of these solutions we refer the reader to the book by Zakrzewski [436]. Note that for $m > 1$, all loops in the field configuration space are contractible, because of the homotopy group relation $\pi_3(\mathbb{CP}^m) = 0$. However, $\pi_4(\mathbb{CP}^m) = \mathbb{Z}$, so it is probably the existence of non-contractible spheres which underlies the existence of these saddle point solutions.

We have found various saddle point solutions in the Skyrme model, for example, the hedgehog solutions of baryon number $|B| > 1$. Also there is the octahedral $B = 5$ solution, and the cubic $B = 7$ solution. However, there is no systematic topological classification of these. Since the configuration space of the Skyrme model is Maps$(S^3 \mapsto S^3)$, there are non-contractible loops, because $\pi_1(\text{Maps}(S^3 \mapsto S^3)) = \pi_4(S^3) = \mathbb{Z}_2$. Bagger *et al.* have attempted to construct sphalerons systematically, using these non-contractible loops [26, 153]. For $B = 0$, they considered creating a Skyrmion-antiSkyrmion pair from the vacuum, separating them, rotating the Skyrmion by 2π, and annihilating them again. This follows Taubes' discussion of the monopole-antimonopole loop, but here the energy argument is not clear-cut. It is essential that the energy along the loop remains less than $2M = 2.464$, twice the mass of a single Skyrmion. However, no matter how the Skyrmion is rotated, the energy, as estimated in the dipole-dipole approximation (and in the absence of electromagnetic corrections), always reaches $2M$. It is possible that a higher order calculation will lead to a loop whose maximal energy is less than $2M$, but this has not been established. So far, then, there is no convincing evidence for a Skyrmion-antiSkyrmion pair forming a sphaleron. There definitely are some unstable solutions of the Skyrme equation in the $B = 0$ sector. Any soliton of the Skyrme-Faddeev model can be embedded in the Skyrme model (by embedding the target S^2 as the equator of S^3) and is a solution there. The lowest energy solution of this type has energy 4.4.

Finally, a challenging problem is to find saddle point solutions of the pure Yang-Mills equation on $\mathbb{R}^4$ or on S^4. The equation (and not just the self-dual equation for instantons) is conformally invariant, so these two problems are the same. The existence of saddle point solutions was established by Sibner, Sibner and Uhlenbeck [375]. They exploited the fact that the imposition of $SO(2)$ symmetry reduces the Yang-Mills equation to equations for hyperbolic monopoles. Taubes' arguments can be applied here, to again show the existence of a solution representing a monopole-antimonopole pair in unstable equilibrium. This solution is then also an $SO(2)$-invariant sphaleron of the Yang-Mills theory. A more concrete method for constructing solutions was subsequently discovered by Sadun and Segert [357]. They noted that $SO(5)$ has an interesting $SO(3)$ subgroup, and that by imposing this symmetry on fields on S^4 the Yang-Mills equation is reduced to ODEs. Solutions of these ODEs satisfying appropriate boundary conditions have been rigorously proved to exist, and they give solutions of the Yang-Mills equation, by the principle of symmetric criticality. They have also been obtained numerically [358]. As a result, a solution is known with energy (action) $5.43 \times \pi^2$ and topological charge zero. There are probably many more unstable solutions of the Yang-Mills equation, but there is no systematic topological understanding of them.

References

[1] F. Abdelwahid and J. Burzlaff, Existence theorems for 90° vortex-vortex scattering, *J. Math. Phys.* **35**, 4651 (1994).

[2] M. J. Ablowitz and P. A. Clarkson, *Solitons, Nonlinear Evolution Equations and Inverse Scattering*, Cambridge University Press, 1991.

[3] M. Abolfath, J. J. Palacios, H. A. Fertig, S. M. Girvin and A. H. MacDonald, Critical comparison of classical field theory and microscopic wave functions for Skyrmions in quantum Hall ferromagnets, *Phys. Rev.* **B56**, 6795 (1997).

[4] E. Abraham, Nonlinear sigma models and their Q-lump solutions, *Phys. Lett.* **B278**, 291 (1992).

[5] M. Abramowitz and I. A. Stegun, *Handbook of Mathematical Functions*, New York, Dover, 1965.

[6] A. A. Abrikosov, On the magnetic properties of superconductors of the second group, *Sov. Phys. JETP* **5**, 1174 (1957).

[7] G. S. Adkins, C. R. Nappi and E. Witten, Static properties of nucleons in the Skyrme model, *Nucl. Phys.* **B228**, 552 (1983).

[8] S. L. Adler, Axial-vector vertex in spinor electrodynamics, *Phys. Rev.* **177**, 2426 (1969).

[9] T. Akiba, H. Kikuchi and T. Yanagida, Free energy of the sphaleron in the Weinberg-Salam model, *Phys. Rev.* **D40**, 588 (1989).

[10] S. L. Altmann and P. Herzig, *Point-Group Theory Tables*, Oxford, Clarendon Press, 1994.

[11] J. Ambjørn, T. Askgaard, H. Porter and M. E. Shaposhnikov, Lattice simulations of electroweak sphaleron transitions in real time, *Phys. Lett.* **B244**, 479 (1990); Sphaleron transitions and baryon asymmetry: A numerical, real-time analysis, *Nucl. Phys.* **B353**, 346 (1991).

[12] H. Aratyn, L. A. Ferreira and A. H. Zimerman, Exact static soliton solutions of (3+1)-dimensional integrable theory with nonzero Hopf numbers, *Phys. Rev. Lett.* **83**, 1723 (1999).

[13] P. Arnold and L. McLerran, Sphalerons, small fluctuations, and baryon-number violation in electroweak theory, *Phys. Rev.* **D36**, 581 (1987).

[14] K. Arthur and J. Burzlaff, Existence theorems for π/n vortex scattering, *Lett. Math. Phys.* **36**, 311 (1996).

[15] M. F. Atiyah, Magnetic monopoles in hyperbolic spaces, in *M. Atiyah: Collected Works, vol. 5*, Oxford, Clarendon Press, 1988.

[16] M. F. Atiyah and R. Bott, The Yang-Mills equations over Riemann surfaces, *Phil. Trans. R. Soc. Lond.* **A308**, 523 (1983).

[17] M. F. Atiyah and N. J. Hitchin, *The Geometry and Dynamics of Magnetic Monopoles*, Princeton University Press, 1988.

[18] M. F. Atiyah, N. J. Hitchin, V. G. Drinfeld and Yu. I. Manin, Construction of instantons, *Phys. Lett.* **A65**, 185 (1978).

[19] M. F. Atiyah, N. J. Hitchin and I. M. Singer, Deformations of instantons, *Proc. Natl. Acad. Sci. USA* **74**, 2662 (1977).

[20] M. F. Atiyah and J. D. S. Jones, Topological aspects of Yang-Mills theory, *Commun. Math. Phys.* **61**, 97 (1978).

[21] M. F. Atiyah and N. S. Manton, Skyrmions from instantons, *Phys. Lett.* **B222**, 438 (1989).

[22] M. F. Atiyah and N. S. Manton, Geometry and kinematics of two Skyrmions, *Commun. Math. Phys.* **153**, 391 (1993).

[23] M. F. Atiyah and R. S. Ward, Instantons and algebraic geometry, *Commun. Math. Phys.* **55**, 117 (1977).

[24] S. J. Avis and C. J. Isham, Vacuum solutions for a twisted scalar field, *Proc. R. Soc. Lond.* **A363**, 581 (1978).

[25] J. Baacke, Fluctuations and stability of the 't Hooft-Polyakov monopole, *Z. Phys.* **C53**, 399 (1992).

[26] J. Bagger, W. Goldstein and M. Soldate, Static solutions in the vacuum sector of the Skyrme model, *Phys. Rev.* **D31**, 2600 (1985).

[27] F. A. Bais and D. Wilkinson, Exact $SU(N)$ monopole solutions with spherical symmetry, *Phys. Rev.* **D19**, 2410 (1979).

[28] D. Bak and C. Lee, Scattering of light by a BPS monopole, *Nucl. Phys.* **B403**, 315 (1993).

[29] D. Bak, C. Lee and K. Lee, Dynamics of Bogomol'nyi-Prasad-Sommerfield dyons: Effective field theory approach, *Phys. Rev.* **D57**, 5239 (1998).

[30] A. P. Balachandran, A. Barducci, F. Lizzi, V. G. J. Rodgers and A. Stern, Doubly strange dibaryon in the chiral model, *Phys. Rev. Lett.* **52**, 887 (1984).

[31] J. M. Baptista, Some special Kähler metrics on $SL(2, \mathbb{C})$ and their holomorphic quantization, *J. Geom. Phys.* **50**, 1 (2004).

[32] I. V. Barashenkov and A. O. Harin, Nonrelativistic Cherns-Simons theory for the repulsive Bose gas, *Phys. Rev. Lett.* **72**, 1575 (1994); Topological excitations in a condensate of nonrelativistic bosons coupled to Maxwell and Chern-Simons fields, *Phys. Rev.* **D52**, 2471 (1995).

[33] J. Bardeen, L. N. Cooper and J. R. Schrieffer, Theory of superconductivity, *Phys. Rev.* **108**, 1175 (1957).

[34] C. Barnes, K. Baskerville and N. Turok, Normal modes of the $B = 4$ Skyrme soliton, *Phys. Rev. Lett.* **79**, 367 (1997); Normal mode spectrum of the deuteron in the Skyrme model, *Phys. Lett.* **B411**, 180 (1997).

[35] W. K. Baskerville, Making nuclei out of the Skyrme crystal, *Nucl. Phys.* **A596**, 611 (1996).

[36] W. K. Baskerville, Vibrational spectrum of the $B = 7$ Skyrme soliton, hep-th/9906063 (1999).

[37] G. K. Batchelor, *An Introduction to Fluid Dynamics*, Cambridge University Press, 1967.

[38] L. Bates and R. Montgomery, Closed geodesics on the space of stable two-monopoles, *Commun. Math. Phys.* **118**, 635 (1988).

[39] R. A. Battye, *String Radiation, Interactions and Cosmological Constraints*, Ph.D. thesis, Cambridge University, 1995.

[40] R. A. Battye and P. M. Sutcliffe, Multi-soliton dynamics in the Skyrme model, *Phys. Lett.* **B391**, 150 (1997).

[41] R. A. Battye and P. M. Sutcliffe, Symmetric Skyrmions, *Phys. Rev. Lett.* **79**, 363 (1997).

[42] R. A. Battye and P. M. Sutcliffe, A Skyrme lattice with hexagonal symmetry, *Phys. Lett.* **B416**, 385 (1998).

[43] R. A. Battye and P. M. Sutcliffe, Knots as stable soliton solutions in a three-dimensional classical field theory, *Phys. Rev. Lett.* **81**, 4798 (1998); Solitons, links and knots, *Proc. R. Soc. Lond.* **A455**, 4305 (1999).

[44] R. A. Battye and P. M. Sutcliffe, Q-ball dynamics, *Nucl. Phys.* **B590**, 329 (2000).

[45] R. A. Battye and P. M. Sutcliffe, Solitonic fullerene structures in light atomic nuclei, *Phys. Rev. Lett.* **86**, 3989 (2001); Skyrmions, fullerenes and rational maps, *Rev. Math. Phys.* **14**, 29 (2002).

[46] A. A. Belavin and A. M. Polyakov, Metastable states of two-dimensional isotropic ferromagnets, *JETP Lett.* **22**, 245 (1975).

[47] A. A. Belavin, A. M. Polyakov, A. S. Schwarz and Yu. S. Tyupkin, Pseudoparticle solutions of the Yang-Mills equations, *Phys. Lett.* **B59**, 85 (1975).

[48] J. S. Bell and R. Jackiw, A PCAC puzzle: $\pi^0 \to \gamma\gamma$ in the σ-model, *Nuovo Cim.* **60**, 47 (1969).

[49] G. Benettin, L. Galgani and A. Giorgilli, Realization of holonomic constraints and freezing of high-frequency degrees of freedom in the light of classical perturbation theory. 1, *Commun. Math. Phys.* **113**, 87 (1987).

[50] M. S. Berger and Y. Y. Chen, Symmetric vortices for the Ginzburg-Landau equations of superconductivity and the nonlinear desingularization phenomenon, *J. Func. Anal.* **82**, 259 (1989).

[51] F. Bethuel, H. Brezis and F. Hélein, *Ginzburg-Landau Vortices*, Boston, Birkhäuser, 1994.

[52] L. M. A. Bettencourt and R. J. Rivers, Interactions between $U(1)$ cosmic strings: An analytical study, *Phys. Rev.* **D51**, 1842 (1995).

[53] R. Bielawski, Existence of closed geodesics on the moduli space of k-monopoles, *Nonlinearity* **9**, 1463 (1996).

[54] R. Bielawski, Monopoles and the Gibbons-Manton metric, *Commun. Math. Phys.* **194**, 297 (1998).

[55] A. I. Bochkarev and M. E. Shaposhnikov, Anomalous fermion number nonconservation at high temperatures: Two-dimensional example, *Mod. Phys. Lett.* **A2**, 991 (1987); (Erratum) **A4**, 1495 (1989).

[56] E. B. Bogomolny, The stability of classical solutions, *Sov. J. Nucl. Phys.* **24**, 449 (1976).

[57] E. B. Bogomolny and M. S. Marinov, Calculation of the monopole mass in gauge theory, *Sov. J. Nucl. Phys.* **23**, 355 (1976).

[58] J. Boguta, Can nuclear interactions be long ranged?, *Phys. Rev. Lett.* **50**, 148 (1983).

[59] M. Born and L. Infeld, Foundations of the new field theory, *Proc. R. Soc. Lond.* **A144**, 425 (1934).

[60] R. Bott and L. W. Tu, *Differential Forms in Algebraic Topology*, New York, Springer-Verlag, 1982.

[61] C. P. Boyer, J. C. Hurtubise, B. M. Mann and R. J. Milgram, The topology of instanton moduli spaces, I. The Atiyah-Jones conjecture, *Ann. Math.* **137**, 561 (1993).

[62] P. J. Braam and D. M. Austin, Boundary values of hyperbolic monopoles, *Nonlinearity* **3**, 809 (1990).

[63] P. J. Braam and P. van Baal, Nahm's transformation for instantons, *Commun. Math. Phys.* **122**, 267 (1989).

[64] E. Braaten and L. Carson, Deuteron as a toroidal Skyrmion, *Phys. Rev.* **D38**, 3525 (1988).

[65] E. Braaten, S. Townsend and L. Carson, Novel structure of static multisoliton solutions in the Skyrme model, *Phys. Lett.* **B235**, 147 (1990).

[66] H. W. Braden and P. M. Sutcliffe, A monopole metric, *Phys. Lett.* **B391**, 366 (1997).

[67] S. B. Bradlow, Vortices in holomorphic line bundles over closed Kähler manifolds, *Commun. Math. Phys.* **135**, 1 (1990).

[68] G. E. Brown (ed.), *Selected Papers, with Commentary, of Tony Hilton Royle Skyrme*, Singapore, World Scientific, 1994.

[69] S. A. Brown, H. Panagopoulos and M. K. Prasad, Two separated $SU(2)$ Yang-Mills-Higgs monopoles in the Atiyah-Drinfeld-Hitchin-Manin-Nahm construction, *Phys. Rev.* **D26**, 854 (1982).

[70] J. Burzlaff, A classical lump in $SU(2)$ gauge theory with a Higgs doublet, *Nucl. Phys.* **B233**, 262 (1984).

[71] C. G. Callan, R. F. Dashen and D. J. Gross, Toward a theory of the strong interactions, *Phys. Rev.* **D17**, 2717 (1978).

[72] D. K. Campbell, J. F. Schonfeld and C. A. Wingate, Resonance structure in kink-antikink interactions in ϕ^4 theory, *Physica* **D9**, 1 (1983).

[73] S. M. Carroll, S. Hellerman and M. Trodden, Domain wall junctions are 1/4-BPS states, *Phys. Rev.* **D61**, 065001 (2000).

[74] L. Carson, $B = 3$ nuclei as quantized multi-Skyrmions, *Phys. Rev. Lett.* **66**, 1406 (1991).

[75] L. Castillejo, P. S. J. Jones, A. D. Jackson, J. J. M. Verbaarschot and A. Jackson, Dense Skyrmion systems, *Nucl. Phys.* **A501**, 801 (1989).

[76] L. Castillejo and M. Kugler, The interaction of Skyrmions, Weizmann Institute preprint WIS-87/66-Ph (1987).

[77] J. M. Cervero, Exact monopole solution and Euclidean Yang-Mills field, Harvard University preprint HUTP-77/A011 (1977).

[78] A. Chakrabarti, Harrison-Neugebauer-type transformations for instantons: Multicharged monopoles as limits, *Phys. Rev.* **D25**, 3282 (1982).

[79] A. Chakrabarti, Construction of hyperbolic monopoles, *J. Math. Phys.* **27**, 340 (1986).

[80] G. Chalmers, Multi-monopole moduli spaces for $SU(n)$ gauge group, hep-th/9605182 (1996).

[81] S. J. Chapman, Nucleation of superconductivity in decreasing fields. I, *Euro. J. Appl. Math.* **5**, 449 (1994); Nucleation of superconductivity in decreasing fields. II, *ibid* **5**, 469 (1994).

[82] S. J. Chapman, Nucleation of vortices in type-II superconductors in increasing magnetic fields, *Appl. Math. Lett.* **10**, 29 (1997).

[83] S. J. Chapman, S. D. Howison and J. R. Ockendon, Macroscopic models for superconductivity, *SIAM Review* **34**, 529 (1992).

[84] N. H. Christ, E. J. Weinberg and N. K. Stanton, General self-dual Yang-Mills solutions, *Phys. Rev.* **D18**, 2013 (1978).

[85] S. Coleman, Quantum sine-Gordon equation as the massive Thirring model, *Phys. Rev.* **D11**, 2088 (1975).

[86] S. Coleman, *Q*-balls, *Nucl. Phys.* **B262**, 263 (1985).

[87] S. Coleman, *Aspects of Symmetry*, p. 218, Cambridge University Press, 1985.

[88] *ibid* p. 258.

[89] S. Coleman, S. Parke, A. Neveu and C. M. Sommerfield, Can one dent a dyon?, *Phys. Rev.* **D15**, 544 (1977).

[90] J. E. Colliander and R. L. Jerrard, Ginzburg-Landau vortices: Weak stability and Schrödinger equation dynamics, *J. d'Anal. Math.* **77**, 129 (1999).

[91] S. A. Connell, The dynamics of the $SU(3)$ charge $(1, 1)$ magnetic monopole, University of South Australia preprint (1994).

[92] E. Corrigan, Recent developments in affine Toda field theory, hep-th/9412213 (1994).

[93] E. Corrigan and D. B. Fairlie, Scalar field theory and exact solutions to a classical $SU(2)$ gauge theory, *Phys. Lett.* **B67**, 69 (1977).

[94] E. F. Corrigan, D. B. Fairlie, S. Templeton and P. Goddard, A Green function for the general self-dual gauge field, *Nucl. Phys.* **B140**, 31 (1978).

[95] E. F. Corrigan, D. B. Fairlie, R. G. Yates and P. Goddard, The construction of self-dual solutions to $SU(2)$ gauge theory, *Commun. Math. Phys.* **58**, 223 (1978).

[96] E. Corrigan and P. Goddard, An n monopole solution with $4n - 1$ degrees of freedom, *Commun. Math. Phys.* **80**, 575 (1981).

[97] E. Corrigan and P. Goddard, Construction of instanton and monopole solutions and reciprocity, *Ann. Phys.* **154**, 253 (1984).

[98] A. S. Dancer, Nahm data and $SU(3)$ monopoles, *Nonlinearity* **5**, 1355 (1992).

[99] A. S. Dancer, Nahm's equations and hyperkähler geometry, *Commun. Math. Phys.* **158**, 545 (1993).

[100] A. Dancer and R. Leese, Dynamics of $SU(3)$ monopoles, *Proc. R. Soc. Lond.* **A440**, 421 (1993).

[101] A. S. Dancer and R. A. Leese, A numerical study of $SU(3)$ charge-two monopoles with minimal symmetry breaking, *Phys. Lett.* **B390**, 252 (1997).

[102] R. F. Dashen, B. Hasslacher and A. Neveu, Nonperturbative methods and extended-hadron models in field theory. II. Two-dimensional models and extended hadrons, *Phys. Rev.* **D10**, 4130 (1974).

[103] R. F. Dashen, B. Hasslacher and A. Neveu, Nonperturbative methods and extended-hadron models in field theory. III. Four-dimensional non-abelian models, *Phys. Rev.* **D10**, 4138 (1974).

[104] P. G. De Gennes, *Superconductivity in Metals and Alloys*, New York, Benjamin, 1966.

[105] P. G. De Gennes and J. Prost, *The Physics of Liquid Crystals*, Oxford, Clarendon Press, 1993.

[106] S. Demoulini and D. Stuart, Gradient flow of the superconducting Ginzburg-Landau functional on the plane, *Commun. Anal. Geom.* **5**, 121 (1997).

[107] G. H. Derrick, Comments on nonlinear wave equations as models for elementary particles, *J. Math. Phys.* **5**, 1252 (1964).

[108] A. M. Din and W. J. Zakrzewski, Skyrmion dynamics in 2+1 dimensions, *Nucl. Phys.* **B259**, 667 (1985).

[109] P. A. M. Dirac, Quantized singularities in the electromagnetic field, *Proc. R. Soc. Lond.* **A133**, 60 (1931).

[110] S. K. Donaldson, Nahm's equations and the classification of monopoles, *Commun. Math. Phys.* **96**, 387 (1984).

[111] S. K. Donaldson and P. B. Kronheimer, *Geometry of Four-manifolds*, Oxford, Clarendon Press, 1990.

[112] N. Dorey, T. J. Hollowood, V. V. Khoze and M. P. Mattis, The calculus of many instantons, *Phys. Reports* **371**, 231 (2002).

[113] N. Dorey, V. V. Khoze, M. P. Mattis, D. Tong and S. Vandoren, Instantons, three-dimensional gauge theory, and the Atiyah-Hitchin manifold, *Nucl. Phys.* **B502**, 59 (1997).

[114] A. T. Dorsey, Vortex motion and the Hall effect in type-II superconductors: A time-dependent Ginzburg-Landau theory approach, *Phys. Rev.* **B46**, 8376 (1992).

[115] P. G. Drazin and R. S. Johnson, *Solitons: An Introduction*, Cambridge University Press, 1989.

[116] G. V. Dunne and B. Tekin, Calorons in the Weyl gauge, *Phys. Rev.* **D63**, 085004 (2001).

[117] W. E, Dynamics of vortices in Ginzburg-Landau theories with applications to superconductivity, *Physica* **D77**, 383 (1994).

[118] J. Eells and J. C. Wood, Restrictions on harmonic maps of surfaces, *Topology* **15**, 263 (1976).

[119] G. Eilam, D. Klabucar and A. Stern, Skyrmion solutions to the Weinberg-Salam model, *Phys. Rev. Lett.* **56**, 1331 (1986).

[120] F. Englert and R. Brout, Broken symmetry and the mass of gauge vector mesons, *Phys. Rev. Lett.* **13**, 321 (1964).

[121] N. Ercolani and A. Sinha, Monopoles and Baker functions, *Commun. Math. Phys.* **125**, 385 (1989).

[122] O. Espinosa, High-energy behaviour of baryon- and lepton-number violating scattering amplitudes in the standard model, *Nucl. Phys.* **B343**, 310 (1990).

[123] M. J. Esteban, A direct variational approach to Skyrme's model for meson fields, *Commun. Math. Phys.* **105**, 571 (1986).

[124] L. C. Evans, *Partial Differential Equations (Graduate Studies in Mathematics, vol. 19)*, Providence, American Mathematical Society, 1998.

[125] L. D. Faddeev, Quantization of solitons, Princeton preprint IAS-75-QS70 (1975).

[126] L. D. Faddeev, Some comments on the many dimensional solitons, *Lett. Math. Phys.* **1**, 289 (1976).

[127] L. Faddeev and A. J. Niemi, Stable knot-like structures in classical field theory, *Nature* **387**, 58 (1997).

[128] L. Faddeev and A. J. Niemi, Partially dual variables in $SU(2)$ Yang-Mills theory, *Phys. Rev. Lett.* **82**, 1624 (1999).

[129] E. Farhi, K. Rajagopal and R. Singleton Jr., Gauge-invariant variables for spontaneously broken $SU(2)$ gauge theory in the spherical ansatz, *Phys. Rev.* **D52**, 2394 (1995).

[130] A. L. Fetter, Vortices in an imperfect Bose Gas. I. The condensate, *Phys. Rev.* **138**, A429 (1965).

[131] M. Fierz, Zur Theorie magnetisch geladener Teilchen, *Helv. Phys. Acta* **17**, 27 (1944).

[132] D. Finkelstein and J. Rubinstein, Connection between spin, statistics and kinks, *J. Math. Phys.* **9**, 1762 (1968).

[133] H. Flanders, *Differential Forms*, New York, Academic Press, 1963.

[134] P. Forgács and Z. Horváth, Topology and saddle points in field theories, *Phys. Lett.* **B138**, 397 (1984).

[135] P. Forgács, Z. Horváth and L. Palla, Exact multimonopole solutions in the Bogomolny-Prasad-Sommerfield limit, *Phys. Lett.* **B99**, 232 (1981); Generating monopoles of arbitrary charge by Bäcklund transformations, *ibid* **B102**, 131 (1981); Soliton theoretic framework for generating multimonopoles, *Ann. Phys.* **136**, 371 (1981); Non-linear superposition of monopoles, *Nucl. Phys.* **B192**, 141 (1981); Solution-generating technique for self-dual monopoles, *ibid* **B229**, 77 (1983).

[136] P. Forgács and N. S. Manton, Space-time symmetries in gauge theories, *Commun. Math. Phys.* **72**, 15 (1980).

[137] P. W. Fowler and D. E. Manolopoulos, *An Atlas of Fullerenes*, Oxford, Clarendon Press, 1995.

[138] E. Fradkin, *Field Theories of Condensed Matter Systems*, Redwood City, Addison-Wesley, 1991.

[139] R. Froese and I. Herbst, Realizing holonomic constraints in classical and quantum mechanics, *Commun. Math. Phys.* **220**, 489 (2001).

[140] N. Ganoulis, P. Goddard and D. Olive, Self-dual monopoles and Toda molecules, *Nucl. Phys.* **B205** [FS5] 601 (1982).

[141] M. García Pérez, A. González-Arroyo, C. Pena and P. van Baal, Nahm dualities on the torus – a synthesis, *Nucl. Phys.* **B564**, 159 (2000).

[142] O. García-Prada, A direct existence proof for the vortex equations over a compact Riemann surface, *Bull. London Math. Soc.* **26**, 88 (1994).

[143] C. L. Gardner, The 't Hooft-Polyakov monopole near the Prasad-Sommerfield limit, *Ann. Phys.* **146**, 129 (1983).

[144] G. Gat, $SU(3)$ Skyrmions from instantons, *Phys. Lett.* **B257**, 357 (1991).

[145] J. P. Gauntlett and D. A. Lowe, Dyons and S-duality in $N = 4$ supersymmetric gauge theory, *Nucl. Phys.* **B472**, 194 (1996).

[146] H. Georgi and S. L. Glashow, Unified weak and electromagnetic interactions without neutral currents, *Phys. Rev. Lett.* **28**, 1494 (1972).

[147] H. Georgi and S. L. Glashow, Unity of all elementary-particle forces, *Phys. Rev. Lett.* **32**, 438 (1974).

[148] G. W. Gibbons and N. S. Manton, Classical and quantum dynamics of BPS monopoles, *Nucl. Phys.* **B274**, 183 (1986).

[149] G. W. Gibbons and N. S. Manton, The moduli space metric for well-separated BPS monopoles, *Phys. Lett.* **B356**, 32 (1995).

[150] G. W. Gibbons, P. Rychenkova and R. Goto, Hyperkähler quotient construction of BPS monopole moduli spaces, *Commun. Math. Phys.* **186**, 581 (1997).

[151] G. W. Gibbons and P. K. Townsend, Bogomol'nyi equation for intersecting domain walls, *Phys. Rev. Lett.* **83**, 1727 (1999).

[152] V. L. Ginzburg and L. D. Landau, On the theory of superconductivity (in Russian), *Zh. Eksp. Teor. Fiz.* **20**, 1064 (1950); *Collected Papers of L. D. Landau* (ed. D. ter Haar), Oxford, Pergamon, 1965.

[153] T. Gisiger and M. B. Paranjape, Recent mathematical developments in the Skyrme model, *Phys. Reports* **306**, 109 (1998).

[154] D. Giulini, On the possibility of spinorial quantization in the Skyrme model, *Mod. Phys. Lett.* **A8**, 1917 (1993).

[155] J. Gladikowski and M. Hellmund, Static solitons with nonzero Hopf number, *Phys. Rev.* **D56**, 5194 (1997).

[156] P. Goddard, J. Nuyts and D. Olive, Gauge theories and magnetic charge, *Nucl. Phys.* **B125**, 1 (1977).

[157] P. Goddard and D. I. Olive, Magnetic monopoles in gauge field theories, *Rep. Prog. Phys.* **41**, 1357 (1978).

[158] J. N. Goldberg, P. S. Jang, S. Y. Park and K. C. Wali, Interactions between 't Hooft-Polyakov monopoles, *Phys. Rev.* **D18**, 542 (1978).

[159] A. S. Goldhaber, Role of spin in the monopole problem, *Phys. Rev.* **140**, B1407 (1965).

476   *References*

[160] A. S. Goldhaber and N. S. Manton, Maximal symmetry of the Skyrme crystal, *Phys. Lett.* **B198**, 231 (1987).

[161] A. S. Goldhaber and W. P. Trower, *Magnetic Monopoles*, College Park, American Assoc. Physics Teachers, 1990.

[162] J. Goldstone, Field theories with "superconductor" solutions, *Nuovo Cim.* **19**, 154 (1961).

[163] H. Gomm, F. Lizzi and G. Sparano, Multibaryons in Skyrme and quark models, *Phys. Rev.* **D31**, 226 (1985).

[164] L. P. Gorkov and G. M. Eliashberg, Generalization of the Ginzburg-Landau equations for non-stationary problems in the case of alloys with paramagnetic impurities, *Sov. Phys. JETP* **27**, 328 (1968).

[165] D. Yu. Grigoriev, V. A. Rubakov and M. E. Shaposhnikov, Sphaleron transitions at finite temperatures: Numerical study in (1+1) dimensions, *Phys. Lett.* **B216**, 172 (1989); Topological transitions at finite temperatures: A real-time numerical approach, *Nucl. Phys.* **B326**, 737 (1989).

[166] D. Groisser and T. H. Parker, The Riemannian geometry of the Yang-Mills moduli space, *Commun. Math. Phys.* **112**, 663 (1987).

[167] S. Gustafson and I. M. Sigal, The stability of magnetic vortices, *Commun. Math. Phys.* **212**, 257 (2000).

[168] A. H. Guth, Inflationary universe: A possible solution to the horizon and flatness problems, *Phys. Rev.* **D23**, 347 (1981).

[169] C. Hagen, A new gauge theory without an elementary photon, *Ann. Phys.* **157**, 342 (1984).

[170] K. Hagiwara *et al.*, Review of particle physics, *Phys. Rev.* **D66**, 010001 (2002).

[171] J. L. Harden and V. Arp, The lower critical field in the Ginzburg-Landau theory of superconductivity, *Cryogenics* **3**, 105 (1963).

[172] B. J. Harrington and H. K. Shepard, Periodic Euclidean solutions and the finite-temperature Yang-Mills gas, *Phys. Rev.* **D17**, 2122 (1978).

[173] R. Hartshorne, Stable vector bundles and instantons, *Commun. Math. Phys.* **59**, 1 (1978).

[174] M. Hassaïne and P. A. Horváthy, The symmetries of the Manton superconductivity model, *J. Geom. Phys.* **34**, 242 (2000).

[175] M. Hassaïne, P. A. Horváthy and J.-C. Yera, Non-relativistic Maxwell-Chern-Simons vortices, *Ann. Phys.* **263**, 276 (1998).

[176] M. Hellmund and J. Kripfganz, The decay of the sphaleron, *Nucl. Phys.* **B373**, 749 (1991).

[177] R.-M. Hervé and M. Hervé, Etude qualitative des solutions réelles d'une équation differentielle liée à l'équation de Ginzburg-Landau, *Ann. Inst. H. Poincaré (Anal. Non Lin.)* **11**, 427 (1994).

[178] K. F. Herzfeld and M. Goeppert-Mayer, On the states of aggregation, *J. Chem. Phys.* **2**, 38 (1934).

[179] J. Hietarinta and P. Salo, Faddeev-Hopf knots: dynamics of linked unknots, *Phys. Lett.* **B451**, 60 (1999).

[180] P. W. Higgs, Broken symmetries and the masses of gauge bosons, *Phys. Rev. Lett.* **13**, 508 (1964).

[181] P. J. Hilton, *An Introduction to Homotopy Theory*, Cambridge University Press, 1953.

[182] M. Hindmarsh and M. James, Origin of the sphaleron dipole moment, *Phys. Rev.* **D49**, 6109 (1994).

[183] N. J. Hitchin, Monopoles and geodesics, *Commun. Math. Phys.* **83**, 579 (1982).

[184] N. J. Hitchin, On the construction of monopoles, *Commun. Math. Phys.* **89**, 145 (1983).

[185] N. J. Hitchin, Integrable systems in Riemannian geometry, in *Surveys in Differential Geometry, vol. 4*, Cambridge, International Press, 1998.

[186] N. J. Hitchin, A. Karlhede, U. Lindström and M. Roček, Hyperkähler metrics and supersymmetry, *Commun. Math. Phys.* **108**, 535 (1987).

[187] N. J. Hitchin, N. S. Manton and M. K. Murray, Symmetric monopoles, *Nonlinearity* **8**, 661 (1995).

[188] Z. Hlousek and D. Spector, Understanding solitons algebraically, *Mod. Phys. Lett.* **A7**, 3403 (1992).

[189] T. Hollowood, Solitons in affine Toda theory, *Nucl. Phys.* **B384**, 523 (1992).

[190] J. Hong, Y. Kim and P. Y. Pac, Multivortex solutions of the abelian Chern-Simons-Higgs theory, *Phys. Rev. Lett.* **64**, 2230 (1990).

[191] C. J. Houghton, Instanton vibrations of the 3-Skyrmion, *Phys. Rev.* **D60**, 105003 (1999).

[192] C. J. Houghton, N. S. Manton and N. M. Romão, On the constraints defining BPS monopoles, *Commun. Math. Phys.* **212**, 219 (2000).

[193] C. J. Houghton, N. S. Manton and P. M. Sutcliffe, Rational maps, monopoles and Skyrmions, *Nucl. Phys.* **B510**, 507 (1998).

[194] C. J. Houghton and P. M. Sutcliffe, Tetrahedral and cubic monopoles, *Commun. Math. Phys.* **180**, 343 (1996).

[195] C. J. Houghton and P. M. Sutcliffe, Octahedral and dodecahedral monopoles, *Nonlinearity* **9**, 385 (1996).

[196] C. J. Houghton and P. M. Sutcliffe, Monopole scattering with a twist, *Nucl. Phys.* **B464**, 59 (1996).

[197] C. J. Houghton and P. M. Sutcliffe, Inversion symmetric 3-monopoles and the Atiyah-Hitchin manifold, *Nonlinearity* **9**, 1609 (1996).

[198] C. J. Houghton and P. M. Sutcliffe, $SU(N)$ monopoles and Platonic symmetry, *J. Math. Phys.* **38**, 5576 (1997).

[199] J. Hurtubise, $SU(2)$ monopoles of charge 2, *Commun. Math. Phys.* **92**, 195 (1983).

[200] J. Hurtubise, Monopoles and rational maps: A note on a theorem of Donaldson, *Commun. Math. Phys.* **100**, 191 (1985).

[201] J. Hurtubise and M. K. Murray, On the construction of monopoles for the classical groups, *Commun. Math. Phys.* **122**, 35 (1989).

[202] D. Huterer and T. Vachaspati, Gravitational lensing by cosmic strings in the era of wide-field surveys, *Phys. Rev.* **D68**, 041301 (2003).

[203] T. Ioannidou, Soliton dynamics in a novel discrete $O(3)$ sigma model in (2+1) dimensions, *Nonlinearity* **10**, 1357 (1997).

[204] T. Ioannidou, $SU(N)$ Skyrmions from instantons, *Nonlinearity* **13**, 1217 (2000).

[205] T. Ioannidou, B. Piette and W. J. Zakrzewski, Spherically symmetric solutions of the $SU(N)$ Skyrme models, *J. Math. Phys.* **40**, 6223 (1999).

[206] T. Ioannidou, B. Piette and W. J. Zakrzewski, $SU(N)$ Skyrmions and harmonic maps, *J. Math. Phys.* **40**, 6353 (1999).

[207] T. Ioannidou and P. M. Sutcliffe, Monopoles and harmonic maps, *J. Math. Phys.* **40**, 5440 (1999).

[208] T. Ioannidou and P. M. Sutcliffe, Monopoles from rational maps, *Phys. Lett.* **B457**, 133 (1999).

[209] P. Irwin, $SU(3)$ monopoles and their fields, *Phys. Rev.* **D56**, 5200 (1997).

[210] P. Irwin, Zero mode quantization of multi-Skyrmions, *Phys. Rev.* **D61**, 114024 (2000).

[211] P. W. Irwin and N. S. Manton, Gradient flow for well-separated Skyrmions, *Phys. Lett.* **B385**, 187 (1996).

[212] I. T. Ivanov and M. Roček, Supersymmetric σ-models, twistors, and the Atiyah-Hitchin metric, *Commun. Math. Phys.* **182**, 291 (1996).

[213] R. Jackiw, K. Lee and E. J. Weinberg, Self-dual Chern-Simons solitons, *Phys. Rev.* **D42**, 3488 (1990).

[214] R. Jackiw and N. S. Manton, Symmetries and conservation laws in gauge theories, *Ann. Phys.* **127**, 257 (1980).

[215] R. Jackiw, C. Nohl and C. Rebbi, Conformal properties of pseudoparticle configurations, *Phys. Rev.* **D15**, 1642 (1977).

[216] R. Jackiw and S. Templeton, How super-renormalizable interactions cure their infrared divergences, *Phys. Rev.* **D23**, 2291 (1981).

[217] R. Jackiw and E. J. Weinberg, Self-dual Chern-Simons vortices, *Phys. Rev. Lett.* **64**, 2234 (1990).

[218] A. Jackson, A. D. Jackson and V. Pasquier, The Skyrmion-Skyrmion interaction, *Nucl. Phys.* **A432**, 567 (1985).

[219] A. D. Jackson, N. S. Manton and A Wirzba, New Skyrmion solutions on a 3-sphere, *Nucl. Phys.* **A495**, 499 (1989).

[220] A. D. Jackson and M. Rho, Baryons as chiral solitons, *Phys. Rev. Lett.* **51**, 751 (1983).

[221] A. D. Jackson and J. J. M. Verbaarschot, Phase structure of the Skyrme model, *Nucl. Phys.* **A484**, 419 (1988).

[222] L. Jacobs and C. Rebbi, Interaction energy of superconducting vortices, *Phys. Rev.* **B19**, 4486 (1979).

[223] A. Jaffe and C. Taubes, *Vortices and Monopoles*, Boston, Birkhäuser, 1980.

[224] M. E. R. James, The sphaleron at non-zero Weinberg angle, *Z. Phys.* **C55**, 515 (1992).

[225] S. Jarvis, A rational map for Euclidean monopoles via radial scattering, *J. reine angew. Math.* **524**, 17 (2000).

[226] S. Jarvis and P. Norbury, Zero and infinite curvature limits of hyperbolic monopoles, *Bull. London Math. Soc.* **29**, 737 (1997).

[227] R. L. Jerrard and H. M. Soner, Dynamics of Ginzburg-Landau vortices, *Arch. Rational Mech. Anal.* **142**, 99 (1998).

[228] C. A. Jones and P. H. Roberts, Motions in a Bose condensate: IV. Axisymmetric solitary waves, *J. Phys.* **A15**, 2599 (1982).

[229] B. Julia and A. Zee, Poles with both magnetic and electric charges in non-abelian gauge theory, *Phys. Rev.* **D11**, 2227 (1975).

[230] L. B. Kapitanski and O. A. Ladyzenskaia, On the Coleman's principle concerning the stationary points of invariant functionals, *Zap. Nauchn. Semin, LOMI* **127**, 84 (1983).

[231] J. B. Ketterson and S. N. Song, *Superconductivity*, Cambridge University Press, 1999.

[232] A. Khare, Charged vortices and Q-balls in an abelian Higgs model exhibiting a first order phase transition, *Phys. Lett.* **B255**, 393 (1991).

[233] T. W. B. Kibble, Symmetry breaking in non-abelian gauge theories, *Phys. Rev.* **155**, 1554 (1967).

[234] T. W. Kirkman and C. K. Zachos, Asymptotic analysis of the monopole structure, *Phys. Rev.* **D24**, 999 (1981).

[235] I. Klebanov, Nuclear matter in the Skyrme model, *Nucl. Phys.* **B262**, 133 (1985).

[236] B. Kleihaus and J. Kunz, Monopole-antimonopole solution of the $SU(2)$ Yang-Mills-Higgs model, *Phys. Rev.* **D61**, 025003 (2000).

[237] F. Klein, *Lectures on the Icosahedron*, London, Kegan Paul, 1913.

[238] W. H. Kleiner, L. M. Roth and S. H. Autler, Bulk solution of Ginzburg-Landau equations for type II superconductors: Upper critical field region, *Phys. Rev.* **133**, A1226 (1964).

[239] F. R. Klinkhamer, Sphalerons, deformed sphalerons and configuration space, *Phys. Lett.* **B236**, 187 (1990).

[240] F. R. Klinkhamer, Construction of a new electroweak sphaleron, *Nucl. Phys.* **B410**, 343 (1993).

[241] F. R. Klinkhamer and N. S. Manton, A saddle-point solution in the Weinberg-Salam theory, *Phys. Rev.* **D30**, 2212 (1984).

[242] O. Kobayashi and K. Takahashi, Instability of classical solution in an $SU(2)$ gauge model, *Prog. Theor. Phys.* **60**, 311 (1978).

[243] S. Kobayashi and K. Nomizu, *Foundations of Differential Geometry, vol. 1*, New York, Interscience, 1963.

[244] V. B. Kopeliovich and B. E. Stern, Exotic Skyrmions, *JETP Lett.* **45**, 203 (1987).

[245] V. E. Korepin and S. L. Shatashvili, Rational parametrization of the three-instanton solutions of the Yang-Mills equations, *Sov. Phys. Dokl.* **28**, 1018 (1983).

[246] S. Krusch, S^3 Skyrmions and the rational map ansatz, *Nonlinearity* **13**, 2163 (2000).

[247] S. Krusch, Homotopy of rational maps and the quantization of Skyrmions, *Ann. Phys.* **304**, 103 (2003).

[248] M. Kugler and S. Shtrikman, A new Skyrmion crystal, *Phys. Lett.* **B208**, 491 (1988); Skyrmion crystals and their symmetries, *Phys. Rev.* **D40**, 3421 (1989).

[249] A. Kundu and Yu. P. Rybakov, Closed-vortex-type solitons with Hopf index, *J. Phys.* **A15**, 269 (1982).

[250] J. Kunz and Y. Brihaye, New sphalerons in the Weinberg-Salam theory, *Phys. Lett.* **B216**, 353 (1989).

[251] J. Kunz, B. Kleihaus and Y. Brihaye, Sphalerons at finite mixing angle, *Phys. Rev.* **D46**, 3587 (1992).

[252] V. A. Kuzmin, V. A. Rubakov and M. E. Shaposhnikov, On anomalous electroweak baryon-number non-conservation in the early universe, *Phys. Lett.* **B155**, 36 (1985).

[253] K. Lee, E. J. Weinberg and P. Yi, Electromagnetic duality and $SU(3)$ monopoles, *Phys. Lett.* **B376**, 97 (1996).

[254] K. Lee, E. J. Weinberg and P. Yi, Moduli space of many BPS monopoles for arbitrary gauge groups, *Phys. Rev.* **D54**, 1633 (1996).

[255] K. Lee, E. J. Weinberg and P. Yi, Massive and massless monopoles with non-abelian magnetic charges, *Phys. Rev.* **D54**, 6351 (1996).

[256] T. D. Lee, *Particle Physics and Introduction to Field Theory*, Chur, Harwood, 1981.

[257] R. A. Leese, Discrete Bogomolny equations for the nonlinear $O(3)$ sigma model in 2+1 dimensions, *Phys. Rev.* **D40**, 2004 (1989).

[258] R. A. Leese, Low energy scattering of solitons in the $\mathbb{CP}^1$ model, *Nucl. Phys.* **B344**, 33 (1990).

[259] R. A. Leese, Q-lumps and their interactions, *Nucl. Phys.* **B366**, 283 (1991).

[260] R. A. Leese and N. S. Manton, Stable instanton-generated Skyrme fields with baryon numbers three and four, *Nucl. Phys.* **A572**, 575 (1994).

[261] R. A. Leese, N. S. Manton and B. J. Schroers, Attractive channel Skyrmions and the deuteron, *Nucl. Phys.* **B442**, 228 (1995).

[262] R. A. Leese, M. Peyrard and W. J. Zakrzewski, Soliton stability in the $O(3)$ σ-model in (2+1) dimensions, *Nonlinearity* **3**, 387 (1990).

[263] R. A. Leese, M. Peyrard and W. J. Zakrzewski, Soliton scatterings in some relativistic models in (2+1) dimensions, *Nonlinearity* **3**, 773 (1990).

[264] A. N. Leznov and M. V. Saveliev, Representation of zero curvature for the system of nonlinear partial differential equations $x_{\alpha,z\bar{z}} = \exp(kx)_\alpha$ and its integrability, *Lett. Math. Phys.* **3**, 489 (1979); Representation theory and integration of nonlinear spherically symmetric equations to gauge theories, *Commun. Math. Phys.* **74**, 111 (1980).

[265] J.-Q. Liang, H. J. W. Müller-Kirsten and D. H. Tchrakian, Solitons, bounces and sphalerons on a circle, *Phys. Lett.* **B282**, 105 (1992).

[266] F. H. Lin, Some dynamical properties of Ginzburg-Landau vortices, *Commun. Pure Appl. Math.* **49**, 323 (1996); A remark on the previous paper "Some dynamical properties of Ginzburg-Landau vortices", *ibid* **49**, 361 (1996).

[267] L. A. Ljusternik, *The Topology of the Calculus of Variations in the Large (Amer. Math. Soc. Transl. 16)*, Providence, American Mathematical Society, 1966.

[268] M. A. Lohe, Two- and three-dimensional instantons, *Phys. Lett.* **B70**, 325 (1977).

[269] H. A. Lorentz, *Theory of Electrons (2nd edn, 1915)*, New York, Dover, 1952.

[270] I. G. Macdonald, Symmetric products of an algebraic curve, *Topology* **1**, 319 (1962).

[271] A. Maciocia, Metrics on the moduli spaces of instantons over Euclidean 4-space, *Commun. Math. Phys.* **135**, 467 (1991).

[272] R. MacKenzie, Remarks on gauge vortex scattering, *Phys. Lett.* **B352**, 96 (1995).

[273] S. F. Magruder, Classical interactions of 't Hooft monopoles, *Phys. Rev.* **D17**, 3257 (1978).

[274] S. Mandelstam, Soliton operators for the quantized sine-Gordon equation, *Phys. Rev.* **D11**, 3026 (1975).

[275] N. S. Manton, The force between 't Hooft-Polyakov monopoles, *Nucl. Phys.* **B126**, 525 (1977).

[276] N. S. Manton, Complex structure of monopoles, *Nucl. Phys.* **B135**, 319 (1978).

[277] N. S. Manton, Instantons on a line, *Phys. Lett.* **B76**, 111 (1978).

[278] N. S. Manton, An effective Lagrangian for solitons, *Nucl. Phys.* **B150**, 397 (1979).

[279] N. S. Manton, A remark on the scattering of BPS monopoles, *Phys. Lett.* **B110**, 54 (1982).

[280] N. S. Manton, Topology in the Weinberg-Salam theory, *Phys. Rev.* **D28**, 2019 (1983).

[281] N. S. Manton, Monopole interactions at long range, *Phys. Lett.* **B154**, 397 (1985); (Erratum) **B157**, 475 (1985).

[282] N. S. Manton, Geometry of Skyrmions, *Commun. Math. Phys.* **111**, 469 (1987).

[283] N. S. Manton, Is the $B = 2$ Skyrmion axially symmetric?, *Phys. Lett.* **B192**, 177 (1987).

[284] N. S. Manton, Unstable manifolds and soliton dynamics, *Phys. Rev. Lett.* **60**, 1916 (1988).

[285] N. S. Manton, Statistical mechanics of vortices, *Nucl. Phys.* **B400** [FS], 624 (1993).

[286] N. S. Manton, Skyrmions and their pion multipole moments, *Acta Phys. Pol.* **B25**, 1757 (1994).

[287] N. S. Manton, First order vortex dynamics, *Ann. Phys.* **256**, 114 (1997).

[288] N. S. Manton and S. M. Nasir, Conservation laws in a first-order dynamical system of vortices, *Nonlinearity* **12**, 851 (1999).

[289] N. S. Manton and S. M. Nasir, Volume of vortex moduli spaces, *Commun. Math. Phys.* **199**, 591 (1999).

[290] N. S. Manton and B. M. A. G. Piette, Understanding Skyrmions using rational maps, *Prog. Math.* **201**, 469 (2001).

[291] N. S. Manton and P. J. Ruback, Skyrmions in flat space and curved space, *Phys. Lett.* **B181**, 137 (1986).

[292] N. S. Manton and T. M. Samols, Sphalerons on a circle, *Phys. Lett.* **B207**, 179 (1988).

[293] N. S. Manton and T. M. Samols, Radiation from monopole scattering, *Phys. Lett.* **B215**, 559 (1988).

[294] N. S. Manton, B. J. Schroers and M. A. Singer, The interaction energy of well-separated Skyrme solitons, *Commun. Math. Phys.* **245**, 123 (2004).

[295] N. S. Manton and J. M. Speight, Asymptotic interactions of critically coupled vortices, *Commun. Math. Phys.* **236**, 535 (2003).

[296] N. S. Manton and P. M. Sutcliffe, Skyrme crystal from a twisted instanton on a four-torus, *Phys. Lett.* **B342**, 196 (1995).

[297] L. J. Mason and N. M. J. Woodhouse, *Integrability, Self-duality and Twistor Theory*, Oxford, Clarendon Press, 1996.

[298] M. R. Matthews, B. P. Anderson, P. C. Haljan, D. S. Hall, C. E. Wieman and E. A. Cornell, Vortices in a Bose-Einstein condensate, *Phys. Rev. Lett.* **83**, 2498 (1999).

[299] R. A. Matzner, Interaction of $U(1)$ cosmic strings: numerical intercommutation, *Computers in Physics* **2** (Sep/Oct), 51 (1988).

[300] H. P. McKean, The sine-Gordon and sinh-Gordon equations on the circle, *Commun. Pure Appl. Math.* **34**, 197 (1981).

[301] J. B. McLeod and C. B. Wang, Existence of the solution for the 't Hooft-Polyakov monopole, math-ph/9902002 (1999).

[302] U.-G. Meissner, Toroidal solitons with unit Hopf charge, *Phys. Lett.* **B154**, 190 (1985).

[303] J. Milnor, *Morse Theory (Annals of Math. Studies 51)*, Princeton University Press, 1969.

[304] C. Montonen and D. Olive, Magnetic monopoles as gauge particles?, *Phys. Lett.* **B72**, 117 (1977).

[305] J. N. Moore, *Gauged Vortices: Dynamics, Radiation and Cosmological Networks*, Ph.D. thesis, Cambridge University, 2000.

[306] K. J. M. Moriarty, E. Myers and C. Rebbi, Dynamical interactions of cosmic strings and flux vortices in superconductors, *Phys. Lett.* **B207**, 411 (1988).

[307] M. Moshir, Soliton-antisoliton scattering and capture in $\lambda\phi^4$ theory, *Nucl. Phys.* **B185**, 318 (1981).

[308] E. Mottola and A. Wipf, Unsuppressed fermion-number violation at high temperature: An $O(3)$ model, *Phys. Rev.* **D39**, 588 (1989).

[309] M. Murray, Stratifying monopoles and rational maps, *Commun. Math. Phys.* **125**, 661 (1989).

[310] M. K. Murray, A note on the $(1, 1, \ldots, 1)$ monopole metric, *J. Geom. Phys.* **23**, 31 (1997).

[311] M. K. Murray and M. A. Singer, On the complete integrability of the discrete Nahm equations, *Commun. Math. Phys.* **210**, 497 (2000).

[312] W. Nahm, The interaction energy of 't Hooft monopoles in the Prasad-Sommerfield limit, *Phys. Lett.* **B79**, 426 (1978).

[313] W. Nahm, The construction of all self-dual multimonopoles by the ADHM method, in *Monopoles in Quantum Field Theory*, eds. N. S. Craigie, P. Goddard and W. Nahm, Singapore, World Scientific, 1982.

[314] H. Nakajima, Monopoles and Nahm's equations, in *Einstein Metrics and Yang-Mills Connections (Sanda, 1990), Lecture Notes in Pure and App. Math.* **145**, New York, Dekker, 1993.

[315] C. Nash, Geometry of hyperbolic monopoles, *J. Math. Phys.* **27**, 2160 (1986).

[316] S. M. Nasir, Vortices and flat connections, *Phys. Lett.* **B419**, 253 (1998).

[317] J. Neu, Vortices in complex scalar fields, *Physica* **D43**, 385 (1990).

[318] H. B. Nielsen and P. Olesen, Vortex line models for dual strings, *Nucl. Phys.* **B61**, 45 (1973).

[319] P. Nozières and W. F. Vinen, The motion of flux lines in type II superconductors, *Phil. Mag.* **14**, 667 (1966).

[320] H. Oda, K. Ito, M. Naganuma and N. Sakai, An exact solution of BPS domain wall junction, *Phys. Lett.* **B471**, 140 (1999).

[321] Yu. N. Ovchinnikov and I. M. Sigal, The Ginzburg-Landau equation III. Vortex dynamics, *Nonlinearity* **11**, 1277 (1998); Long-time behaviour of Ginzburg-Landau vortices, *ibid* **11**, 1295 (1998); Dynamics of localized structures, *Physica* **A261**, 143 (1998).

[322] R. S. Palais, The principle of symmetric criticality, *Commun. Math. Phys.* **69**, 19 (1979).

[323] G. Papadopoulos and P. K. Townsend, Solitons in supersymmetric sigma models with torsion, *Nucl. Phys.* **B444**, 245 (1995).

[324] N. Papanicolaou and T. N. Tomaras, Dynamics of magnetic vortices, *Nucl. Phys.* **B360**, 425 (1991).

[325] N. Papanicolaou and T. N. Tomaras, On the dynamics of vortices in a nonrelativistic Ginzburg-Landau model, *Phys. Lett.* **A179**, 33 (1993).

[326] R. D. Parkes (ed.), *Superconductivity*, New York, Marcel Dekker, 1969.

[327] H. Pedersen and Y. S. Poon, Hyper-Kähler metrics and a generalization of the Bogomolny equations, *Commun. Math. Phys.* **117**, 569 (1988).

[328] L. Perivolaropoulos, Asymptotics of Nielsen-Olesen vortices, *Phys. Rev.* **D48**, 5961 (1993).

[329] J. K. Perring and T. H. R. Skyrme, A model unified field equation, *Nucl. Phys.* **31**, 550 (1962).

[330] B. M. A. G. Piette, B. J. Schroers and W. J. Zakrzewski, Multisolitons in a two-dimensional Skyrme model, *Z. Phys.* **C65**, 165 (1995).

[331] B. M. A. G. Piette, B. J. Schroers and W. J. Zakrzewski, Dynamics of baby Skyrmions, *Nucl. Phys.* **B439**, 205 (1995).

[332] B. Piette and W. J. Zakrzewski, Shrinking of solitons in the (2+1)-dimensional S^2 sigma model, *Nonlinearity* **9**, 897 (1996).

[333] B. Plohr, Ph.D. thesis, Princeton University, 1980; The behavior at infinity of isotropic vortices and monopoles, *J. Math. Phys.* **22**, 2184 (1981).

[334] H. Poincaré, Remarques sur une expérience de M. Birkeland, *Comptes Rendus Acad. Sc.* **123**, 530 (1896).

[335] J. Polchinski, *String Theory, vol. 2*, Cambridge University Press, 1998.

[336] A. M. Polyakov, Particle spectrum in quantum field theory, *JETP Lett.* **20**, 194 (1974).

[337] M. K. Prasad, Instantons and monopoles in Yang-Mills gauge field theories, *Physica* **D1**, 167 (1980).

[338] M. K. Prasad, Yang-Mills-Higgs monopole solutions of arbitrary topological charge, *Commun. Math. Phys.* **80**, 137 (1981).

[339] M. K. Prasad and P. Rossi, Construction of exact Yang-Mills-Higgs multimonopoles of arbitrary charge, *Phys. Rev. Lett.* **46**, 806 (1981).

[340] M. K. Prasad and C. M. Sommerfield, Exact classical solution for the 't Hooft monopole and the Julia-Zee dyon, *Phys. Rev. Lett.* **35**, 760 (1975).

[341] J. P. Preskill, Cosmological production of superheavy magnetic monopoles, *Phys. Rev. Lett.* **43**, 1365 (1979).

[342] R. Rajaraman, Intersoliton forces in weak-coupling quantum field theories, *Phys. Rev.* **D15**, 2866 (1977).

[343] R. Rajaraman, *Solitons and Instantons*, Amsterdam, Elsevier Science, 1982.

[344] F. H. Ree and W. G. Hoover, Fifth and sixth virial coefficients for hard spheres and hard disks, *J. Chem. Phys.* **40**, 939 (1964); Seventh virial coefficients for hard spheres and hard disks, *ibid* **46**, 4181 (1967).

[345] A. Ringwald, High energy breakdown of perturbation theory in the electroweak instanton sector, *Nucl. Phys.* **B330**, 1 (1990).

[346] V. N. Romanov, A. S. Schwarz and Yu. S. Tyupkin, On spherically symmetric fields in gauge theories, *Nucl. Phys.* **B130**, 209 (1977).

[347] N. M. Romão, Quantum Chern-Simons vortices on a sphere, *J. Math. Phys.* **42**, 3445 (2001).

[348] N. M. Romão, *Classical and Quantum Aspects of Topological Solitons*, Ph.D. thesis, Cambridge University, 2002.

[349] G. Rosen, Particlelike solutions to nonlinear complex scalar field theories with positive-definite energy densities, *J. Math. Phys.* **9**, 996 (1968).

[350] C. Rosenzweig and A. M. Srivastava, Towards a qualitative understanding of the scattering of topological defects, *Phys. Rev.* **D43**, 4029 (1991).

[351] P. J. Ruback, Sigma model solitons and their moduli space metrics, *Commun. Math. Phys.* **116**, 645 (1988).

[352] P. J. Ruback, Vortex string motion in the abelian Higgs model, *Nucl. Phys.* **B296**, 669 (1988).

[353] P. J. Ruback, unpublished (see T. M. Samols, *Soliton Scattering*, Ph.D. thesis, Cambridge University, 1990).

[354] V. A. Rubakov and M. E. Shaposhnikov, Electroweak baryon number non-conservation in the early universe and in high-energy collisions, *Phys. Usp.* **39**, 461 (1996).

[355] B. Rüber, *Eine axialsymmetrische magnetische Dipollösung der Yang-Mills-Higgs-Gleichungen*, Diplomarbeit, Universität Bonn, 1985.

[356] H. Rubin and P. Ungar, Motion under a strong constraining force, *Commun. Pure Appl. Math.* **10**, 65 (1957).

[357] L. Sadun and J. Segert, Non-self-dual Yang-Mills connections with non-zero Chern number, *Bull. Am. Math. Soc.* **24**, 163 (1991).

[358] L. Sadun and J. Segert, Stationary points of the Yang-Mills action, *Commun. Pure Appl. Math.* **45**, 461 (1992).

[359] L. A. Sadun and J. M. Speight, Geodesic incompleteness in the $\mathbb{CP}^1$ model on a compact Riemann surface, *Lett. Math. Phys.* **43**, 329 (1998).

[360] P. M. Saffin, Tiling with almost-BPS-invariant domain-wall junctions, *Phys. Rev. Lett.* **83**, 4249 (1999).

[361] A. D. Sakharov, Violation of CP invariance, C asymmetry, and baryon asymmetry of the universe, *JETP Lett.* **5**, 24 (1967).

[362] G. Salmon, *Lessons Introductory to the Modern Higher Algebra*, Dublin, Hodges and Smith, 1866.

[363] T. M. Samols, Vortex scattering, *Commun. Math. Phys.* **145**, 149 (1992).

[364] J. F. Schonfeld, A mass term for three-dimensional gauge fields, *Nucl. Phys.* **B185**, 157 (1981).

[365] B. J. Schroers, Dynamics of moving and spinning Skyrmions, *Z. Phys.* **C61**, 479 (1994).

[366] B. J. Schroers, Bogomol'nyi solitons in a gauged $O(3)$ sigma model, *Phys. Lett.* **B356**, 291 (1995).

[367] B. J. Schroers, On the existence of minima in the Skyrme model, *JHEP* proceedings PRHEP-unesp2002/034 (2002).

[368] A. S. Schwarz, On regular solutions of Euclidean Yang-Mills equations, *Phys. Lett.* **B67**, 172 (1977).

[369] J. Schwinger, A magnetic model of matter, *Science* **165**, 757 (1969).

[370] G. Segal, The topology of spaces of rational maps, *Acta Math.* **143**, 39 (1979).

[371] G. Segal and A. Selby, The cohomology of the space of magnetic monopoles, *Commun. Math. Phys.* **177**, 775 (1996).

[372] A. Sen, Dyon-monopole bound states, self-dual harmonic forms on the multi-monopole moduli space, and $SL(2, \mathbb{Z})$ invariance in string theory, *Phys. Lett.* **B329**, 217 (1994).

[373] P. A. Shah and N. S. Manton, Thermodynamics of vortices in the plane, *J. Math. Phys.* **35**, 1171 (1994).

[374] E. P. S. Shellard and P. J. Ruback, Vortex scattering in two dimensions, *Phys. Lett.* **B209**, 262 (1988).

[375] L. M. Sibner, R. J. Sibner and K. Uhlenbeck, Solutions to Yang-Mills equations that are not self-dual, *Proc. Natl. Acad. Sci. USA* **86**, 8610 (1989).

[376] M. A. Singer and P. M. Sutcliffe, Symmetric instantons and Skyrme fields, *Nonlinearity* **12**, 987 (1999).

[377] T. H. R. Skyrme, A nonlinear field theory, *Proc. R. Soc. Lond.* **A260**, 127 (1961).

[378] T. H. R. Skyrme, Particle states of a quantized meson field, *Proc. R. Soc. Lond.* **A262**, 237 (1961).

[379] T. H. R. Skyrme, A unified field theory of mesons and baryons, *Nucl. Phys.* **31**, 556 (1962).

[380] S. L. Sondhi, A. Karlhede, S. A. Kivelson and E. H. Rezayi, Skyrmions and the crossover from the integer to fractional quantum Hall effect at small Zeeman energies, *Phys. Rev.* **B47**, 16419 (1993).

[381] J. M. Speight, Low-energy dynamics of a $\mathbb{CP}^1$ lump on the sphere, *J. Math. Phys.* **36**, 796 (1995).

[382] J. M. Speight, Static intervortex forces, *Phys. Rev.* **D55**, 3830 (1997).

[383] J. M. Speight and I. A. B. Strachan, Gravity thaws the frozen moduli of the $\mathbb{CP}^1$ lump, *Phys. Lett.* **B457**, 17 (1999).

[384] I. A. B. Strachan, Low-velocity scattering of vortices in a modified abelian Higgs model, *J. Math. Phys.* **33**, 102 (1992).

[385] D. Stuart, Dynamics of abelian Higgs vortices in the near Bogomolny regime, *Commun. Math. Phys.* **159**, 51 (1994).

[386] D. Stuart, The geodesic approximation for the Yang-Mills-Higgs equations, *Commun. Math. Phys.* **166**, 149 (1994).

[387] D. Stuart, Interaction of superconducting vortices and asymptotics of the Ginsburg-Landau flow, *Appl. Math. Lett.* **9**(5), 27 (1996).

[388] P. M. Sutcliffe, The interaction of Skyrme-like lumps in (2+1) dimensions, *Nonlinearity* **4**, 1109 (1991).

[389] P. M. Sutcliffe, Sine-Gordon solitons from $\mathbb{CP}^1$ instantons, *Phys. Lett.* **B283**, 85 (1992).

[390] P. M. Sutcliffe, Instanton chains with soliton limits, *Phys. Lett.* **B302**, 237 (1993).

[391] P. M. Sutcliffe, Monopole zeros, *Phys. Lett.* **B376**, 103 (1996).

[392] P. M. Sutcliffe, Seiberg-Witten theory, monopole spectral curves and affine Toda solitons, *Phys. Lett.* **B381**, 129 (1996).

[393] P. M. Sutcliffe, Cyclic monopoles, *Nucl. Phys.* **B505**, 517 (1997).

[394] P. M. Sutcliffe and W. J. Zakrzewski, Skyrmions from harmonic maps, in *Integrable Hierarchies and Modern Physical Theories*, eds. H. Aratyn and A. S. Sorin, *NATO Science Series II, vol. 18*, Dordrecht, Kluwer, 2001.

[395] M. Takahasi, Ph.D. thesis, University of Tokyo.

[396] C. H. Taubes, Arbitrary N-vortex solutions to the first order Ginzburg-Landau equations, *Commun. Math. Phys.* **72**, 277 (1980).

[397] C. H. Taubes, On the equivalence of the first and second order equations for gauge theories, *Commun. Math. Phys.* **75**, 207 (1980).

[398] C. H. Taubes, The existence of a non-minimal solution to the $SU(2)$ Yang-Mills-Higgs equations on $\mathbb{R}^3$: Part I, *Commun. Math. Phys.* **86**, 257 (1982); The existence of a non-minimal solution to the $SU(2)$ Yang-Mills-Higgs equations on $\mathbb{R}^3$: Part II, *ibid* **86**, 299 (1982).

[399] M. Temple-Raston, BPS two-monopole scattering, *Phys. Lett.* **B206**, 503 (1988); Two-magnetic-monopole Julia sets, *ibid* **B213**, 168 (1988).

[400] W. Thomson, On vortex atoms, *Trans. R. Soc. Edin.* **6**, 94 (1867).

[401] G. 't Hooft, Magnetic monopoles in unified gauge theories, *Nucl. Phys.* **B79**, 276 (1974).

[402] G. 't Hooft, unpublished.

[403] G. 't Hooft, Symmetry breaking through Bell-Jackiw anomalies, *Phys. Rev. Lett.* **37**, 8 (1976); Computation of the quantum effects due to a four-dimensional pseudoparticle, *Phys. Rev.* **D14**, 3432 (1976).

[404] D. Tong, NS5-branes, T-duality and worldsheet instantons, *JHEP* **0207**, 013 (2002).

[405] A. F. Vakulenko and L. V. Kapitanski, Stability of solitons in S^2 in the nonlinear σ-model, *Dokl. Akad. Nauk USSR* **246**, 840 (1979).

[406] J. J. M. Verbaarschot, Axial symmetry of bound baryon number two solution of the Skyrme model, *Phys. Lett.* **B195**, 235 (1987).

[407] A. Vilenkin and E. P. S. Shellard, *Cosmic Strings and other Topological Defects*, Cambridge University Press, 1994.

[408] R. Vinh Mau, M. Lacombe, B. Loiseau, W. N. Cottingham and P. Lisboa, The static baryon-baryon potential in the Skyrme model, *Phys. Lett.* **B150**, 259 (1985).

[409] T. Waindzoch and J. Wambach, Stability of the $B = 2$ hedgehog in the Skyrme model, *Nucl. Phys.* **A602**, 347 (1996).

[410] N. R. Walet and T. Weidig, Full 2D numerical study of the quantum Hall Skyrme crystal, cond-mat/0106157 (2001).

[411] T. S. Walhout, Quantizing the four-baryon Skyrmion, *Nucl. Phys.* **A547**, 423 (1992).

[412] R. S. Ward, On self-dual gauge fields, *Phys. Lett.* **A61**, 81 (1977).

[413] R. S. Ward, A Yang-Mills-Higgs monopole of charge 2, *Commun. Math. Phys.* **79**, 317 (1981).

[414] R. S. Ward, Two Yang-Mills-Higgs monopoles close together, *Phys. Lett.* **B102**, 136 (1981).

[415] R. S. Ward, Deformations of the embedding of the $SU(2)$ monopole solution in $SU(3)$, *Commun. Math. Phys.* **86**, 437 (1982).

[416] R. S. Ward, Slowly-moving lumps in the $\mathbb{CP}^1$ model in (2+1) dimensions, *Phys. Lett.* **B158**, 424 (1985).

[417] R. S. Ward, Integrable and solvable systems, and relations among them, *Phil. Trans. R. Soc. Lond.* **A315**, 451 (1985).

[418] R. S. Ward, Hopf solitons on S^3 and $\mathbb{R}^3$, *Nonlinearity* **12**, 241 (1999).

[419] R. S. Ward, The interaction of two Hopf solitons, *Phys. Lett.* **B473**, 291 (2000).

[420] R. S. Ward and R. O. Wells, *Twistor Geometry and Field Theory*, Cambridge University Press, 1990.

[421] T. Weidig, The baby Skyrme models and their multi-skyrmions, *Nonlinearity* **12**, 1489 (1999).

[422] E. J. Weinberg, Parameter counting for multimonopole solutions, *Phys. Rev.* **D20**, 936 (1979).

[423] E. J. Weinberg, Fundamental monopoles and multimonopole solutions for arbitrary simple gauge groups, *Nucl. Phys.* **B167**, 500 (1980).

[424] J. Wess and B. Zumino, Consequences of anomalous Ward identities, *Phys. Lett.* **B37**, 95 (1971).

[425] F. Wilczek, Geometry and interactions of instantons, in *Quark Confinement and Field Theory*, eds. D. R. Stump and D. H. Weingarten, New York, Wiley, 1977.

[426] J. G. Williams, Topological analysis of a nonlinear field theory, *J. Math. Phys.* **11**, 2611 (1970).

[427] E. Witten, Some exact multipseudoparticle solutions of classical Yang-Mills theory, *Phys. Rev. Lett.* **38**, 121 (1977).

[428] E. Witten, Global aspects of current algebra, *Nucl. Phys.* **B223**, 422 (1983); Current algebra, baryons, and quark confinement, *ibid* **B223**, 433 (1983).

[429] E. Witten and D. Olive, Supersymmetry algebras that include topological charges, *Phys. Lett.* **B78**, 97 (1978).

[430] M. P. Wojtkowski, Bounded geodesics for the Atiyah-Hitchin metric, *Bull. Am. Math. Soc.* **18**, 179 (1988).

[431] G. Woo, Pseudoparticle configurations in two-dimensional ferromagnets, *J. Math. Phys.* **18**, 1264 (1977).

[432] T. T. Wu and C. N. Yang, Concept of nonintegrable phase factors and global formulation of gauge fields, *Phys. Rev.* **D12**, 3845 (1975).

[433] L. G. Yaffe, Static solutions of $SU(2)$-Higgs theory, *Phys. Rev.* **D40**, 3463 (1989).

[434] C. N. Yang and R. L. Mills, Conservation of isotopic spin and isotopic gauge invariance, *Phys. Rev.* **96**, 191 (1954).

[435] Y. Yang, Coexistence of vortices and antivortices in an abelian gauge theory, *Phys. Rev. Lett.* **80**, 26 (1998).

[436] W. J. Zakrzewski, *Low Dimensional Sigma Models*, Bristol, Institute of Physics Publishing, 1989.

[437] W. J. Zakrzewski, Soliton-like scattering in the $O(3)$ sigma model in (2+1) dimensions, *Nonlinearity* **4**, 429 (1991).

[438] Ya. B. Zeldovich and M. Yu. Khlopov, On the concentration of relic magnetic monopoles in the universe, *Phys. Lett.* **B79**, 239 (1978).

[439] S. C. Zhang, T. H. Hansson and S. Kivelson, Effective-field-theory model for the fractional quantum Hall effect, *Phys. Rev. Lett.* **62**, 82 (1989).

[440] D. Zwanziger, Point monopoles in quantum field theory, in *Monopoles in Quantum Field Theory*, eds. N. S. Craigie, P. Goddard and W. Nahm, Singapore, World Scientific, 1982.

Index